HISTORY AS A SCIENCE:
THE PHILOSOPHY OF R.G. COLLINGWOOD

MARTINUS NIJHOFF PHILOSOPHY LIBRARY

VOLUME 3

Volumes in this series:

1. D. Lamb, Hegel – From Foundation to System.
   ISBN 90-247-2359-0
2. I.N. Bulhof, William Dilthey – A Hermeneutic Approach to the Study of History and Culture.
   ISBN 90-247-2360-4
3. W.J. van der Dussen, History as a Science: The Philosophy of R.G. Collingwood.
   ISBN 90-247-2453-8
4. M. Chatterjee, The Language of Philosophy.
   ISBN 90-247-2372-8
5. E.-H. Kluge, The Metaphysics of Gottlob Frege.
   ISBN 90-247-2422-8
6. D. Dutton and M. Krausz, eds., The Concept of Creativity in Science and Art
   ISBN 90-247-2418-X

ISBN series 90-247-2344-2

HISTORY AS A SCIENCE: THE PHILOSOPHY OF R.G. COLLINGWOOD

by

W.J. VAN DER DUSSEN

1981

MARTINUS NIJHOFF PUBLISHERS

THE HAGUE/BOSTON/LONDON

Distributors

for the United States and Canada
Kluwer Boston, Inc.
190 Old Derby Street
Hingham, MA 02043
USA

for all other countries
Kluwer Academic Publishers Group
Distribution Center
P.O. Box 322
3300 AH Dordrecht
The Netherlands

Permission to quote from W. Debbins ed., *R.G. Collingwood: Essays in the Philosophy of History* (Austin, 1965), was kindly granted by the University of Texas Press.

Permission to quote from Collingwood's *Roman Britain* (1923, 1932), *Speculum Mentis* (1924), *An Essay on Philosophical Method* (1933), *Roman Britain and the English Settlements* (1936), *The Principles of Art* (1938), *An Autobiography* (1939), *An Essay on Metaphysics* (1940), *The New Leviathan* (1942), *The Idea of History* (1946), was kindly granted by the Oxford University Press.

ISBN 90-247-2453-8 (this volume)
ISBN 90-247-2344-2 (series)

PRINTED IN THE NETHERLANDS

For ARDA

Teaching and research are alike valueless unless they are based on a reasoned conviction as to what it is that we are teaching and what it is that we are trying to find out.
(R.G. Collingwood, *The English Historical Review* 46 (1931), 465).

Photo taken at the Roman fort at Birdoswald on Hadrian's Wall, in 1929. From right to left: R.G. Collingwood, S. Applebaum, K. Stade, H.S. Addison, F.G. Simpson, R. Turner, E.B. Birley, J. Birley, J. Charlton. At the background on the far left stands Th. Hepple, for many decades foreman-excavator of F.G. Simpson and later of E.B. Birley. (Photo in the possession of Dr. Grace Simpson).

R.G. Collingwood (the original picture of this photograph was owned by the late Sir Malcolm Knox).

CONTENTS

PART TWO

PART THREE

ACKNOWLEDGMENTS

The present study has its origins in a doctoral thesis presented at the University of Leiden. It is also the product of a number of people, who have supported me in various ways. I would like to use this opportunity to express my thanks to them.

I am in the first place obliged to Professor G. Nuchelmans and Professor C.J.M. Schuyt for their encouragement and interest in my study from the very beginning. I remember with particular gratitude the valuable discussions with them on the composition of the book.

It is difficult to express in brief the great indebtedness I owe to Professor W.H. Dray, since it is of such a special nature, both in kind and degree. He introduced me to Collingwood's philosophy of history, both in his writings and – during a year's stay at Toronto in 1969-1970 – through his teaching. At a later date I had the opportunity to discuss with him on several occasions special aspects of Collingwood's philosophy of history. Having studied Collingwood's unpublished manuscripts I was more than once impressed by the degree to which his interpretation of various aspects of Collingwood's thought was corroborated by them. It was finally extremely kind of him to be willing to read through the complete manuscript of the book, making many valuable suggestions for improving the English and saving me from a serious error in chapter 8. For the errors that remain I am of course responsible. The same holds for Professor L.O. Mink, who read the first two chapters, for which I would like to express my special thanks. I also thank Dr. A.J. Vanderjagt for his critical reading of parts of the manuscript.

I recall with gratitude the meeting I had with the late Sir Malcolm Knox. Not only was the discussion I had with him on Collingwood's work of great importance for me, but also the way he spoke of his beloved teacher and friend is a precious memory.

Not being an archaeologist, my discussion of this aspect of Collingwood's work was an hazardous enterprise. I am most grateful, therefore, to the archaeologists Professor E.B. Birley, Professor J.E. Bogaers and Dr. Grace Simpson for their willingness to read chapter 5. I also highly appreciated the meetings I had with Professor Birley and Mr. R.P. Wright, and I thank Professor Birley as well for allowing me to publish extracts from a letter written by Collingwood to him.

I am especially indebted to Dr. Grace Simpson. For she has greatly encouraged me to pay special attention to Collingwood's contributions to archaeology. She assured me that they are not only of interest for philosophers, but for archaeologists as well. Besides this, she has always been a most willing source of information and provided me with many valuable suggestions for relevant literature. I also thank her for permission to publish extracts from letters written by Collingwood to her father and to reproduce the photograph of the excavation at Birdoswald.

I am most obliged to the staff of Oxford University Press for their kind co-operation in my search for relevant information on Collingwood. In this connection I would also like to thank Mrs. M.W. Cordy, Deputy Librarian of Pembroke College.

I am grateful to the staff of the Department of Western Manuscripts at the Bodleian Library in Oxford for their cordial assistance during the many hours I spent with them studying Collingwood's manuscripts.

I would also like to express grateful acknowledgment to Mr. T.H. Aston, Keeper of the Archives of the University of Oxford, for his permission to publish Collingwood's reports of 1932 and 1935 to the Faculty of Literae Humaniores.

The person to whom I am indebted most for the final version of the book is undoubtedly Mrs. Teresa Smith. For she has undertaken the difficult task of correcting my English. She has not rewritten the original text, but limited herself to making corrections. I am conscious of the fact that for this reason this book is far from a literary masterpiece. If the English is tolerable, it is the work of Mrs. Smith, and if it is not, the fault is mine.

The discussion of Collingwood's unpublished manuscripts plays an important role in this study. Without the decision of the late Mrs. Kate Collingwood to make these manuscripts available this would have been impossible. This decision will certainly be welcomed by the growing circle of Collingwood scholars, and I am sure the understanding of the many aspects of Collingwood's thought will be greatly enhanced by the availability of the manuscripts. Having limited myself to the study of the manuscripts related to the philosophy of history I am grateful to Mrs. Collingwood for her permission to publish the quotations I used, and also for allowing me to publish extracts of letters by Collingwood.

Finally I would like to thank the Netherlands Organization for the Advancement of Pure Research (Z.W.O.) for the grant they offered me to study Collingwood's manuscripts.

Nijmegen
December 1980

W.J.v.d.D.

ABBREVIATIONS

In the text:

RPh:	*Religion and Philosophy* (1916).
RuPh:	*Ruskin's Philosophy* (1922).
SM:	*Speculum Mentis* (1924).
EPhM:	*An Essay on Philosophical Method* (1933).
PA:	*The Principles of Art* (1938).
Aut:	*An Autobiography* (1939).
EM:	*An Essay on Metaphysics* (1940).
NL:	*The New Leviathan* (1942).
IH:	*The Idea of History* (1946).

From W. Debbins ed., *R.G. Collingwood: Essays in the Philosophy of History* (Austin, 1965):

HSc:	"Are History and Science Different Kinds of Knowledge?" (1922).
NAPhH:	"The Nature and Aims of a Philosophy of History" (1925).
SHC:	"Oswald Spengler and the Theory of Historical Cycles" (1927).
THC:	"The Theory of Historical Cycles" (1927).
LHK:	"The Limits of Historical Knowledge" (1928).
PhP:	"A Philosophy of Progress" (1929).
PhH:	"The Philosophy of History" (1930).

In notes and bibliography:

LM:	List of Manuscripts.
CW:	*Transactions of the Cumberland and Westmorland Antiquarian and Archaeological Society*, New Series.
PSAN:	*Proceedings of the Society of Antiquaries of Newcastle upon Tyne.*
JRS:	*The Journal of Roman Studies.*
RB:	*Roman Britain* (1923, 1932).
RBES:	*Roman Britain and the English Settlements* (1936).
CAH:	*The Cambridge Ancient History.*

CHAPTER ONE

INTRODUCTION

1.1. COLLINGWOOD'S RECEPTION

Collingwood's scholarly reputation is a complicated and variegated affair. For one has not only to make a distinction between his reputation during his life and after his premature death in 1943, but also between his reputation as a philosopher and as an archaeologist and historian. Collingwood himself considered philosophy as his primary occupation and his work in archaeology and history as that of an amateur. This work, however, reached the highest standards and his contributions to archaeology and history have always been appreciated accordingly. Though Collingwood's reputation as the main expert on Roman Britain in the period between the two wars remains unchallenged, modern developments in this field have inevitably superseded his contributions and made them primarily voices from a past period. Philosophy was the other half of Collingwood's scholarly life. In his own thinking there was always a close relationship between philosophy and archaeological and historical practice. His interpreters have not always recognized this connection. I have met archaeologists who were surprised to hear that Collingwood was a philosopher as well, and philosophers who either did not know that he had been a practising archaeologist and historian, or thought it no more than a private hobby.

Collingwood's reputation as a philosopher was very different from the one he gained in archaeology and history. For in the philosophical climate at Oxford between the wars he was always an isolated figure. There were several reasons for this: not only his resistance to prevailing realism, but also his interest in the historical dimension of philosophy and in continental philosophers like Vico, Hegel and Croce. One should add perhaps, his unconventional involvement in archaeological and historical research. Though looked at askance by his colleagues for his ideas, Collingwood was highly appreciated by his pupils for his teaching. "He was one of the spell-binders of my own undergraduate days at Oxford", Dorothy Emmet declares; "and for some of us our delight in his lectures was spiced by knowing that our tutors disapproved of him"[1]. Likewise, the historian Wedgwood asserts that Collingwood's personality

"... in the lecture hall, impressed itself indelibly on the rising generation of historians and philosophers"[2], and Beloff speaks of "... the thrill and stimulus", which he "like so many of his Oxford contemporaries got as an undergraduate from hearing Collingwood lecture... "[3].

Collingwood's reputation as a philosopher began a new career after the posthumous publication of *The Idea of History.* Though it was considered an intriguing and sometimes even an obscure book, it was hailed at least as an important contribution to a field of study neglected until then by English philosophy. It was discussed by the Oxford philosophers Walsh and Gardiner[4], among others, but their assessments did not make clear that Collingwood's ideas were of more than passing interest. Harris, at least, contended in 1952 that "[i]t would appear that Collingwood's work is in the process of being forgotten in this country"[5]. That this did not prove to be true is especially due to Collingwood's reception in North America. The first person to have made fruitful use of Collingwood's ideas on history was Dray in his influential book *Laws and Explanation in History*[6]. He still thought it necessary, though, to declare that he made "no apology" for his attempt to "make sense" of these ideas[7]. Donagan's *The Later Philosophy of R.G. Collingwood*[8] was the first study on Collingwood of a more systematic nature, including not only his philosophy of history, but also a discussion of his philosophy of mind, logic of question and answer, philosophy of art and metaphysics. The next landmarks in Collingwood interpretation were: Mink's *Mind, History, and Dialectic: The Philosophy of R.G. Collingwood*[9] and Rubinoff's *Collingwood and The Reform of Metaphysics: A Study in The Philosophy of Mind*[10].

These studies have contributed to a growing interest in Collingwood's ideas, to the point that one could even speak of a Collingwood revival. Mink's observation that "[s]o many people have called Collingwood an 'unduly neglected' thinker that he is coming to be surely the best known neglected thinker of our time"[11], is already outdated, since Collingwood can hardly any longer be called a neglected thinker. Recent studies by Goldstein[12], Martin[13] and Munz[14] show that Collingwood's views still stimulate new developments in philosophy of history; but social scientists also seem to be increasingly aware of the importance of Collingwood's ideas[15]. The relevance of Collingwood's views is increased, moreover, by the fact that it is realized more and more that many of them are remarkable anticipations of influential modern theories. Donagan and Hayden White, for instance, have pointed out similarities between the views of Collingwood and Wittgenstein[16], Mink and Toulmin have noted those between the theory of "absolute presuppositions" and Kuhn's theory of "paradigms"[17], while Mink has also referred to Collingwood's affinities with pragmatism and existentialism[18]. I would add to this list the similarity of Collingwood's view of science as "problem-solving" (using the "logic of question and answer") to Popper's theory of science[19], as well as the use by both of an idea of "situational analysis" in historical explanation[20].

These modern interpretations of Collingwood differ sharply from the traditional ones, which consider him primarily an idealist and a follower of Hegel or Croce. Though these thinkers – to whom Vico, Kant, Whitehead and Alexander should be added – certainly have influenced Collingwood, he is no "follower" of any of them. As with all important philosophers, he absorbed certain past and present ideas into a vision of his own. It is not possible to assign Collingwood to any philosophical "school", which may partly explain his isolation during his life. Toulmin observes that "[i]n part, the trouble was that Collingwood needed a bigger pond than the Oxford of his time provided"[21], but remarks also that his "... criticism of 'realism' and the 'sense datum theory' has by now been accepted by most of Collingwood's successors at Oxford"[22]. Yet the acceptance of Collingwood's ideas may also be based on a misinterpretation. An example of this is the way the German philosopher Gadamer hailed Collingwood as an adherent of the German tradition of thought[23]. Collingwood's philosophy of history is both too rational and empirical for this contention to be justified. The only way to do justice to Collingwood's views is to study them in themselves, without trying to bring them under simple classification.

1.2. COLLINGWOOD'S DEVELOPMENT

In all the major studies of Collingwood the discussion of his development plays an important role. I do not think, though, that this issue, as it has been discussed until now, is as important as generally believed. I shall therefore simply make some brief comments on it.

The topic of Collingwood's development has become an issue primarily because of the discussion by Knox in his "Editor's Preface" to *The Idea of History*. Knox divides Collingwood's philosophical writings into the following three groups:

> The first consists of what he came to regard as juvenilia, *Religion and Philosophy* (1916) and *Speculum Mentis* (1924). The second begins with the *Essay on Philosophical Method* (1933) and continues with *The Idea of Nature* (which dates, except for its Conclusion, from 1934) and much (1936) of *The Idea of History*. The last comprises the *Autobiography* (1939), the *Essay on Metaphysics* (1940), and *The New Leviathan* (1942). *The Principles of Art* (1938) is akin in part to the second group, in part to the third (IH,VII).

In the second period, from 1933 to 1936, Collingwood was, in Knox's view, at the zenith of his powers; *An Essay on Philosophical Method* is especially valued by him. Between 1936 and 1938 an important change took place in Collingwood's ideas, Knox contends. For while in 1936 Collingwood still believed in the possibility of metaphysics as a separate study, he argues, in 1938 he had given up this idea and, in his *Essay on Metaphysics* and the *Autobiography*, had reduced metaphysics to history. Knox is highly critical of this change of view and considers it a relapse into an extreme form of historicism implying complete scepticism. Though he is of the opinion that its roots are already to be found in the writings of the first group, Knox explains Collingwood's drastic change of mind especially by "one decisive factor which cast a dark shadow over all his later work: his ill health". At about the time the *Essay on Philosophical Method* was prepared for publication during the spring of 1932, he asserts,

> ...Collingwood's health began to give trouble and he was given a term's leave of absence from his college work. It was not then realized that this was the beginning of the ill health against which the rest of his life was to be an heroic struggle. What started to happen at some point during the following years was that tiny blood-vessels began to burst in the brain, with the result that the small parts of the brain affected were put out of action. It was only an intensification of this process when in 1938 he had the first of a series of strokes which eventually reduced him to helplessness.. (IH,XXI).

Since Knox attaches great importance to the influence of Collingwood's illness on the development of his views – a suggestion which has been echoed by others as well – some observations on it are perhaps called for. It should be observed in the first place that the illness which was to trouble Collingwood the rest of his life started at an earlier date than Knox asserts. For it all began with the complications of chicken-pox which Collingwood contracted in April 1931[24]. For a whole year he was seriously ill and in the first term of 1932 he took leave of absence, travelling to the Mediterranean. Secondly, *An Essay on Philosophical Method* was not finished, as Knox contends, in the spring of 1932, but more than a year later. For in a "list of work done" Collingwood writes about this book: "Written November 1932-June 1933. Press June 1933"[25]. So the book was written after a period of serious illness and not, as Knox asserts, before it. Since Knox values highly *An Essay on Philosophical Method* and even considers it Collingwood's most important book, this fact seriously weakens his suggestion that Collingwood's judgment was marred by his illness.

I think, however, that this suggestion may also be challenged for more fundamental reasons. For even if it were conceded that Collingwood's illness marred his judgment, it is far from clear why this should have expressed itself in the form of an extreme historicist and sceptical viewpoint. The suggestion of such a necessary relation seems more illustrative, therefore, of Knox's negative assessment of the latter. I believe, moreover, that Collingwood's alleged historicism is much less disastrous than Knox suggests[26]. In "What 'Civilization' means"[27], written in preparation for *The New Leviathan*, Collingwood also shows that he was well aware of the dangers of an extreme form of historicism[28].

Finally, there is evidence that Collingwood had already developed a theory of metaphysics similar to the one of "absolute presuppositions" – considered by Knox the clearest example of a complete historicism – some years before 1938. For in a lecture on "The Nature of Metaphysical Study"[29], delivered in January 1934 (thus in the midst of Collingwood's "second period"), Collingwood is already developing the view that 17*th* century and modern science are based on certain metaphysical presuppositions[30].

For these reasons, I would contend, Knox's view that Collingwood's development shows after 1936 a sharp change towards an extreme form of historicism is ill founded. This view has also been challenged by Donagan. In his opinion, Collingwood's works after 1933 should be seen as a unity. He sees these works, however, too much as a planned series. The manuscript on cosmology, for instance, which is written in 1933-'34 and is among the unpublished manuscripts[31], would not seem to be "the first version of *The Idea of Nature*", as Donagan contends[32]. Of *The New Leviathan* it can certainly be said that it was not planned by Collingwood as part of an overall plan[33].

While Donagan deals only with Collingwood's works after 1933, both Mink and Rubinoff discuss all his writings. They interpret these as a unity exemplifying a dialectical philosophy. The only change in Collingwood's development, Mink argues, was the one to a dialectical viewpoint, realized after his first book *Religion and Philosophy*. Rubinoff considers *Speculum Mentis* crucial in the interpretation of Collingwood's writings and even sees it as a program laid out by Collingwood for his subsequent works.

I mention these views briefly in order to show how various interpretations have been given of Collingwood's development after Knox's and in reaction to his. My own view, limited to the development of Collingwood's views on history, will be discussed later.

1.3. DESIGN OF THE BOOK

Collingwood's philosophy of history is undoubtedly the most widely discussed aspect of his thought. As we have seen, it was the publication of *The Idea of History* that gave Collingwood's reputation a new dimension. His views on history have been discussed in dozens of articles. In the major books on Collingwood mentioned above, his philosophy of history is also extensively dealt with. Taking into account the great amount of attention paid to Collingwood's views on history and the important role they play in modern discussions on philosophy of history, it is surprising, however, that there has not yet appeared a systematic study on Collingwood's philosophy of history. The present study is an attempt to undertake this task.

Not without reason I have emphasized at the beginning that

Collingwood was not only a philosopher, but also a practising archaeologist and historian. With regard to Collingwood's philosophy of history it is surprising that this fact has hardly been taken into account by his interpreters. It is the more surprising, since Collingwood himself explicitly asserts in his *Autobiography* that his philosophical ideas on history were influenced by his practice. I think it is justified, therefore, to pay special attention to the latter. But while this aspect of Collingwood's work has been mostly unknown, and could have been studied before, the latter cannot be said of Collingwood's unpublished manuscripts. For these have only been available for consultation since 1978. They are extremely important and throw considerable light on many aspects of Collingwood's thinking that either remained obscure or were unknown before. I will limit myself to their bearing on Collingwood's philosophy of history and will not discuss, therefore, their implications for his philosophy of art, philosophy of nature or theory of absolute presuppositions.

By means of a study of the new information provided by Collingwood's archaeological and historical writings and his unpublished manuscripts, I will try to give a clearer picture of – to use Mink's words– "the figure in the carpet"[34] of Collingwood's philosophy of history. To fill in this picture I shall document my arguments thoroughly. I have thought this necessary not only because of the new information involved in these arguments, but also because of the new light this information often sheds upon Collingwood's published writings. I am conscious of the fact that I am, for this reason, rather lavish with quotations. In view of the importance of the subject and its special nature I hope, however, that this will be acceptable to the reader.

The book consists of three parts. The first is based upon Collingwood's published philosophical writings, and discusses, with reference to them, both his development and the reactions which *The Idea of History* roused. The second part deals with the hitherto unknown aspects of Collingwood's work, that is, the unpublished manuscripts and his archaeological and historical writings. The third part offers a general survey of the various aspects of Collingwood's philosophy of history.

Collingwood's development is only discussed from the point of view of his ideas on history. I will argue for the thesis that there is a dividing line in his development between the writings before 1926 and those after. For until that date Collingwood had a realist view of history, while after it he developed an explicitly idealist theory of the past. There is a complicating factor, though, that the realist theory of knowledge was already rejected by Collingwood long before 1926. And, paradoxically, his historical practice was instrumental in this. For, as Collingwood asserts in his *Autobiography*, he used his experience in archaeology and the teaching of the history of philosophy as a "flank attack" on the realist position (Aut, 23-28). This paradox may be resolved, however, if one keeps in mind the different meanings of the word "history" in Collingwood's writings. In *Speculum Mentis* history is primarily dealt with as a "form of experience", along with those of art, religion, science and philosophy, and history as a science is not discussed explicitly. However, between 1926 and 1930 Collingwood directed his attention towards the science of history and developed in his lectures an idealist theory of history. With this theory Collingwood gave an answer to the Kantian question how historical knowledge is possible, a question which had been neglected by him until then, since he was involved in other aspects of history. It is in this context, then, that the re-enactment doctrine was developed and the thesis that all history is the history of thought. They are answers to a purely philosophical question and in Collingwood's historical practice we do not, therefore, see a corresponding change of view from 1926.

The interpretation given above is in line with the description Collingwood himself gives of his development in his *Autobiography*. For he declares that he found in 1928 the "solution" to the problem of the epistemological foundation of history and that his train of thought was completed by about 1930 (Aut,107,115). I do not, therefore, endorse the views of Knox, Donagan and Mink, who all disbelieve the description Collingwood gives of his own development[35].

Besides the change to an idealist theory of history from 1926, Collingwood made between 1916 and 1919 another important change of view, namely to a dialectical position. Mink has rightly pointed out the importance of this fact[36]. For the dialectical viewpoint is all important, not only for a correct understanding of *Speculum Mentis*, but also for Collingwood's philosophy of mind, his theory of concepts and his view of history as a process. Since the latter aspect has, in my view, been too much neglected by Collingwood's interpreters, I pay special attention to it in chapters 2, 3 and 6.

The treatment of *The Idea of History* and its discussion in chapter 3 is based on Collingwood's principle that in studying a subject the history of thought about it should be included (Aut,132). The history of Collingwood's interpretations proves the value of this principle. For there is no aspect of Collingwood's philosophy of history about which his interpreters agree. This fact makes the question of a correct interpretation all the more urgent. The apparent diversity of interpretations is especially due to the fact that they have nearly all been exclusively based on *The Idea of History*, without sufficiently realizing that it is hardly possible to give an accurate account of Collingwood's philosophy of history resting only on this book. It is the aim of this study, then, to put Collingwood's ideas on the subject into a broader context.

The unpublished manuscripts provide this context in a most relevant way. Collingwood lectured on the philosophy of history from 1926 to 1932 and his notes have survived. The lectures of 1926 and 1928 are undoubtedly the most important ones. They are completely written out and could be correctly described as brilliant essays rather than mere notes. They are not only of importance for the considerable light they shed on Collingwood's views and their development, but also because topics are discussed in them which are still of great current interest. The same can be said of many parts of other manuscripts. The manuscripts also contain the products of two large projects on which Collingwood worked concerning cosmology and folklore, written respectively in 1933-'34 and 1936-'37. I shall only discuss the latter, however, in so far as they are relevant for a better understanding of Collingwood's philosophy of history.

The discussion of Collingwood's archaeological and historical practice is based on the exploration of Collingwood's own thesis that his philosophy of history rests to a large degree on this practice. Though it is not possible to give an exact indication of the relation between Collingwood's theoretical and practical work, I would contend that a study of the latter often clarifies the former. To give some examples: the method of selective excavation well illustrates the logic of question and answer and the principle that one should look for the purposes expressed in archaeological remains can be considered a "practical" exemplification of the re-enactment doctrine. I do not think it suffices, though, simply to give some examples from Collingwood's archaeological and historical practice. To put the latter in its proper context it is necessary also to have a general view of Collingwood's contributions in this field. I discuss briefly, therefore, the books and most important articles Collingwood wrote.

In the final part of this study an attempt is made to give a general interpretation of the most important aspects of Collingwood's philosophy of history. This is based on the discussions in the previous parts and therefore often refers to them. In reaction to the existing interpretations of the various aspects of Collingwood's philosophy of history, I develop my own. All the subjects discussed in chapter 3 are accordingly dealt with again in this part. Taking into account, however, Collingwood's unpublished manuscripts, as well as his writings on archaeology and history, one is forced to the conclusion that there are many aspects of his philosophy of history that have insufficiently been discussed until now or have not been considered at all. Examples are Collingwood's philosophy of mind and his views on the historical process, the principle of rationality, historical evidence, the logic of question and answer, the relation of history to natural and social science and the practical dimension of history. For this reason many new aspects of Collingwood's philosophy of history come up for discussion, which have not been dealt with in the previous parts. Though others will have to judge the cogency of the interpretations given, I think the conclusion may safely be drawn that Collingwood was a more "complete" and accordingly a more important philosopher of history than has been generally realized hitherto.

PART ONE

CHAPTER TWO

THE DEVELOPMENT OF COLLINGWOOD'S THOUGHT ON HISTORY

2.1. FROM *RELIGION AND PHILOSOPHY* TO *SPECULUM MENTIS*

Religion and Philosophy, (Oxford, 1916) was Collingwood's first book[1]. He had already begun writing it in 1912, the year he both took a first class degree in *literae humaniores* and was elected to a philosophy fellowship at Pembroke College (in fact he was elected before taking his degree)[2]. In 1914 the book was finished[3]. In this work, Collingwood develops a highly intellectualised view of religion, which is stated right at the beginning as follows: "This book is the result of an attempt to treat the Christian creed not as dogma but as a critical solution of a philosophical problem" (RPh, XIII); and further on: "In the first place, religion is undoubtedly an affair of the intellect, a philosophical activity. Its very centre and foundation is creed, and every creed is a view of the universe, a theory of man and the world, a theory of God" (RPh, XV).

It is clear that Collingwood considers the relation between religion and philosophy a very close one, even to the point of being identical. The same can be said of the relation between religion and history. In *Religion and Philosophy* Collingwood displays a limited view with regard to history. He equates it with "the gradual and cumulative experience of facts" (RPh, 48). This "pure history", however, cannot exist without philosophy. "There is no such thing as an entirely non-philosophical history", Collingwood says:

> History cannot proceed without philosophical presuppositions of a highly complex character. It deals with evidence, and therefore makes epistemological assumptions as to the value of evidence; it describes the actions of historical characters in terms whose meaning is fixed by ethical thought; it has continually to determine what events are possible and what are not possible, and this can only be done in virtue of some general metaphysical conclusions... It is equally certain that philosophy is impossible without history; for any theory must be a theory of facts, and if there were no facts there would be no occasion for

theory (RPh, 46-47).

So history and philosophy are from the epistemological point of view closely related to each other: history needs philosophy for a real understanding, and philosophy needs history for the facts. As for the objects of both intellectual activities, Collingwood arrives at a conclusion that goes even further and declares them to be identical:

> History, like philosophy, is the knowledge of the one real world ... History *a parte objecti* – the reality which historical research seeks to know – is nothing else than the totality of existence; and this is also the object of philosophy. History *a parte subjecti* – the activity of the historian – is investigation of all that has happened and is happening; and this is philosophy too ... History and philosophy are therefore the same thing (RPh, 51).

In this phase of his thought Collingwood's ideas on history were not yet fully developed. For instance, no definition is given of any special field of interest for history, history being directed towards the totality of existence; there is no mention of the specific epistemological problems concerning the study of the past and moreover no reference is made to the past as a distinguishing characteristic of history. In his later development Collingwood was to deal extensively with precisely these topics. Of course, we must keep in mind that in *Religion and Philosophy* history is only dealt with obliquely and that his main interest lies elsewhere. In this context the only role assigned to history as such is the ascertaining and collection of facts.

On August 8th, 1919, Collingwood delivered an address at the Ruskin Centenary Conference, held at Coniston, under the title *Ruskin's Philosophy*[4]. This address is an interesting document in the evolution of Collingwood's thought. Here we find history discussed in a way which illustrates certain characteristics of his ideas on history which were to play a pivotal role in his future philosophy. It seems paradoxical to talk about philosophy and history in connection with Ruskin, because however versatile Ruskin's mind, he was neither a philosopher nor a historian. Collingwood clearly recognises this and it is, indeed, his starting-point in his discussion of Ruskin's ideas. We come here to a fundamental element in Collingwood's thought – we might call it a fundamental principle. When he talks about religion, history, art, science or philosophy, Collingwood's primary frame of reference is the treatment of these subjects as common aspects of the human mind. Already in *Religion and Philosophy* he says:

> Just as every man has some working theory of the world which is his philosophy, some system of ideals which rule his conduct, so every one has to some degree that unified life of all the faculties which is a religion (RPh,XVII).

In *Ruskin's Philosophy* Collingwood defines this implicit philosophy of a man as "certain central principles which the man takes as fundamental and incontrovertible, which he assumes as true in all his thinking and acting" and calls it a "ring of solid thought" (RuPh,10)[5]. The task of the philosopher, then, is to discover what people's philosophy is (RuPh,11). The central topic of Collingwood's address is Ruskin's "ring of thought", which he characterizes as "the historical habit of mind" (or thought) and is described by him – interestingly enough – as "historicism"[6]. Collingwood contrasts this type of thought with "logicism". To the latter he ascribes the tendency to look for general laws and to subsume facts under them, which results in a "contempt for facts", "a habitual intolerance" and "a tendency towards monotony and rigidity in all kinds of mental work" (RuPh,13). Historicism, by contrast, shows a respect for facts in preference to theories and laws, and possesses a "natural inclination" toward tolerance, "for it respects facts to such an extent as to suppose that nothing can ever have existed unless it had something to say for itself. This induces a broad outlook and a readiness to study, not without sympathy, ideals which differ widely from one's own" (RuPh,15). Collingwood finds the starting point for this way of thinking in the philosophy of Hegel. As another example he mentions the belief in the unity of mind – in contrast with faculty-psychology – which is exemplified by Ruskin's work. Though Ruskin himself was hardly acquainted with Hegel's philosophy Collingwood also refers to historicism as "Hegelism".

It is clear from Collingwood's discussion of Ruskin's ideas that he fully agrees with his historicism. It would, perhaps, be more correct to say that Collingwood projects his own ideas on to Ruskin's. For this reason, it is interesting how Collingwood deals with the concept of history as expressing a habit of mind. This habit will form the essential background to the rest of his intellectual career. It is also apparent that Collingwood highly appreciates certain aspects of Hegel's thought. The best example of this is to be found, however, in *Speculum Mentis*, which appeared a few years later, in 1924.

Speculum Mentis (Oxford, 1924), with *An Essay on Philosophical Method* (Oxford, 1933), is the best worked out and also most important of Collingwood's philosophical books. Nevertheless, I do not follow Rubinoff in his argument that in *Speculum Mentis* we find the expression of Collingwood's master plan and that the whole of his subsequent work can be seen "as a projection of the programme outlined in *Speculum Mentis*"[7]. Though there are certainly continuities in Collingwood's thought, his mind is too versatile and always too much engaged in both empirical and speculative studies, for a view of Collingwood's development as the working-out of a single plan, however complex, to be other than artifical.

Mink's opinion of *Speculum Mentis* is less pretentious, but he too regards it as "in certain ways the most illuminating of his books"[8]. This judgment by Mink is based on the fact that in *Speculum Mentis*

Collingwood develops for the first time a consistent dialectical philosophy. He is certainly right in emphasizing the difference from *Religion and Philosophy*. We have seen how Collingwood in this earlier work ends up postulating an identity between religion and philosophy on the one hand, and history and philosophy on the other – certainly not a product of dialectical thought[9].

We can only guess at the "moment of *kairos*"[10] through which Collingwood became a dialectical thinker for the rest of his life. It must have been before his address on *Ruskin's Philosophy* in 1919, because one cannot support Hegelism without thinking dialectically. Most probably Collingwood's close acquaintance with the Italian idealists Croce, Gentile and de Ruggiero played a vital role in this development. It is beside the mark, however, to describe Collingwood as a "pupil" of Croce as is so often done. Though he translated some works of Croce and was undoubtedly influenced by his philosophy, especially on art, he was also critical of certain aspects of his thought[11]. According to H.S. Harris, Collingwood was mainly influenced by Gentile[12]. There exists little evidence for this view. The extensive correspondence of Collingwood with de Ruggiero, however, proves that there was at least from 1920 a close relation between them[13].

In Oxford, J.A. Smith, from 1910 to 1935 Waynflete Professor in Moral and Metaphysical Philosophy, whom Collingwood was to succeed in this chair, was deeply influenced by the Italian idealists[14]. It is most probable that Smith inspired Collingwood and that their common interest was hardly accidental. In his *Autobiography*, Collingwood mentions him as "an intimate and beloved friend"[15].

Published in 1924, *Speculum Mentis* was written between December 1922 and August 1923[16]. It is surely the most Hegelian of Collingwood's books. On 2 October 1920 he had already written to de Ruggiero, after a visit from the latter to Oxford: "You have left me full of new thoughts and new hopes, and quite ready to undertake the task you assign to me, of being the only English neo-Hegelian"[17]. Fortunately, *Speculum Mentis* differs from the works of Hegel in being extremely readable. Like all his books, Collingwood wrote *Speculum Mentis* for a general public in the belief "that a philosophy which cannot be written in plain terms, without reliance on the jargon of any school, must be a false philosophy" (SM, 11). This does not mean that the subjects he writes about are always easy to understand. Collingwood's dislike, however, of a "philosopher's philosophy" is clearly shown in a book review he wrote while he was working on *Speculum Mentis*, in which he says:

> Few of our leading philosophers understand how to write a book. As a rule, when they publish, they either reprint a scratch collection of technical essays on points of detail, or else expand one such essay into a volume. In either case they write as specialists for specialists, and seem hardly to suspect that a

> printed book addressed only to specialists is thrown away[18].

The theme of *Speculum Mentis* is essentially the self-knowledge of mind, the question of the unity and diversity of mind being the crucial problem. In the Middle Ages mind was still a unity, religion, art and philosophy being closely related to each other. This unity was disrupted by the Renaissance, when mind became fragmentized and the relation between the different aspects became more and more problematic. Collingwood distinguishes art, religion, science, history and philosophy as "forms of experience". Their sequence is important, because they are not five species of a genus, but form a natural order of their own. This is the case for both the development of individual man and the history of mankind (SM, 50-54). The relation between the various forms of experience is dialectical. Its fundamental characteristic is the correlation between the implicit and explicit: what is implicit at one level, becomes explicit at the next, with the turning of an error into a truth. This process comes to an end in philosophy, when mind has only itself as object and arrives in this way at explicit self-consciousness. Collingwood himself made clear that he considered the dialectical relation between the implicit and the explicit the main difference between *Speculum Mentis* and *Religion and Philosophy*, saying of the latter:

> With much of what that book contains I am still in agreement; but there are certain principles which I then overlooked or denied, in the light of which many of its faults can be corrected. The chief of these principles is the distinction between implicit and explicit. I contended throughout that religion, theology, and philosophy were identical, and this I should now not so much withdraw as qualify by pointing out that the 'empirical' (i.e. real but unexplained) difference between them is that theology makes explicit what in religion as such is always implicit, and so with philosophy and theology (SM, 108).

Art is the lowest form of experience. It is pure imagination, ignoring the question of the reality of its imaginary products; it falsely conceives itself as merely intuitive. In religion, the figments of the imagination are asserted as real. The distinction between symbol and meaning is only implicit here, and is made explicit by theology. In science, for the first time the mind becomes explicitly rational. Science is abstract and in its purest form a priori and deductive. The abstract, however, must rest on the concrete, because it cannot rest on the more abstract. So science becomes empirical and bases itself on historical fact. In history, then, the individual fact is made explicit. It breaks down, however, because the totality of facts can never be reached. This can only be done by philosophy, which sees the infinite whole of fact as "the nature of the knowing mind as such" (SM, 241).

Of course this survey is much too short to do justice either to the substance of *Speculum Mentis* or to the often fine pieces of philosophising about the different forms of experience and their mutual relations. In a letter to de Ruggiero of 24 August 1923, Collingwood writes about his book most illuminatingly because he shows very briefly its fundamentals:

> I have just finished the book which I call *SPECULUM MENTIS*, which is my Philosophy of the Spirit, and I must write and tell you about it. It began from two ends at once – (i) elaborate empirical studies of art, religion, science, history and philosophy and their various modifications and forms: these being my empirically chosen 'forms of the spirit' (ii) the principle that all these must be identical (in the sense in which you proved the identity of science and philosophy in your *Scienza)*. The problem was to find a principle which would serve to articulate without destroying the unity of the spirit. I found this by reflecting that the spirit was not an infinite given whole but a process of self-discovery and self-creation, and therefore the principle required must be simply self-knowledge with its negative self-ignorance. I actually reached this from the empirical side, by discovering (a) that all religion *is* metaphor, but cannot *admit* that it is metaphor without ceasing to be religion: (b) that all science *is* hypothesis, but similarly cannot admit it: and so on: thus each form contains in its definition a negative element, viz. 'in this form, the mind is *an sich* such and such, but not *für sich*'. Thus in each form there is a contradiction between its own view of itself and an outside observer's view of it (artist's theory of art and philosopher's theory of art etc.) and this contradiction arises from self-ignorance[19].

I propose to concentrate now on Collingwood's concept of history in *Speculum Mentis.* This concept is difficult to grasp and has been sometimes misunderstood. But a clear grasp is essential for a better comprehension of the subsequent development of Collingwood's thought on history.

In *Speculum Mentis* history is in the first place one of the five forms of experience. As Rotenstreich has rightly pointed out, history however is also "the background against which all the realms of mind are made manifest"[20]. For a dialectic of Hegelian type it is of course natural to see history playing an essential role in the whole system. Hegel, however, did not consider history as a separate category in his system, as Collingwood does. For this reason Rotenstreich criticises the latter for not clarifying the double meaning of history[21].

Rubinoff points to another distinction within the concept of history as developed in *Speculum Mentis* – history as a form of experience and history as a science. According to Rubinoff "[t]his ... difficult question ... draws attention to what is perhaps the greatest single weakness in Collingwood's system"[22]. This is a severe judgment from someone who is apparently trying to solve as many contradictions and uncertainties as possible in Collingwood's work. Further on he says that Collingwood "therefore writes at different levels in different places without explicity acknowledging it, and it is unhappily left to the reader to decide on which occasions he is referring to history as a general habit of mind and on which occasions he is referring to it as a special discipline, that is, as a science"[23].

So one can distinguish three meanings of the concept of history in *Speculum Mentis*: a) as a form of experience, b) as a science, c) as the background of the system. One can discern, however, a fourth meaning in the view that history plays (implicitly) a role in all forms of experience "before" the one of history. For instance, Collingwood says that "all art, religion, and science rest on perception or history, as the earlier terms of any dialectical series on the later" (SM, 207-208); "[t]here is... no feature of experience, no attitude of mind towards its object, which is alien to history" (SM,218); and "the work of art, God, and the abstract concept are all attempts on the part of thought to reach the organized individuality of history" (SM,220).

In my view the last two meanings do not present serious problems. What is essential is the relation between history as a form of experience and as a science. History as a form of experience is variously described by Collingwood as "historical consciousness" (SM,208), "the historical spirit" (SM,212, 219), "historical thought" (SM,217), or "the historical conception of reality" (SM,214). It is crucial to note, however, that the "historical consciousness" etc. forms a scale of different levels, an idea which was to play a pivotal role in Collingwood's later *Essay on Philosophical Method*. History is characterized by perception. Collingwood calls it "an ultimate form of historical thought which is the most rudimentary of all" (SM,204). "In perception", he says a few lines further on, "we are immediately aware of our object, which is a concrete and therefore historical fact: perception and history are thus identical". Perception must not be confused with sensation, Collingwood warns us, because perception always contains thought and therefore an element of mediation. If history is based on the activity of perception, its object is "fact as such" (SM,211).

Perception and attention to facts are characteristics of history as a form of experience and as such they must not be associated with history as a science. The perception of objects in the present is in this sense historical. Historical consciousness develops, however; the *cogito ergo sum* of Descartes, for instance, Collingwood interprets as saying that the concrete historical fact, the fact of my actual present awareness, is the root of science (SM,202). The laboratories of modern science since the Renaissance are another example of the "historising" of science. Having said this, Collingwood arrives at a crucial passage, where he turns from the concept of history as a form of experience to the concept of history as a science:

The form of thought which we call *specifically* history came to its

> maturity in the course of this process. History *in the special sense of the word* came into being in the eighteenth century and shot up to a gigantic stature in the nineteenth. It is an absolutely new movement in the life of mankind. In the sense in which Gibbon and Mommsen were historians, there was no such thing as an historian before the eighteenth century (SM,203) (italics mine).

The transition from one meaning of the concept of history to the other is made clear some pages further on:

> The object of history is fact as such. To determine facts far distant in space and time is not the essence of history but its climax, the very heroism and bravado of the historical spirit in its defiance of empirical limitations. There is a growth of history, as we have already seen that there is a growth of art and religion and science, which proceeds from perception through annals and memoir to history in this highest sense (SM,211).

As said before, perception is not limited to the immediately given:

> To perceive is to see what we do not see, to grasp the object as a whole in a synthesis of front and back, top and bottom, past, present, and future; all this is implied in my perception of the ink-pot I see before me. Thus in perception we have that very identical process of reconstruction from data which is the essence of history. In the work of the annalist this is extended (SM,212).

If Collingwood says that "[i]t is only when the concept of fact has become explicit that we can ... reach, for the first time, historians' history" (SM,216), it is implicitly asserted that there is also a non-historians' history. By "historians' history" he means "the revolution in historical thought which took place in the eighteenth century" (SM,216). While "[t]he whole past and present universe is the field of history, to its remotest parts and in its most distant beginnings" (SM,217), history as a form of experience is itself constantly in a process of development. Only when it becomes conscious of its own characteristics, as they become explicit, is the stage of "real" history reached: "historians' history" or history in its highest sense. It is the modern science of history, as it has developed since the eighteenth century, which embodies this stage. So the science of history is a natural product of history as a form of experience – indeed, its ideal realization.

Collingwood also speaks about the study of history as "empirical". In the letter quoted above to de Ruggiero of 24 August 1923, he says that one of the starting points for *Speculum Mentis* was "elaborate empirical studies" of history, among other things. In the book itself, at one point he translates a difficulty "into empirical terms" and then says that he "insists upon this difficulty not as a hostile and unsympathetic critic of historians, but as an historian himself, one who takes a special delight in historical research and inquiry" (SM,235), moving on to discuss certain problems concerning importance and selection in writing history (SM, 235-237).

So Goldstein is not completely right when he claims that "one absolutely fundamental way in which what he [Collingwood] says about history in *Speculum Mentis* differs from what he says in *The Idea of History* is that in the former there is not the slightest evidence of attention to the actual practice of history. History in that book is only a dialectically motivated stage in a series of modes of experience and it has the character it does, not because of anything historians do but because *Speculum Mentis* requires that it does ..."[24]. In another article Goldstein claims that for Collingwood "[a]t this stage of his thought, history... is not the discipline real historians work at", and that in *Speculum Mentis* history is discussed "with almost no attention to the actual activities of historians"[25]. It is undoubtedly true that Collingwood in *Speculum Mentis* does not focus his attention on the science of history as such. History is there primarily treated as one of the forms of experience. The science of history, however, is not in contradiction with this concept of history. On the contrary, it must be seen as the highest stage of it. Because Collingwood in later life paid so much explicit attention to the science of history – his fame, indeed, is based largely on this aspect of his thinking – the treatment of history in early works such as *Speculum Mentis* is easily misunderstood. This had another starting-point, and dealt with the science of history obliquely and in a special context. Hence it is not correct to condemn Collingwood for his "failure" to distinguish between the two concepts of history and even less to consider it "perhaps the greatest single weakness in Collingwood's system", as does Rubinoff[26].

In *Ruskin's philosophy* Collingwood considered the modern science of history as the best example of the "historical habit of mind" or thought. Contrasting it with the logical, he says: "In calling these two types of thought the logical and the historical respectively I do not mean to imply that the first has no dealings with history nor the second with logic. There were historians in the seventeenth and eighteenth centuries; but their notion of what history ought to be was a trivial and elementary thing compared with the vast development of historical knowledge and historical method which took place in the nineteenth century" (RuPh,16). We have seen how Collingwood, in his address on Ruskin, identified the "historical habit of mind" with Hegel's thinking. As an example he gives the attempt "to live up to the maxim that in every conflict or dispute there is right on both sides", from which he concludes that "[t]he history of a struggle – and all history is the history of struggles – cannot be written by a man who believes that one party must have been simply right and the other simply wrong" (RuPh,23).

From these examples it is clear that Collingwood, whether dealing with history as a habit of mind in *Ruskin's philosophy*, or as a form of experience in *Speculum Mentis*, is far from excluding the science of history. There is no sharp distinction between the two concepts of history. The science of history must rather be considered the logical outcome of the more fundamental concept of history.

2.2. COLLINGWOOD AND REALISM

The two concepts of history discussed above will hold our attention, because after *Speculum Mentis* Collingwood shifted emphasis in his treatment of history, concentrating from then on on the science of history. This shift was combined with a completely different epistemological attitude towards history. Until *Speculum Mentis*, Collingwood's view of history can be characterized as realistic, while afterwards he turned to an explicitly anti-realistic conception. We will see that these shifts are closely related to each other and that Collingwood's special attention to the problems of the science of history is accompanied by an anti-realistic view. We have not discussed until now Collingwood's attitude towards realism. For a better understanding of his thinking about history this is necessary, however, especially as, according to Collingwood himself, the questions were related.

The topic of realism runs like a continuous thread through Collingwood's *Autobiography*, written in 1938 and published in 1939. For this reason it is one of the best known aspects of his thought. One should add, not only because the *Autobiography* is perhaps after *The Idea of History* Collingwood's most widely read book, but especially because of the vehement and often emotional way he attacks realism in it. He even goes so far as to conclude the review of his intellectual career with the extremely harsh judgment that the realists ("the minute philosophers of my youth"), "for all their profession of a purely scientific detachment from practical affairs, were the propagandists of a coming Fascism" (Aut,167).

I shall try to explain Collingwood's attitude towards realism, by paying special attention to its relation with the concept of history. In his *Autobiography*, Collingwood tells us that he was "thoroughly indoctrinated with [the "realist" school's] principles and methods" during the time he studied at Oxford, adding that "though I called myself a 'realist', it was not without some reservations" (Aut,22). Here it is interesting to compare what he has to say about the subject in *Religion and Philosophy*, which he finished in 1914. Discussing the question of the distinction between an object like a table and the thought about it, he adds the following note:

> I believe that the argument I have tried to express contains little if anything which contradicts the principles of either Realism or Idealism in their more satisfactory forms. There is an idealism with which I feel little sympathy, and there is a so-called realism which seems to me only distinguishable from that idealism by its

> attempt to evade its own necessary conclusions. But I do not wish to appear as a combatant in the battle between what I believe to be the better forms of the theories. Indeed, if they are to be judged by such works as Joachim's *Nature of Truth* on the one hand and Prichard's *Kant's Theory of Knowledge* and Carritt's *Theory of Beauty* on the other, I hope I have said nothing with which both sides would not to some extent agree; though I can hardly expect to avoid offending one or other – or both – by the way in which I put it (RPh,101).

This passage clearly demonstrates the "diffidence of youth" (Aut,22) he felt by then. In his discussion of history one discerns however in *Religion and Philosophy* a plainly realistic approach, when he says: "History must be regarded ... simply as *objectivity*; as the real fact of which we are conscious. History is that which actually exists; fact, as something independent of my own or your knowledge of it" (RPh,49). In "The Devil", an essay published in 1916, we find the same approach: "Intellectual evil consists in setting up that which I believe as the standard of truth, whereas I ought rather to test and if necessary reject my beliefs by comparing them with reality ... [E] vil arises when man takes himself, exactly as he stands, for the measure of all things"[27]. Further on he says: " ... we must bear in mind that truth, reality, God, are real things existing quite independently of our individual life and private opinions ... [Man's nature] is incapable of being the standard of anything. It is itself in need of a standard, and that standard, which for science is reality, for religion is God"[28].

According to his *Autobiography*, by 1914 Collingwood already had reservations about the realists. The passages quoted seem to contradict this. We should note, however, that Collingwood's resistance to the realists was characterized initially by a dislike of only certain aspects of this movement. In the beginning there was no all-out attack on its stronghold, but only a "flank attack" (Aut,23,28). This flank attack was directed against the mistaken approach, in Collingwood's eyes, of the realists towards history. It consisted of two manoeuvres. In the first place archaeology had taught him that for every excavation a concrete question had to be asked. Collingwood extends this lesson to all knowledge, on the principles of Bacon and Descartes, that "knowledge comes only by answering questions, and that these questions must be the right questions and asked in the right order" (Aut,25). The realists, by contrast, "talked as if knowing were a simple 'intuiting' or a simple 'apprehending' of some reality" (Aut,25). On this view, reality is just waiting, as it were, to be grasped by a knowing mind, while according to Collingwood knowledge is essentially an activity consisting of both questions and answers.

The second aspect of Collingwood's "flank attack" concerned the fact that the realists used to treat philosophers merely as targets on which to practise their philosophical wit. The question of whether a certain philosopher really held the ideas imputed to him did not interest them at all. Collingwood was critical of this attitude, because in his eyes the ideas of other philosophers were not only distorted in this way, but also misused. He proposed instead a historical approach, teaching his pupils "that they must never accept any criticism of anybody's philosophy which they might hear or read without satisfying themselves by first-hand study that this was the philosophy he actually expounded" (Aut,27). That Collingwood really held this attitude in philosphy (it is interesting to note that he calls it a "habit of mind" (Aut,26)) can be demonstrated by a review of a book on Nietzsche (written in 1918) which begins as follows:

> Early in 1915, when everybody in this country was talking of Nietzsche as the greatest of the three arch-fiends who had preached Germany into madness, it seemed strange that no one should come forward and in plain words tell the public what he actually said, and how far it justified the title with which an ingenious bookseller in Piccadilly has immortalized the present conflict – 'the Euro-Nietzschean War'[29].

Collingwood's offensive against realism was completed by his rejection of their propositional logic, and his offering a so-called "logic of question and answer" instead (Aut,36-37). This logic may be considered the philosophical elaboration of the principle mentioned above of the question and answer activity of knowledge. According to the newly coined logic, questions of the meaning, agreement and contradiction or truth and falsehood of propositions cannot be answered in the abstract, but must be seen in relation to the questions the propositions concerned were meant to answer (Aut,33). The identification of the original question is the result of historical research; so we see here again an illustration of the importance of the historical approach in dealing with philosophical questions, an aspect grossly neglected, in Collingwood's eyes, by the realists. His logic of question and answer Collingwood worked out in 1917 in a book called *Truth and Contradiction*, but he could not find a publisher for it (Aut,42-43)[30].

By the end of the war Collingwood seems to have rounded off his objections against realism. They were based on an extensive study of realist literature, judging by the amount of it he mentions in a letter of 2 October 1920 to de Ruggiero[31]. He ends the same letter by saying about the movement of "new-realism": "if I can, as I hope to do, devote some time this winter to studying the history of the movement, you shall hear more".

In a letter to de Ruggiero of 4 November of the same year Collingwood writes:

> ... I am to read a paper to the Oxford Philosophical Society at the end of November on the collapse of modern Realism, which is the first occasion on which I have put my views before the professors and tutors in philosophy here. I think, to judge by what I hear, that

> it comes at the right moment, when most people in Oxford who were realists are giving up their old position and the younger men have broken away from that school: so it may be that I shall find some people willing to listen to me. If that is so, I shall think about publishing the 'Libellus'[32].

The experience of reading his paper to his colleagues must have made a great impression on Collingwood, because 18 years later he writes about it in his *Autobiography*:

> The War ended, I came back to Oxford an opponent of the 'realists'. I had not yet learnt the uselessness of reading papers and holding discussions on philosophical subjects; so, with the intention of putting my cards on the table, I read a paper to my colleagues, trying to convince them that Cook Wilson's central positive doctrine, 'knowing makes no difference to what is known', was meaningless (Aut,44).

His estimation of the impact of his attack on realism, as expressed in the letter of November 1920, turned out to be beside the mark. On 20 March 1921 he writes again to de Ruggiero:

> I read my paper on Realism last November and no one seemed much interested; but afterwards a small and rather exclusive body of philosophers which meets every week under J.A. Smith's presidency asked me to join them, they being all men much older than myself and evidently intending to do me an honour. But it doesn't do much good really[33].

Perhaps it was not very tactful to call the realist movement "the undischarged bankrupt of modern philosophy" (Aut,45) to an audience of realists. In any case, this incident may be considered a dividing-line in Collingwood's intellectual development and probably was the beginning of his isolation in the philosophical arena. Collingwood himself apparently thought about it this way: "So far as my philosophical ideas were concerned, I was now cut off not only from the 'realist' school to which most of my colleagues belonged, but from every other school of thought in England, I might almost say in the world" (Aut,53).

The feeling of isolation, but also a certain air of resignation, is demonstrated by a letter to de Ruggiero of 4 October 1927, when Collingwood says:

> The official philosophy of the day is the realism of Moore and the rest; a very bad philosophy it is, in my opinion, as in yours; but it prevails at present, and those who disagree with it are either abused or merely neglected[34].

Collingwood thought the doctrine of realism that "knowing makes no difference to what is known" especially disastrous in the field of moral philosophy, as it taught that practice had to be divorced from theory. In his view, there was no better way "to train up a generation of Englishmen and Englishwomen expressly as the potential dupes of every adventurer in morals or politics, commerce or religion, who should appeal to their emotions and promise them private gains which he neither could procure them nor even meant to procure them ..."(Aut,48-49).

According to the philosophy textbooks, idealism is the view opposed to realism. It is no surprise, therefore, to find Collingwood always labelled as an idealist. Without further explanation this is not correct, however. We have seen that his rejection of realism was based on different arguments. It was in fact a brand of thought distinctively his own. In my opinion dialectics played a more important role in it than idealism[35]. The concept of dialectic was hardly used, however, in England in the period between the wars, so it is understandable that Collingwood's protests at being labelled an idealist proved to be ineffective. He could accept severe criticism of *Speculum Mentis*, he said, but not that it was "the usual idealistic nonsense", as one reviewer had said – it being in his opinion neither "usual", nor "idealistic" (Aut, 56-57).

One of the rare occasions when Collingwood gives his opinion about idealism is in a letter to de Ruggiero of 20 March 1921:

> I find myself now rather inclined to react against the English idealists because they imported so much of what was *bad* in Hegelism into England; and I find their present successors a real nuisance and my chief enemies. I am even becoming tolerant of Mill, in that he did *try* to get a concept of thought *in fieri*; but the result of the idealist tradition has been to solidify thought into a pure Platonic being[36].

To see the idealists of the time characterized by Collingwood as his "chief enemies" certainly has an air of paradox, to put it mildly. Did he fight a war on two fronts, against both the idealists and the realists? Has Collingwood's philosophy nothing to do with idealism? It certainly has. We shall see later in this study that from the late twenties his concept of history became explicitly idealistic, turning from an – at least implicitly – realistic position. While holding this latter view on history, he had however already for some years been an avowed anti-realist.

Collingwood's position, or rather his development, is a complicated affair. He cannot be classified as belonging to any specific movement, either Idealism or Realism. This does not mean that there are no idealistic or realistic *aspects* to be discerned in his thought. To track these down we will have to go back to Collingwood's work itself.

2.3. HISTORY: FROM REALISM TO IDEALISM

Between 1921 and 1930 Collingwood published some articles on history[37]. They are of great interest, because we can see in them the development of his thought on this subject, including some important changes in it. In the article "Croce's Philosophy of History" Collingwood gives a critical treatment of Croce's *Theorie und Geschichte der Historiographie* (Tubingen, 1915)[38]. His objections are directed against what Croce himself calls "naturalism" or "transcendence". The transcendence of Croce is expressed, according to Collingwood, by his dualism of thought and will, truth and error, thought and life and the idea of the positivity of history. The transcendent attitude he defines as "asserting the existence of a criterion outside the historian's mind by which the points of view which arise within that mind are justified and condemned"[39]. With his objections against Croce Collingwood implicitly criticizes realism. At one point he even characterizes Croce as "the realist, dualist, empiricist, or naturalist, who delights in formal distinctions ..."[40]. That this aspect of Croce's thought is balanced by an idealistic tendency, which is exemplified in his historical writings, does not, however, concern us here.

We have seen that in *Speculum Mentis*, in the sequence of art, religion, science, history and philosophy, history has the distinguishing features of being concerned with facts and being based on perception. In our discussion on history we concentrated on the two concepts of history: as a form of experience and as a science. We will now deal with the aspect of its relation to realism.

In *Speculum Mentis* Collingwood gives a plainly realistic picture of history. "[T]he historical consciousness asserts concrete fact", he says (SM,208) and" [history] does not come to the facts with a ready-made law in its hand and try to force them into it, throwing them away in disgust when they are too hard; it rejoices in their hardness and finds its satisfaction in their very diversity and uniqueness" (SM,210). Other examples are such statements as "[t]he object of history is fact as such" (SM,211), "[a]n historian must state the facts as they happened" (SM,216) or "[h]istory is the knowledge of the infinite world of facts" (SM,231). This idea of history "breaks down" however. To quote the words of Collingwood in a crucial passage:

> History is the knowledge of the infinite world of facts. It is therefore itself an infinite whole of thought: history is essentially universal history, a whole in which the knowledge of every fact is included. This whole, universal history, is never achieved. All history is fragmentary. The historian – he cannot help it – is a specialist, and no one takes all history for his province unless he is content to show everywhere an equal ignorance, an equal falsification of fact. But this is a fatal objection to the claims of historical thought as we have, without favour or exaggeration,

> stated them. History is the knowledge of an infinite whole whose parts, repeating the plan of the whole in their structure, are only known by reference to their context (SM,231).

So the context is always essential to facts and they cannot be studied in abstraction. This has the serious consequence of a complete scepticism: "[i]f history exists, its object is an infinite whole which is unknowable and renders all its parts unknowable" (SM,234). A few pages further Collingwood gives his final judgment: "As long as we pretend to write history, we must claim access to the fact as it really was. This fact, we have seen, is inaccessible. History as a form of knowledge cannot exist" (SM,238). With the addition that "history is the crown and the *reductio ad absurdum* of all knowledge considered as knowledge of an objective reality independent of the knowing mind", Collingwood nevertheless shows the way out of this dilemma. For philosophy supersedes the distinction between subject and object, as practised in history, and realizes that "[t]he world of fact which is explicitly studied in history is ... implicitly nothing but the knowing mind as such" (SM,245).

It is important to notice in this connection that Collingwood distinguishes two definitions of philosophy, namely a dogmatic and a critical one. The first defines philosophy by reference to its object and is self-conscious thought about the way we are aware of a certain object. The second is "the investigation by thought itself of the limitations of its capacity" (SM,254). Philosophy is defined in this case by reference to its method and regarded "as the self-liberation of thought from uncriticized assumptions" (SM,247). Because dogmatic philosophy is defined by its object, each form of experience has its own philosophy: art, religion, science and history, philosophy itself being in essence self-referential.

These dogmatic philosophies must be seen as the philosophical affirmations of the different forms of experience and from the standpoint of critical philosophy they are to be considered erroneous (SM,250-252). In his treatment of the dogmatic philosophy of history, called by him "historical philosophy", Collingwood, in the very first sentence, gives its essence: "The historical form of dogmatism is that represented by modern realism" (SM,281). The substance of it is that "[h]istorical dogmatism is the assertion of fact as ultimately real, and fact means not only the facts of 'history' but the facts of perception" (SM,282). So it starts from the principle of the positivity of the object and the separateness of subject and object from each other, which implies the "denial that the object is conditioned or affected by becoming known to any thinking mind" (SM,282-283).

Collingwood's criticism of historical dogmatism (philosophy) is presented in the form of a severe criticism of realism itself: "... modern realism is essentially inconsistent. It is a halt, or rather a confused running to and fro, between two principles, the abstract concept and the concrete fact" (SM,285).

Each form of experience thinks itself the only valid one. Dogmatic philosophy reinforces this tendency, basing itself on the false abstraction between subject and object and is for this reason a philosophical error. The only way to master this error is to criticize the false assumptions on which the different forms of experience are based and to realize that the true object studied by them is always mind itself. The distinction between subject and object then disappears. We are here at the level of "absolute mind" and philosophy is at the stage of "absolute knowledge": "The various countries on our initial map, then, turn out to be variously-distorted versions of one and the same country ... To explore that country is the endless task of the mind; and it only exists in being explored. Of such a country there is no map, for it is itself its own map. The explorer, the country explored, and the map are one and the same thing" (SM,309).

In concluding our survey of *Speculum Mentis* it is important to note that Collingwood explicitly renounces realism; his discussion of historical philosophy makes this clear (SM,281-287). It is also clear, however, that at the same time his conception of history is a plainly realistic one. History has a non-valid foundation and as a form of knowledge cannot exist. Critical philosophy constitutes the only possibility for overcoming this defect, realizing also that in history mind studies only itself and not an external object. *Speculum Mentis* therefore leaves us in the situation that history on its own cannot find its way out of the realistic fly-bottle. Only philosophy can do the job. The consequence is that in the sphere of absolute knowledge we have left the realm of history as such. For an historian this is hardly acceptable[41]. In *Speculum Mentis* this consequence is mitigated by the fact that history is primarily, though not exclusively, dealt with as a form of experience. When Collingwood, having written this book, turned to the actual practice of the science of history, his treatment of this subject in *Speculum Mentis* – or better, the place he gave it in his system – must have been unsatisfactory to him too. For this reason we see him developing a new approach towards history in the following years.

In "The Nature and Aims of a Philosophy of History"[42], published in 1925, the concept of history that Collingwood develops is still closely linked to that of *Speculum Mentis*. He begins by explaining what the philosophy of history is not. He comments on the traditional meanings of this concept, being either the search for general laws in history or the working-out of a plot, in which the facts are connected by a plan or certain fundamental forces. Collingwood's objections to the first of these so-called philosophies of history are that a) it is based on assumed certainty of facts which is unfounded, b) its conclusions are not really universal, but only valid for certain periods in history, c) it ignores the individuality of history. In his later writings Collingwood will stick to these views on history, opposing the idea of searching for general laws. Collingwood is sympathetic towards the notion of history as expressing a certain plan or plot, but refuses to call it a philosophy of history. It is just history showing nothing apart from its details: "[t]he real plot of history ... is coincident with universal history in all its extent and with all its profusion of detail" (NAPhH,39). If a plot is seen in the form of certain fundamental historical forces, for instance economic ones, Collingwood is less tolerant and calls it "simply a philosophical blunder" (NAPhH,40).

Concentrating on what a philosophy of history should be it is striking that Collingwood begins by mentioning "actual historians" (NAPhH,41). Compared with his treatment of history in *Speculum Mentis* the science of history gets much more emphasis, though the links with this book are still clear. In this respect "The Nature and Aims of a Philosophy of History" represents an intermediate position between *Speculum Mentis* and his later work on the subject.

The emphasis on facts and perception is seen again as the main characteristic of history (NAPhH,42,49). From the first aspect also a sceptical conclusion is drawn, but this time of a different type. We have seen in *Speculum Mentis* that facts can never be studied in abstraction, but only by reference to their context, this having as consequence that the real object of history is nothing less than the infinite whole. This leads to a complete scepticism concerning historical knowledge, and the breakdown of history. Philosophy as absolute knowledge, then, is brought forward in *Speculum Mentis* as the only possible escape from this dilemma.

The scepticism concerning our knowledge of historical facts in "The Nature and Aims of a Philosophy of History" is of a different nature than the one in *Speculum Mentis*. Collingwood gives as an example the battle of Hastings being fought in 1066. What does this fact include?, he asks. One can refer to the tactics of the battle or its political significance, but no one will ever suppose that any one particular account of the battle will be a complete one. "In other words", Collingwood concludes: "when we speak of the battle of Hastings we are speaking not of something known but of something partly known and partly unknown" (NAPhH,42). "[N]o one ever will know what exactly it was that happened", he continues a few lines further, concluding his analysis of historical facts as follows:

> In other words, no fact ever has been wholly ascertained, but a fact may be progressively ascertained; as the labour of historians goes forward, they come to know more and more about the facts, and to reject with greater and greater confidence a number of mistaken accounts of them; but no historical statement can ever express the complete truth about any single fact.
> This is perfectly well known to all historians. No historian imagines that he knows any single fact in its entirety, or that any historian ever will (NAPhH,43).

Collingwood is reluctant to call this doctrine scepticism, "for scepticism implies that no one opinion is preferable to any other; and it is certainly possible to choose between different historical views ... " (NAPhH,43). This is true indeed, when scepticism is defined in the way Collingwood does here. If however complete knowledge is put forward as criterion and all historical knowledge is found falling short of this, one has to come to a sceptical conclusion concerning the latter. And this is exactly what Collingwood had done in *Speculum Mentis*. So according to the criterion of this book the analysis of our knowledge of historical facts, as in "The Nature and Aims of a Philosophy of History", gives certainly no less reason for scepticism. There is an important difference, though, between the two scepticisms. In *Speculum Mentis* history is seen as the knowledge of the infinite whole of facts, which can never be reached, while in "The Nature and Aims of a Philosophy of History" it is stated that no fact can be wholly ascertained. So we have here a scepticism concerning the totality of facts versus the total fact. The first finds its place in the system of *Speculum Mentis* and is the product of a specific theory, while the second is the result of reflection on the work of actual historians. The crucial difference between the two is that according to "The Nature and Aims of a Philosophy of History", the problem of this fundamental scepticism can be overcome by historical thinking itself. No historian will claim to have complete knowledge of a single fact and "[a]n historian speaking to historians speaks on the basis of an assumed agreement on this point ... he does not perpetually qualify his statement with 'in my opinion', 'probably', 'so far as the available evidence goes', just because a qualification of this kind is assumed as a standing order in all historical thinking" (NAPhH,43).

In contrast with this, the analysis of history as given in *Speculum Mentis* leads to a fatal scepticism, a real breakdown of history, which can only be overcome by philosophy. While history in *Speculum Mentis* is, when left to itself, in an epistemologically helpless state, in "The Nature and Aims of a Philosophy of History" it proves to be able to stand on its own feet, at least with regard to this problem of fundamental scepticism. One could consider it a first indication of the autonomy of history – so often emphasized in Collingwood's later works.

"The Nature and Aims of a Philosophy of History", however, is far from having no links with *Speculum Mentis*. History as a form of experience and as a science are now called respectively ideal and actual history. As in *Speculum Mentis* they are closely linked, the difference being the possibility of realizing historical knowledge. Collingwood says about it:

> Ideally, historical thought is the apprehension of a world of fact. Actually, it is the presentation by thought to itself of a world of half-ascertained fact: a world in which truth and error are at any given moment inextricably confused together. Thus the actual object of actual historical thinking is an object which is not 'given' but perpetually in process of being given ... If there is to be a philosophy of history, it can only be a philosophical reflexion on

> the historian's effort to attain truth, not on a truth which has not been attained. The philosophy of history, therefore, is the study of historical thinking: not only the psychological analysis of its actual procedure, but the analysis of the ideal which it sets before itself (NAPhH,44).

Speculum Mentis did not speak of a philosophy of history, but of "historical philosophy", the latter being dogmatic. In contrast, "The Nature and Aims of a Philosophy of History" states, explicitly taking Kant as example, that the philosophy of history must not take a dogmatic attitude, "but a critical attitude, which undertakes the task of inquiring not only into the results of a certain type of thought but into the nature and value, the presuppositions and implications, of that type of thought itself" (NAPhH,45). We find here a different relation between history and philosophy to the one exposed in *Speculum Mentis*, where they are reconciled only at the level of absolute knowledge and history as such is left behind.

The ideal history which is critically studied by the philosophy of history is in fact identical with history as a form of experience as elaborated in *Speculum Mentis*. This is clear when Collingwood says for instance that "[h]istorical thought is one among a number of attitudes taken up by the mind towards the objective world" (NAPhH,44). Collingwood discusses several characteristics of "[t]he historical consciousness in its ideal nature" (NAPhH,45) and mentions in this connection "the knowledge of the individual"; that history is a specification of art, because it tells a (true) story; that it resembles science in using generalizations and that "[h]istory in its fundamental and elementary form is perception" (NAPhH, respectively 45,48,49).

Dealing with perception, Collingwood moves on to a discussion of "the historian in the higher sense", as "the man who is not content to accept what he is told but endeavours to criticise his sources in order to discover, so far as he can, whether they tell the truth" (NAPhH,51). He then emphasizes the need for a critical interpretation of the sources and the impossibility of avoiding taking up a point of view. This analysis winds up with a perspective-theory of historical knowledge:

> [The historian] can only travel from one perspective to another. He can never get outside his own point of view and see it as a monad among monads. He is a monad, not a monadologist; that is to say, he is a necessary victim of the 'egocentric predicament' which holds good of all perception (NAPhH,55).

It is only at the level of philosophy that we can transcend this monadism of historical thought, "to desert monadism for monadology, to see not merely a perspective but the space of perspectives. History is finite thinking, because in its concentration upon its object it suppresses the question of its relation to that object" (NAPhH,55-56). An historian thinks about his object, but not about his thought about his object: "he thinks not about his point of view but from his point of view" (NAPhH,55).

In "The Nature and Aims of a Philosophy of History", we still find the same realistic view concerning historical facts as in *Speculum Mentis*, as the following passages make clear:

> History assumes that there is a world of fact independent of the knowing mind, a world which is only revealed and in no sense constituted by the historian's thought ... [The historian] seeks to study the activities of the human spirit ... by apprehending them in their full actuality, as they really exist in the world of fact. But these actual happenings are always the object of his thought, and never his thought itself ... Consequently he is always the spectator of a life in which he does not participate: he sees the world of fact as it were across a gulf which, as an historian, he cannot bridge ... The historian is thus always thinking of an object other than his own historical thinking ... (NAPhH, respectively 46,47,48).

As in *Speculum Mentis* the conclusion that "[f]act, in its reality, is unknowable" (NAPhH,55), is also put forward. We have seen, however, that there is one fundamental difference. According to "The Nature and Aims of a Philosophy of History" the historian is implicitly conscious of this scepticism inherent in historical knowledge, but is not deterred by it from aiming at such knowledge[43]. Philosophy makes this explicit at a theoretical level: "The world of fact, which for history is an external presupposition of thought, becomes for philosophy a world of perspectives each having at its centre an historical consciousness; a world of worlds of thought each relative to its thinker" (NAPhH,56).

After 1925 Collingwood concentrated on the epistemological problems of the study of history. This aspect of his thought is not only the best known part of Collingwood's work, but was also considered by himself his chief work. There is ample evidence that from 1926 till at least 1930 he was mainly occupied with this problem. On 18 August 1926 he writes to de Ruggiero: "For myself, I am trying to clear up my conception of History – helped greatly, but not wholly satisfied, by both Croce and Gentile, and developing further the view expressed in *Speculum Mentis*. And always pursuing the study of history itself"[44].

To Croce he writes on 5 January 1928:

> ... and the leisure which the chair would give me ... would enable me to pursue the work on the philosophy of historical method which I regard as my chief task in philosophy.
>
> I have learnt from you to regard philosophy as primarily the methodology of history; and, alone (I think) among English philosophers, I have devoted much time to specialised historical work, in order to train myself for a treatment of philosophy from this point of view – I am now 38 years old, and am ready to begin

> the work of systematising the problems of historical methodology as they present themselves to me; but I shall never be able to do this until I get free from college teaching[45].

According to the *Autobiography* the final results of his thoughts on the methodology of history were put on paper while on holiday in Die, in France, in 1928 (Aut,107). This manuscript, together with one of 1926 on the same subject, is still in existence and they constitute the most valuable evidence we have concerning the development of his ideas on history. We will deal with these manuscripts at a later point in this study and will here discuss only Collingwood's published writings.

In "The Limits of Historical Knowledge"[46], published in 1928 but written in 1927[47], we see already a drastic change in Collingwood's approach to the epistemological foundations of historical science as compared with what he wrote about it before. We have seen that in "The Nature and Aims of a Philosophy of History" Collingwood still held a realistic view concerning history. This was a paradoxical situation, not only because his general position was already clearly anti-realist, but even more so because it was, according to the *Autobiography*, his historical and archaeological experience which had played an essential role in his opposition towards the realists[48].

The central topic of "The Limits of Historical Knowledge" is the problem of realism and scepticism in relation to historical knowledge. These subjects were discussed before by Collingwood, but in this article they are dealt with together for the first time. Until now Collingwood had focussed his attention on historical facts, finding these unknowable either "in the whole" or "wholly". In "The Limits of Historical Knowledge" this subject is not brought up again by Collingwood. This time he speaks about "fact" in a more direct sense, meaning by it the evidence or sources on which an historical study has to be based. He now considers this the essential distinguishing feature of history as compared with science, which can make use of experiments. Collingwood is fully aware that this makes history vulnerable to scepticism, being so much at the mercy of the sources which have come down to us. The easy argument that scepticism is contradicting itself, because it claims to possess the knowledge which it formally denies, he does not want to use (LHK,96). Collingwood takes the matter up in a surprising way by appealing to the actual practice of historians. This approach may be quite acceptable and even common in our day, but this was certainly not the case in the philosophical climate at the time the article was written. It was characteristic, however, of Collingwood's philosophy of history – at least after 1925 – that he always tried to keep contact with the actual practice of historians.

Discussing the problem of historical scepticism in "The Limits of Historical Knowledge" Collingwood gives the following argument:

> [W]hen one takes up the study of some difficult historical question as yet unsettled, and enters with well-equipped and honest

> opponents into the *concordia discors* of learned controversy, there is one thing which one cannot fail to observe. This is the existence of what I may call rules of the game. One rule – the first – runs thus: 'You must not say anything, however true, for which you cannot produce evidence'. The game is won not by the player who can reconstitute what really happened, but by the player who can show that his view of what happened is the one which the evidence accessible to all players, when criticised up to the hilt, supports (LHK,97).

The comparison of the study of history with a game is considered by Collingwood to be a real description of what this study is like. "The so-called rules of the game", he says, "are really the definition of what historical thinking is; the winner of the game is the historian proper – the person who thinks historically, whose thought fulfils the ideal of historical truth. For historical thinking means nothing else than interpreting all the available evidence with the maximum degree of critical skill" (LHK,98-99). This being dependent on the evidence is what is meant by the limit of historical knowledge. An historian can never go further than the evidence allows him to do.

In pointing to this fundamental feature of historical knowledge Collingwood also attacks the false illusion that it is the "real" past the historian is aiming at: "[historical thinking] does not mean discovering what really happened, if 'what really happened' is anything other than 'what the evidence indicates'" (LHK,99). Collingwood takes here the important step of leaving aside all realism with regard to our knowledge of the past. We have seen that until then he himself had taken a realistic attitude concerning historical knowledge. In "The Limits of Historical Knowledge", however, he is definite in his rejection of it.

Collingwood acknowledges that all thought has the tendency to consider the objects of our knowledge as real. This "plain man's realism" sees the past as "real" in one way or the other. The necessary counterpart to this historical realism is scepticism: according to this theory a historian ought to know nothing less than the past in its totality, but he will realize at the same time that this is impossible (LHK,100). This realistic attitude towards the past is considered by Collingwood a fatal misunderstanding. The past as such has finished happening and is non-existent. The only objects which are real are present and consist of the evidence a historian has at his disposal. While for a historical realist the only limit to historical knowledge is the past as past – which in fact is limitless – for Collingwood the limit consists of the evidence. Historical problems therefore cannot be separated from historical evidence: "The historian does not first think of a problem and then search for evidence bearing on it; it is his possession of evidence bearing on a problem that alone makes the problem a real one" (LHK,102).

With his anti-realist theory Collingwood solves the otherwise persistent problem of historical scepticism. At the end of "The Limits of Historical Knowledge" he tersely formulates it in a way which is important enough to be quoted fully:

> It thus appears that history is not doubtful at all. It seemed doubtful, to say the least, so long as we imagined its object to be the past as past; but though the question 'what really happened', where 'what happened', and 'what the evidence proves' are assumed as distinct, is necessarily doubtful, the question 'what the evidence proves' is not doubtful. Granted a training in historical methods, and equipment of historical scholarship, without which no one can fairly judge, it is possible to take a particular problem, to study the solution of that problem advanced by a particular historian on a particular review of the evidence, and within the limits of this problem, as stated, to raise the question whether he has or has not proved his case. That question can be answered, by a competent scholar, with no more doubt than must attend any man's answer to any question that can be asked in any department of knowledge. And in the certainty of that answer lies the formal dignity, the logical worth, the scientific value in the highest sense of that word, of historical studies (LHK,102-103)[49].

We have said that after 1925 Collingwood changed his views concerning the epistemology of history. In "The Limits of Historical Knowledge" we find the first element of this in his shift to an explicitly anti-realist position. This is a substantive improvement in the field of the theory of history and still of current interest. For historians still show a tendency, at least implicitly, towards the "plain man's realism", seeing it as their ideal aim to know as much as possible of "the real past". With the more reflective historians this tendency, then, is of course accompanied by the necessary scepticism[50].

The second element of Collingwood's change of viewpoint on history is related to the question of how historical knowledge could be soundly based in a philosophically satisfactory way. This question – the problem of how knowledge of a non-existent past is possible – was not discussed by him in "The Limits of Historical Knowledge". A solution to this problem was found in 1928 and described by Collingwood in his *Autobiography* (Aut,Ch.X). It was on that occasion that he came to his well-known theory that "all history is the history of thought" (Aut,110) and that "historical knowledge is the re-enactment in the historian's mind of the thought whose history he is studying" (Aut,112).

It is surprising indeed that hardly any of the interpretors of the development of Collingwood's thought on history have noticed this clear dividing-line. Debbins even goes so far as to say that "[i]t is my contention that from the earlier essays to the later works there is no significant change in Collingwood's conception of history or philosophy of history. I mean simply that there is nothing in the earlier essays which is inconsistent with what Collingwood says about history or philosophy of history in the later works"[51]. In a review of Debbin's edition of Collingwood's essays on philosophy of history Errol E. Harris says that "[s]ome of the doctrines which were to be developed more fully later appear here, often in embryonic form, not yet fully thought out, and sometimes apparently incompatible one with another"[52]. He does not dwell however on these incompatibilities and the change from realism to anti-realism is not noticed. In an earlier article on Collingwood's theory of history Harris had already tried to show "that the various accounts of history can be welded together into a single coherent theory"[53].

Though Rotenstreich observes "the fundamental change ... in the later stage of his thought"[54], he does not locate it in "The Limits of Historical Knowledge"[55]. Rubinoff did notice the alteration in Collingwood's view: "The transition from the realism of 'The Nature and Aims of a Philosophy of History' to the idealism of the *Idea of History* is represented ... by 'The Limits of Historical Knowledge' (1928) and 'The Philosophy of History' (1930). In the 1928 essay Collingwood seems virtually to be criticizing the very stand he took earlier. Realism with its concept of the historical past regarded as a thing-in-itself is explicitly repudiated"[56]. This transition, however, is then worked out by Rubinoff in a not altogether satisfactory way by interpreting it within the context of a theory about different stages of the distinction between subject and object in Collingwood's thought[57]. In another place Rubinoff develops in this connection the theory that Collingwood's concept of perception has to be interpreted as a dialectically developing scale of forms, a conception which is taken from Collingwood's *An Essay on Philosophical Method* [58].

Goldstein is aware of the change which took place in Collingwood's views on history. "The direction Collingwood's historical thought was taking during the period from *Speculum Mentis* to *The Idea of History* is now clear", he writes. "Increasingly he came to think that the way to avoid the pitfalls of skepticism was to move away from an unknowable past-in-itself and pay rather close heed to what historians do"[59]. But though Goldstein mentions "The Limits of Historical Knowledge" in this connection, his view that Collingwood "increasingly" came to an anti-realistic standpoint, suggests that he did not see this essay as the watershed in Collingwood's development that I consider it to be.

How did Collingwood come to his anti-realist view of history? In "The Limits of Historical Knowledge" he not only gives us no explanation of the change of view, but also provides no epistemological foundation for it. He does provide such a foundation, however, in an article called "Some Perplexities About Time: With an Attempted Solution", an address to the *Aristotelian Society* on 15 February 1926[60]. I shall not follow the complete argument of this article, but rather concentrate on the conclusions in the last pages. These are of great importance with regard to Collingwood's anti-realistic position, which from then on was to be so characteristic of his thinking about history.

Collingwood begins by distinguishing *being* from *existing* and then makes within being the distinction between the actual and the ideal.

"The ideal is that which is thought", Collingwood continues, "but not thought as real or existing; and in this class fall the future, which is possible but not necessary, and the past, which is necessary but not possible"[61]. The present, however, conceived as "the union of present and past", is real:

> Within this present there are, as really as you like, two elements (necessity and possibility), each of which taken singly or in isolation characterizes a being which is not real but ideal – the past and future respectively. Thus the past *as past* and the future *as future* do not exist at all, but are purely ideal; the past as living in the present and the future as germinating in the present are wholly real and indeed are just the present itself[62].

A thing may be both ideal and real. As an example Collingwood mentions duty, which is real in spite of the fact that it only exists for mind. Past and future however are merely ideal. Collingwood, then, continues as follows:

> Hence, if there were no mind, there would at any given moment be no past and no future; there would only be a present in which the past survived transformed and in which the future was present in germ. The past *as past* and the future *as future*, in contradistinction from their fusion in the present, have being for mind and only so. We do call the past, *as such*, into being by recollecting and by thinking historically; but we do this by disentangling it out of the present in which it actually exists, transformed, and re-transforming it in thought into what it was[63].

It is in this theory of the ideality of the past implicitly existing in the present that the epistemological foundation of Collingwood's rejection of historical realism must be found. The theory is also illuminating with regard to the next step Collingwood was to take, namely the notion of all history as the history of thought and of the historian as having to re-enact in his mind the thought he is studying. These notions are to be connected with the idea that the past cannot be seen as separated from mind[64].

"The Philosophy of History", which appeared in 1930[65], is the last of the series of essays Collingwood wrote on history which were not included in *The Idea of History*. In a nutshell, it gives the conclusions he had reached in his thinking about history and it also outlines the programme of future work he still intended to do in this field. For this reason it is worth our attention.

In "The Limits of Historical Knowledge" Collingwood dealt with the science of history. In "The Philosophy of History" he discusses the same subject, but he also returns to the subject of history as a "habit of mind" or a "form of experience", an approach he had used before.

These terms are not used by Collingwood this time, but the meaning is the same when he discusses history as "a universal and necessary human interest", a potential object for philosophical reflection for this reason alone. In this connection it is of interest to refer to a distinction Collingwood had made elsewhere between empirical and philosophical concepts: Mink has drawn attention to the importance of this distinction for a better understanding of Collingwood's concept of history[66].

In "Economics as a Philosophical Science"[67], Collingwood deals with the philosophical concept of economy, distinguishing it from the empirical concept. According to Collingwood, economic concepts such as value and wealth "are various aspects of, or various attempts to describe, a certain form of action which, for the sake of a provisional name, we shall call economic action"[68]. Philosophical thought, then, conceives its object as activity, while empirical thought conceives it as "substance or thing". Empirical economics may be regarded, according to Collingwood, as a branch of psychology. Taking a philosophical approach to economics, on the other hand, for Collingwood means to analyse the essential features of economic action. The fundamental fact, then, of economic action is in his opinion its being utilitarian[69]. In contrast with the philosophical concept, the empirical one is classificatory: " ... we must leave to empirical economics the question of what particular actions are economic or expedient in particular circumstances, and the attempt to classify, in a manner none the less useful for being arbitrary, these particular economic actions"[70]. A philosophical distinction on the other hand "is made not because we choose to make it, but because we cannot think about the subject at all without making it"[71]. Collingwood then makes a philosophical distinction between moral, political and economic actions.

In the same way he gives in "Political Action"[72] a philosophical analysis of the concept of political action, in which for instance the state is not thought of "as a thing but as the collective name for a certain complex of political actions"[73]. Political action, then, is seen by Collingwood as "essentially regulation, control, the imposition of order and regularity upon things"[74].

In "The Philosophy of History" Collingwood does not speak about philosophical and empirical concepts, but about the difference between philosophy and science in general: "philosophy studies the universal and necessary characteristics of things: science their particular and contingent characteristics" (PhH, 121). We can only speak about the philosophy of art or religion, according to Collingwood, if art and religion can be considered as "a universal and necessary characteristic of things". This Kantian phraseology is equivalent to a "form of experience" in *Speculum Mentis*. So for a philosophy of history to be possible, history must be "a universal and necessary human interest". "Historians must live", he continues, "and therefore history must be a trade; but unless history were a universal and necessary human interest the historian's trade would be of less value than the plumber's, because, whereas we pay plumbers to save us having to do our own plumbing, we pay historians to help us to become historians ourselves" (PhH, 123).

It is not difficult for Collingwood to demonstrate the universality and necessity of history. For everything has a past, which is relevant for its present situation. Hence anyone who is really interested in something must be interested in its historical aspect too: "History, as the study of the past, is therefore a universal and necessary human interest – interesting to anybody who is interested in anything – and not the affair of a special professional group" (PhH,124).

The conclusion of this argument, convincing in its simplicity, seems to exclude the professional historians as such, that is, the science of history. This impression appears to be a false one, however, the transition from the "philosophical" concept of history to the "empirical" one being surprisingly smooth. For history is a special kind of interest, Collingwood argues: it is an intellectual interest, a form of knowledge. It is the task of the philosophy of history, then, "to discover the essential characteristics of this form of knowledge" (PhH,124). History as a form of knowledge also has a history of its own, which of course is reflected by the philosophy of history. Collingwood ends the section on "History and the Philosophy of History" as follows:

> It is, therefore, natural that the philosophy of history should follow a course of development parallel to that of history itself. Where historical knowledge exists only in a desultory and casual form, there will be only a very crude and shallow philosophy of history. Where historical knowledge is a highly-organised thing, involving a technique of its own and a consciousness of its own peculiar aims and methods, the philosophy of history will be a definite and individual philosophical science, whose importance in philosophy as a whole will more or less correspond to the importance of history in human thought as a whole. With this clue we can turn to the past and describe the way in which the philosophy of history has developed *pari passu* with the development of history itself (PhH, 125).

The foregoing discussion shows the ambiguity of Collingwood's concept of history. In the first place this concept may refer to the idea of history as the historical consciousness in general, history as "a universal and necessary human interest". Besides this rather vague concept of history (the "philosophical" one), there is history conceived as a science (the "empirical" concept). History in the second sense must be conceptually distinguished from the one in the first sense, but is certainly not separate from it. On the contrary, the science of history must be considered the most developed expression of the historical consciousness: one could describe it even as the "paradigmatic" example of it. The philosophy of history, then, may refer either to the philosophical concept of history or the empirical one. It will understandably concentrate on the latter, that is, on history as an organised form of knowledge. It is important to analyse the two meanings of the concept of history, their mutual relation and their relation to the philosophy of history, as Collingwood worked it out in "The Philosophy of History", because this essay can be considered a sort of miniature of *The Idea of History*. Certain aspects of his theory, however, are put more clearly in the essay than in the subsequent book.

Collingwood continues "The Philosophy of History" with an analysis of the development of historical thought (he calls it a "History of the Idea"). It is interesting to note that in contrast with *The Idea of History*, which starts with antiquity, Collingwood refuses to go back beyond the seventeenth century and begins his survey with Bacon. He gives the following reason:

> Before there can be a philosophy of history there must be a sustained and systematic attempt to build up a body of historical knowledge. It would, therefore, be more reasonable to look for a philosophy of history in Herodotus or Thucydides, Polybius or Livy or Tacitus. But even here we find nothing that can properly be so called; all that we find is statements of fact, even when these are broad and sweeping in their scope, like Polybius's picture of Rome standing in the centre of universal history. After all, the philosophy of history is nothing but the deliberate attempt to answer the question 'what *is* history?' and none of the ancient historians raised the question (PhH, 126).

This argument is not very convincing and even puzzling when compared with what Collingwood himself contends a few pages before about history and its philosophy. History, as "a sustained and systematic attempt to build up a body of historical knowledge", must not be equated with the philosophy of history. The latter tries to find the essential characteristics of the first and in Collingwood's view also has to show how it developed. In this connection the ancient historians he mentions, from Herodotus to Tacitus, certainly deserve to be dealt with as attempts to build up a body of historical knowledge. Collingwood here confuses history with the philosophy of history. In *The Idea of History* this inconsistency is removed and "the history of the idea of history" is seen to begin in antiquity.

I shall not discuss Collingwood's survey of the development of historical thought from Bacon to Croce. He pays special attention to people like Vico, who contributed to a better understanding of the methodological principles of history. In this connection he says about historical knowledge that "it has to be built up by each historian for himself, using the universal and necessary principles of historical thought to interpret the data which the past has left behind it" (PhH, 128).

In the last section of "The Philosophy of History", called "Outline of a Philosophy of History", Collingwood tries to give in only a few pages the conclusions he had reached with regard to some essential characteristics of the study of history. It is much too short to be satisfactory and has therefore hardly been noticed before, but it is still full of fertile ideas. In "The Limits of Historical Knowledge" Collingwood had emphasized the importance of evidence for history. In the last section of "The Philosophy of History" the need of principles is stressed:

> ... data are not enough. They must be interpreted. This requires principles, and the body of principles constitutes historical method or technique. Some of these principles are scientific in character, that is, they concern particular groups of evidence, and compose the special sciences of archaeology, palaeography, numismatics, and so forth. Some are philosophical, that is, they apply universally to all evidence whatever, and compose the logic of historical method. It is to this that we must refer such problems as, the nature and limits of negative evidence, the possibility of analogical argument, and so forth (PhH, 136-137).

This passage shows that within the field of the "empirical" concept of history (history as a science), Collingwood discerns certain principles, some of which are scientific in character and some philosophical. The latter are universal features of historical knowledge (the logic of historical method), the first concern only particular groups of evidence. The analogy with the two concepts of history is clear. So we may conclude that on the one side history may be philosophically seen as "a universal and necessary human interest" and empirically as a science, while on the other hand the science of history itself must be seen as subject to certain philosophical or universal principles and certain empirical or scientific ones.

The principles of historical knowledge and its data must be seen in close relation: "Data, on the one hand, and principles of interpretation, on the other, are the two elements of all historical thought. But they do not exist separately and then undergo a combination. They exist together or not at all" (PhH, 137). Other points which Collingwood mentions are the need to ask particular questions and not to try to collect so-called crude facts, the impossibility of a universal history, and the need to re-interpret and re-write history (PhH, 137-139). Again he is emphatic in his rejection of historical realism:

> ...[S]ince the past in itself is nothing, the knowledge of the past in itself is not, and cannot be, the historian's goal. His goal, as the goal of a thinking being, is knowledge of the present; to that everything must return, round that everything must revolve. But, as historian, he is concerned with one special aspect of the present – how it came to be what it is. In that sense, the past is an aspect or function of the present; and that is how it must always appear to the historian who reflects intelligently on his own work, or, in other words, attempts a philosophy of history (PhH, 139).

With "The Philosophy of History", Collingwood had come to the end of a long road of reflection on history. In the beginning his thought was dominated by the effort to come to grips with the different spheres of human life, combined with his struggle against realism. This culminated in *Speculum Mentis*. History was assigned a certain role in the system, but his conception of it was as yet undeveloped and therefore unsatisfactory. It was after *Speculum Mentis* that Collingwood focussed his attention on the special problems of history and it was only then that he progressively came to a better understanding of it. "The Nature and Aims of a Philosophy of History" may be considered transitional. The great change was accomplished by "The Limits of Historical Knowledge", in which for the first time a clear anti-realist position was taken concerning historical knowledge and the role of evidence was emphasized. Finally in "The Philosophy of History" the need was brought to the fore of using certain principles in historical research. In the same article the relation between the philosophical and empirical concepts of history on the one side, and the philosophy of history on the other, was made explicit and clarified. One could say that in 1930 the basic principles of Collingwood's ideas on history had been developed[75]. What appeared afterwards on the subject may be considered elaborations of these basic principles. On one important topic, however, I have not been able to give evidence for this interpretation from his printed work: Collingwood's theory of history as the re-thinking of past thoughts. This theory was worked out by him in the so-called Die-manuscript of 1928 (Aut, 107) and will be discussed later in this study.

Before moving on to Collingwood's work after 1930, it is necessary to deal with two topics, neglected until now, but important for a better understanding of Collingwood's view on history: the relation between history and science and history as process.

2.4. HISTORY AND SCIENCE

One of the best known aspects of Collingwood's thought is his defence of the autonomy of the science of history and his vindication of the epistemological soundness of historical knowledge against the dominating pretensions of natural science and with it the positivistically conceived social sciences. Here, too, one can discern a development in his thinking, which justifies our attention.

In 1922 Collingwood wrote an article entitled "Are History and Science different Kinds of Knowledge?"[76]. He begins by giving the answer to the question posed in the title, and declares that he will argue that the distinction is in his view illusory (HSc, 23). He is aware that this is contrary to the traditional view. From Aristotle till the nineteenth century science and history were seen to be in opposition, science being directed towards the universal and history towards the particular. At the end of the nineteenth century, with people like Bergson and Croce, a reaction took place in the sense that history was valued more highly than science, but the dichotomy between the two forms of knowledge was maintained.

Collingwood objects that this distinction between science and history is based on a wrong conception of what their essence really is. Science is usually held to deal in generalizations. According to Collingwood, however, the *application* of generalizations is a more essential feature of science. This application necessarily stands in close relation to individual facts: the latter are interpreted by the former. "Science is this interpretation", Collingwood says, and continues:

> To live the life of a scientist consists in the understanding of the world around one in terms of one's science. To be a geologist is to look at landscape geologically: to be a physiologist is to look at organisms physiologically, and so on. The object which the scientist cognises is not 'a universal', but always particular fact, a fact which but for the the existence of his generalising activity would be blank meaningless sense-data. His activity as a scientist may be described alternatively as the *understanding* of sense-data by concepts, or the *realising* of concepts in sensation, 'intuiting' his thoughts or 'thinking out' his intuitions (HSc, 28).

This definition of science finds its counterpart in the definition of history. History, Collingwood maintains, is not exempt from generalizations, which, as in science, are used to interpret individual facts. Here he points to sciences such as palaeography, numismatics and archaeology, which play an important role in history. As with science their generalizations, however, must not be looked upon as an aim in themselves; they must rather be seen on the analogy of "the interest of a workman in the improvement of his tools" (HSc, 31).

The logical foundation of the epistemological convergence of science and history is found in Collingwood's views on the relation between the particular, the individual and the universal. We can only have knowledge of the individual, in which the particular and universal are united: " ... the sense-datum (pure particular) and concept (pure universal) are false abstractions when taken separately which yet, as elements in the one concrete object of knowledge, the individual interpreted fact, are capable of being analytically distinguished" (HSc, 29). This is true of both science and history; so neither is science aimed at the universal, nor history at the particular. That science and history, however, are traditionally seen as having precisely these orientations is explained by Collingwood by their historical background. From the Greeks onwards science developed faster than history, with the consequence that philosophical reflection concentrated on this intellectual endeavour, to the neglect of history. This resulted in an uneven balance between the epistemology of science and history. For this reason: " ... in the theory of science attention has always been drawn to the concepts of principles of interpretation according to which the active work of thought proceeds, while the theory of history has contented itself with attending to the finished product of thought, the fully-compiled historical narrative". This leads to history being seen as "complete and immovable, while what we arbitrarily call science (the mere abstract generalisation) is an early stage in the process of thought which looks forward to its own completion in what inductive logic calls verification" (HSc, 32). The alleged distinctions between science and history, Collingwood concludes, " ... are the result of comparing an inside view of science with an outside view of history – science as an actual process of thought with history as a dead, finished article. When both are regarded as actual inquiries, the difference of method and of logic wholly disappears" (HSc, 33)[77].

This explicit rejection of a dualism between science and history is not Collingwood's final view on the subject. His later views include certain conceptions, which clearly distinguish science from history, but were not mentioned in his first article on the subject. They could not have been mentioned, because his views on history were not yet fully developed, as we have seen before. On the other hand, we should also consider the fact that Collingwood was a dialectical thinker, which is especially apparent when he is concerned to distinguish different aspects of a subject. This is the case with the problem of the relation between science and history. For having written "Are History and Science different Kinds of Knowledge? he sent de Ruggiero a copy on 21 September 1922, calling it "a trifle which I contributed to one of those Congresses, held this summer at Manchester". He is not much pleased with it, because "the rules of the game called a 'symposium' ... oblige the disputant to take up an exaggerated position and defend his thesis". He continues: "Of course I should really distinguish history and science by a distinction between the categories under which the historian and the scientist think the object: the logical formula of the thinking (sense, category, individualisation of the object) being identical in each case"[78]. In a later paragraph in the same letter, Collingwood writes:

> *Somehow* there is a real identity and a real distinction between things like art, history, religion, action, science, philosophy etc. How to formulate the identity and the differences? Croce analyses the differences between his four 'elements', but misses their identity – he gives them the wrong kind of identity, the identity (merely generic) of four sails of a windmill. Your *Szienza*, and this paper of mine on history, seize the identity but say nothing about the differences[79].

The differences between science and history are discussed by Collingwood in an article, entitled "Science and History"[80], which appeared shortly after the one emphasizing their identity. That Collingwood takes here an opposite view is not the only example of his dialectical approach. For the article, written in a charming way, is partly in the form of a Socratic dialogue. It is of great interest, too, because it makes clear how the transition came about between Collingwood's views on the relation between science and history in the article of 1922 and in *Speculum Mentis*.

The article consists of two parts, the first being a discussion between two men about the characteristics of science and history, while in the second Collingwood gives his comments on the subject. Of course the dialogue between the two men, taking place in a lonely inn, is also the expression of Collingwood's own ideas. It is symbolic in this connection that the "scientific" point of view is put by an "older man" and the "historical" one by a "younger man".

The latter begins the discussion by stating that scientific methods are all right " ... for stars and electrons and prime numbers ... but they don't seem to fit when you come to men and women". "Every fresh person is a fresh fact", he argues, "and you have to study him afresh from the very beginning. A planet isn't a fresh fact in that sense, it is just another instance of a known law. But there aren't any laws of human conduct; or if there are, they have so many exceptions that they aren't any use". Of man therefore there cannot be any science, only history, the laws of anthropology, psychology and economics being only "rough generalizations". The reason that man cannot be studied in a scientific way is that "when you get to human beings you get to mind and every mind is unique and unlike every other"[81].

According to the older man sciences like psychology and economics are capable of explaining human actions. "The historian", however, he says, "as I look at him, is only a harmless necessary drudge collecting facts for the scientist. When he has got the facts, the scientist will produce the theory which makes them intelligible. Meantime, please go on with your history: it will all come in handy one day – like butterfly-collecting, you know"[82]. The younger man retorts that "[t] hat's all nonsense":

> You talk about this fine future science that is going to be brewed somehow out of historical facts, when it is settled what the facts are; but there are heaps of historical facts settled already, quite enough to form the basis of any science. Newton wanted *some* celestial motions accurately observed before he could work out his theory of gravitation, but he didn't say to the astronomers 'first discover every fact about every heavenly body, and then I'll tell you why it all happens'. The moon was good enough for him. Why isn't Julius Caesar good enough for the sociologist, or whatever you are going to call your scientist of human life? I'll tell you: it is because the moon faithfully exemplifies the general rule about bodies, but there isn't any general rule about

> people, so Julius Caesar can't exemplify it – and neither would all the facts of all history, past, present and to come[83].

The participants then converge to the point where they agree that on the one hand science is also concerned with facts, and history on the other hand also aims to give explanations. The younger man summarizes the conclusion as follows:

> We both try to find out the why of things, and we are both more interested in facts than in anything else. It is all nonsense to say that what a scientist cares for is mere generalities, and what a historian cares for is mere facts. Neither is any use by itself. Generalizations are ways of grouping facts, and facts aren't facts at all until they are grouped. All the ordinary talk about the difference between science and history is on the wrong lines. Still, I do think there's an important difference somewhere ...[84].

Except for the last sentence, we have arrived here at the same conclusion as the one to which Collingwood had come in his article of 1922. The articles differ, though, in that in his 1922 paper the alleged difference between science and history was discussed in terms of their logical base, turning on the question of the use of the universal, the individual or the particular; while in the later article attention focusses on the question of explanation versus mere establishment of facts. The main difference, however, from the earlier article is that this time Collingwood is not content with denying any difference between science and history. The debate between the two men having reached deadlock, he picks up the discussion himself and continues the argument:

> There is a real difference between the scientific and historical points of view, I thought, if only one could grasp it: and I felt sure that it was the clue to a right understanding of human life. The average scientist thinks of man as a complex machine, evolved by degrees out of other machines as the universe dances its unceasing and unmeaning dance ...
> History, I thought, valued man differently. Its concern is with seeing man as he is, in the full flush of his momentary existence; conscious of the universe and of himself, acting and reacting in a world that is not dust at all but mind, a world of moral ideals, political systems, scientific discoveries, hopes and fears[85].

A generalizing science like psychology, Collingwood contends, always tries to reduce mental phenomena to abstract concepts. A historian, in contrast with this, will always aim at the concrete manifestations of these concepts: " ... for though he may be told that man has this or that faculty, his problem remains the same: how have these faculties actually been used on special occasions?"[86]. Collingwood develops his view on the relation between the abstract and the concrete in a way which is worth quoting in full:

> The historian is concerned with the individual, with the wholly real. All generalization is abstract, and to that extent arbitrary and partly untrue; the concrete individual alone is real. Queen Elizabeth's fit of temper is ultimate reality, and we travel further and further from the truth when we progressively generalize it into a fit of temper, a state of consciousness, a thing that happened. Thus, *the least true thing that can be said about a man* is that he is a product of nature, the latest fruit of evolution, a fleeting shape in the dust. 'Is it true?' Yes, it is true: but it is so little of the truth that if it is presented as the whole truth it turns into a great bouncing lie. 'What then is the whole truth about man?' Well, a truer truth would be to describe what this natural product, man, is in general like: to leave vague metaphysics and plunge into comparative anatomy, physiology, psychology. But that would still be mere abstraction, truth mutilated and falsified. At most it would be about ten per cent., instead of about one per cent. of the facts. The whole, final and ultimate truth about man is this or that historical fact. Here, in the concrete acts and achievements of man, we really for the first time know him as he is, and in knowing, understand...
>
> History and science both deal with reality – the only reality in existence – concrete fact. Now to understand a fact is simply to see it in its true perspective, in its connexion with other facts: and this is what both history and science attempt to do. But the scientist tries to understand the facts by building up round them a network of abstractions and generalizations, which come between him and the facts, and lead him to fancy that what he is really studying is not the facts at all but the generalizations. Hence he misunderstands his own purpose, and comes to believe that the object which he is investigating is the 'system of laws of nature', a system which does not exist at all in the real world and is only a metaphysical name for an error in logical analysis.
>
> Now the historian is really doing the same thing as the scientist, that is, trying to understand the facts. But he differs from the scientist in realizing what he is doing. He knows that the object of his investigation is no abstract 'law of nature', but the facts, and that the generalizations which he uses are only a scaffolding of thought round the object, not the object itself. But this difference of point of view has important consequences. It leads the scientist to think that he has given us the ultimate truth when he has really given us the ultimate abstraction: to aim at exhausting all the reality from the object and reducing it to a mere formula, which, as a half-truth masquerading as a whole truth, becomes a positive falsehood[87].

Collingwood's conclusion is that " ... science, even at its best, always falls short of understanding the facts as they really are"[88]. Only history is able to realize this. It is another example of the realistic view Collingwood held at that time with regard to history as concerned with "the facts as they really are".

Strictly speaking, Collingwood does not deny that there are similarities between science and history. He even says that they are "really doing the same thing" by trying to understand the facts[89]. History, however, differs from science in realizing what it is doing and this is emphasized in "Science and History". This is a different argument from that of the earlier article "Are History and Science different Kinds of Knowledge?", in which the similarity between science and history was emphasized.

"Science and History" is also interesting for other reasons. It demonstrates, for instance, that already Collingwood had a keen interest in the epistemological problems of history. Certain themes emerge, which play an important role in Collingwood's later thought – for example, that the object of history is human actions, that mind is the essential feature of man and that it is strictly unique. In 1922, however, Collingwood's main interest was not yet focussed on history as such, but on the forms of human experience in general, with history as one of these forms. This resulted in the system set out in *Speculum Mentis*. "Science and History" was written immediately before Collingwood began working on *Speculum Mentis*[90]. The way he ends the article makes the connection between the two very clear. For he refers there to the fact that modern science is a product of the sixteenth and seventeenth centuries, while since the end of the eighteenth century the importance of history had been growing (this theme is referred to again and again by Collingwood in his work), and then concludes: "I seemed to see the human mind wandering in its search for knowledge from one resting-place to another, from magic to religion, from theology to science, and now, its latest migration, from science to history"[91].

We have seen that in *Speculum Mentis* the difference between science and history is accentuated, science being "the assertion of the abstract concept" (SM, 158) and history "the assertion of fact" (SM, 201). In this conception history, however, is implicit in science and develops in a dialectical way out of it.

In "The Nature and Aims of a Philosophy of History" (1925) Collingwood repeats that scientifc thinking is abstract and historical thinking concrete. Scientific judgments are hypothetical, taking the form "if A, then B", while historical judgments are categorical (NAPhH,46). Collingwood argues, however, that history uses generalizations too. The distinction between science and history lies in the different emphasis: " ... the special activity of the scientist is to generalise; but the historian, as we have seen, generalises too, only he generalises not for the sake of generalising, like the scientist, but for the sake of helping himself to determine historical fact" (NAPhH,48). It is concerning this emphasis on the generalizing nature of science that Collingwood changed his view, because in "Are History and Science different Kinds of Knowledge?" (1922) he had said: "The scientist generalises, certainly: but generalisation is subordinate to his real work as a scientist, the interpretation of individual fact" (HSc,30). We have seen that already in "History and Science" (1923) the generalizing character of science was emphasized against history[92].

Collingwood's views on the relation between science and history are of importance for a better understanding of his views on history. For they are the basis of his opposition to the "positivistic" approach in history, which tries to ape scientific methods. A clear example of this approach is provided for Collingwood by Spengler's *Der Untergang des Abendlandes* (1918). In "Oswald Spengler and the Theory of Historical Cycles" (1927)[93] it is precisely the confusion of scientific with historical thinking that is condemned by Collingwood:

> History deals with the individual in all its individuality; the historian is concerned to discover the facts, the whole facts, and nothing but the facts. Now comparative anatomy is not history but science; and Spengler's morphology is simply the comparative anatomy of historical periods. The historical morphologist is concerned not to discover what happened, but, assuming that he knows what happened, to generalise about its structure as compared with the structure of other happenings. His business is not to *work at* history, but to *talk about* it, on the assumption that someone else has already done the work – the work, that is, of finding out what the facts are, the historian's work. In this sense, Spengler nowhere shows the slightest desire to do a piece of historical work, or the slightest sign of having done one. His history consists of ready-made facts which he has found in books; and what he wants to do is to arrange these in patterns (SHC, 67).

That Collingwood had reflected on the concept of time[94] is demonstrated by the way in which he uses this concept in distinguishing science from history. Spengler had claimed that his morphology enabled him to foretell the future[95] -an easy target for Collingwood's scorn:

> The touchstone of the historical sense is the future. Science determines the future, foretells an eclipse or the like, just because the object of science is Nature and 'Nature has no history'. The laws of Nature are timeless truths. For history, time is the great reality; and the future is the infinite well-spring of those events which, when they happen, become present, and whose traces left upon the present enable us to

> reconstruct them when they are past. We cannot know the future, just because the future has not happened and therefore cannot leave its traces in the present. The historian who tries to forecast the future is like a tracker anxiously peering at a muddy road in order to descry the footsteps of the next person who is going to pass that way (SHC,68).

In "The Philosophy of History" (1930) again "universal" historians like Hegel, Marx, Comte and Spencer are criticized for confusing science with history. As with Spengler, facts are not taken seriously enough: "When they plugged a hole in their knowledge by inventing a fact, what they invented was not an individual but simply any instance of some general law" (PhH,132). When discussing the problem of whether it is possible to have reasoned knowledge of the individual, let alone of a scientific nature, Collingwood returns to the view expressed in "Are History and Science different Kinds of Knowledge?", in which the distinction between science and history was denied. The argument is the same as in this earlier article, but this time Croce is mentioned as its author (PhH,135). Its essence lies in the character of the individual judgment, which is reflected both in science and history. In science "'All S is P' means 'This S, in its character as S, is P'. When the element of individuality is taken away, we have, not a universal judgment 'All S is P', but nothing at all" (PhH,135). The universality lies in the predicate, "which is a concept, a universal idea properly thought out". The same, however, is true for history, which contains in its individual judgments universalities in the form of the predicate used. Calling someone "unscrupulous" is given by Collingwood as example (PhH,136).

It is not necessary to interpret this position as another surprising switch in Collingwood's ideas on the relation between science and history. It is rather that in the articles of 1922 and 1930 the analysis is in terms of the *logic* of science and history, in which their similarity is emphasized as basically consisting of individual judgments. In the other articles, however – and mainly too in his later work – attention is focussed on the difference in *practice* between science and history, especially in the categories they use. This latter aspect is made clear again in Collingwood's comments on an article by de Ruggiero.

On 9 January 1931 Collingwood writes to de Ruggiero that he has translated one of his articles, entitled "Science, History and Philosophy", which was to be published in the same year[96]. In his letter Collingwood writes:

> [Y]our paper interests me very much indeed, and I think it very valuable – perhaps the best *compte rendu* that anyone has yet produced on the progress hitherto made towards a Critique of Historical Reason. It clears up a great many difficulties and sets the problems in a new and improved light, and the time which I

> have spent in translating it has been extremely well spent from the point of view of my own progress in the same subject ... The general tendency of English philosophy at the present day – and American still more so – is, as you quite rightly say, based on copying science and assimilating philosophy to scientific thought with the result that it merely negates history – either ignoring it or openly denying its claims. And I find that my own attempts to introduce a slender thread of historical thought into English philosophy are met everywhere with a blank refusal ... Therefore a historically-minded philosopher here is a vox clamantis in deserto[97].

De Ruggiero's starting-point in discussing science and history was to see them as expressions of two "mental categories". English thought is, according to him, traditionally orientated towards the problems of natural science and Italian thought to those of history. In his article de Ruggiero makes a comparative study of the categories of both, from a philosophical point of view. Starting from Kant he considers "being in space" and "becoming in time" as two fundamental "cosmic frames". Kant's analysis is especially relevant for the spatially orientated natural science and "it must be revised and broadened if it is to do justice to the wider horizon of a theoretic spirit including historical as well as scientific knowledge"[98].

According to de Ruggiero, we can ask ourselves the following questions about anything which is presented to us in space or experienced in time: 1) whether it exists, 2) what it is, 3) why it exists, 4) what it is worth, these being the categories of modality, substance, causality and value. These categories are used both by science and history, " ... but they use them in different ways, so that the same categorial functions are specified ... according to the direction taken by consciousness in each case"[99]. Summarized very briefly, the different categories have the following characteristics in science and history: 1) modality: scientific judgments are hypothetical, historical ones express factual reality; 2) substance: science thinks in terms of things, history in terms of activity; 3) causality: scientific thought is deterministic, historical thought intentional; 4) value: truth is seen in science in terms of the coherence of a system, while in history it is the individual that counts[100].

"In short", de Ruggiero concludes this part of his argument, "between science and history there is an inverse relation. The one is *abstracting* thought, which tends towards uniformity and identity, and cancels all the differences which it meets in its path. The other is *individuating* thought, tending towards multiformity and diversity, and enriching itself as it advances by collecting to itself all the individual determinations of activity that emerge in the course of time. On the one side we have a static projection of the world in space, on the other a dynamic progression in time"[101]. De Ruggiero does not want to consider this opposition between science and history as ultimate and irreducible. It is the task of philosophy to show " ... that these two opposed forms of thought spring, in spite of their opposition, from a common root and must end in a final reunion...". Philosophy is assigned the role of judge of the claims of science and history[102]. More specifically, it is only dialectical thought, which is able to realize "... the synthetic point of view, which combines the opposites into an articulated and living untity"[103].

It is interesting to note the differences between the solutions offered by de Ruggiero, and by Collingwood in *Speculum Mentis*. For Collingwood, philosophy was denied the role of judge of the different forms of experience[104]: they were "left behind" as such by philosophy and elevated ("aufgehoben") to its level. Collingwood has some criticisms of de Ruggiero's position, but these do not concern the role of philosophy. He regards the latter's analysis of the relation between science and history as "an understatement of the case" and says the following about it:

> In the antithesis between science and history, I find that science negates history in a way in which history does *not* negate science: science dismisses history as a vain fable, while history accepts science as a necessary element in truth, but not the whole truth. A scientifically trained man tends to despise history: a historically trained man does not tend to despise science, he regards it as an instrument necessary for his own use. Scientific philosophy, as in England, regards historical philosophy as pure nonsense; but historical philosophy, as in Italy, regards scientific philosophy as an element dialectically contained and transcended in itself. Hence the relation between science and history is not, I think, so symmetrical as your paper allows a reader to suppose. You say that philosophers must abandon their hostility to dialectic: I agree; but this hostility does not exist, I think, among the historical philosophers of Italy, but only among the scientific philosophers of England. Science is hostile to dialectic, history is not; and if a philosopher acquires a dialectical point of view this means that he is acquiring a historical point of view. Therefore I think, when you say 'abandon your hostility to dialectic', you mean 'abandon this English preoccupation with natural science and learn from us Italians to think historically'. I agree that this is what my countrymen ought to do; but I'm not sure that this interpretation does not disturb the symmetry of the relation which you establish between science and history[105].

We find Collingwood here in the role of an ardent defender of history against the claims of science. The passage quoted also demonstrates that Collingwood refers to the *practice* of scientists and scientific philosophy, especially in England.

Since Collingood opposed the dominance of "scientific" thought it is understandable that his criticism should be levelled especially at cases where history itself is affected by this mode of thought. This "positivism" can take different shapes and we have seen that Collingwood attacks one of them in Spengler. In a review of the "Selected Essays" of the historian Bury, Collingwood shows himself the defender of history against the claims of positivism in another form[106].

Originally Bury held a positivistic view of history, treating individual events as instances of universal laws. In his inaugural lecture "The Science of History" (1903), he has moved some way from this conception, according to Collingwood. But he still calls history "a science, no less and no more" and this is considered by Collingwood "the formula of positivism"[107]. In the following years, Bury moved further and further away from positivism and became aware of the special characteristics of history, particularly its individuality. He could not get to grips properly, however, with this essential feature of history and the only place he could assign to individuality in his theory was to interpret it as chance or contingency. Collingwood sums up Bury's dilemma, which was, according to him, the result of his positivistic background, as follows:

> He had discovered the fundamental element of individuality in history; but his early training had taught him that only the universal was intelligible: we could understand why this *kind* of thing happened, namely because this *kind* of cause was operative: but why *this* thing happened, rather than any other thing of the same kind, could not be understood at all. The universal is the rational; the individual is the accidental, the irrational, the meaningless; so that the element of individuality in history might be called the element of chance, irrationality, or contingency[108].

Collingwood is strongly opposed to the theory that only the universal is intelligible and that the individual is forever condemned to remain unintelligible. This positivistic theory leaves for history no alternative but to choose to be assimilated to the sciences in one way or another, or to acquiesce in remaining irrational and incapable of making things intelligible. Bury could not find a way out of this dilemma, which is demonstrated by his theory of the contingency of history, although he suggested on the other hand that the role of the individual and the accidental might become less[109].

We have come here to a crucial element of Collingwood's mature view on history: he is strongly convinced that, although history deals with the individual, it is none the less rational and capable of making things intelligible. As with the other aspects of Collingwood's view on history, its main features took shape in the twenties.

2.5. HISTORY AS PROCESS

In his essay "Collingwood's Historicism: a Dialectic of Process"[110], Mink has argued that according to the traditional interpretation of Collingwood's ideas on history some features are dominant, while others remain recessive. The reason for this, he contends, must be found in the fact that Collingwood's views on history are usually interpreted on the basis of *The Idea of History*, to the neglect of Collingwood's other philosophical writings, particularly in so far as they are relevant for his ideas on history. According to Mink, one of the "recessive" elements of Collingwood's thought on history is the concept of process[111]. This aspect is indeed almost wholly neglected by interpreters of Collingwood's philosophy of history. One of the few interpreters who have paid attention to the idea of process in Collingwood's thought is Rotenstreich. In his view, Collingwood tries to deny time in his philosophy of history. He considers this to be his "main trend" and asserts that it is only in his later works – *An Essay on Metaphysics* (Oxford, 1938) and *The New Leviathan* (Oxford, 1942) – that Collingwood takes " ... refuge in the common-sense meaning of history *qua* process"[112]. According to Rotenstreich, " ... [t]his time-factor that has been neglected in the realm of history comes back into history *via* the realm of politics"[113], in *The New Leviathan*.

I do not think this interpretation is correct. Collingwood always had a keen interest in the "object" of historical thought, that is, in history as "res gestae". In his *Autobiography* he tells us that his "first principle of a philosophy of history" was related to this topic (Aut,97). This was about 1920. The principle was "that the past which an historian studies is not a dead past, but a past which in some sense is still living in the present". "At the time, I expressed this by saying", Collingwood continues, "that history is concerned not with 'events' but with 'processes'; that 'processes' are things which do not begin and end but turn into one another; and that if a process P_1 turns into a process P_2, there is no dividing line at which P_1 stops and P_2 begins; P_1 never stops, it goes on in the changed form P_2, and P_2 never begins, it has previously been going on in the earlier form P_1" (Aut,97-98).

Collingwood wrote down these ideas in an essay called *Libellus de Generatione*, which was not published but given to de Ruggiero (Aut,99). He calls it "... primarily a study of the nature and implications of process or becoming" (Aut,99). He also intended this to be an attack on "realism", "... showing how the *non possumus* of 'realists' towards a theory of history arose from their refusal to admit the reality of becoming, and from their analysis of the true proposition 'P_1 becomes P_2' into the complex of propositions 'P_1 is P_1', 'P_1 is not P_2', 'P_1 ends where P_2 begins', 'P_2 is P_2' and 'P_2 is not P_1', all of them either tautologous or false" (Aut,99).

This analysis of history as process is, of course, thoroughly dialectical in nature. The importance that Collingwood apparently attached to it is demonstrated by his discussion, at the end of the

Autobiography, of a historical problem. The problem was that Celtic arts and crafts had flourished in Britain before the Roman conquest, that after the conquest they were replaced by Roman fashions, but – and this was the puzzling fact – towards the end of the Roman period and after it, a "Celtic Revival" took place (Aut,137). Collingwood rejects three attempts that had been made to explain the Celtic revival, based on the following theories: a) the Celtic tradition had never been broken (rejected by Collingwood for lack of evidence); b) not all the Celts were subjected to the Romans (rejected for the same reason); c) Celtic art was the product of the "Celtic temperament", which only blossomed under certain conditions (rejected by Collingwood, because it depended on an "occult entity") (Aut,138-139).

Collingwood took the problem seriously, because it focussed, in his opinion, not only "the whole problem of Romanization", but also "the whole problem of art-history and indeed of what the Germans call history of culture" (Aut,140). The solution of the problem was, according to Collingwood, not to consider the Celtic revival as something exceptional. It was seen as such, " ... due to the fact that the nature of historical process was misconceived" (Aut,140). It is in this context that he refers again to his views on the historical process, as analysed before: "As I had long ago proved in the *Libellus de Generatione*, any process involving an historical change from P_1 to P_2 leaves an unconverted residue of P_1 incapsulated within an historical state of things which superficially is altogether P_2. This, I thought, might prove the key to my problem" (Aut,140-141).

Collingwood shows the importance which he attached to the application of this theory to his historical work in the passage with which he concludes his discussion of the topic:

> This was the idea which I expressed in the chapter on 'Art' in the *Oxford History of England*; a chapter which I would gladly leave as the sole memorial of my Romano-British studies, and the best example I can give to posterity of how to solve a much-debated problem in history, not by discovering fresh evidence, but by reconsidering questions of principle. It may thus serve to illustrate what I have called the *rapprochement* between philosophy and history, as seen from the point of view of history (Aut,144-145).

It is paradoxical indeed, that Collingwood's views on the historical process, extensively dealt with in his *Autobiography* and put forward by him as his best example of the use of a theoretical principle in the study of history, have hardly been noticed by any interpreter of Collingwood's views on history and have had to be brought to the fore by Mink as one of the "recessive" elements of Collingwood's ideas on history.

There are other places in Collingwood's published work where ample evidence can be found of his views on history as process. In *Speculum Mentis*, for instance, he says:

> History ... is not a sheer flux of unique and disconnected events, each absolutely new and unprecedented. And, on the other hand, it is not a barren cyclical repetition of the same pattern over and over again, still less a shuffling of rearranged units like repeated throws of dice, every new event an arbitrary selection from a given number of possibilities. It is a process in which method or regularity does not exclude novelty; for every phase, while it grows out of the preceding phase, sums it up in the immediacy of its own being and thereby sums up implicitly the whole of previous history. Every such summation is a new act, and history consists of this perpetual summation of itself (SM,56).

In "Oswald Spengler and the Theory of Historical Cycles" Collingwood argues against Spengler that the self-consistent character of a given culture is not static but dynamic, " ... not a single unchanged thing, ... but a process of spiritual development ..." (SHC,73). "Better historical thinking", Collingwood says, in opposition to the one of Spengler, "would show us within the heart of classical culture, not a single unchanged idea, but a dynamic interplay of ideas, containing elements which, even quite early, prepare it for its conversion into Magian ... It is truer to say that the classical is not a style but an age, a process, a development, which led to the Magian by its own inner logic. Thus the Pantheon is *both* Magian *and* classical; it is classical in the act of *turning into* Magian. And this conception of 'turning into', the conception of becoming, is (as Spengler himself industriously asserts, and industriously forgets) the fundamental idea of all history" (SHC,74).

A process can be seen from one point of view as decay or, from another, as progress. The "decline of Hellas", Collingwood points out, "will figure as the movement leading up to the Hellenistic world. Was it, then, 'really' a decline or an advance? Neither, because both; it was a becoming, a change, a development; and the historian's highest task is to discover *what* developed, through *what* phases, into *what*. If anyone is not interested in that question, he is not interested in history" (SHC,75). In "The Theory of Historical Cycles" (1927)[114] the same idea is put forward: "In history, *tout lasse, tout passe, tout casse*; everything decays, and all movement is a movement away from something, a loss of something won, a withering, a death. The growth of the steamship is the passing-away of that splendid thing, the sailing-ship; the rise of fire-arms is the decadence of archery" (THC,81).

A process contains within itself elements of identity or continuity and difference. "Anybody would admit that Plato's *Republic* and Hobbes's *Leviathan* are about two things which are in one way the same thing and in another way different", Collingwood says in the *Autobiography*. "That is not in dispute. What is in dispute is the kind of sameness and the kind of difference ... The sameness is the sameness of an historical process, and the difference is the difference between one thing which in the course of that process has turned into something else, and the other thing into which it has turned" (Aut,62). Collingwood rejects the "realist" idea of eternal problems in philosophy, but adds in a note: "If 'eternal' is used in its vulgar and inaccurate sense, as equivalent to 'lasting for a considerable time', the phrase 'eternal problem' may be used to designate collectively a series of problems connected by a process of historical change, such that their continuity is discernible even by the presumably rather unintelligent eye of the person who thus misuses the word, but the differences between them not so discernible" (Aut,68). The differences in a process, however, may also be more apparent than its continuity. An example of this is given by Collingwood in "A Philosophy of Progress" (1929)[115], when he discusses our political institutions and says about them: " ... they have been evolved through a process in the course of which they have incorporated into themselves portions of primitive law, of the Greek city-state, of Roman imperialism, of feudal organisation, and so forth. All these elements have gone to make modern political life what it is" (PhP,120).

This short discussion demonstrates not only the value which Collingwood attaches to the view of history as process, but shows also that it was elaborated by him at an early date. Similarly it can be said that the other aspects of his thought on history, which have been dealt with in this chapter, were developed in the twenties: this is the case with regard to his change from realism to idealism, the epistemological foundation of history and its special character as compared with science. My reason for discussing the development of Collingwood's ideas is that considerable light is thereby thrown on certain aspects of his mature and more generally familiar views, especially as worked out in *The Idea of History*. It is on the latter book and the reactions it aroused, that attention will be focussed in the next chapter.

CHAPTER THREE

THE IDEA OF HISTORY AND ITS DISCUSSION

3.1. THE PHILOSOPHY OF HISTORY IN COLLINGWOOD'S LATER YEARS

An Essay on Philosophical Method was written by Collingwood from November 1932 to June 1933[1]. Both Knox and Collingwood himself consider this book his best. According to Knox it is the "only one book of Collingwood's which could be called great" and in a review of it he had even called it "a philosophical classic" (IH,XX), while Collingwood says about it: "It is my best book in matter; in style, I may call it my only book, for it is the only one I ever had the time to finish as well as I knew how, instead of leaving it in a more or less rough state" (Aut,118).

Collingwood deals in this book with the question what philosophy is. He does this by contrasting its method with that of mathematics and science. In the latter concepts are classified as precisely as possible into mutually exclusive classes. With philosophy, however, this is not the case, as its class-concepts overlap and are not mutually exclusive. Moreover, the species of a philosophical genus differ not only in degree, but also in kind. Collingwood supplements his theory of the overlap of classes in the field of philosophy with the theory that a philosophical concept exemplifies a "scale of forms", in which its generic essence is realized in varying degrees. The concept of the good in ethics, for instance, can be distinguished in this way into the pleasant, the expedient and the right, which are not to be considered mutually exclusive (EPhM,41). In a philosophical scale of forms there is not only a synthesis of kind and degree, but also of distinction and opposition:

> The higher term is a species of the same genus as the lower, but it differs in degree as a more adequate embodiment of the generic esssence, as well as in kind as a specifically different embodiment; it follows from this that it must be not only distinct from it, as one specification from another, but opposed to it, as a higher specification to a lower, a relatively adequate to a relatively inadequate, a true embodiment of the generic essence to a false embodiment; as true, it possesses not only its

own specific character, but also that which its rival falsely claimed. The higher thus negates the lower, and at the same time reaffirms it: negates it as a false embodiment of the generic essence, and reaffirms its content, that specific form of the essence, as part and parcel of itself (EPhM,88).

The relation between the different terms is also described by Collingwood "by the metaphor of promising and performing", or by saying "that the higher is the reality of which the lower is the appearance, or the ideal to which the lower is an approximation, or the truth of which the lower is a perversion" (EPhM,87). "Each term in the scale, therefore, sums up the whole scale to that point", Collingwood says. "Wherever we stand in the scale, we stand at a culmination. Infinity as well as zero can thus be struck out of the scale, not because we never reach a real embodiment of the generic concept, but because the specific form at which we stand is the generic concept itself, so far as our thought yet conceives it" (EPhM,89).

Certain words can refer both to scientific and philosophical concepts. As examples, Collingwood gives the philosophical concept of matter as used by materialism, and the scientific concept of Newtonian physics; mind for the spiritualistic philosopher and for the psychologist; and the concept of evolution for the biologist, and the philosopher who sees it as a "cosmic process" (EPhM,34). Again, a concept like art is for the critic "a highly specialized thing, limited to a small and select body of works outside which lie all the pot-boilers and failures of artists, and the inartistic expressions of everyday life", while "for the aesthetic philosopher, these too are art, which becomes a thread running all through the fabric of the mind's activity" (EPhM,35). When words are used in two different ways, they are, according to Collingwood, not on that account equivocal: "they undergo a regular and uniform change in meaning when they pass from one sphere to the other, and this change leaves something fundamental in their meaning unaltered, so that it is more appropriate to speak of two phases of a concept than two senses of a word" (EPhM,33).

In *An Essay on Philosophical Method* little attention is paid to history and this subject is only mentioned at the end of the book, when history and philosophy are compared as branches of literature. The theory of philosophical concepts as elaborated in this book is nevertheless important also for the concept of history. We have seen that, in Collingwood's view, there is both a philosophical and an empirical or scientific concept of history. The first refers to a universal aspect of mind, necessarily having a certain vagueness, while the second designates what historians are involved in. The relation between the two types of concept is expressed by Collingwood as follows:

> It appears ... that when a concept has a dual significance, philosophical and non-philosophical, in its non-philosophical phase

> it qualifies a limited part of reality, whereas in its philosophical it leaks or escapes out of these limits and invades the neighbouring regions, tending at last to colour our thought of reality as a whole. As a non-philosophical concept it observes the rules of classification, its instances forming a class separate from other classes; as a philosophical concept it breaks these rules, and the class of its instances overlaps those of its co-ordinate species (EPhM,35).

In the preceding chapter we have dealt extensively with the idea of two concepts of history in Collingwood's thought and the passages quoted above make clear how this idea is further clarified in *An Essay on Philosophical Method.*

The *Essay* was intended by Collingwood as a starting-point for his future work. In a letter to de Ruggiero he mentions it "as a programme for future work rather than a conclusion or final theoretical position"[2]. In his *Autobiography* Collingwood says that he "planned a series, to begin with an *Essay on Philosophical Method*", and that he wrote in 1937 "the second book of my series, *The Principles of Art*" (Aut,117-118). The *Autobiography* itself was written in 1938. The overwhelming productivity Collingwood exhibited in the following years and the extraordinary speed with which he could work – he already knew that he had not many years to live – meant that in 1939 the information about his planned series, as given in the *Autobiography*, was out of date. For in that year *An Essay on Metaphysics* was finished, together with one-third of a book he intended to write, with the title *The Principles of History.* He further prepared the publication of his lectures on the philosophy of history and philosophy of nature, which were to be titled *The Idea of History* and *The Idea of Nature.* All these works were to be published by the Oxford University Press, and in a correspondence between the Press and Collingwood, dated 18 and 19 October 1939, it is revealed that publication of the following series and items was intended: I "Philosophical Essays", with *An Essay on Philosophical Method* and *An Essay on Metaphysics*, II "Philosophical Principles", with *The Principles of Art* and *The Principles of History*, III "Studies in the History of Ideas", with *The Idea of History* and *The Idea of Nature*[3].

Of these planned works, *The Principles of History* was not finished and *The Idea of History* and *The Idea of Nature* have only been posthumously published. We will confine our attention to the works Collingwood intended to write on philosophy of history and the way their publication eventually took shape. It is first of all important to note that *The Idea of History*, as it was published posthumously, does not completely correspond to the book Collingwood intended to publish under the same title.

In 1936 Collingwood lectured on philosophy of history. The lectures consisted of two parts: a "history of the idea of history" and a "metaphysical epilegomena", with the subtitles "Re-enactment of Past Experience the Essence of History", "The Subject-matter of History" and "Progress"[4]. In a letter to de Ruggiero of 12 June 1937 Collingwood writes about his views on history and his lectures on the subject. The way this topic is discussed by him is important enough to be extensively quoted. Collingwood refers in his letter to his paper "Human Nature and Human History" (reprinted in IH,205-231), saying: "A year ago I published a paper on the conception of 'human nature', arguing that what went by that name in the 18th century was really human history, falsely crystallized into a special case of 'nature', and implying that the so-called sciences of mind were faulty in so far as they treat mind as something given, to be analysed like a natural object, instead of something whose only reality is its historical process". Collingwood, then, continues as follows:

> This thesis involves a programme of recasting the science of mind (including the Crocian science of spirit) into the form of history; not into the form of history *wie sie staht und geht*, but into a new form of history, not merely philological but philosophical. The philosophy in it is not, as Croce has said, simply its methodology. The absorption is mutual: the product is not philosophy based on history nor history based on philosophy, it is both these things at once. I think you will understand what I mean, and will very likely say you have heard it, and thought it, long ago.
> I am trying now to carry out a part of that programme, in two ways. (I) I am lecturing on the philosophy of history, and mean to publish my lectures a few years hence. Their ground-plan is based on the idea that the concept of history, as it now controls the work of historians, is the product of a long process of historical development. It is not the same as the concept of history current 50 or 100 or 200 or 500 or 1000 or 2000 years ago. It is a new and peculiar concept, and we can trace its rise by tracing the history of historiography and of historical criticism and methodology. So I do this, beginning with Herodotus and coming down to the present day: and hoping that in this way I can give people a *concrete* methodology of present-day historiography, integrated in its own past, which will show them what the problems and methods of present-day historiography are. It is as if one took Croce's *Storiografia*, and turned it inside out: greatly expanding Part II, and absorbing all the subject-matter of Part I into its interstices. I find this method *effective* in practice. People who are rendered suspicious or contradictious by Croce's method of presentation, where his own central idea is stated *first*, become quite tame and intelligent if the very same idea is led up to by showing how it developed out of a determinate situation. Like Hegel, they are annoyed by having the Absolute fired at them out of a pistol![5].

The other way of carrying out his programme was related to Collingwood's studies in aesthetics, since he mentions in the same letter that he worked in this field "on similar principles". Though *The*

Principles of Art does not show a particularly historical approach he says in the letter that "[t]he problem of 'the nature of Art' should become the problem of the nature of *modern art*; and the solution of this problem is the historical problem (but philosophical as much as historical) of the genesis of modern art"[6].

Collingwood does not mention in his letter to de Ruggiero the "metaphysical epilegomena" of his lectures of 1936. So probably the book he intended to publish under the title *The Idea of History* would only consist of his "history of the idea of history".

From October 1938 to April 1939 Collingwood made a trip to the East Indies to recover from ill-health. From 24 October to 13 November 1938 he wrote the first draft of *An Essay on Metaphysics* on board[7]. A few days after his arrival at Batavia (the present-day Jakarta) he travelled to the east of the island of Java and from there to the island of Bali, where he stayed from 8 to 30 December[8]. In the following weeks he visited different islands in the archipelago on various ships. Returning to Java again Collingwood writes in his diary on 9 February 1939: "Wrote sketch contents for *Principles of History*". Having arrived at Surabaya the next day he writes: "Began writing *Principles of History*". During the following days he worked intensively on it till 21 February, when he took the train from Yogya, which he had reached on 14 February, to Batavia. After that date he worked on the proofs of the *Autobiography* he received there (on 19 March he rewrote the last chapter on board the ship returning home) and revised *An Essay on Metaphysics*. The only occasions on which the *Principles of History* are mentioned again are on 26 March: "Playing with *Principles of History*" and 27 March: "Tried to begin ch. IV of *Principles of History* in morning – stuff wouldn't flow".

Collingwood's plan for *The Principles of History* was to divide it into 3 books, dealing with: "1) a simple account of the most obvious characteristics of history *as a special science* 2) Relation between this and others 3) Relation of history as thought to practical life"[9]. Of these books only the first one was finished, consisting of four chapters, dealing respectively with the concept of evidence, action, re-enactment and history as the self-knowledge of mind. From the evidence of letters it is clear that Collingwood himself considered this book a very important one, in fact his most important. For on 14 February 1939 he writes to his son: "...I have begun writing *The Principles of History*, which will go down to posterity as my masterpiece. I suddenly began it, quite unexpectedly, as my boat was approaching Soerabaja, and spent my whole time in that damnably hot town writing it as hard as I could". (The day before Collingwood had written in his diary: "All these days, 82° in my room at dawn, rising to 90° in the late afternoon"). On 18 February he writes to his wife "...I began writing the book I have been working on up to 30 years, *The Principles of History*" and on 22 February: "...My new book goes well. I have spent only nine days, not whole days, on it and I have now written 41,000 words, which I think is one third of the bulk of *Principles of Art*, which I do not wish greatly to exceed. This looks as if there would be two third of *Principles of History* written by the time I sail"[10].

That Collingwood eventually only managed to finish one third of the book is explained by the fact that he received on 22 February the proofs of the *Autobiography* and worked the following weeks on these – on board the ship returning home – and also on the revision of parts of *An Essay on Metaphysics*. That he had not forgotten *The Principles of History*, however, is made clear by a letter he wrote to the archaeologist Simpson after his return on "Easter Monday 1939": "The third [book], of which I wrote some 40,000 words in Java, is called *The Principles of History* and is the book which my whole life has been spent in preparing to write. If I can finish that, I shall have nothing to grumble at".[11] In his preface to *The Idea of History* Knox quotes Collingwood using the same wording in his authorization to publish the finished part of *The Principles of History*, with a preface "explaining that it is a fragment of what I had, for twenty-five years at least, looked forward to writing as my chief work" (IH,VI).

In view of the extreme value Collingwood himself apparently attached to *The Principles of History* the question arises as to why this book was never finished. The question is the more pressing, because he began working on a new book, *The New Leviathan*. According to Knox the reason must be found in the extreme historicism Collingwood had arrived at by that time, with the opinion that "philosophy as a separate discipline is liquidated by being converted into history" (IH,X). This brings Knox to the following theory:

> Why did he never finish *The Principles of History*? Diminished physical strength, and preoccupation with *The New Leviathan*, are two obvious answers. But the true answer is that his project had become either impossible or unnecessary. *The Principles of History* was either a philosophical work, an attempt to describe what history is and to explain how historical knowledge is possible, or else it was no more than autobiography, an account of how the author as a matter of fact proceeded in his own historical work. For Collingwood by 1939 it could not be the former, because philosophy had been absorbed by history; and it was useless for him to write the latter because his *Autobiography* was already in print. By this time it was not even open to him to distinguish his practice as an historian from his philosophical theory about his practice, because in the *Autobiography* theory and practice had been indentified (IH,XVII).

In my opinion this contention may be seriously questioned. There is no evidence that Collingwood decided not to complete *The Principles of History*. On the contrary, we have seen that in a letter to the Oxford University Press of October 1939 he still mentions his plans to finish the book[12]. The real reason why *The Principles of History* was never finished is in fact mentioned by Knox at the beginning of the quotation given above. It is Collingwood's preoccupation with *The New Leviathan*. When the war broke out, he considered it his duty to work on this latter book. In a letter on April 1941 he writes to the archaeologist Crawford: "When the war broke out I saw that the whole business was due to the fact that everybody concerned was in a completely muddled condition about the first principles of politics and, examining my own mind, I saw that I had plenty of ideas which it would be a public service to state"[13].

3.2. *THE IDEA OF HISTORY*

In 1927 Collingwood reviewed *Roman Society in Gaul in the Merovingian Age*, by S. Dill (London, 1926). This work had appeared posthumously and had been edited by C.B. Armstrong. Collingwood says about it:

> It was a task involving much labour, and requiring, as is evident, tact and judgment; the provision of notes and references to such a work is arduous, and the polishing of an unfinished manuscript by another hand is a task which no one would undertake except as a labour of love. Both tasks have been well done. It is true that there are still roughnesses of style, repetitions of incident, and here and there a judgment, an emphasis, or even a statement of fact which second thoughts might have altered; but to have smoothed away all these things would have involved robbing the book of just that personal quality which makes it visibly the work of its author[14].

Exactly the same could have been said of his own book *The Idea of History*, posthumously edited by his friend and, one could say, only real pupil, Sir Malcolm Knox. It is a strange coincidence that Collingwood, who had published during his life such an enormous quantity of work, attained most of his fame after his death from a book which is in fact a patchwork put together from lectures, printed work and parts of a first draft for a book. That Knox managed to make more or less a unity of it is a considerable achievement. Knox also wrote an "Editor's Preface", in which he gave a short sketch of Collingwood's background and development[15].

When Knox edited *The Idea of History* not all of Collingwood's manuscripts were available to him. Of the ones he did not see, the most important were the lectures on philosophy of history of 1926 and 1928. For this reason it is clear that Knox's first sentence in his Preface, "During the first six months of 1936 Collingwood wrote thirty-two lectures on *The Philosophy of History*", must now be seen in another light, given our knowledge that Collingwood lectured from 1926 on the subject and that his lectures of 1936 are the result of a long development. We have seen that the book Collingwood intended to publish with the title *The Idea of History* was to be based on the lectures of 1936, parts of which were revised in 1940. The lectures dealt mainly with the history of the idea of history. It is probable that Collingwood did not intend to publish the "metaphysical epilegomena" to the lectures, as in *The Principles of History* he made a fresh start on the same subject.

The Idea of History, as it was edited by Knox, consists of four elements: 1) The lectures on the history of the idea of history of 1936, partly revised in 1940 (IH, 1-122, 126-147, 151-204)[16] 2) Three essays on history, which formed Part II of the lectures of 1936 under the head "Metaphysical Epilegomena", included in *The Idea of History* under the titles "History as Re-enactment of Past Experience" (IH, 282-302), "The Subject-matter of History" (IH,302-315) and "Progress as created by Historical Thinking" (IH, 321-334) 3) Two essays which had been published before: "Human Nature and Human History" (IH,205-231)[17] and "Historical Imagination" (IH,231-249)[18] 4) Parts of *The Principles of History*,given in *The Idea of History* the titles "Historical Evidence" (IH,249-282) (apparently being chapter 1 of book I of *The Principles of History*), "History and Freedom" (IH,315-320) and "Hegel and Marx" (IH,122-126) (the last two probably being taken from chapter 4 of book I of *The Principles of History*). *The Idea of History* was divided by Knox in such a way that the lectures of 1936 on the history of the idea of history formed the first four parts of the book (except the sections on "Hegel and Marx" and on Bury), while the rest was put together by him as Part V with the title "Epilegomena".

The fate of *The Principles of History* has not been a very fortunate one – not only because Collingwood never finished it, but especially because the manuscript containing what he did finish is lost. Collingwood wrote about ninety pages of *The Principles of History*. More than half of this manuscript was not used by Knox because he did not think it attained a standard fit for publication. It is very unfortunate that this particular portion of the manuscript has been lost[19].

Before reviewing the discussion on *The Idea of History*, it is first necessary to state the argument. The problem with this approach is that it is unavoidable that an outline of Collingwood's theory as developed in *The Idea of History* will also be an interpretation. However, we will try to let Collingwood speak as much as possible for himself. The subject is Collingwood's philosophy of history and for this reason – though it is not the usual practice – we will not confine ourselves to *The Idea of History* alone, but also discuss what Collingwood has to say about the subject in the *Autobiography*. For, together with what is left of *The Principles of History*, the *Autobiography* expresses Collingwood's latest views on philosophy of history.

In the preceding chapter we have seen that Collingwood's "flank attack" on the realists was based on the "laboratory of knowledge" of his archaeological experience. In the chapter "The Need for a

Philosophy of History" Collingwood explicitly returns in the *Autobiography* to the subject of the relation between history and philosophy, beginning as follows:

> My life's work hitherto, as seen from my fiftieth year, has been in the main an attempt to bring about a *rapprochement* between philosophy and history. In the preceding chapter I have described one aspect of this *rapprochement*, namely my demand that when philosophers thought about the history of their own subject they should recognize that what they were thinking about was history, and should think about it in ways which did not disgrace the contemporary standards of historical thinking. From the first, however, I saw that more than this was involved. I was also demanding a philosophy of history (Aut,77).

Collingwood then turns in a crucial passage to his idea of what a philosophy of history should be about:

> This meant, in the first instance, a special branch of philosophical inquiry devoted to the special problems raised by historical thinking. Epistemological problems, such as one might group together under the question 'how is historical knowledge possible?' Metaphysical problems, concerned with the nature of the historian's subject-matter: the elucidation of terms like event, process, progress, civilization, and so forth (Aut,77).

What Collingwood does not say in this passage is that the epistemological problems of history and its metaphysical problems are closely related to each other and in fact cannot be separated in his theory of history. We will see that this is made clear by what he has to say about the subject, both in the *Autobiography* and *The Idea of History*. It is important to note this, because it is nowhere explicitly stated by Collingwood.

With regard to the "metaphysical" aspect of history – history *a parte objecti* – by about 1920 it was his "first principle of a philosophy of history" that the past studied by an historian is not a dead past, "but a past which in some sense is still living in the present". Collingwood expressed this by saying "that history is concerned not with 'events' but with 'processes'" (Aut,97).[20]

In the chapter "History as the Self-knowledge of Mind" Collingwood deals with the epistemological problems of history. He begins by distinguishing "history proper" from "pseudo-history". By the latter he refers to "such things as the narratives of geology, palaeontology, astronomy, and other natural sciences which in the late eighteenth and the nineteenth centuries had assumed a semblance at least of historicity" (Aut,107). On the other hand, the essence of a historical object like archaeological remains lies in the fact that, unlike a natural object, it embodies a purpose. "[I]n pseudo-history there is no conception of purpose", Collingwood says, "there are only relics of various kinds, differing among themselves in such ways that they have to be interpreted as relics of different pasts which can be arranged on a time-scale" (Aut,109).

We will not discuss the details of Collingwood's succeeding argument, but only mention the three important conclusions (he calls them "propositions") which he reaches (the first one immediately after the passage quoted above): 1) "I expressed this new conception of history in the phrase: 'all history is the history of thought'. You are thinking historically, I meant, when you say about anything, 'I see what the person who made this (wrote this, used this, designed this etc.) was thinking.' Until you can say that, you may be trying to think historically but you are not succeeding. And there is nothing else except thought that can be the object of historical knowledge" (Aut, 110). 2) The answer to the question on which conditions it is possible to know the history of a thought is: a) "the thought must be expressed", b) "the historian must be able to think over again for himself the thought whose expression he is trying to interpret" (Aut,111). This leads to the proposition: "historical knowledge is the re-enactment in the historian's mind of the thought whose history he is studying" (Aut,112). 3) The solution of the problem how far the re-thought thought is the same as the original one is expressed in the propositon: "Historical knowledge is the re-enactment of a past thought incapsulated in a context of present thoughts which, by contradicting it, confine it to a plane different from theirs" (Aut,114).

The discussion of Collingwood's ideas on history usually rest on the Epilegomena of *The Idea of History.* The corollary of this is that certain aspects which come to the fore in the first four parts of the book but are not dealt with so explicitly in the Epilegomena, risk being neglected. One could say that this is the case with Collingwood's view on history *a parte objecti,* which in his eyes first of all has the characteristic of being a process. Collingwood's view on history as a process is made clear by his explicit rejection of certain ideas which do not take this into account. For instance, Collingwood criticizes the "substantialism" of Greco-Roman historiography, based on the "substantialistic metaphysics" of Plato and Aristotle, which "implies a theory of knowledge according to which only what is unchanging is knowable" (IH,42). According to Collingwood, this theory is reflected for example in the history of Rome as described by Livy: "Rome is the heroine of his narrative. Rome is the agent whose actions he is describing. Therefore Rome is a substance, changeless and eternal. From the beginning of the narrative Rome is ready-made and complete. To the end of the narrative she has undergone no spiritual change ... hence the origin of Rome, as he describes it, was a kind of miraculous leap into existence of the complete city as it existed at a later date" (IH,44). Collingwood declares that a modern historian, in contrast, would have interpreted the history of Rome as "a history of how Rome came to be what it is, a history of the process which brought into existence the characteristic Roman institutions and moulded the typical Roman character" (IH,43). The same defect he finds in Tacitus, who sees human characters as fixed substances, because "the idea of development in a character, an idea so familiar to ourselves, is to him a metaphysical impossibility" (IH,44).

In another context Rickert is criticized for not achieving the idea of historical process – a defect which this time is seen as the product of a positivistic remnant in his thought: "Rickert regards nature, after the positivistic manner, as cut up into separate facts and he goes on to deform history by regarding it in a similar way as an assemblage of individual facts supposed to differ from the facts of nature only in being vehicles of value. But the essence of history lies not in its consisting of individual facts, however valuable these facts may be, but in the process or development leading from one to another" (IH,169). The same criticism is leveled against Dilthey: "In the *Introduction to the Sciences of Mind* he took up the position ... that history deals with concrete individuals and natural science with abstract generalizations. But this never led him to a satisfactory philosophy of history, because the individuals of which he was thinking were conceived as isolated past facts and were not integrated into a genuine process of historical development" (IH,172). Again, Toynbee's *Study of History* is attacked as "a restatement of the positivistic view" (IH,159): "His whole scheme is really a scheme of pigeon-holes elaborately arranged and labelled, into which ready-made historical facts can be put ... This act, become habitual, leads to an obsession: one forgets that the historical fact, as it actually exists and as the historian actually knows it, is always a process in which something is changing into something else. This element of process is the life of history" (IH,163). A few lines further, summing up his criticism, Collingwood asserts on Toynbee: " ... [H]e regards history itself, the historical process, as cut up by sharp lines into mutually exclusive parts, and denies the continuity of the process in virtue of which every part overlaps and interpenetrates others. His distinction between societies or civilizations is really a distinction between focal points in the process: he has misunderstood it as a distinction between chunks or lumps of fact into which the process is divided" (IH,164).

According to Collingwood, the first person to realize that history forms a process was Vico: "Vico regards the historical process as a process whereby human beings build up systems of language, custom, law, government, etc.: i.e. he thinks of history as the history of the genesis and development of human societies and their institutions. Here we reach for the first time a completely modern idea of what the subject-matter of history is" (IH,65).

However, the claim that history is essentially a process does not suffice. For Darwin's great accomplishment was exactly that he proved nature to be a process. In Collingwood's eyes it is a grave misconception, however, to equate historical processes with natural ones. It is here that we come to an essential element in Collingwood's theory of history, since the latter depends on the distinction between the two types of process or between nature and history in general. Collingwood expresses this distinction in various ways, which can be reduced to his view that history deals with human beings with their distinguishing characteristic of having thoughts and being able to express them, that is, of being rational. This brings Collingwood to his principle that all history is the history of thought. This principle constitutes the basis for three other formulations of the difference between nature and history: 1) nature consists of events, while history consists of (human) actions; 2) natural events are seen from the outside, actions essentially from the inside; 3) the past of a natural process is dead, while the past of an historical process is living in the present.

Again, the best way to discern how Collingwood develops these ideas is, in the first instance, to follow his discussion of other philosophers of history. Hegel is given credit for having stated the distinction between nature and history, though he was wrong in denying the doctrine of the evolution of nature. "But it remains true that the process of nature is different from the process of history – that, for example, the succession of geological periods is not a truly historical succession", Collingwood adds. "Geology presents us with a series of *events*, but history is not history unless it presents us with a series of *acts*. Thus Hegel's conclusion is right, that there is no history except the history of human life, and that, not merely as life, but as rational life, the life of thinking beings" (IH,115). "[F]ollowing immediately from this", Collingwood continues, "all history is the history of thought. In so far as human actions are mere events, the historian cannot understand them; strictly, he cannot even ascertain that they have happened. They are only knowable to him as the outward expression of thoughts" (IH,115). A few pages further on Collingwood returns to Hegel's distinction between events and actions: " ... [H]istory consists of actions, and actions have an inside and an outside; on the outside they are mere events, related in space and time but not otherwise; on the inside they are thoughts, bound to each other by logical connexions" (IH,118).

The German philosophers of history are not only criticized by Collingwood because they did not grasp the idea of historical process, but also because they did not see the past as living in the present. Of Rickert Collingwood writes:

> [He] fails to see that the peculiarity of historical thought is the way in which the historian's mind, as the mind of the present day, apprehends the process by which this mind itself has come into existence through the mental development of the past. He fails to see that what gives value to past facts is the fact that they are not mere past facts, they are not a dead past but a living past, a heritage of past thoughts which by the work of his historical consciousness the historian makes his own. The past cut off from

> the present, converted into a mere spectacle, can have no value at all; it is history converted into nature (IH,169-170).

Of Simmel Collingwood contends that "because he has not sufficiently grasped the nature of the historical process he does not realize that the historian's own mind is heir to the past and has come to be what it is through the development of the past into the present, so that in him the past is living in the present" (IH,171). Similarly he says about Spengler's approach, that "at every point the idea of historical process as a mental process, where the past is conserved in the present, is elaborately denied" (IH,182), and about Toynbee, that "history is converted into nature, and the past, instead of living in the present, as it does in history, is conceived as a dead past, as it is in nature" (IH,164). We have seen in the preceding chapter that Collingwood criticizes Spengler and Bury as examples of a positivistic approach; in a separate section he deals with this theory in general and blames it for implying "that history could only be the history of external events, not the history of the thought out of which these events grew" (IH,132).

The characteristics of history which distinguishes it from nature, developed by Collingwood in his lectures of 1936, are dealt with more systematically in the Epilegomena of *The Idea of History*, especially in "Human Nature and Human History", also dated 1936. Collingwood's ideas on the subject, as developed in the latter essay, are much better known. For this reason most of the passages that will be quoted here have become more or less classic. I will only pick out those topics which relate to the present discussion, for instance when Collingwood says:

> There is a certain analogy between the archaeologist's interpretation of a stratified site and the geologist's interpretation of rock-horizons with their associated fossils; but the difference is no less clear than the similarity. The archaeologist's use of his stratified relics depends on his conceiving them as artifacts serving human purposes and thus expressing a particular way in which men have thought about their own life; and from his point of view the palaeontologist, arranging his fossils in a time-series, is not working as an historian, but only as a scientist thinking in a way which can at most be described as quasi-historical (IH,212).

One of the passages quoted most often is the one in which Collingwood illustrates the difference between an action – the object of an historian – and mere event with the metaphor of the outside and inside:

> The historian, investigating any event in the past, makes a distinction between what may be called the outside and the inside of an event. By the outside of the event I mean everything belonging to it which can be described in terms of bodies and their movements: the passage of Caesar, accompanied by certain men,

> across a river called the Rubicon at one date, or the spilling of his blood on the floor of the senate-house at another. By the inside of the event I mean that in it which can only be described in terms of thought: Caesar's defiance of Republican law, or the clash of constitutional policy between himself and his assassins. The historian is never concerned with either of these to the exclusion of the other. He is investigating not mere events (where by a mere event I mean one which has only an outside and no inside) but actions, and an action is the unity of the outside and inside of an event. He is interested in the crossing of the Rubicon only in its relation to Republican law, and in the spilling of Caesar's blood only in its relation to a constitutional conflict. His work may begin by discovering the outside of an event, but it can never end there; he must always remember that the event was an action, and that his main task is to think himself into this action, to discern the thought of its agent (IH,213).

It is obvious that the differences between nature and history will be reflected in the activities of the scientist and the historian. The way this is described by Collingwood is so crucial and the subject of so many discussions that its quotation in full is justified:

> To the scientist, nature is always and merely a 'phenomenon', not in the sense of being defective in reality, but in the sense of being a spectacle presented to his intelligent observation; whereas the events of history are never mere phenomena, never mere spectacles for contemplation, but things which the historian looks, not at, but through, to discern the thought within them.
>
> In thus penetrating to the inside of events and detecting the thought which they express, the historian is doing something which the scientist need not and cannot do. In this way the task of the historian is more complex than that of the scientist. In another way it is simpler: the historian need not and cannot (without ceasing to be an historian) emulate the scientist in searching for the causes or laws of events. For science, the event is discovered by perceiving it, and the further search for its cause is conducted by assigning it to its class and determining the relation between that class and others. For history, the object to be discovered is not the mere event, but the thought expressed in it. To discover that thought is already to understand it. After the historian has ascertained the facts, there is no further process of inquiring into their causes. When he knows what happened, he already knows why it happened.
>
> This does not mean that words like 'cause' are necessarily out of place in reference to history; it only means that they are used there in a special sense. When a scientist asks 'Why did that piece of litmus paper turn pink?' he means 'On what kinds of occasions do pieces of litmus paper turn pink? ' When an historian asks 'Why did

> Brutus stab Caesar? ' he means 'What did Brutus think, which made him decide to stab Caesar?' The cause of the event, for him, means the thought in the mind of the person by whose agency the event came about: and this is not something other than the event, it is the inside of the event itself.
> The processes of nature can therefore be properly described as sequences of mere events, but those of history cannot. They are not processes of mere events but processes of actions, which have an inner side, consisting of processes of thought; and what the historian is looking for is these processes of thought. All history is the history of thought (IH,214-215).

Collingwood also deals again with the idea of a living past:

> The past, in a natural process, is a past superseded and dead. Now suppose the historical process of human thought were in this sense an evolutionary process. It would follow that the ways of thinking characteristic of any given historical period are ways in which people must think then, but in which others, cast at different times in a different mental mould, cannot think at all. If that were the case, there would be no such thing as truth: according to the inference correctly drawn by Herbert Spencer, what we take for knowledge is merely the fashion of present-day thought, not true but at the most useful in our struggle for existence (IH,225).

The starting-point of our discussion of Collingwood's ideas on history was his view of history *a parte objecti*. We have seen, however, that the activity of historians – history *a parte subjecti* – crept in the discussion. We may think this is perfectly natural, in the sense that a correct view of history as object is a precondition of being a good historian. In Collingwood's view, however, the two are also related to each other in another and more essential way. For the possibility of having knowledge of the past is directly related to and dependent on the nature of the historical process as a process of thought and therefore living in the present. The reverse, however, is also true: the historical process only exists in our knowledge of it. "There is not, first, a special kind of process, the historical process, and then a special way of knowing this, namely historical thought". Collingwood says. "The historical process is itself a process of thought, and it exists only in so far as the minds which are parts of it know themselves for parts of it ... It would therefore be sophistical to argue that, since the historical process is a process of thought, there must be thought already present, as its presupposition, at the beginning of it, and that an account of what thought is, originally and in itself, must be a non-historical account. History does not presuppose mind; it is the life of mind itself, which is not mind except so far as it both lives in historical process and knows itself as so living" (IH,226-227). The close interrelation between the historical process, as seen by Collingwood, and our knowledge of it is made clear again by the passage, where he says: "Thought is therefore not the presupposition of an historical process which is in turn the presupposition of historical knowledge. It is only in the historical process, the process of thoughts, that thought exists at all; and it is only in so far as this process is known for a process of thoughts that it is one. The self-knowledge of reason is not an accident; it belongs to its essence" (IH,227).

This passage is especially relevant to Collingwood's view on the function of historical knowledge, which in his eyes "is no luxury, or mere amusement of a mind at leisure from more pressing occupations, but a prime duty ... of reason itself" (IH,227-228). At the moment, however, we will not discuss this aspect, but direct our attention to history *a parte subjecti* – and in particular the celebrated doctrine of history as the re-enactment of past thought. But we should remind ourselves, as a preliminary, of the relation in Collingwood's theory between the thought of the historian and the historical process as a living process of thought, taking the lectures of 1936 as evidence. This conception is made clear, for instance, when Collingwood criticizes Toynbee for failing "to see that the historian is an integral element in the process of history itself, reviving in himself the experiences of which he achieves historical knowledge. Just as the various parts of the process are misconceived as placed outside one another, so the process as a whole and the historian are placed outside one another" (IH,164). Discussing the dilemma arrived at by Simmel and Dilthey (the former seeing history as a mere projection into the past of present states of mind and the latter seeing history in the last resort as the object of psychological analysis) Collingwood develops the same contention: "The answer to both doctrines is that since the past is not a dead past but lives on in the present, the historian's knowledge is not exposed to the dilemma at all: it is not either knowledge of the past and therefore not knowledge of the present, or else knowledge of the present and therefore not knowledge of the past; it is knowledge of the past in the present, the self-knowledge of the historian's own mind as the present revival and reliving of past experiences" (IH,175).[21]

As far as I know there are only two places in his published works before 1936 where Collingwood alludes to the idea of re-thinking. It is characteristic that they are expressed in the context of a consideration of the idea of a living past. In "Croce's Philosophy of History" (1921) he says: "When a man is dead, the world has judged him, and my judgment does not matter; but the mere fact that I am rethinking his history proves that he is not dead, that the world has not yet passed its judgment. In my person, indeed, it is now about to pass judgment"[22]. In "Oswald Spengler and the Theory of Historical Cycles" (1927) he states: "If history is possible, if we can understand other cultures, we can do so only by re-thinking for ourselves their thoughts, cherishing within us the fundamental idea which framed their lives; and in that case their culture lives on within ours, as Euclidean geometry lives on within modern geometry and Herodotean history within the mind of the modern historian"[23]. The concept of re-thinking past thought as an essential element in historical thinking is developed by Collingwood for the first time in the Die manuscript of 1928 and will be dealt with in chapter 4 of this study. In the published work, it is Collingwood's expression of his theory in *The Idea of History* that has attracted most attention, though the subject is also treated in the *Autobiography*. We will start again by giving a few examples from the lectures of 1936.

The concept of re-thinking (Collingwood also speaks of re-enactment) is mentioned for the first time in *The Idea of History* when discussing Tacitus: "History cannot be scientifically written unless the historian can re-enact in his own mind the experience of the people whose actions he is narrating. Tacitus never tried to do this: his characters are seen not from inside, with understanding and sympathy, but from outside, as mere spectacles of virtue or vice" (IH,39). Kant is criticized for seeing history, like nature, as a spectacle. He was wrong in this, Collingwood contends, "because history is not a spectacle. The events of history do not 'pass in review' before the historian. They have finished happening before he begins thinking about them. He has to re-create them inside his own mind, re-enacting for himself so much of the experience of the men who took part in them as he wishes to understand" (IH,97). Dealing with the difference between nature and history Collingwood says that "it is peculiar to history that the historian re-enacts in his own mind the thoughts and motives of the agents whose actions he is narrating, and no succession of events is an historical succession unless it consists of acts whose motives can, in principle at least, be thus re-enacted" (IH,115).

In the lectures of 1936 the re-enactment doctrine is confined to some casual remarks, though it is already clear that Collingwood considers it essential for history. As with his general theory of history, the doctrine is more systematically worked out in the Epilegomena, especially in the essays "Human Nature and Human History" and "History as Re-enactment of Past Experience", both from 1936. After stating in the first essay that "all history is the history of thought" Collingwood continues:

> But how does the historian discern the thoughts which he is trying to discover? There is only one way in which it can be done: by re-thinking them in his own mind. The historian of philosophy, reading Plato, is trying to know what Plato thought when he expressed himself in certain words. The only way in which he can do this is by thinking it for himself. This, in fact, is what we mean when we speak of 'understanding' the words. So the historian of politics or warfare, presented with an account of certain actions done by Julius Caesar, tries to understand these actions, that is, to discover what thoughts in Caesar's mind determined him to do them. This implies envisaging for himself the situation in which Caesar stood, and thinking for himself what Caesar thought about the situation

> and the possible ways of dealing with it. The history of thought, and therefore all history, is the re-enactment of past thought in the historian's own mind (IH,215).

The conclusion of this passage does not imply, however, according to Collingwood, that all human actions are the subject-matter of history, and he adds that "indeed historians are agreed that they are not":

> But when they are asked how the distinction is to be made between historical and non-historical human actions, they are somewhat at a loss how to reply. From our present point of view we can offer an answer: so far as man's conduct is determined by what may be called his animal nature, his impulses and appetites, it is non-historical; the process of those activities is a natural process. Thus, the historian is not interested in the fact that men eat and sleep and make love and thus satisfy their natural appetites; but he is interested in the social customs which they create by their thought as a framework within which these appetites find satisfaction in ways sanctioned by convention and morality (IH,216).

Again, the irrational aspects of the human mind are not accepted by Collingwood as subject-matter for history: these aspects are referred by him to the science of psychology. "They are the blind forces and activities in us", Collingwood contends, "which are part of human life as it consciously experiences itself, but are not parts of the historical process: sensation as distinct from thought, feelings as distinct from conceptions, appetite as distinct from will. Their importance to us consists in the fact that they form the proximate environment in which our reason lives, as our physiological organism is the proximate environment in which they live. They are the basis of our rational life, though no part of it. Our reason discovers them, but in studying them it is not studying itself" (IH,231).

Referring to the idea of re-enactment in "History as Re-enactment of Past Experience", Collingwood says: "What we must now do is to look more closely at this idea, and see what it means in itself and what further consequences it implies" (IH,282). When an historian has a written document or relic of the past before him, Collingwood continues, "he has to discover what the person who wrote those words meant by them. This means discovering the thought ... which he expressed by them. To discover what this thought was, the historian must think it again for himself" (IH, 282-283). Now, if we are to understand Collingwood's concept of re-thinking past thoughts we must first of all understand his concept of thought, especially as the common sense notion of thought does not suffice.

According to Collingwood, one aspect of thought is that it "can never be mere object". This statement is clarified by him as follows: "To know someone else's activity of thinking is possible only on the assumption that this same activity can be re-enacted in one's own mind. In that sense, to know 'what someone is thinking' (or 'has thought') involves thinking it for oneself" (IH,288)[24]. It is obvious that for a historian the activity of re-thinking a past thought cannot be accidental, but must be consciously done: "unless he knows that he is thinking historically, he is not thinking historically" (IH,289). Thought is part of our "total" mental life, our "stream of consciousness". It has, however, a peculiar characteristic: " ... if mere experience is conceived as a flow of successive states", Collingwood says, "thought must be conceived as something that can apprehend the structure of this flow and the forms of succession which it exhibits: that is, thought is able to think the past as well as the present" (IH,292-293).

So a thought is not "wholly entangled in the flow of experience" (IH,296), an idea Collingwood elaborates on in the following way:

> An act of thought is certainly a part of the thinker's experience. It occurs at a certain time, and in a certain context of other acts of thought, emotions, sensations, and so forth. Its presence in this context I call its immediacy; for although thought is not mere immediacy it is not devoid of immediacy. The peculiarity of thought is that, in addition to occurring here and now in this context, it can sustain itself through a change of context and revive in a different one. This power to sustain and revive itself is what makes an act of thought more than a mere 'event' or 'situation' ... It is because, and so far as, the act of thought is misconceived as a mere event that the idea of re-enacting it seems paradoxical and a perverse way of describing the occurrence of another, similar, event. The immediate, as such, cannot be re-enacted. Consequently, those elements in experience whose being is just their immediacy (sensations, feelings, &c. as such) cannot be re-enacted; not only that, but thought itself can never be re-enacted in its immediacy. The first discovery of a truth, for example, differs from any subsequent contemplation of it, not in that the truth contemplated is a different truth, nor in that the act of contemplating it is a different act; but in that the immediacy of the first occasion can never again be experienced: the shock of its novelty, the liberation from perplexing problems, the triumph of achieving a desired result, perhaps the sense of having vanquished opponents and achieved fame, and so forth (IH,297-298).

" ... [A]n act of thought", Collingwood says a few pages further on, "in addition to actually happening, is capable of sustaining itself and being revived or repeated without loss of its identity" (IH,300). Then, giving the re-thinking of a thought of Plato as an example he declares: " ... in their immediacy, as actual experiences organically united with the body of experience out of which they arise, Plato's thought and mine are different. But in their mediation they are the same" (IH,301). The latter is called by Collingwood the aspect of universality of thought, meaning by this phrase "the way in which thought, transcending its own immediacy, survives and revives in other contexts" (IH,303).

So history being the history of thought, the gulf of time between the historian and his object can be bridged. The gulf must be bridged however from both ends:

> The object must be of such a kind that it can revive itself in the historian's mind; the historian's mind must be such as to offer a home for that revival. This does not mean that his mind must be of a certain kind, possessed of an historical temperament; nor that he must be trained in special rules of historical technique. It means that he must be the right man to study that object. What he is studying is a certain thought: to study it involves re-enacting it in himself; and in order that it may take its place in the immediacy of his own thought, his thought must be, as it were, pre-adapted to become its host (IH,304).

In "The Subject-matter of History" the concept of thought is further analysed by Collingwood. He considers "its power of recognizing the activity of the self as a single activity persisting through the diversity of its own acts", which is self-consciousness, to be" [t]he positive peculiarity which distinguishes thought from mere consciousness" (IH,306). In an obscure and difficult passage Collingwood then makes a distinction between different "levels" of thought, the last being called by him reflection, which means being conscious of thinking. "Historical thinking is always reflection", Collingwood concludes, "for reflection is thinking about the act of thinking, and we have seen that all historical thinking is of that kind. But what kind of thinking can be its object?" (IH,307). His answer to this question is as follows:

> In order ... that any particular act of thought should become subject-matter for history, it must be an act not only of thought but of reflective thought, that is, one which is performed in the consciousness that it is being performed, and is constituted what it is by that consciousness. The effort to do it must be more than a merely conscious effort. It must not be the blind effort to do we know not what, like the effort to remember a forgotten name or to perceive a confused object; it must be a reflective effort, the effort to do something of which we have a conception before we do it. A reflective activity is one in which we know what it is that we are trying to do, so that when it is done we know that it is done by seeing that it has conformed to the standard or criterion which was our initial conception of it. It is therefore an act which we are enabled to perform by knowing in advance how to perform it (IH,308).

There is one element of the re-enactment doctrine which we have not yet dealt with, and that is the "third proposition" to which Collingwood had come in the *Autobiography* in his treatment of the characteristics of historical knowledge. In this proposition it is said that historical knowledge is the re-enactment of past thought "incapsulated in a context of present thoughts which, by contradicting it, confine it to a plane different from theirs" (Aut,114). This is the answer to a problem Collingwood apparently considered a difficult one, for he says about it that "[n]o question in my study of historical method ever gave me so much trouble; and the answer was not complete until some years later" (Aut,112). An example he discusses in this connection is that of understanding what Nelson meant by saying "in honour I won them, in honour I will die with them", when he was advised to take off his decorations to make himself a less conspicuous target for his enemies. "Understanding the words", Collingwood says, "means thinking for myself what Nelson thought when he spoke them ... Unless I were capable – perhaps only transiently – of thinking that for myself, Nelson's words would remain meaningless to me" (Aut,112). He continues, then, as follows:

> But this re-enactment of Nelson's thought is a re-enactment with a difference. Nelson's thought, as Nelson thought it and as I re-think it, is certainly one and the same thought; and yet in some way there is not one thought, there are two different thoughts. What was the difference?...
> The difference is one of context. To Nelson, that thought was a present thought; to me, it is a past thought living in the present but (as I have elsewhere put it) incapsulated, not free. What is an incapsulated thought? It is a thought which, though perfectly alive, forms no part of the question-answer complex which constitutes what people call the 'real' life, the superficial or obvious present, of the mind in question (Aut,112-113).

Collingwood then makes a distinction between a "primary" and a "secondary" series of thoughts, the first being that of the historian and the second that of his object, saying about their relation:

> No question that arises in this primary series, the series constituting my 'real' life, ever requires the answer 'in honour I won them, in honour I will die with them'. But a question arising in that primary series may act as a switch into another dimension. I plunge beneath the surface of my mind, and there live a life in which I not merely think about Nelson but am Nelson, and thus in thinking about Nelson think about myself. But this secondary life is prevented from overflowing into my primary life by being what I call incapsulated, that is, existing in a context of primary or surface knowledge which keeps it in its place and prevents it from thus overflowing (Aut,113).

So the same act of thought can happen twice (or more) in different "primary series". This is important to note, because it means that by re-thinking a past thought a historian is not totally committed to that thought, but keeps his present mental life as his primary one. In *The Idea of History* Collingwood also refers to this principle, for instance when he says:

> When, as an historian, I relive in my own mind a certain experience of Julius Caesar, I am not simply being Julius Caesar; on the contrary, I am myself, and know that I am myself; the way in which I incorporate Julius Caesar's experience in my own personality is not by confusing myself with him, but by distinguishing myself from him and at the same time making his experience my own. The living past of history lives in the present; but it lives not in the immediate experience of the present, but only in the self-knowledge of the present (IH,174).

By being incapsulated in a present thought a past thought may be judged or criticized. After explaining his conception of re-enactment, with the thoughts of Plato and Caesar as examples, Collingwood maintains:

> This re-enactment is only accomplished, in the case of Plato and Caesar respectively, so far as the historian brings to bear on the problem all the powers of his own mind and all his knowledge of philosophy and politics. It is not a passive surrender to the spell of another's mind; it is a labour of active and therefore critical thinking. The historian not only re-enacts past thought, he re-enacts it in the context of his own knowledge and therefore, in re-enacting it, criticizes it, forms his own judgement of its value, corrects whatever errors he can discern in it. This criticism of the thought whose history he traces is not something secondary to tracing the history of it. It is an indispensable condition of the historical knowledge itself (IH,215).

In another context the same principle is put forward by Collingwood, when he says that "[w]hat is required, if I am to know Plato's philosophy, is both to re-think it in my own mind and also to think other things in the light of which I can judge it" (IH,301).

This survey of Collingwood's philosophy of history is far from complete. I have not dealt, for instance, with the question and answer activity, the function of history as the self-knowledge of mind, the role of evidence and Collingwood's conception of scientific history against so-called "scissors-and-paste" history. It is not the aim, however, of this chapter to give an all-round picture of Collingwood's ideas on history, and those aspects which have been neglected until now will come up for discussion elsewhere in this study. The only purpose of this survey of certain aspects of Collingwood's thought on history is to provide the necessary background to make the many comments on Collingwood's ideas on the subject understandable. Since an "objective" survey of Collingwood's thought on history would be an idle endeavour, no attempt has been made to evade a certain interpretation. My intention has been to find a coherence in Collingwood's fundamental views on history. Against this background we will turn now to the various comments given on them.

3.3. THE DISCUSSION OF *THE IDEA OF HISTORY*

3.3.1. INTRODUCTION

A whole history could be written of the reception of *The Idea of History* and the discussion of the views expressed in it. Such a history would not only be intrinsically valuable, but would also form one of the essential elements in a history of the philosophy of history after World War II in general. For it is no exaggeration to say that during this period there has hardly been one study in this field – at least in the English speaking world – in which Collingwood's contribution has not come up for discussion in one way or another. Nevertheless, no attempt will be made in this study to review the many comments on *The Idea of History*. Only those topics will be discussed which in my opinion are the most essential and are still of current interest.

Immediately after its appearance the importance of *The Idea of History* was recognized, both by historians and philosophers. Though the more detailed comments subsequently came almost exclusively from philosophers, the first reviews were written, interestingly enough, by historians. Max Beloff, for instance, in a review titled "The Historian's Philosopher"[25] says about *The Idea of History* that not "... a better introduction to historical studies [could] be imagined than a combination of the *Autobiography* and this latest work". C.V. Wedgwood and A.L. Rowse did not endorse all aspects of Collingwood's views, but the latter calls it "... in spite of its defects, ... one of the most important works ever devoted to the subject"[26]. Since Collingwood had dealt with Oakeshott's *Experience and its Modes* (IH, 151-159) it is interesting to note the latter's assessment of *The Idea of History*: "... unfinished and scrappy though it is, it is enough to show that if he [Collingwood] had been unhindered by ill-health and early death he could have done for historical knowledge something like what Kant did for natural science", and further: "The last hundred pages of *The Idea of History* is all that remains of Collingwood's projected work on the principles of historical knowledge, but it is enough to put him ahead of every other writer on the subject"[27].

It is understandable that historians were most interested in the first four parts of *The Idea of History*, which gives an historical survey of the idea of history[28]. Philosophers, on the other hand, have focussed their attention almost exclusively on the Epilegomena. The first occasion when some aspects of Collingwood's views were discussed was at a joint session of *The Aristotelian Society* and *The Mind Association* in July 1947, with contributions by MacIver, Walsh and Ginsberg[29]. From that date till the present day there has been a never ceasing flow of publications on Collingwood's philosophy of history. In the fifties, especially with contributions by Dray and Donagan, the philosophical discussion became increasingly detailed and sophisticated. This has undoubtedly advanced the understanding of Collingwood's ideas, but has also had the effect that few historians contribute to it any more, and one even wonders if the discussion could always be followed by them. I am not sure that this development would have been appreciated by Collingwood himself.

I propose to confine myself to certain topics that have attracted most attention in the discussions of Collingwood's philosophy of history. With regard to history *a parte objecti* we will begin with Collingwood's thesis that all history is the history of thought and the related issue of the neglect of objective conditions alleged by some interpreters. On the methodological aspects of history we will begin with the interpretation of Collingwood as an intuitionist and a treatment of the re-enactment doctrine. Then the topics of understanding and explanation, the use of generalizations and historical objectivity will be dealt with.

On these subjects a separate study could easily be written and perhaps it should. The following survey makes no pretence at completeness.

3.3.2. ALL HISTORY IS THE HISTORY OF THOUGHT

We have seen in the preceding section that Collingwood's first thesis with regard to history as object consisted of his process view of history. This aspect of his thought was hardly noticed and consequently never contested. We should not, however, conclude that this notion is simply a common-sense one: Collingwood did not consider it so and we have seen how he developed an explicit theory of his own.

Collingwood's definition of the historical process as essentially a process of thought has been the object of many critical comments. It is no surprise that the few remarks made by historians on Collingwood – mostly in an oblique way – are especially directed towards this aspect of his thought. For it can easily be interpreted as a directive for their research, and as such does not fit with their practice. It will thus be rejected as an unacceptable limitation on their activities, and at best be considered the product of an apparently extravagant philosophical theory. "What warrant is there for this limitation of the idea of history?", asks Renier. "The reasonable, indeed the rational attitude, would be to accept as history that which has been recognized as such through the ages, or, if we are to be meticulous in our use of words, that which has been called 'history' through the ages ... [T]o construct an abstraction, to annex for it the common name of 'history' ... why should this provide anyone with a criterion for the excommunication of recognized historians?"[30]. According to Smith "[t]he characterization of *all* history as the history of thought will strike the practicing historian as an excessively idealistic as well as arbitrary concept ..."[31], while Tholfsen says about Collingwood's theory that " [h]is conception of 'thought' ... is much too narrow to do justice to the actual subject matter of historical writing"[32]. Collingwood's view is considered by Fischer to be an example of the "idealist fallacy", which "consists in interpretations of human conduct which rest upon a conception of man as *Homo Sapiens* in a narrow and exclusive sense". He regards this as an absurd reduction: "To isolate merely the rational component of human existence is to falsify both humanity and rationality"[33]. Elton speaks of "... the fatal suggestion that ideas are the only realities in history"[34], while according to the German historian Faber "... die Interpretation der Geschichte als Geschichte der menschlichen Gedanken durch *Collingwood* würde eine unzulässige Verengung des Gegenstandes Geschichte bedeuten"[35]. Collingwood's theory is strongly rejected by Marwick in his widely read book *The Nature of History*. Discussing Collingwood's contention that what is described in military history is "thinking about strategy and thinking about tactics", not "weary marches in heat or cold, or the thrills and chills of battle or the long agony of wounded men" (Aut, 110), he goes so far as to call this passage "absolute rubbish, well illustrating what can happen when a highly refined mind pushes a pet theory too far". A few lines later he speaks of "odd mystical outbursts"[36].

Among historians, Toynbee has been the most outspoken and comprehensive critic of Collingwood's theory of history[37]. "'[T]he acts which we do on purpose'", he contends, "include acts of other kinds – for instance, acts of will – besides acts of reflective thought; and acts of other kinds – for instance, acts of impulse as well as acts of will – play a much larger part than any acts of reflective thought in the action that is the subject-matter of History as we find this 'in real life' when we look at the actual practice of historians without allowing a philosopher's *a priori* dictum to hypnotize us into ignoring the realities"[38]. "If he [the historian] is to participate in other people's experiences", Toynbee declares a few pages further on, "he must participate, not only in their thoughts, but also in their emotions and in their volitions ... The historian must obey Collingwood's commandment over a wider field of experience than the intellectual allotment of which, alone, Collingwood takes cognizance"[39]. Toynbee's conclusion is that "[a]n idolization of thought is the philosopher's idolatrous sacrifice on the altar of his professional patriotism"[40].

Though usually in a more subtle way, philosophers too have raised objections to Collingwood's definition of history. Murphy calls "Collingwood's identification of all history with the history of thought" an "unhappy overstatement"[41] and Grant speaks of "Collingwood's exaggerated emphasis upon the rationality of the historical object"[42]. Winch calls Collingwood's conception "no doubt an exaggeration"[43], while the German philosopher Apel refers to "der extremen These, dass der Historiker es nur mit den 'Gedanken' der Menschen zu tun habe"[44]. Walsh was the first to endorse in principle Collingwood's notion of all history being the history of thought. In his view, however, an analysis in terms of thoughts or purposes is "only a *prima facie* analysis". "We are not claiming that all action is through and through the product of reason, but only that it has, as it were, a rational superstructure"[45].

Many others beside Fischer equate Collingwood's theory with an idealistic position, like that of Dilthey, Croce or Oakeshott[46].

The most obvious objection to Collingwood's concept of history as history of thought is that it seriously distorts the historical object. At best, the argument usually goes, it could be considered relevant for intellectual history. The theory is defective because: a) the irrational aspects of human actions are not taken into account; b) social and economic history fall outside its sphere; and c) it is only relevant for individual actions, but not for group or mass-behaviour. We will deal with these objections separately.

We have seen that Toynbee's criticism is focussed on the first objection. Others have followed him. "[E] vents have other insides than the acts of thought which Collingwood regarded as alone the objects of historical knowledge", Child contends, "besides thought, there is feeling or emotion"[47]. And the theologian Roberts: "[N] ot all human behaviour is purposive; sometimes it is instinctive, impulsive, reflexive, or intuitive, and sometimes such actions, which cannot be explained in terms of their agents' conscious purposes or intentions, have made a tremendous impact on the course of human history"[48]. Another theologian, Hartt, also criticizes Collingwood for denying irrational factors "any real potency or at least any potency before which reason on its throne need tremble"[49]. According to Gardiner not all human activity is "thought out": "it may be routine, skilled, or impulsive"[50] and Fischer contends that "Collingwood's method, strictly applied, would exclude not merely the nonintellectual problems in which historians are actually interested but also many intellectual problems, which are characteristically neither rational nor irrational, but transrational"[51], whatever may be meant by this last concept.

"When it is said that history is essentially concerned with 'thought', what is being referred to?", asks Walsh. "The term is capable of both a wider and a narrower meaning, and the ambiguity is reflected in an important division among supporters of the idealist theory"[52]. He then contrasts Collingwood's theory with that of Dilthey. While the latter interprets the subject-matter of history to include men's feelings, emotions and sensations, Collingwood opts for a narrow view: "When *he* said that all history was the history of thought, he meant that it was properly concerned with intellectual operations. All thinking, he explained, took place against a background of feeling and emotion, but it was not with that that the historian was concerned". "The reader may well be puzzled to know what led Collingwood to maintain so apparently extreme and paradoxical a theory as this", Walsh then comments[53]. A few pages further on he explicitly rejects Collingwood's "very narrow definition of the field of history": "The historian ... does try to resurrect the thought of the past; but he is interested not solely in ideas proper, but also in the background of feeling and emotion which those ideas had. When he attempts to uncover the spirit of an age, it is not merely its intellectual life he hopes to penetrate: he wants to get at its emotional life too"[54].

The sociologist Baumann's comments on Collingwood illustrate another criticism: it is not applicable to the social and economic features of history:

> The motives of the historical actors can hardly be dignified as 'thoughts', because they are too complex and obscure, the field of causation too multiple, the stuff and dramatic pattern of history too multifarious to venture intellectualistic reductions and generalizations. Moreover, the thoughts even of the 'greatest individuals', the most eminent actors in the historical drama (even if they are 'thoughts' in Collingwood's sense) are unimportant in comparison with the concatenation of social and historical forces. The reading of Hitler's *Mein Kampf* is hardly the primary source for understanding the tragedy of Germany under Nazi domination. Further we have to distinguish between political, military and cultural history, the proper field of Collingwood's concern, on one side, and economic history on the other. In analysing, say, the history of prices, are we able to rethink the thoughts of the agents who did them?[55].

This criticism is of course closely related to the criticism that Collingwood's theory is only relevant for individual behaviour. It is taken separately because the latter point brings in the question of methodological individualism, and this cannot be equated with the question of the possibility of social and economic history.

That Collingwood's theory would only be applicable to individuals had been realized by Ginsberg in 1947:

> The view that history consists in the discovery of the thought which is the inner side of the event to be explained is only plausible, if at all, with reference to certain kinds of history such as diplomatic history, which is often concerned with specific acts of individuals ... But it is almost meaningless when applied to the history of large-scale massive interactions such as is found in economic history or the history of language, or more generally in the history of institutions. ... The view that the data with which history or social science have to deal are only ascertainable by re-thinking the thought expressed in them would seem to make any explanation of human affairs highly precarious. For it is the interactions between human minds and the consequences that follow from them that are important and these are not present to any one mind. If recourse be had to the extremely dubious notion of a group mind over and above individual minds it would still be necessary to show how this mind could ever be accessible to the individual mind. Consider, for example, the work of historians of languages in tracing long-range trends or drifts of phonetic changes ... So it is with all the major social institutions, the changes which they undergo may ultimately be traceable to changes in the minds of the individuals sustaining them but the institutions have, so to

say, a structure of their own, which reacts on individual minds, and which must be studied as such[56].

Gardiner and White express the same criticism[57] and the historian Carr calls it "[o] ne of the serious errors of Collingwood's view of history... that the thought behind the act, which the historian was called on to investigate, was the thought of the individual actor"[58]. Leff, also an historian, castigates Collingwood for "narrowing down the area of meaningful historical experience to individual thoughts and intentions" and "its excessive reliance upon the autonomy of individual conduct"[59]. Fischer, however, sees the subject from another angle: "One problem for an idealist epistemology is the group phenomenon", he says: "Can one rethink the thought of a collectivity? Only, it seems, by conjuring up the fiction of a 'corporate mind', as Collingwood called it"[60].

This claim is plainly contradicted by Donagan, who maintains that Collingwood "was a methodological individualist, in the strongest sense of that disputable term", adding a few lines further on: "Historians interest themselves in processes which take place in groups: the expansion and contraction of populations, the rise and fall of empires, wars, religious movements, economic booms and depressions, struggles between parties and factions, constitutional changes, and so forth. Collingwood maintained not only that all such processes are in principle analysable in terms of the acts of individuals and the relations between them, but also that all historical explanations of such processes are in terms of acts of individuals or of classes of such acts"[61]. Mink has taken Donagan's interpretation further by emphasizing that Collingwood's theory allowed for thoughts of anonymous individuals. According to him the link between the thought of individuals and human institutions is provided by a *priori* conceptual systems, a notion he derives from Collingwood's theory of absolute presuppostions as worked out in *An Essay on Metaphysics*[62].

Gardiner discusses the important concept of thought in Collingwood's theory in his article "The 'Objects' of Historical Knowledge"[63]. Though the re-enactment doctrine forms the background, the more general question of the status of the concept of thought is discussed by Gardiner. His starting-point is Collingwood's division of historical events into an "outside" and "inside". Gardiner has no objection against using this metaphor if it is not taken too literally. In his view, however, this is not the case with Collingwood and consequently his theory of historical knowledge is based on an unsatisfactory epistemology. Collingwood's notion of the revival of the object of historical knowledge in the historian's mind, the latter offering "a home" for the former, is the particular target of Gardiner's criticism. "In the light of this interpretation, an everyday occurrence begins to look very mysterious"[64], he comments. The "outside – inside" metaphor, "with its quasi-spatial associations" treats thoughts as "entities ... which two different persons may 'have' at different times"[65]. According to Gardiner, this also implies the unacceptable notion of crossing the "boundaries" of the "outsides" of events to their "insides", containing the realm of thoughts[66]. This theory, then, is intended to provide the basis for our knowledge of past thoughts, because they would be reduced to the priviliged model of "introspective acquaintance"[67].

In his book *The Nature of Historical Explanation* the same subject is discussed by Gardiner in a similar way. The distinction between the outside and the inside of events is called by him "both artificial and misleading":

> Artificial, because we do not, for example, talk of human actions having 'insides' and 'outsides': the distinction is normally put in terms of what was done and why it was done. Misleading, because the introduction of a spatial metaphor gives the impression that what are called the 'insides' of events are queer objects, invisible engines that make the wheels go round. And it is only too easy to move from this to the supposition that, in order to 'know' the insides of historical events (where 'knowing' is knowing by acquaintance) some peculiar technique for looking at these is required, analogous to the use by bacteriologists and astronomers of microscopes and telescopes, although, of course, at the same time subtly different. Thus a picture is presented which depicts the historian as a man who examines difficult, recalcitrant entities – thoughts and intentions, plans and 'mental processes' – by means of 'intuition', 're-enactment of past experience', and so forth[68].

Though he was not acquainted with the views of Gardiner, Cohen takes the same line, in an article which also appeared in 1952. "Collingwood's use of the terms 'inner' and 'outer', as names for two categories into which all human action can be divided", he maintains, "illustrated his adherence to a metaphysics of the type which G. Ryle has attacked as 'the dogma of the ghost in the machine'"[69].

In his article "R.G. Collingwood and the Acquaintance Theory of Knowledge"[70] Dray put the interpretation of Collingwood's philosophy of mind on quite another track, explicitly repudiating Gardiner's view. He makes clear that Collingwood was in fact an avowed opponent of certain ideas imputed to him by Gardiner. For he opposed both "metaphysical" and "positive" theories of mind. By the former Collingwood meant the idea of a "mental substance", distinguishing between "what mind is and what it does", and by the latter he meant the failure to take account of the distinction between a thought and a mere "flow of consciousness"[71]. Dray emphasizes that thoughts were regarded by Collingwood neither as entities nor as occurrences, but were seen as belonging to the mind's activities. "What makes activities 'mental' is not their occurrence in consciousness, but a pattern, course or direction in the activities themselves", Dray then contends. But it is not "simply whatever pattern or direction the flow of immediate experience may exhibit": "What makes a pattern mental is that familiar but elusive characteristic which we might call 'falling under standards or criteria'; for, as Collingwood puts it, somewhat paradoxically, 'all thinking is critical thinking'. It is of the very essence of mind that those activities said to be a part of it be open to assessment, evaluation, criticism"[72]. The view that Collingwood's theory would imply that the agent has priviliged access to his private thoughts, as directly introspectible and known by acquaintance, is, according to Dray, beside the mark: "Indeed, I cannot see why Collingwood could not, on his theory, allow the possibility that at the time an agent's overt activity showed that he had a certain thought (e.g. opinion, belief, intention), there should have been no introspectible happenings whatever which were relevant to the question whether or not he had it. Collingwood's claim that the agent himself may be unaware of, or mistaken about, his own thought supports such a view of the implications of his account"[73]. Having given these arguments Dray rightly concludes that "[i]n view of such considerations, Collingwood's conception of mind begins to look a good deal more like the one popularized by Professor Ryle than Gardiner seems to have realized"[74].

One aspect of Dray's argument is worth special attention, that is, his contention that Collingwood's theory not only warrants the possibility that an agent is not (fully) aware of his thoughts, but also that this does not preclude them being known by an observer. This conception is extremely important with regard to the relevance of Collingwood's theory for historical practice. Walsh realized this clearly in his discussion of two possible objections to Collingwood's view, viz. that it would only be relevant with regard to deliberate human actions and it would only be plausible so long as certain types of history were considered, economic history for instance being obviously excluded. "Of these two objections", Walsh comments, "the first can perhaps be met with the reflection that much action which is impulsive and, to that extent, seems 'thoughtless', can none the less be shown on further investigation to be the expression of thought. If I strike a man in a fit of passion my action is certainly not deliberate; but it would be idle to deny that there was, as we say, an idea behind it"[75]. In this connection he draws a comparison with the practice of the psycho-analyst, "whose success in revealing carefully worked out plans behind apparently irrational actions is surely relevant to the subject we are considering". The same principle applies, according to Walsh, to economic history, where

> [t]he actions with which economic history deals are the actions of innumerable agents – in fact, all those who take part in the economic processes under investigation. And the thoughts which the economic historian tries to get at are expressed, often enough, in complicated series of actions carried out by different persons over long stretches of time, few, if any, of whom are aware of the direction of the whole movement. It may well be impossible to

> detect any deliberate plan here; but is that an insuperable objection to the idealist theory? Surely there is nothing very revolutionary in the suggestion that an idea can be persistently influential without its being continuously before anyone's mind: it can have, as it were, a background effect, being assumed unconsciously by persons who have never explicitly thought about it[76].

Walsh concludes that we must not equate what a person has *in mind* with what he has *before his mind.*

Though, according to Walsh, "Collingwood's language in this context ... is not free from ambiguity", Dray is of the opinion that Collingwood's theory of mind does clearly provide the possibility for historians to discover thoughts "which were unknown, not only to any contemporary eyewitness of the actions concerned, but even to the agents themselves", adding: "... even thought*less* actions may express a thought, and hence be properly regarded as having a thoughtside. For to act thoughtlessly is not necessarily to act to no purpose ..."[77].

3.3.3. OBJECTIVE CONDITIONS

Another aspect of Collingwood's theory which is often criticized is its not taking into account the role of objective conditions in history. Among these conditions may be reckoned both "objective" social ones (for instance institutions) and natural conditions (or events). We will concentrate on the latter, not only because the former have already been discussed, but especially because natural conditions constitute an "objective force" in a more obvious way. It is they, therefore, which have particularly attracted attention in the discussion.

Collingwood has not dealt with this subject extensively, but it is nevertheless an important aspect of this thought. Expressed in a simple way, Collingwood's view is that natural conditions in themselves never determine human conduct, but only influence it through the way they are interpreted by men. It is obvious that this view may be taken as a concrete illustration of the theory that all history is the history of thought. As early as "The Theory of Historical Cycles" (1927) Collingwood clearly contends that it is essentially man himself that moulds his destiny:

> Man is not confronted by changing circumstances outside himself; or if he is, that belongs to the mere externals of his life. The essential change is within himself; it is a change in his own habits, his own wants, his own laws, his own beliefs and feelings and valuations; and this change is brought about by the attempt to meet a need itself arising essentially from within. It is because man is not content to react automatically to the stimulus of nature that he is man, and not a plant or a mere animal; his humanity

> consists in his self-consciousness, his power to mould his own nature, which comes simultaneously with his awareness of that power[78].

Collingwood rejects theories which conceive human life to be a reflection of geographical and climatic conditions, as for instance Montesquieu: "History so conceived would become a kind of natural history of man, or anthropology, where institutions appear not as free inventions of human reason in the course of its development, but as the necessary effects of natural causes" (IH, 79). A few lines further on he gives his own interpretation: "To be sure, there is an intimate relation between any culture and its natural environment; but what determines its character is not the facts of that environment, in themselves, but what man is able to get out of them; and that depends on what kind of man he is".

After repeating elsewhere that all history is the history of thought Collingwood continues: "and when an historian says that a man is in a certain situation this is the same as saying that he thinks he is in this situation. The hard facts of the situation, which it is so important for him to face, are the hard facts of the way in which he conceives the situation" (IH, 317). If a man does not dare to cross the mountains, because he is frightened of the devils in them, Collingwood declares, it would be folly for the historian to call it sheer superstition and not in accordance with the facts: "The compulsion which the devil-haunted mountains exercise on the man who would cross them consists in the fact that he cannot help believing in the devils. Sheer superstition, no doubt: but this superstition is a fact, and the crucial fact in the situation we are considering" (IH, 317). That on Collingwood's view thought is ascribed the primary function within human conduct is also demonstrated by his description of a poor man's action as not being determined "by the fact of his children's unsatisified hunger, the fact, the physiological fact, of empty bellies and wizened limbs, but by his thought of that fact" (IH,315-316).

This view has been criticized by many interpreters of Collingwood's thought as obviously neglecting the "objective" facts that man has to cope with. Rotenstreich, for instance, puts it as follows:

> Collingwood could place history in the realm of purposive activities since in the later phase of his development he did not take into account the objective circumstances in which the purposive activity takes place, for instance, the geographical data essential for purposive planning of an action, or the stamina and endurance of a people or a society which is called upon to act, etc. Collingwood – and this is the main criticism of his view with reference to his contraction of history to purposive activities – placed the activity, as it were, in a vacuum; he understood it is having meaning only when related from the outset to meaningful activity. The only meaningful activity which he took into account

> was that of sponsoring an action with a purpose in view. But in history there are meanings assigned to given facts through responses to circumstances: an earthquake, although by no means a purposive activity created within the human realm, has a historical meaning through its impact on the human realm, that is to say through the meaning connected with this disaster after the event and not in anticipation of it[79].

Renier, too, castigates Collingwood for his neglect of natural conditions in history. As an example of the real influence of such conditions he cites a change which occurred in the Gulf Stream in the fifteenth century, resulting in a boom in herring-fishery in the Low Countries, the expansion of vessel-building etc.[80]. Roberts is of the same opinion when he says that "there are surely occasions when the historian must take account of natural changes in the environment which have profoundly influenced human societies or states, and these natural changes cannot possibly be explained in terms of human motives or intentions"[81]. Grant considers Collingwood's restriction to the study of human action an "over-simplification": "For man does not only act in history, he suffers also. Many natural events, like the Lisbon earthquake or the explosion of Krakatoa, are also historical events – not because they are caused by minds but because they affect people's minds". He criticizes Collingwood for "ignoring the importance of human *reactions* as well as *actions*. In other words, an historical event is to be defined as a past event in which a human mind is involved either as agent or patient"[82].

In his recent book *Knowledge and Explanation in History*[83] Atkinson is of the same opinion: "Much of what goes on in history is significantly influenced, not to say determined, by factors outside human control; by the natural environment, and the non-willed social environment too. Intentional action is but one of the influences upon what comes about. A great deal of human behaviour is, moreover, reactive and irrational or non-rational, rather than rational in the means/end sense"[84].

On Collingwood's example of the man who believes the mountains to be inhabited by devils he argues:

> There is here a curious mixture of exaggeration and refusal to face the obvious. It is, no doubt, all one whether the mountains really are or are only thought to be uncrossable, to the extent anyway that the people concerned cannot distinguish these two possibilities. Necessarily what people think, they think to be true; our beliefs may be false, but having them is holding them as true. And one can see how an historian, concerned to explain why people acted as they did, should be tempted to feel no obligation to distinguish the facts as they saw them from the facts as they are ... But it is still not literally the case that the historian can ignore the truth – what *he* takes to be the truth – about the mountains. It

must make a difference whether, in the then state of technology, they were or were not unclimbable. There is, at the very least, more to explain, if the beliefs which influenced people's actions were false in ways which could have been detected at the time. And Collingwood presumably cannot have meant to suggest that beliefs that the mountains were climbable, when they were not, would lead to their actually being climbed[85].

Not all interpreters, however, have rejected Collingwood's emphasis on the thought-side of human conduct. Walsh, as one of the first of them, has defended Collingwood's position against possible Marxist objections. According to him Collingwood is not denying that natural conditions have an influence on human beings, because actions necessarily take place within a physical world. "What he is doing", Walsh says, "is denying that the natural background to men's actions has a constant or inevitable effect on them ... Certainly human beings are affected by the material conditions in which they live; but what is important in considering this situation is not so much the external conditions themselves as the attitudes men take up to them. These vary from case to case, and so we cannot speak (as historical materialists, despite their parade of dialectical terminology, in fact do) of a causal determination which is simple and constant"[86].

Collingwood's notion that human thoughts as the insides of historical events are the proper object of history is fully endorsed by E.E. Harris: "No other sort of event is of interest to an historian. Other events, like earthquakes or plagues, are of historical interest only so far as they affect human action; that is to say, only so far as human beings perceive, apprehend and intelligently respond to them. This being so, no historical event is historically known unless the thought of which it is the outward expression is known and understood"[87].

Donagan takes up an intermediate position. Though he considers Collingwood's view that an historical situation consists entirely of thoughts "flagrantly opposed to the facts" and calls it an "eccentric doctrine", he argues that an act cannot be caused by a physical force *per se*: "Pure physical force, e.g. a push, can produce only a pure physical effect, e.g. a fall, but not an action, e.g. a walk or a run". Donagan sees this as analogous to Toynbee's conception of human actions as responses to challenges[88].

Collingwood's view is accepted without reservation by Mink. It is a misunderstanding, the latter contends, to object that Collingwood neglects natural facts; he merely says that these "are relevant to history *only* to the extent that they enter the consciousness of men". Consequently they are only mentioned "in the mode of indirect discourse; that is, if 'p' is a proposition describing natural fact, then historical discourse proper never contains statements of the form 'p' but only of the form, 'It was known that p', 'It was believed that p', 'X said to Y that p', and the like"[89].

Because of their impact, natural disasters such as earthquakes and volcanic eruptions may be considered paradigm cases of the objective forces of nature. We have seen that they are mentioned in this connection both by Rotenstreich and Grant in their arguments against Collingwood. What these authors have apparently missed is that it was Collingwood himself who has discussed this type of events in a note in the *Autobiography*:

> Some 'events' of interest to the historian are not actions but the opposite, for which we have no English word: not *actiones* but *passiones*, instances of being acted upon. Thus the eruption of Vesuvius in A.D. 79 is to the historian a *passio* on the part of the people affected by it. It becomes an 'historical event' in so far as people were not merely affected by it, but reacted to this affection by actions of various kinds. The historian of the eruption is in reality the historian of these actions (Aut, 128).

Although this passage is not mentioned by him, the theologian Bultmann in his book *Geschichte und Eschatologie* develops a point of view which is directly relevant in his discussion of the role of natural conditions. He declares:

> Im Gegensatz zu den Handlungen könnte man diesen Bereich der Naturgegebenheiten und Naturereignisse, sofern sie für die menschliche Geschichte etwas bedeuten, als *Widerfahrnisse*, *Erleidungen* bezeichnen. Zur Geschichte gehört nicht nur das Handeln der Menschen, sondern auch ihr Erleiden. Mann könnte fragen, ob nicht das Erleiden immer erst das Handeln in Gang bringt. Doch bringt das Erleiden das Handeln nicht nur in Gang, sondern es ist als menschliches im Unterschied vom bloss natürlichen oder mechanischen Widerfahrnis auch in gewissem Sinne ein Handeln, eine actio als reactio[90].

Apart from the fact that Bultmann is not consistent in his phrasing (he first equates "Widerfahrnisse" with "Erleidungen" and then distinguishes "das Erleiden" from "bloss natürlichen oder mechanischen Widerfahrnis"), his contention is not in accordance with Collingwood's view. For the latter clearly states that a *passio* becomes only an "historical event" "in so far as people were not merely affected by it, but reacted to this affection by actions of various kinds" (Bultmann speaks of a "reactio" only in the limited sense of an "Erleiden", that is, what Collingwood calls "affection")[91].

Thus, with regard to the relation of natural conditions to human actions, Collingwood apparently makes, in his example of the eruption of the Vesuvius, a distinction between two levels: the one of "affections" and actions, the first one serving, one could say, as a link between the natural event itself and human action. One may regret that this view was not properly worked out by Collingwood, because it is far from clear, especially on the question of where to put the borderline between the two levels of causal relationship.

Dray has rightly noticed the importance of Collingwood's short discussion of the eruption of the Vesuvius as an historical event. He interprets it as a concession on Collingwood's part that human thoughts are not to be considered the only causal factor in history:

> En se servant du terme 'affecter' dans un tel cas, Collingwood emploie évidemment une expression causale pour relier aux actions ce qui ne constitue ni leur intérieur ni leur extérieur; et cette concession demeure même quand il ajoute que l'événement n'entre dans l'histoire que si une réaction humaine se produit. En effet, en admettant de cette facon que les événements naturels ont un rôle explicatif en histoire, Collingwood accorde aux historiens un sens du mot 'cause' qui s'ajoute à celui qu'il avait d'abord tenu pour le seul sens historique légitime. Car ce qu'il dit par rapport au Vésuve revient à ceci: l'environnement physique peut être cause du fait que les agents historiques le regardent d'une certaine façon; et ces façons de regarder, à leur tour, peuvent être cause du fait que ces agents agissent d'une certaine manière. Dans cette séquence, le dernier emploi du mot 'cause' illustre bien le sens spécialement historique que possède ce mot dans la théorie originelle de Collingwood; mais il n'en est pas ainsi du premier emploi[92].

Notwithstanding this conclusion Dray concedes with regard to historical explanations that the reasons of the agents, that is, their thoughts, may be considered explanations by themselves of their actions. "Ce qui ne peut pas ... survivre", he adds however, "c'est cette autre prétention plus forte, à savoir: que les situations réelles ne sont jamais causes de telles actions – il en est ainsi, du moins, si la relation causale s'avère transitive, comme le suppose Collingwood lui-même..."[93].

One wonders, though, how a human action may be seen on the one side as sufficiently explained by the reasons of the agent, while on the other side the possibility should be left open that "real situations" could also be the cause of it. This question is especially acute with historical events such as the eruption of the Vesuvius and its human aftermath. Though the passage referring to this event is unsufficiently worked out by Collingwood, his other statements with regard to the role of natural conditions in history are clear enough. The preceding discussion of some reactions to his views shows how differently they may still be interpreted.

3.3.4. THE INTUITIVE VERSION OF THE RE-ENACTMENT DOCTRINE

Moving on from the discussion of the historical object to the activities of the historian – history a *parte subjecti* – one inevitably comes to the doctrine of the re-enactment of past thought as Collingwood's central conception. Before dealing with the doctrine itself we will first discuss a special interpretation of it, which has been particularly influential. Collingwood argued – to quote Walsh – "that history involves a unique and direct form of understanding which raises it above other kinds of knowledge"[94]. This special kind of understanding is usually described as intuition. Cohen for instance speaks of "a quasi-Spinozist *scientia intuitiva*[95], Hayden White of Collingwood's "resolution of historical knowledge into intuition"[96] and Marwick of a history "which turned out to depend solely on the historian's intuition"[97].

When an intuitionistic conception is attributed to Collingwood, this is without exception with disapproval. One may distinguish three points of criticism: a) Collingwood's pretensions to a theory of truth of a subjective nature; b) his theory's being non-inferential and unscientific; c) its not taking into account general knowledge.

Gallie mentions the fact that Collingwood's doctrine "has been criticised chiefly on the ground that it offers yet another intuitionist theory of truth, i.e. a theory which rests upon some entirely subjective criterion of truth, in this case the feeling or sensation of successful problem-solving", adding that "there can be no doubt that it does offend on this score"[98]. Munz talks in one breath of "a subjective and intuitionist factor"[99]. According to the theologican Harvey, what the re-enactment doctrine holds is that "the historian does not infer what the agent is thinking or feeling but grasps it immediately and directly"[100]. D'Arcy is of the same opinion[101], and Bagby speaks of "philosophers, like Croce and Collingwood, [who] have supposed that only some 'unscientific' or, at best, semi-rational mode of thought, some sort of direct intuition, was suited to the understanding of historical events"[102]. Shoemaker is more cautious in his judgment that intuition for Collingwood is not a sufficient, but only a necessary condition for historical knowledge. According to him "Collingwood does not say that the historian has *no* use for inference on *any* level", though he concedes on the other hand that "Collingwood *does* claim that there is more to historical knowledge than whatever is gained through inference"[103].

The intuitive version of Collingwood's theory has been familiarized by Walsh's interpretation in the well-known *Introduction to Philosophy of History*. He agrees with Collingwood that an historian has to penetrate behind the historical phenomena to the thought-side of them, but "to hold that such penetration is achieved by an intuitive act is something very different"[104]. What he particularly objects to is the fact that this would be done without referring to any general truth. Walsh calls this "Collingwood's main thesis" and says that it "will not bear examination": "It is not true that we grasp and understand the thought of past persons in a single act of intuitive insight. We have to discover what they were thinking, and find out why they thought it, by interpreting the evidence before us, and this process of interpretation is one in which we make at least implicit reference to general truths"[105].

In his important article "The Verification of Historical Theses"[106] Donagan strongly opposes the intuitive version of Collingwood's re-enactment doctrine, which is called by him "the received interpretation". His "alternative interpretation" is based not only on passages in "The Idea of History", which explicitly refer to the inferential nature of history (esp. IH,252), but also on Collingwood's practice as an historian. He argues that Collingwood – for instance in *Roman Britain and the English Settlements* – "does not profess to have penetrated Caesar's thoughts by intuition"[107], and that, especially in the sections "The Historical Imagination" and "Historical Evidence" of *The Idea of History*, Collingwood elaborated "a thoroughly inferential and non-intuitive theory of historical verification"[108].

Donagan emphasizes as well the "Baconian", that is, the scientific spirit of Collingwood's conception of history, in his book *The Later Philosophy of R.G. Collingwood.* He ends a section called "Re-enactment and Intuition" with the conclusion that "if Collingwood held it to be by interpreting evidence that historians come to know that 'the thoughts they create' are in fact re-creations of past thoughts, it is quite impossible that he should also have held that evidence is interpreted by means of the re-creation of past thoughts, or that such re-creation is a matter of self-certifying intuition"[109].

Dray has applauded Donagan's alternative interpretation[110] and in a review of his book he remarks "that it should scotch once and for all the objection so often heard that Collingwood offers an 'intuitive' account of historical enquiry"[111].

Goldstein too is of the opinion that re-thinking "requires a good deal of preparation" and that it "certainly [is] not the immediate, intuitive grasping that so many critics have imputed to Collingwood". It is "systematic and conceptual" and, he adds, "[a] ll those critics who have treated rethinking as a species of empathy or intuition have simply no idea at all of what it involves"[112].

The opposition to the intuitive version seems to have gained ground among the interpreters of Collingwood's philosophy of history, because in his recent study Atkinson too is explicit in its rejection:

> Collingwood's insistence that actions involve thoughts, which can somehow be re-thought by the historians, has led some to interpret him as crediting historians with a species of intuitive insight into the minds of the dead. It cannot be denied that his phraseology occasionally lends support to this interpretation, but it is none the less beyond doubt that it was not his considered view. Not only does he always take it for granted that historians should continue to go about their business in the way everybody knows they do and must, i.e. by working from documents and other records, he actually insists that history is inferential not intuitive[113].

In another recent study, *The Shapes of Time* by P. Munz, the alleged intuitive aspect of Collingwood's re-enactment doctrine is dealt with rather ambiguously. He first says that "[t]he only thing that suggests that the historian in the last instance must fall back upon irrational empathy is Collingwood's use of the word 'reenactment' and its associations", but adds that "[o]ur analysis clearly shows that the process is far more intellectual and rational than Collingwood led us to believe"[114]. He then remarks in a footnote that "[i]t is a moot point, again to be resolved by exegesis, whether Collingwood did or did not mean by reenactment something like 'intuition'"[115], while a few lines further on he says that "Collingwood's reenactment does not have the irrational associations it conjures up in the minds of many readers"[116]. On the following page, however, "a subjective and intuitionist factor" and "Collingwood's intuitionism" are mentioned again, to which a note is added that "there is room for disagreement as to the exact meaning of Collingwood's intuitionism"[117]. His final assessment, however, is clear in maintaining that Collingwood's "choice of the word 'reenactment' is indicative of his intention to depsychologise the process involved and to get away from various forms of intuition ... He ... chose the term 'reenactment' as the next best thing to describe the thoroughly intellectual nature of the process"[118].

3.3.5. HISTORY AS THE RE-ENACTMENT OF PAST THOUGHT

The re-enactment doctrine is undoubtedly the most discussed aspect of Collingwood's philosophy of history. Though most comments have been critical, Cebik's statement that "[t]oday, we inform practically no one when we record that the doctrine of re-enactment makes little if any sense"[119] will certainly not be generally endorsed. On the contrary, one may rather discern a recent revival of interest in the doctrine, developing new interpretations.

The first reactions to Collingwood's re-enactment doctrine were nearly all sceptical or even negative. We have seen that it was judged by many as unduly restrictive with regard to the historical object and giving an untrue picture of the historian's activities. It is in this connection typical that the few approvals were only made in connection with a limited aspect of history. The theologian Sykes for instance goes so far as to call the passage where Collingwood states that it is the historian's main task to discern the thought of the agent "the *Magna Carta* of the historian, and more particularly of the historian of the origins of Christianity"[120]. Though he is critical of other aspects of his theory Murphy contends that the "profundity" of Collingwood's conception "is summed up in the words of wisdom which proclaim that to know what someone has thought is to think it for oneself", this being especially relevant for the "historians of ideas"[121].

These comments are seldom to be found, however. As said, the majority were critical. A unique one is given by Strauss when he criticizes Collingwood for looking to the past too much with present eyes: "He did not attempt to look at scientific history, for once, from the point of view of the earlier thinkers"[122]. Collingwood's contention that an historian can and must criticize the thoughts he rethinks is not acceptable to Strauss, because "Collingwood therefore rejected the thought of the past as untrue in the decisive respect"[123]. Strauss claims in this respect to be more Collingwoodian than Collingwood himself, saying that "Collingwood's attitude towards the thought of the past was in fact that of a spectator who sees from the outside the relation of an earlier thought to its time" and that he "lacked the incentive for re-enacting the thought of the past: he did not re-enact the thought of the past"[124].

Roughly speaking one may distinguish two interpretations of Collingwood's re-enactment doctrine: one which takes it as a methodological recommendation and one which considers it an element of Collingwood's analysis of the a priori characteristics of history. We will first deal with the methodological interpretation, of which the intuitive version is the most popular form (We shall see, however, that it is not necessary to accept this version in order to interpret Collingwood's re-enactment doctrine in a methodological way).

A clear example is given in Toynbee's discussion of Collingwood's views. Mentioning the theory of re-enactment he speaks of Collingwood's "rule", "instruction", "admonishment", "commandment" or "prescription" to re-enact the past[125]. He strongly rejects this as completely impossible, since the historian would have to make "Tamerlane's experience 'an integral part of his own ... by re-enacting' it 'for himself'; and at this point, if the historian is a truly conscientious workman, the shadow of the madhouse once again falls athwart his thorny path"[126]. Though he himself highly values imagination, Toynbee asks with regard to Collingwood's theory: "But what is this faculty of Imagination which makes it possible, after all, for an historian to participate in Timur Lenk's experience without his having to re-experience it in real life?"[127]. It is understandable that historians like Toynbee in the first instance focus their attention on the possible methodological consequences of Collingwood's theory. It is equally understandable that they cannot find much inspiration in it and so arrive at a hostile reaction, expressing, as it were, their disappointment. Another example is the very unfriendly criticism by Renier, who calls the re-enactment of past thought an "exquisite symbiosis ... by going into a trance". "In Collingwood's nomenclature", he sums this up, "the resulting clairvoyance is called 'the historical imagination'"[128].

Though using a different language, philosophers interpreting the re-enactment doctrine methodologically have not been less critical. We have seen how Gardiner blames Collingwood for taking the metaphor of "insides" and "outsides" of events too literally. "[I]f we follow our metaphor through", he says, "we must somehow be able to *look into* the agent's mind to detect the thoughts occurring there"[129]. In his book *The Nature of Historical Explanation* Gardiner speaks of "the suggestion of some sort of telepathic communication with past thoughts", which would be connected with Collingwood's "desire to assimilate the past to the present so that the requirements of the acquaintance theory of knowledge may be satisfied ..."[130]. Further on he talks about the study of thoughts and intentions "by means of 'intuition', 're-enactment of past experience', and so forth" and even of "esoteric methods"[131].

Many interpreters of Collingwood have adopted the methodological version of the re-enactment doctrine. To give some examples: Mandelbaum describes it as holding that "the method which alone can accomplish... [the] task [of understanding human actions in the past] is to re-think the thoughts of the past"[132]; Flenley speaks about the "reconstruction or re-enactment the historian performs by rethinking the thoughts of the past"[133]; while Buchdahl maintains that, on Collingwood's view, "construction is ... being turned into reconstruction, and this into re-enactment ... [I]f you should somehow be able to reconstruct, to re-live, that thought, then you will have recaptured the exact flow of the connected tissue of events"[134]. In discussing Collingwood's description of Caesar's activities in *Roman Britain and the English Settlements* K.M. Martin speaks of Collingwood's "technique of re-enacting the past in his own mind"[135]. However, this technique of the historian, in Martin's eyes, is apparently not very effective: "Whereas he wishes, theoretically, to see into the mind of Caesar, his practice often does not permit this ... This inability to see the thought behind the act means that Collingwood can only supply, in many cases, a statement of the event and not of the action he wishes to narrate"[136]. Roberts too interprets Collingwood as contending that "the historical method ...demands a special technique, not required in the natural sciences, to penetrate the inside of events, thus detecting the thought they express"[137].

According to Grant "Collingwood's conception of re-thinking can be interpreted as a heuristic principle of some value", though he criticizes Collingwood for ignoring that "the task of the historian is not only to re-think past thoughts", but also, "in at least equal measure, the re-feeling, or re-living, of past desires"[138]. The "re-living of the inside of the action" Grant interprets as supervening upon "the inferences as to what was done ... carried out in accordance with the general principles of historical methodology". "These two complementary stages of inference and understanding", he then concludes, "could be described as the scientific and aesthetic aspects of historiography respectively"[139].

Though he values highly Collingwood's contribution to the philosophy of history, Cebik also gives a methodological interpretation of the re-enactment doctrine, and strongly rejects the latter. He speaks of "the means invented for penetrating human beings and capturing the thought behind their actions"[140]. His assessment of Collingwood's theory is as follows: "Putting aside his idealism, which sired the deformed notions of reenactment and historical imagination, one may find in Collingwood perhaps the nearest approach to a correct analysis of doing history ... The seeds of his error arise from too stringently holding the historian's attention to human action, from interpreting statements about human thought as statements about something hidden and demanding special talent and skill to uncover, and from transforming certain aspects of the logic of historical inquiry from pragmatic into metaphysical considerations"[141].

A recent and original version of the methodological interpretation is that given by Goldstein[142]. According to him the re-enactment of past thought must be seen as a central element in the constitution of the past by historical research. With this interpretation Goldstein objects to re-enactment as a part of historical explanation: "Collingwood's own intention was to use the idea of re-enactment in his account of how the historical past is known, not explained"[143] (the subject of explanation will be dealt with in the next section). Making use of examples from Collingwood's historical work, Goldstein develops the theory that it is by re-enactment that evidence is interpreted as the result of certain human actions, establishing in this way what these actions were. This re-enactment requires a good deal of empirical research and for this reason the intuitive interpretation is rejected by Goldstein. "What, after all, does the statement, 'That is the Antonine Wall', tell us about the object it designates?", he asks. "To know what it is as something historical is to know what purpose it served, what thoughts – policies – it embodies. Short of that, one really does not know what that thing stretched across Scotland from sea to sea is ... The answer to the... question, 'What was Antoninus doing here? 'turns out not to be 'Building a wall', to which we then respond 'Why?' It turns out, rather, to be carrying out a certain defensive policy, on the basis of these and those assumptions, requiring, *inter alia*, the construction of a wall having certain minimal characteristics, and so on. All this is established by means of the re-enactment of thought – together, of course, with evidence, the wall itself, to be sure, but all manner of other sources about Roman administrative and military policy"[144].

That Goldstein adheres to an explicitly methodological interpretation of the re-enactment doctrine is also made clear by other statements, for instance when he says that what Collingwood tries to do is "present an account of what took place, based upon evidence and using the method of re-enacting past thought". A few lines further on he avers: "Until the historian has gone through the course of re-enactment, he knows what evidence has survived the depradations of time, but he does not know what human actions have taken place. With re-enactment, we get a conception of what human action must have or might have ... taken place"[145]. Goldstein speaks in this connection of the re-enactment of past thought as "the central technique used by Collingwood in his attempt to constitute the historical past"[146].

The first fundamental attack on the methodological interpretation of the re-enactment doctrine was made by Donagan in his article "The

Verification of Historical Theses", which has been mentioned already in connection with his criticism of the intuitive interpretation[147]. We have seen that he considered this interpretation an element of the "received" one, setting his "alternative interpretation" against it. In his article Donagan develops a sustained argument against the methodological version in general, basing himself on a careful reading of certain crucial passages in *The Idea of History*. Even before analysing these he maintains that "[t]o re-think significant past thoughts is part of the end an historian strives to accomplish; it is not even the whole of it, for he must also both demonstrate that he has re-thought them, and use them to explain past actions. Collingwood's interpreters have mistaken his descriptions of an element in the goal of historical inquiry for descriptions of historical method; in so doing, they have defied his explicit warning, 'the re-enactment of past thought is not a pre-condition of historical knowledge, but an integral element in it' (IH,290)"[148].

The passages Donagan uses as evidence for his interpretation are taken from *The Idea of History*, where Collingwood illustrates his concept of re-thinking with the example of understanding the meaning of the Theodosian Code and "a passage of an ancient philosopher". Collingwood remarks: "Thus he is re-enacting in his own mind the experience of the emperor; and only in so far as he does this has he any historical knowledge ... of the meaning of the edict", and with the latter: "This means re-thinking for himself the thought of his author, and nothing short of that will make him the historian of that author's philosophy" (IH,283). Donagan's comment on these passages is the following:

> Both examples suggest that Collingwood's subject may be, not historical method, but what historical method achieves. The sentences ... read less like prescriptions of what to do in order to acquire historical knowledge than like descriptions of what historical knowledge consists in. In that case, Collingwood's question would be 'On what conditions is historical knowledge possible?' which is not a methodological question, but a philosophical one. And his answer would not tell historians how to make and verify historical theses, but would specify what conditions must obtain if such theses are to be put forward at all: unless past thoughts can be re-enacted, history is impossible[149].

These passages by Collingwood serve to introduce his discussion of "the criticism of an imaginery objector" (IH, 283). Their summing up is quoted by Donagan: "I am considering how history, as the knowledge of past thoughts (acts of thought), is possible; and I am only concerned to show that it is impossible except on the view that to know another's act of thought involves repeating it for oneself" (IH,288). Donagan concludes: "This unmistakably declares that the passage we are considering is not concerned with how historians verify their reconstructions of past thoughts; after all, that has been discussed at length in the previous section, 'Historical Evidence'. Wittgenstein would have described Collingwood's problem as a 'conceptual' one, and it arises as serious conceptual problems commonly do, with a philosophical objection to the possibility of something which nevertheless obdurately exists"[150].

Another passage that Donagan considers crucial is one in which Collingwood asks: "How does the historian discern the thoughts which he is trying to discover?" (IH, 215). He contends that "discerning" must be seen here as distinguished from "discovering" and preliminary to it. "On this interpretation", Donagan says, "to discern a past thought would be to reconstruct it – as a preliminary to proof or disproof ... [A]n historical reconstruction of a thought ... does not differ from e.g. discerning the sense of a complicated argument; and, just as an historical reconstruction must be demonstrated, so must a claim to have discerned the sense of an argument". Collingwood's question "would then be equivalent to, 'How is it possible even to form a notion of the inner side of another's actions?' and his answer would state the conditions on which it is possible to put forward theses about that inner side, not the methods by which such theses may be verified"[151]. When Collingwood says, in his example of understanding Caesar's actions, "This implies envisaging for himself the situation in which Caesar stood, and thinking for himself what Caesar thought about the situation and the possible ways of dealing with it" (IH, 215), Donagan maintains:

> 'Imply' is often used in a looser sense, but Collingwood's statement is true if it is taken in its strongest sense, namely, of logically necessary implication. To discover Caesar's thoughts logically implies envisaging his situation and what he thought of it, just as bisecting an angle logically implies dividing it. By this example Collingwood draws attention to something implicit in the very concept of historical knowledge[152].

Though he concedes that Collingwood's question with regard to the possibility of historical knowledge is formulated in "infelicitous and misleading ways"[153], Donagan is strongly convinced that the methodological interpretation is wrong. While the latter interprets the re-enactment doctrine, one could say, as a necessary and sufficient condition for obtaining historical knowledge, Donagan is of the opinion that it only serves as a description of a necessary condition for such knowledge, re-enactment being logically implied by it. The methodological aspects, according to him, are discussed by Collingwood elsewhere in *The Idea of History*.

After Donagan it has been Dray who has made a major contribution to the non-methodological interpretation of the re-enactment doctrine, particularly with respect to historical understanding and explanation. His theory of the "rationale of actions", as developed in *Laws and Explanation in History*[154], has become very influential, but is in fact based on his interpretation of Collingwood's re-enactment doctrine.

In his first article on the subject Dray attacks Gardiner's interpretation of Collingwood's theory as an example of the acquaintance theory of knowledge[155]. Against Gardiner's objection that, according to the re-enactment doctrine, the thoughts of the historical agent and the historian would have to be the same, Dray emphasizes that this sameness does not require a qualitative similarity: "In terms of experience, there are many ways of thinking the 'same thought'"[156]. "What is said to be literally identical in the thought of historian and agent", Dray argues, "is ... the *rational force* of the argument", the thought explaining an agent's acting, when fully articulated, assuming "the form of a practical argument"[157]. Like Donagan, Dray interprets the re-enactment of past thoughts as a necessary condition of understanding:

> [W]hat Collingwood is claiming, when he says the thoughts must be 're-thought', is that they are only explanatory if the historian is able to certify for himself their sufficiency as reasons for doing what the agent did. The theory of 're-thinking' is thus ... the formulation of a condition which must be satisfied for understanding to be claimed. What the theory attempts to do is elicit the *criterion of intelligibility* employed in historical studies[158].

In his article "Historical Understanding as Re-thinking"[159] the same argument is developed by Dray. He maintains that re-thinking is both possible and necessary. He contends:

> In order to grasp the agent's thought, and see that it really does explain his action, the historian must do more than merely reproduce the agent's argument, whether implicit or explicit; he must also *draw his conclusion*. It is not enough merely to examine a report of the agent's 'thought-process'; the historian must, on inspecting the thoughts, and treating them as premisses of practical deliberation, actually *think that* the conclusion follows. The historian's 'seeing' the connection between the agent's 'considerations' and his action entails his *certification* of the connection between them – this entailment being a logical one. If the attempt to re-think, and thus to certify, the agent's thought-action complex breaks down ... then we have a dark spot, an unintelligibility, a failure to understand[160].

"The explanation might thus be said to succeed", Dray says a few lines further on, "to the extent to which it reveals the rationality of the agent. An action is said to be understood, on Collingwood's view, when it is seen to have been rationally necessary"[161]. While Collingwood leaves the impression that a failure to understand must always be attributed to the historian, Dray mentions three possibilities which would make "rational explanation" not realizable: a) "cases where the agent's reasoning about his situation is itself mistaken"[162]; b) "cases where what is done was not intended by the agent"; c) "arbitrary or capricious action"[163].

In the chapter "The Rationale of Actions" in his *Laws and Explanation in History* Dray says: "The discussion to follow may be regarded in part as an attempt to 'make sense' of what Collingwood, in particular, has to say about historical understanding"[164]. It is in this chapter that he develops his well-known theory of "rational explanation" as an alternative to the "covering law" model of historical explanation. There is also a passage where Dray explicitly rejects the methodological version of the re-enactment doctrine and develops independently of Donagan a "necessary condition" interpretation:

> When Collingwood says that historical understanding consists of penetrating to the thought-side of actions – discovering the thought and nothing further – the temptation to interpret this in the methodological way is understandably strong. But there is another way in which the doctrine can be formulated: 'Only by putting yourself in the agent's position can you *understand* why he did what he did'. The point of the 'projection' metaphor is, in this case, more plausibly interpreted as a logical one. Its function is not to remind us of *how we come to know* certain facts, but to formulate, however tentatively, certain *conditions which must be satisfied* before a historian is prepared to say: 'Now I have the explanation'[165].

In a recent article on Collingwood's philosophy of history[166] the same argument is developed by Dray: "Sa thèse, il me semble, c'est que comprendre une action en histoire *consiste* (entre autres choses) à repenser la pensée qui s'y exprime. Il s'agit d'une thèse portant sur le *but* de l'enquête historique, et non sur ses *procédés* ...[N] ous *ne pouvons pas* découvrir ce que fut une pensée dans le passé sans *en même temps* la repenser"[167]. "Telle que je l'ai exposée", Dray concludes his discussion of the re-enactment doctrine, "la doctrine de Collingwood semble requérir, à titre de condition nécessaire pour comprendre une action comme il convient en histoire, que l'historien entretienne à l'égard de l'agent une certaine *empathie* – ceci ne voulant rien dire de plus que la nécessité pour l'historien de tenir compte, dans son évaluation de l'argument pratique de l'agent, de la situation de ce dernier, telle qu'elle se présente à lui"[168]. However, the impression that Collingwood sometimes leaves that a certain sympathy would also be required is not acceptable to Dray, the latter being neither a logical condition nor a psychological necessity for historical understanding.

In more recent publications the interpretations by Donagan and Dray seem to get the upper hand of the methodological version. Shoemaker for instance is very explicit: "'Re-enactment' is not an explanation of how the historian *arrives at* a knowledge of past thoughts, but rather, it explains how or *on what conditions* knowledge of the past is possible. If there is no re-enactment, then there is no scientific history – there is no knowledge of past thoughts ... Re-enactment is the *conditio sine qua non* of historical knowledge"[169]. Toulmin too remarks on Collingwood that he speaks in the section "History as Re-enactment of Past Experience" in *The Idea of History* "explicitly about the *outcome* of the historian's work, not about his *methods*"[170], while Rubinoff mentions the re-enactment of past thoughts in a section where "the transhistorical presuppositions of historical thought" are discussed[171]. Even Gardiner seems to have changed his view, because fourteen years after his obviously methodological (and highly critical) interpretation he writes that what Collingwood did, among other things, "was to focus attention upon the concept of *action*, regarded as a fundamental and ineliminable category of human thought ..."[172]. The "re-living" of the past (Dilthey is mentioned too in this connection) "need not be interpreted in a manner that requires us to suppose that it ascribes to the historian some recondite or 'mystical' mode of cognition which is denied to investigators in other domains", Gardiner maintains. "Instead, it may be regarded as primarily making a point about the kind of conceptual framework to which we commit ourselves when we refer to an event as a human action, such reference carrying certain special ... implications concerning the fashion in which what is referred to can be appropriately characterized and understrood"[173].

In his recent book *Historical Explanation* Rex Martin gives Collingwood's re-enactment doctrine a prominent part in his argument. He is also of the opinion that it should not be interpreted in a methodological way:

> When Collingwood said, for example, that the investigator should 'penetrate' to the 'inside' of an event, he was not hinting at some sort of anti-inductive procedure to get at a peculiarly inaccessible thing. He was stating, metaphorically, the aim of a science of human action: that it should go to the heart of the matter, to the 'thought' of the agent. Collingwood's remark was concerned with what the investigator should be up to, not with how he should proceed to do it... The investigator 'penetrates' to the thought expressed in a deed, not by some queer kind of X-ray perception, but by Baconian questioning of the information he has about the deeds and sayings of an agent[174].

Martin distinguishes two stages in re-enactment, "[t]he inductive reconstruction of the agent's thought through the interpretation of evidence" being the first, while the second is "that the investigator must ascertain that the deed performed is the 'expression' of the reconstituted thought; he must demonstrate that the deed is 'determined' by that thought. It is in this second stage that Collingwood conceived of rethinking under the category of empathy"[175]. Dray's interpretation is endorsed by Martin, when he says that "the deed under consideration ... is explained if and only if the investigator can establish that the deed is plausible, given his own reconstruction of the agent's thought, and also that it was the one, among the several conceivable alternatives, that seems *most* plausible in the light of this reconstruction. The investigator completes his re-enactment by means of a practical deliberation ..."[176].

With regard to the re-enactment of past thoughts Collingwood not only contends that the content of a past thought is rethought, but also the act of thought itself is re-enacted. This view has puzzled Walsh, among others:

> ... [T]here appear to be important ambiguities in this position. In one sense of the word 'thought', that in which it is taken to mean act or process of thinking, my thoughts can never be identical with anyone else's: saying they are mine indicates as much. Yet in another sense, where 'thought' is equated with what a man thinks, two persons can certainly think the same thoughts, and, what is more, can know that they do. But they know it not because their acts of thinking are identical (how could they be?), but because they find they can understand each other[177].

This problem is discussed in a fundamental way by Popper criticizing Collingwood's view, while Skagestad defends Collingwood against Popper. Popper develops his theory in his article "On the Theory of the Objective Mind"[178], in which he distinguishes three worlds: "the first is the physical world or the world of physical states; the second is the mental world or the world of mental states; and the third is the world of intelligibles, or of *ideas in the objective sense*; it is the world of possible objects of thought: the world of theories in themselves, and their logical relations; of arguments in themselves; and of problem situations in themselves"[179]. Within the third world Popper also distinguishes different levels: "Whenever we try to interpret or to understand a theory or a proposition ... we are in fact raising *a problem of understanding*, and this always turns out to be *a problem about a problem*; that is to say, a *higher level problem*"[180]. Popper illustrates this thesis by elaborating the example of the problem of the historical understanding (called by him "p^u") of Galileo's theory of the tides ("P_1"). His conclusion with regard to the relation between the two is that "*our problem of understanding*, P^u, is *on a higher level* than P_1. That is to say, the probelm of understanding is a *metaproblem*"[181]. "[W]e have to distinguish clearly between the metaproblems and metatheories of the historian of science (which are on the P^u level) and the problems and theories of the scientists (which are on the P_1 level)"[182]. "...*[T]here are, in general, no problems common to the different levels*", Popper maintains. "*Thus some of the third-world structural units that constitute the metatheory may be utterly* dissimilar from those that constitute the theory to be interpreted or understood"[183]. He then continues as follows:

> The point is important. It establishes *a fortiori* that even if we could speak at all sensibly (which I am inclined to deny) of such a thing as a *similarity* between third-world *thought-contents* on the one hand and, on the other, those second-world *thought-processes* through which we grasp these contents, even then I should still deny that there actually is, in general, any similarity, on any level of problems, between the contents and the corresponding thought-processes. For the third-world method of historical understanding which I am trying to describe is a method which, wherever possible, replaces psychological explanations by the analysis of *third-world relations*: in place of psychological explanatory principles we make use of third-world considerations mainly of a logical character; and my thesis is that from such analyses our historical understanding can grow[184].

Collingwood's theory is given by Popper as an example of a "psychological explanation": "We part company over the issue of the second and third worlds: the issue of choosing a subjective or an objective method ... Collingwood's psychological way of putting things is by no means merely a matter of formulation. Rather, it is an essential part of his theory of understanding"[185]. Using Collingwood's example of understanding the meaning of the Theodosian Code (IH,283) as an illustration Popper concludes: "Collingwood makes it clear that the essential thing in understanding history is not the analysis of the situation itself, but the historian's mental process of re-enactment, the sympathetic repetition of the original experience"[186]. Against this he puts his own view: "The historian's analysis of the situation is his historical conjecture which in this case is a metatheory about the emperor's reasoning. Being on a level different from the emperor's reasoning, it does not re-enact it, but tries to produce an idealized and reasoned reconstruction of it, omitting inessential elements and perhaps augmenting it"[187].

Besides criticizing Popper for imputing to Collingwood a psychologism, Skagestad disagrees with the distinction within the "third-world" between the object level of the problems of the historical agent and the meta-level of those of the historian: "The rational reconstruction of Galileo's problem, i.e. the search for the presuppositions on which Galileo's answer was a rational one, necessarily involves a serious attempt at solving Galileo's problem. This is essentially Collingwood's position; evidently, it can be stated without any reference to the repetition of psychological processes"[188]. His interpretation of the re-enactment doctrine is accordingly as follows:

> [E]very rational, and hence intelligible, action performed by an

> historical agent has an intellectual content. The agent was faced with a problem, and his action was his attempt to solve that problem. The problem can be expressed as a question, and the action embodies a particular answer. In reconstructing an historical event, the historian must ask the question that confronted the agent and reconstruct his answer to that question. Now, according to Collingwood's monistic philosophy of mind,[189] the assertive content of a thought cannot, even for purposes of analysis, be separated from the act of thinking that thought. It is granted that, say, Euclid's postulates have an objective thought-content, but in grasping that content I actually perform Euclid's act of thinking, not an act 'like' it but that very same act[190].

Giving another example Skagestad maintains:

> The actions of a Caesar or a Napoleon, in so far as they are intelligible, are thoughts as well as actions. Caesar and Napoleon were confronted by problem-situations embodying questions, and the actions by which they sought to solve their problems are the answers to those questions. In writing about Caesar or Napoleon, what the historian tries to do is to ask and answer their questions; that is, to think their thoughts and, to just that extent, actually to be Caesar or Napoleon. In this sense the past which the historian investigates is a living past constituted by thoughts which are his own; and in this sense Collingwood is ready to endorse Croce's notorious dictum, that 'all history is contemporary history'[191].

Shoemaker is of the same opinion: "The distinction between the 'act of thought' and the 'content of thought' is not an actual, but more like a 'formal' or 'metaphorical' distinction. A precise criterion for distinguishing the *content* of the Pythagorean theorem from the *act* of thinking that content would be very difficult indeed to formulate"[192]. With regard to the understanding of Plato's thought he says that "... we must recapture the *argument* – the *process* (in time) by which Plato was led to his conclusions" and continues: "Still, one may insist, the *argument* is also a part of the *content* of the thought. And, ultimately, this must be admitted. It is to be recalled, however, that the distinction between act and content of thought was only a metaphorical one. In reality, there can be no meaningful distinction. The act (the entire process) is co-extensive with the content (the entire content) of the thought"[193].

Debbins too declares that "[t]he more one reflects on the problem, the more one realizes how inadequate the evidence is for distinguishing between the object of thought and the act of thought"[194].

One could remark in this connection that this view is also implied by Dray's interpretation of re-thinking as grasping the rationale of an agent's action in the form of a practical argument.

Collingwood's doctrine that an historian needs to re-think past thoughts has as its preliminary that all history is the history of thought. We have seen that the latter thesis has been generally criticized, often vehemently. It is remarkable in this connection that Munz in his recent book *The Shapes of Time* explicity defends it as a central thesis of his study. His starting-point is that the "raw material" an historian studies, is always a certain interpretation itself: a "Sinngebild", as he calls it, being the basic unit of intelligibility and itself a "mininarrative"[195]. "We must conclude that there is no genuinely raw material at all", Munz contends. "Everything that has come down to us is cooked by somebody for some purpose; and I do not mean 'cooked' in the colloquial sense of 'forged' or 'doctored'. It is simply cooked in the sense that it is an artefact. It is made up by somebody. It is a mininarrative and when we go to work on it critically we must seek first and foremost to find out how it was composed"[196]. The distinction between primary and secondary sources should therefore be abandoned:

> If it is true that even primary sources are not really primary, where then is the raw material of history? The real raw material is present; but it is not 'material' in the sense in which documents are material. The real raw material of history is thought. Precisely, it is the thought that goes into the composition of the mininarratives. The events that actually happened and that cannot be broken down further are the thoughts of the people we are studying. In this sense Croce and Collingwood were right to insist that all history is the history of thought[197].

There remains the problem, however, that feelings, emotions etc. would be excluded, this making the re-enactment doctrine defective. It is of course all-important in this connection what exactly Collingwood means by the concept of thought. According to Mink we have to look for this to Collingwood's theory of mind, as developed by him in *The Principles of Art* and *The New Leviathan:* without this background, he holds, *The Idea of History* is in fact unintelligible. It is not possible to give here a survey of Mink's account of Collingwood's philosophy of mind[198]. Essential to it, Mink contends, is his distinguishing four levels of consciousness. Except for the lowest one, each of these has two aspects, a practical one (action) and a cognitive one (knowledge). After the lowest level of "pure feeling", we get the second of appetite and imagination, the third of desire and perception and the fourth of will and intellect. The relevant point, according to Mink, is that Collingwood called every level above the first a form of "thinking" or "thought". The levels are dialectically related to each other in the sense that at a higher level a lower level becomes conscious and it is this that is meant by "reflection". Only at the fourth (intellectual) level is re-enactment possible. But appetite, for example, can become conscious through desire at this level and consequently be re-enacted.

The second and third levels are only potential objects for a higher level, but when made an object they "survive" at the higher level. The relevance for the re-enactment doctrine is described by Mink as follows:

> Thus it is mental activities of the fourth level which are primarily recoverable and intelligible; but they preserve and carry with them the whole range of activities at lower levels just so far as those activities have become objects of consciousness. Critics of Collingwood who have, understandably enough, argued that historical knowledge includes far more than 're-enactments of acts of reflective thought' have simply not understood that in Collingwood's sense one is performing an 'act of reflective thought' when one orders from a menu, punishes a child, argues about politics, or climbs a mountain[199].

Mink emphasizes that the term "thought" was for Collingwood very wide and referred to all levels of consciousness except the lowest one of the "undifferentiated sensuous-emotional flux". He gives the following example of possible objects to be re-thought by the historian:

> What this means, practically, is that the historian is not limited to reconstructing Caesar's policies while ignoring his ambition. On the other hand, he cannot, as historian, deal with Caesar's ambitiousness as a *psychological* characteristic. 'Upon what meat does this our Caesar feed?' is not an historical question, insofar as it purports to suggest a bio-psychological explanation of individual personality characteristics. *Ambitiousness* belongs to the second level of *appetite* (vague hunger for something); *ambition* belongs to the third level of *desire* (hunger for a specific object); *ambitious decision* belongs to the fourth level of *will*. In re-enacting the latter, the historian can and must re-enact ambitiousness and ambition as far as they survive in it[200].

When we review the many interpretations of the re-enactment doctrine we can easily be overwhelmed by the often widely differing assessments of it. One conclusion, however, can surely be drawn – that it is still of current interest and that the last word will not have been said on it for some time.

3.3.6. EXPLANATION AND UNDERSTANDING

In the philosophy of history the topic of historical explanation is one of the most widely discussed. With regard to this subject too Collingwood's theory is relevant. This is confirmed by Rex Martin's book *Historical Explanation*, which places Collingwood's re-enactment doctrine at the centre of interest. We can distinguish three elements: the relevance of re-enactment for historical explanation in general, Collingwood's dictum that knowing what happened implies knowing why it happened, and his view that only successful actions can be rethought.

One of the objections to the re-enactment doctrine is that the determining conditions of the thoughts in question are neglected. Stover, for instance, contends that Collingwood "mistakenly supposed that when the historian understands, happenings (actions, in this case) thereby become intelligible *in the same way* that deterministic accounts make them intelligible. In other words, Collingwood supposed that the historian, understanding identifyingly, already knows the determining conditions of certain happenings"[201]. According to Stover, he is mistaken in this, and therefore should "consider more circumspectly the possibility, in principle, of deterministic accounts of rational activity, explanations for which nomological universals, far from being superfluous, are indispensable"[202]. Runciman similarly declares that in addition to re-thinking "the investigator must ask himself how his subjects came to have the thoughts and therewith perform the actions which he has succeeded in identifying"[203], while K.M. Martin objects that "although a statement of the agent's thought in one sense explains why the action was executed, it does not explain why the agent thought the thought"[204].

These objections may be criticized in two ways. In the first place it should be remembered that the questioning activity has to stop somewhere, because otherwise one can go on asking "why" as long as one pleases. Thus, according to Dray, no defect in Collingwood's theory can be imputed in this connection: "... [I] f a historian explains Caesar's crossing of the Rubicon by referring to his determination to oust Pompey from the capital, it cannot be taken as a defect of Collingwood's theory that this explanation is incomplete – if by this we mean only that we can still ask why Caesar wanted to get rid of Pompey. It is not a defect of Collingwood's theory because, if a defect at all, it would be a defect of *any* theory of what counts as an answer to the question 'Why?' Explanations can be regarded as given at successive levels of inquiry; and the farther we carry the questioning process, the deeper the explanation may be said to be. What Collingwood has to say, however, can be fairly assessed only as a theory of what counts as an answer to the question 'Why?' at a single level"[205]. Burns makes a similar retort: "it is like saying that the movement of tree branches is not explained by the fact of a wind blowing until we can also explain why the wind is blowing"[206]. Mink too points to the similarity in this respect between historical explanations and explanations of natural events: " ... neither does a causal explanation of a natural event explain how the cause itself came to be ... In science and in history, different questions call for different answers"[207].

In the second place, one has to keep in mind that, according to Collingwood, generalizations which might be used by historians have to be based on ascertained historical facts, that is, on knowledge of thoughts as the inner side of events: "The historian, when he is ready to hand over such a fact to the mental scientist as a datum for generalization, has already understood it in this way from within. If he has not done so, the fact is being used as a datum for generalization before it has been properly 'ascertained'. But if he has done so, nothing of value is left for generalization to do" (IH,222-223). So Collingwood would surely not endorse Stover's recommendation to complete re-thinking with "nomological universals" to find out the determining conditions. If an explanation has to be completed it should be done – in Dray's words – "by rounding out the thought-side, not by adding something of an altogether different kind"[208]. Actions are not necessitated by "laws", either of a natural or historical nature, as claimed by the positivists. They can be necessitated, however, and, according to Dray, Collingwood's answer to the question how this is possible would be "that it can be necessitated in the sense of its being rationally required": "The thoughts or considerations which explain an action in Collingwood's context of discussion, show the action to have been necessary in the sense of being 'the thing to have done, the principles of reason being what they are'"[209]. Instead of a natural necessity one can speak of a rational one, according to Dray: "If something happens in spite of natural necessity, we call it a miracle. If an action is done in spite of rational necessity, we call it a stupidity, a mistake, an irrationality. It is Collingwood's claim that if, and only if, rational necessity can be shown, then we understand what the agent did"[210].

Donagan gives a different argument. His starting-point is the question how, according to Collingwood's theory, an historian can connect his hypotheses with his evidence without making use of general laws[211]. As an illustration he gives an historical problem, taken from Collingwood's own historical writing – the possible aim of Caesar's invasions of Britain[212]. Caesar invaded Britain twice, but withdrew each time. Though he dealt extensively with the expeditions in his *Commentaries* he did not mention their objective. Because Caesar tried in his *Commentaries* to depict himself as favourably as possible, Collingwood comes to the conclusion that he had failed in his intention to conquer at least a part of Britain and consequently did hide this fact. The hypothetical premiss on which Collingwood bases his argument is, according to Donagan, the following: "It is true that there could be no justification for asserting that if Caesar had succeeded in a certain enterprise, and had been writing a book to advertise his successes, and had known that to advertise his success in that enterprise he must make certain things plain in that book, then he would have made those things plain, unless the same assertion could be made of everybody"[213]. Donagan then continues as follows:

> The hypothetical premiss about Caesar does presuppose a general hypothetical about anybody and everybody. Now, is that general hypothetical a law? A moment's reflection will show that it is not.

> A general law must admit of possible empirical falsification, but the general hypothetical presupposed by Collingwood's premiss does not. No conceivable empirical evidence would count against the proposition that if you hold to your intention to bring something about, and believe that you must take certain steps to do so, then you will take those steps if you can. Any evidence which goes to show that you held that belief but did not take those steps, although you could have, also goes to show that you did not stick to your intention. Collingwood's hypothetical premiss therefore rests, not on a general law, but on an analytic truth which derives from the very concept of an intention. It follows that his argument neither contains nor presupposes any general law[214].

Not all hypothetical premisses used by historians, however, are of the nature of analytic truths. Nor are they general laws. Donagan develops the argument that they may be closed hypotheticals, that is, closed in space and time in contrast to the open character of general laws. In both cases an historian refers to the thoughts of historical agents. Summing up his argument Donagan says:

> Collingwood's position, then, was this. Historians follow the Baconian method of systematic questioning, sifting true from false answers to their questions by means of survivals or traces of the past. They connect their conclusions with their evidence as natural scientists do, by means of hypothetical propositions. However, unlike natural scientists, they do not presuppose that every closed hypothetical premiss they employ must be subsumable under a general law, i.e. an open hypothetical. Sometimes their closed hypotheticals are analytic truths, sometimes they are not. When they are, they are about what historical agents will do in certain situations in virtue of what they think, i.e. of their plans, intentions, and so forth; and when they are not, they are derived from independently verifiable propositions about what historical agents think[215].

Dray has subjected Donagan's conception to an ingenious criticism. "Can the action of an historical agent be strictly deduced from what Donagan, following Collingwood, calls its thought-side?" he asks. "Or, more strictly ...: Can the assertion that an action was actually performed at a certain time be deduced from any attribution to the agent of thoughts which are other than those required to constitute it an action of the kind specified in the historian's explanandum?"[216]. Dray strongly doubts it. Thought concepts are sometimes, as Donagan concedes, "logically porous", and they are, according to Dray, also usually vague, making "[w]*hat* counts as falling under them at all, quite apart from whether it must *invariably* fall under them ... often indeterminate"[217]. Other problematic aspects Dray mentions are the possibility of purely private expressions of thought, the necessity of "efficacy" premisses with regard to the agent's powers and opportunities, and the possibility that the agent did not draw the practical conclusion in accordance with the thoughts attributed to him[218].

In his *Historical Explanation* Martin has made an extensive study of the re-enactment doctrine. In the introduction he limits himself to "only *one* of the main types of historical explanation, that whereby we explain actions by referring deeds to the 'thoughts' of individual agents"[219]. (We should note here that it is highly doubtful that this restriction really is implied by Collingwood's re-enactment doctrine. This subject, however, will not be discussed now, but in chapter 8).

Following Dray Martin says that "[w]hen an investigator says of an agent that 'he did x because y', the explanatory force of the 'because ...' part of his statement is derived from the claim that the deed x, in respect of the agent's thought y, is the most appropriate course of action"[220]. This is further worked out in the statement that according to Collingwood's theory an explanation is provided of anyone's action when we can show that: "1) he did perceive himself to be in a certain situation, 2) where he might do some such deed, and 3) he had a purpose relevant to dealing with his situation and 4) could be taken as seeing that that particular deed would accomplish this end"[221]. This model is discussed at length by Martin. He comes to a highly sophisticated solution of the problem of the connection between thoughts and deeds. Re-enactment plays the role, according to Martin, of validating facts for explanatory use. "By this I mean simply", he contends, "that, however the investigator works his evidence, the facts he puts into 'place' in accordance with the schema for explanations must have passed the test of understanding, of re-enactment"[222]. "Validating has to do", he explains, "not with whether facts are evidentially founded or with whether a given piece of evidence 'leads' to a certain conclusion, but with whether facts ... are plausibly or intelligibly related to one another"[223]. This validating is realized by "assertions of appropriateness", which "can be seen as operating in the logical space intermediate between a very high-order but empirically vacuous explanatory schema and a determinate body of available evidence"[224]. "When we explain re-enactively we simply exhibit certain specified connections as holding between particular thought-factors and between these and the particular deed we are explaining"[225], Martin maintains. The assertions of appropriateness realized by re-thinking are based on certain general "hypotheticals of appropriateness"[226]: their status is that they happen "with a high degree of regularity"[227]. Against Hempel's view, however, these general statements are not used for purposes of inference, but of justificaton: "to warrant a connection of particular facts as plausible or intelligible, thereby validating the subsumption of these facts under the schema in a given explanation"[228].

However ingenious Martin's argument, Dray still concludes in a review of his book that "[w]hat is missing is any adequate account of what it is to act for a reason at all, rather than simply to have one and to act in accordance with it", adding to it: "It is discouraging that, after all the philosophical activity (and all the 'sophistication') of recent years, this notion, so crucial for any attempt to provide a rationale for history as a humanistic discipline, and perhaps for the foundations of the social studies generally, remains so murky"[229].

With regard to historical explanation Collingwood has made the following outspoken statement, called by Donagan "[o]ne of [his] best-known sayings about history, and perhaps his boldest"[230]: "When [an historian] knows what happened, he already knows why it happened" (IH,214). The preceding sentences read as follows: "For history, the object to be discovered is not the mere event, but the thought expressed in it. To discover that thought is already to understand it. After the historian has ascertained the facts, there is no further process of inquiring into their causes".

Walsh has interpreted this passage to mean that, according to Collingwood, thoughts are self-explanatory: "It is a ... characteristic of thoughts ... that in re-thinking them we come, *ipso facto*, to understand why they were thought"[231]. According to Donagan, however, this interpretation is incorrect. For the pronoun "it" in "To discover that thought is already to understand it", he says, does not refer to "thought" in the same sentence, but to "event" in the preceding one ("For history, the object to be discovered is not the mere event, but the thought expressed in it"). So what Collingwood contends, according to Donagan, is that by discovering a thought an historian understands a "mere event", knowing "the thought expressed in it"[232].

Though Walsh later revised his interpretation and endorsed Donagan's view[233], Dray is of the opinion that the latter cannot be correct. For what Donagan is saying, Dray argues, is that the outside of an action (the "mere event") is explained by its inside (the thought) and this "would involve denying a major Collingwoodian doctrine: that historical explanation is of actions rather than of 'mere' events"[234]. Dray has always been critical of Collingwood's dictum that when an historian "knows what happened, he already knows why it happened". "For, strictly, speaking", he maintains in his first article dealing with the subject, "an action cannot be explained in terms of its own thought-side ... The thought which is required to make it the action it is ... cannot be considered as something logically distinct from it, by reference to which the action itself can be explained"[235]. The problematic aspect of Collingwood's contention is that the explanandum and the explanans of an explanation are confused. As Dray puts it in a later article: "We are left then with the consequence that full specification of the agent's thought must appear in *both* explanans and explanandum. This leaves the explanation itself open to a charge of circularity"[236]. The solution offered by Dray is to distinguish clearly between explanandum and explanans: "that the historian begins with knowledge of an action performed, including a thought-side, and gives an explanation of it by relating the whole of it to thought of the agent which, in some important sense, is ***not*** a part or 'side' of the action to be explained"[237]. Dray specifies the distinction between explanandum and explanans in the following way:

> I should want to argue that in the analysis of 'thought' explanations of action, we must be prepared to distinguish ... between two different logical roles which knowledge of the agent's thought must play. The first (if I may put it this way) is that of raising the status of a mere physical event to that of an object of historical interest: a human action in the Collingwoodian sense. The place of such reference to thought is exclusively in the historian's explanandum. The second is that of providing an answer to the historian's question why what is specified in the explanandum occurred. Thus reference to Caesar's plan to occupy the country is in no way required to specify his action as an invasion, whereas reference in his intention to land regardless of opposition may be. Of course, if we had wished, we could have asked a why-question about Caesar's action specified as an attempted conquest rather than as an invasion; *what* is to be explained is up to us. In that case, however, reference to his plan of conquest would belong in the historian's explanandum as the thought which helps to *constitute* the action to be explained; and by changing our question we should have ruled out this thought as a possible explanatory factor. A satisfactory explanans would now have to make reference to some further thought of Caesar's[238].

What Dray's analysis also makes clear is that we do not necessarily have to look for "determining conditions" outside the realm of thought, when trying to explain actions at a more fundamental level. We have seen at the beginning of this section, however, that it is exactly this which is required according to certain interpretations.

Though Donagan is inclined to defend Collingwood's thesis that to know what happened is to know why it happened he makes the important restriction that it is only plausible "if the success or failure of the action is excluded from investigation"[239]. It is not at all plausible, according to Donagan, if it is the success or failure that must be explained themselves: "To explain Brutus' killing Caesar as contrasted with his stabbing him, or Napoleon's defeat at Waterloo as opposed to his battle-commands there, it would be necessary to ascertain not only the inside and the outside of what Brutus or Napoleon did, but also their situation"[240]. This argument, though, is more convincing with regard to failures than to successes. Collingwood at least seems to think so, because in his *Autobiography* he says the following: "How can we discover what the tactical problem was that Nelson set himself at Trafalgar? Only by studying the tactics he pursued in the battle. We argue back from the solution to the problem

... Naval historians think it worth while to argue about Nelson's tactical plan at Trafalgar because he won the battle. It is not worth while arguing about Villeneuve's plan. He did not succeed in carrying it out, and therefore no one will ever know what it was. We can only guess. And guessing is not history" (Aut,70). Donagan totally rejects this argument: "No historian could long maintain that history can make sense of success and victory, but not of defeat and disappointment"[241]. Collingwood himself also does not stick to this principle, he says, because in the *Autobiography* (Aut, 131) "he claimed to have discovered what Caesar's purpose had been in invading Britain, even though Caesar 'had failed to achieve it'"[242]. What Donagan does not keep in mind when giving this example, however, is that Collingwood comes to his conclusion with regard to Caesar's intention not from an analysis of Caesar's situation, but from an argument based on an *a priori* hypothetical premiss, as Donagan himself has made clear[243].

In a short article "On Explaining Disaster"[244] Watkins has launched a frontal attack on Collingwood's contention that only successful actions can be reconstructed. He gives a "rational reconstruction" of a disastrous collision of two ships headed by two admirals in 1893, which obviously was not the result of a planned successful action. Having worked out the same example in another article[245], Watkins comes to the conclusion:

> Philosophically speaking, the easiest kind of historical explanation of an action that ended in failure is this. The main components of the agent's decision-scheme have been ascertained to the historian's satisfaction; these point pretty unambiguously to a certain practical conclusion; the action in question was in line with that conclusion; *but* there is a significant discrepancy between the situational appraisal contained in the agent's decision-scheme (as reconstructed by the historian) and the agent's objective problem-situation (as reconstructed by the historian); and the failure of the action can be explained in terms of this discrepancy[246].

What should be noted, however, is that Collingwood was of the opinion that situations consist "altogether of thoughts" (IH,316): "[W]hen an historian says that a man is in a certain situation", he wrote, "this is the same as saying that he thinks he is in this situation. The hard facts of the situation, which it is so important for him to face, are the hard facts of the way in which he conceives the situation" (IH,317). So, according to Collingwood, it would be irrelevant to speak of "the agent's objective problem-situation as reconstructed by the historian" in for instance the case of Villeneuve, for the only thing at issue is his conception of the situation. Certainly one may seriously question Collingwood's view that situations only consist of thoughts. But starting from Collingwood's principle it is at any rate true that Villeneuve's actions would probably require more effort to explain than those of Nelson.

The view that only successful actions can be understood cannot be upheld, of course, as a general principle. That Stauffenberg's bomb-attack on Hitler failed does not imply that we cannot know his plan. It is obvious that certain objective conditions have to be taken into account in a case like this, which cannot possibly be described as thought-situations.

3.3.7. GENERALIZATIONS

"If, by historical thinking, we already understand how and why Napoleon established his ascendancy in revolutionary France", Collingwood says in "Human Nature and Human History", "nothing is added to our understanding of that process by the statement (however true) that similar things have happened elsewhere. It is only when the particular fact cannot be understood by itself that such statements are of value" (IH,223). The idea of a generalizing science of history is therefore rejected by Collingwood: for it would base itself on outward facts, merely to be perceived and not to be understood from within. Consequently "sciences of this type tend systematically to dementalize mind and convert it into nature". As examples Collingwood mentions the "pseudo-history" of Spengler, "where the individual historical facts which he calls 'cultures' are frankly conceived as natural products" and "many psychological theories now fashionable". His second objection is formulated as follows:

> ... [I]f we ask how far the generalizations of such a science hold good, we shall see that its claim to transcend the sphere of history is baseless. Types of behaviour do, no doubt, recur, so long as minds of the same kind are placed in the same kind of situations. The behaviour-patterns characteristic of a feudal baron were no doubt fairly constant so long as there were feudal barons living in a feudal society. But they will be sought in vain (except by an inquirer content with the loosest and most fanciful analogies) in a world whose social structure is of another kind. In order that behaviour-patterns may be constant, there must be in existence a social order which recurrently produces situations of a certain kind. But social orders are historical facts, and subject to inevitable changes, fast or slow (IH,223).

When discussing Bradley Collingwood expresses the same view. Bradley contends in *The Presuppositions of Critical History* (1874) that an historian has to study his authorities critically: when statements are made by the latter about happenings which are according to the historian's experience impossible, Bradley argues, they should be disbelieved. Collingwood objects that this is only possible in the case of natural events: "The laws of nature have always been the same, and what is against nature now was against nature two thousand years ago; but the historical as distinct from the natural conditions of man's life differ so much at different times that no argument from analogy will hold" (IH,239)[247].

It has been Walsh, especially, who has criticized Collingwood for not taking generalizations in history into account. His theory is one, he comments, "which overlooks the part played by general (universal) propositions in all thinking, including historical thinking"[248]. After conceding that "it is not the business of the historian to *arrive at* universal truths", he continues:

> But that does not mean that there is no appeal to such generalizations in history. On the contrary: the historian has at the back of his mind a whole set of principles, based on his own experience, on what he knows of the experience of others, and perhaps on *a priori* considerations too, which sum up his conception of the way human beings of this or that type react to this or that kind of situation. And he uses this body of universal knowledge to a major extent in his historical thinking, sometimes making it explicit, but more often assuming it as the (more or less) common possession of himself and his readers[249].

Walsh describes this body of universal knowledge as "generalizations about human nature"[250]. The understanding the historian has of these generalizations is not the product of psychological or sociological inductions, but should rather be compared, according to Walsh, with the more vague insights of the novelist or the dramatist[251].

In his *Introduction to the Philosophy of History* he develops the same point of view.[252] This time, however, the thesis is added that an historian should also make use of specific generalizations and Collingwood is criticized for neglecting these: "When he tells us that a study of the evidence will enable us to grasp in a single act both what Nelson thought at Trafalgar and why he thought it, and that this knowledge is achieved without reference to any general propositions about the behaviour of admirals, we may well wonder whether he has not been deceived by his own example"[253]. With men like Nelson or Caesar, with whom we feel akin, we easily get the idea that there are no major difficulties in understanding, Walsh argues. "But if we try to apply it to the actions of an African witch-doctor or a Viking chief...we need something more than sympathetic understanding; we need experience, first– or second-hand, of the ways in which they commonly react to the situations in which they find themselves". "And if this applies to unfamiliar cases like that of the witch-doctor", Walsh adds, "should it not apply to familiar cases too? Is it not true that our understanding of Nelson depends in an important way on our knowing something about the conduct of sea battles generally? If we had no such knowledge, should we understand his actions at all?"[254].

Though Walsh is the most serious critic of Collingwood with regard to this subject and has even developed a theory of his own in connection with it, he has not been the only one. Buchdahl for instance calls Collingwood's "playing down of the importance of the discovery of generalisations for the purpose of understanding history" "an important defect"[255], while Acham maintains: "Die Frage, inwiefern singuläre Gedankengänge nicht mit bestimmten Umständen oder Bedingungen regelmässig verknüpft sind, bleibt ausserhalb von Collingwoods Interesse. Individuelle Handlungsgründe und Motive haben bei Collingwood, wie später auch bei Dray, in einem derartigen Ausmass die Weihe des Unwiederholbaren, dass jede Frage nach dem gesetzmässigen Auftreten derartiger psychischer Phänomene als der Sache nach unangemessen angesehen wird"[256].

Another example is Grant, who contends "that we also need to know certain general quasi-psychological propositions about the patterns of behaviour and thought which accompany various types of situation", again using Nelson's strategy at Trafalgar as illustration[257].

According to Martin, however, "Collingwood's position on the character and role of generalization ... is easily and frequently misconstrued"[258]. In his view "he was not committed to asserting that there can be no generalizations in a historical science but, rather, that there could be no universal laws formulated from the data of thought. He did believe, however, that historically localized generalizations, delimited statements about the thought and action of agents at a given stage in historical process, could be framed"[259]. Though Martin thus contends that generalizations must not be seen as altogether excluded from Collingwood's theory of history, he is of the opinion that they have nothing to do with his re-enactment doctrine[260].

3.3.8. HISTORICAL OBJECTIVITY

Walsh has made the remark that the problem of historical objectivity is not only the most important, but also "the most baffling" topic in critical philosophy of history[261]. This is amply demonstrated by the reactions to Collingwood's theory of history on this subject. For it is amazing indeed that the major positions which can be distinguised on the question of the possibility of objective historical knowledge have all been imputed to Collingwood. Collingwood is thus said to have had "a pathetic belief in the possibility of indisputable knowledge"[262], to have "flirt[ed] with a complete relativism"[263], and to have held that "reconstructions of past thoughts are corrigible and, in a sense, hypothetical"[264].

The "objectivist" interpretation of Collingwood is always connected with the methodological version of his re-enactment doctrine. Besides Renier one could mention Cohen, who interprets Collingwood as saying that "[e]very human action has an 'inner side' or thought content, and through his recreation of this thought a historian can learn with certainty the truth about past actions"[265]. According to Walsh it was to avoid scepticism that Collingwood adopted the re-enactment doctrine: "The fact that Collingwood believed that only past thoughts, and not past feelings, could be re-enacted shows his preoccupation with the issue of historical scepticism; it seems hard to avoid the conclusion that his solution to this was to say that thoughts alone could be grasped without possibility of mistake"[266]. Walsh, however, does not agree with this: "If historical scepticism is justified, it applies to thought as well as to feeling"[267]. Fain goes even further in the objectivist interpretation of Collingwood, maintaining not only that re-thinking is a method of obtaining direct knowledge of the past, being as well "a criterion of historical truth", but also that it has the function of circumventing historical evidence[268].

The "subjectivist" interpretation of Collingwood has been expressed in different ways: he is described as an historical sceptic, relativist or historicist, as making historical knowledge exclusively subject to the needs of the present time or even to the whims of the individual historian. The observation should be made, however, that Collingwood's sceptisim, relativism or historicism is sometimes referred to, not in connection with the re-enactment doctrine, but with his theory of metaphysics as the description of the "absolute presuppositions" of the thought of a certain period. According to this theory these presuppositions, which determine the complete "superstructure" of a culture or period, are neither true nor false and consequently cannot be judged, but can only be described. With regard to Collingwood's alleged relativism this theory, of course, is related to the re-enactment doctrine through the notion that each period re-thinks the past in its own way. We will not discuss here, however, the relativism implied by the theory of absolute presuppositions, developed in *An Essay on Metaphysics*.

With regard to Collingwood's theory of historical knowledge Mandelbaum is of the opinion that the thesis of the re-enactment of past actions "must ... inevitably lead to scepticism"[269], while according to Hartt, Collingwood's interpretation of history "carries us ever more deeply into skepticism"[270]. Mentioning Croce's influence on Collingwood, Verhaegen says that the latter shares with Croce "sa conception idéaliste de l'histoire et son subjectivisme absolu qui l'amène à nier toute objectivité dans la connaissance historique et toute possibilité d'atteindre une vérité historique communicable ou définitive"[271].

Walsh discerns two conflicting elements in Collingwood's thought on history. Besides the objectivism of his re-enactment doctrine, Walsh maintains, he has a relativistic view of history: "Collingwood combined (how consistently I do not ask) his theory of history as a re-enactment of past experience with a thoroughgoing historical relativism. Our knowledge of the past, he argued, is itself historically conditioned and hence there is no sense in looking for a final or definitive history of anything: history has to be written anew by every new generation"[272]. As evidence he gives a quotation from Knox's preface to *The Idea of*

History, where Knox quotes from "a manuscript written in 1936" by Collingwood:

> St. Augustine looked at Roman history from the point of view of an early Christian; Tillemont, from that of a seventeenth-century Frenchman; Gibbon, from that of an eighteenth-century Englishman; Mommsen, from that of a nineteenth-century German. There is no point in asking which was the right point of view. Each was the only one possible for the man who adopted it. (IH,XII)[273].

Using the same passage as illustration, Carr contends that "[t]he emphasis on the role of the historian in the making of history tends, if pressed to its logical conclusion, to rule out any objective history al all: history is what the historian makes"[274]. In connection with this argument MacIver links Collingwood's view with Croce's "slogan" that "all history is contemporary history"[275], going even so far as to say that according to "the Idealist philosophers of history ... what happened in the past is nothing but whatever it suits our purposes now to suppose to have happened then"[276].

Strauss and Munz provide another striking example of conflicting interpretations, since they give not only opposite accounts of Collingwood's theory, but also different evaluations. To put it shortly, Strauss criticizes Collingwood for having a "subjectivist" attitude towards the past, while Munz criticizes him for being too "objectivist". To begin with the different accounts, Strauss declares: "Collingwood ... rejected the thought of the past as untrue in the decisive respect. Hence he could not take that thought seriously, for to take a thought seriously means to regard it as possible that the thought in question is true"[277].

Against that Munz contends: "Proceeding from the premises of philosophical idealism, Collingwood never devoted much time to comparisons of thoughts other than comparisons of thoughts that succeeded one another in time and replaced one another. In this way he avoided facing the major issue involved in the idea that history is the history of thought. He seems to have imagined or assumed that in each age or for each person the thoughts entertained were the appropriate thoughts"[278]. Strauss, then, evaluates Collingwood's view as follows:

> Collingwood implicitly denied the possibility of historical objectivity by asserting that criticism of the thought of the past from the point of view of the present is an integral element of understanding the thought of the past ... This conclusion depends in the first place on the premise that there are no unchangeable standards for judging human actions or thoughts. But it depends also on the further premise that the historian's primary task is to pass judgment on the past. Yet before one can pass judgment on the wisdom of, for example, a given policy, one must establish the character of that policy ... The primary task of the political

> historian would then seem to consist in understanding a given situation and given ends as they were understood by those who acted in the situation[279].

Munz, on the other hand, responds to his own account of Collingwood:

> If one accords a privileged status to any thought and takes it to be less subjective than any other, one is an *objectivist*, that is, a person who arbitrarily believes that some thoughts are more objective than others. Collingwood was an objectivist in this sense, for he held that the thoughts entertained by the people of a certain epoch or a certain society are in some sense so appropriate to these people or that society that they must be accorded privileged status ... In this way he avoided the real issue, which arises from the fact that Caesar's thoughts about himself can be presumed to be as erroneous, or hallucinatory, or ideological, or propagandistic, or mendacious as those of any later or of any contemporary observer[280].

The assessment of Collingwood as a subjectivist is sometimes couched in less subtle terms, as for instance by K.M. Martin, when he asserts: "Collingwood employs conjecture in a way which shows that he had adopted a personal and subjective attitude towards history. He interpolates motive and constructs episodes with liberality. However, he cites no documentary evidence but, instead, is content to place fact and conjecture on an equal plane"[281]. Similarly Elton contends that Collingwood's theory of history leads "to the opinion that history is just what the historian dreams up"[282].

Rex Martin is also disturbed by Collingwood's contention that the historian's "criterion" is "simply himself"[283]" "What is disconcerting about Collingwood's suggested criterion is its note of radical subjectivism. If there can be no appeal beyond the investigator's subjective appraisal, then we have subtly shifted the focus of our account of explanation from the apparently objective connection of deed with thought ... to the psychology of the investigator when he asserts a connection of plausibility"[284].

Goldstein has criticized this interpretation of Collingwood. When Collingwood asserts that history is dependent on the historian, he says, we should distinguish two senses of mind-dependence, the idiosyncratic and the non-idiosyncratic: "Mind-dependence in the former sense refers to dependence on the peculiarities of an individual ... In the latter sense, it is like the way in which a good deal of human action depends upon shared meanings, values, and institutional arrangements for its intelligibility. By speaking of it as mind-dependent we mean that it is not merely or directly a creation of nature, but we do not mean to suggest that what we deal with depends upon the idiosyncrasies of particular individuals"[285]. Goldstein concedes that Collingwood failed to distinguish between the two senses of mind-dependence, but there is no reason to believe, in his opinion, that Collingwood did not mean that history is only mind-dependent in the non-idiosyncratic sense[286].

Collingwood left us no criterion to decide whether the thought the historian re-thinks is identical with the agent's. Dray calls it "unfortunate" that he "does not give this question the kind of consideration we should have liked it to get from a philosopher with his experience in historical research"[287]. According to Dray, Collingwood's claim that an historical argument may prove its point "as conclusively as a demonstration in mathematics" (IH,262), was made "in an apparently unguarded moment"[288]. Indeed, Collingwood gives no argument to support this puzzling statement. However, when he gives the example of re-thinking Plato's argument, Collingwood says that " ... the process of argument which I go through is not a process resembling Plato's, it actually is Plato's, so far as I understand him rightly" (IH,301). Besides Dray, Shoemaker too is of the opinion that the final clause of this passage indicates that Collingwood's re-enactment doctrine is not intended to guarantee objective knowledge of the past[289].

PART TWO

CHAPTER FOUR

COLLINGWOOD'S UNPUBLISHED MANUSCRIPTS

4.1. INTRODUCTION

Since March 1978 there have been available for consultation in the Bodleian Library at Oxford no fewer than three thousand pages of Collingwood's unpublished manuscripts[1]. Of the manuscripts that could be consulted at the time of writing, around 2700 pages deal with a great variety of philosophical subjects, the rest with historical and archaeological ones. The former include epistemology, metaphysics, economics, politics, art and (not least) history[2].

The manuscripts are important for two reasons. In the first place, they throw new light on the development of Collingwood's ideas. It is especially fortunate that we now have ample evidence concerning the development of his philosophy of history in his notes for the lectures of 1926 and 1928 (the latter being the "Die manuscript" mentioned by Collingwood in his *Autobiography* (Aut,107)) and in later notes. Also of great interest are his manuscripts on metaphysics of 1934, 1935 and 1937-'38. In the last manuscript, which unfortunately is incomplete, he treats his now much discussed theory of metaphysics as the study of "absolute presuppositions".

The manuscripts also provide evidence of certain aspects of Collingwood's thought which were in fact unknown before. Apparently he had a great interest in folklore and after 1935 he made an extensive study of the subject, of which his manuscripts give proof. He further wrote in 1933-'34 a complete treatise of 130.000 words on cosmology; and his manuscript "Man Goes Mad" of 1936 clearly shows that his concern about the social and political developments of his time was sincere and of an early date; not at all springing from caprice, as the last chapter of his *Autobiography* has led many to believe.

Collingwood always emphasized that we can only talk sensibly about the past if we are in possession of relevant evidence for it. The same is true of course when what is at issue is the ideas of a past philosopher. With all the new material now at our disposal, it is clear that Collingwood will never be the same for us again. I do not hesitate to declare that the new Collingwood is more important than the old one was generally thought to be. This is especially true for his philosophy of history. If his manuscripts on this subject had been known at an early date, the philosophical discussion of historical thinking, as it took shape after World War II, with *The Idea of History* playing such an essential role, would probably have been different. The manuscripts amply show, however, that Collingwood's philosophical interest was not focussed exclusively on history, but covered a wide variety of subjects.

The energy with which Collingwood threw himself into his work is sometimes amazing. To give one example: in his "Log of a Journey in the East Indies" we read for 24 Oct. 1938: "Crossing Bay of Biscay. Began writing Metaphysics in afternoon". The succeeding days are mostly filled with "writing all day", although he also learned Malay (on which the present-day Indonesian language is based). On 13 November he records: "Revising all day; wrote new chapter XXVII (last)". So *An Essay on Metaphysics*, a work which Toulmin has recently called a "powerful and important book" (Aut,XII), was written in three weeks on board a ship, although he did revise parts of it in February and March 1939, mostly on the ship back home.

According to Collingwood, mind is by essence always in development. His own mind well exemplifies this claim, making the interpretation of his ideas not always easy. The manuscripts which are now available for consultation must therefore not be read as a solid block, or as the elaboration of a system. They are more the vivid expression of a mind constantly at work on certain problems, attempts to "think on paper" as he puts it himself[3]. His lectures were usually written down, and those on philosophy of history of 1926 and 1928 are complete essays by themselves: books in embryo. However, the manuscripts are not all of the same nature. Many of them are only notes or treatments of certain problems in a provisional way. To be fair to Collingwood, it is always necessary to realize what the character of a certain piece of writing is; and if quotations are used, the context will be important. Collingwood's reluctance to have his manuscripts published doubtless finds its reasons in these considerations.

No attempt will be made to give a general assessment of all the manuscripts; the sheer bulk of them would prevent this. We will concentrate on the ones on philosophy of history, in which connection the manuscripts on folklore and cosmology are also relevant. Even with regard to these, however, only some aspects can be dealt with. It is the plan of this chapter to describe the manuscripts in general, while in subsequent chapters they will be used again in a wider context.

4.2. HISTORY AND REALISM: THE WRITINGS BEFORE 1926

A FOOTNOTE TO FUTURE HISTORY (1919)

On the "Vigil of All Saints" (i.e. 31 October) 1919 Collingwood wrote a short essay, entitled "A Footnote to Future History", which is interesting as an ilustration of his involvement by that time in the issue of realism. He contrasts the realism of Plato, which had determined European culture until then, with the philosophy of Hegel, which was to determine the future. Of Platonism he says: "It is a realistic philosophy – objective to the core, but preserving its realism and objectivity by recourse to dualism. Its logical foundation is the distinction between the ideal (that is, real) world of timeless and knowable permanence, and the phenomenal (that is, half-real) world of opinionable and changing compromises. This is the definition of Platonism: from which follows the Aristotelian logic of the Three Laws of Thought and the whole of mediaeval culture ... Since the Middle Ages, the common consciousness of humanity has uniformly remained Platonic. Church, state, and universities have together insisted on this". Collingwood makes clear, however, that this is not the whole story: "But we are no longer in the Middle Ages: They have been gradually drawing to a close ever since Descartes pointed the road to a new philosophy – subjective, idealistic, concrete, where Platonism is objective, realistic, abstract"[4].

"It remained for Hegel to go to the root of the matter, to deny the axioms of Aristotle's logic and to assert the unity of the real and the phenomenal, in the synthetic concept of history", Collingwood goes on, "[b]ut Hegel spoke too soon: the world was still too Platonic, and his system appeared to die prematurely with him"[5]. He immediately adds that this did not happen, however:

> Here then lies our point. In this century live Hegelism is struggling with the dead matter of the corpse of Platonism. There can be no doubt which will win: and as Platonism was the philosophy on which all human life was built up to the present time, so Hegelism is that of the next age. But Hegelism does not yet exist. It has had its Plato: it may yet find its Aristotle: in order to become the philosophy of all mankind it must pass through all the stages through which Platonism passed, till a new Age of Faith systematises it into a concrete civilisation founded on Hegel as the Middle Ages were founded on Plato. ... As the Graeco-Roman world was the battlefield of Platonism and Paganism so the world of to-day is the battlefield of Hegelism and Platonism. Hegelism to-day is the property of the learned: the common people are Platonists. We speak of 'the plain man's realism'; but the plain man learnt his realism and his logic at the feet of Aquinas and Duns Scotus, in the Sorbonne and the Great Hall of the University at Oxford. He must learn idealism and the philosophy of contradiction in the unfounded universities of the future – after he has unlearnt his realism[6].

This will lead to nothing less than a complete subversion of our civilization, expressed by Collingwood in the following vision: "In the middle ages speculation and authority were at one: in the Roman Empire, and again to-day, they are at open war. Such a war can have only one issue. Thought cannot be held in leash: but authority can be undermined. And with authority go law and order, church, state, and institutions in one fall"[7]. According to Collingwood a catastrophe awaits our civilization: "Unless history moves faster in these days, it may not come for another nine centuries: our present conflicts, like the agonies of the expiring Roman Republic, are but a ripple on the surface of history. But it must come: and as surely as it comes, so surely it will be followed by the revival of humanity into a new life"[8].

We may see in this vision an anticipation of the – less dramatic – conception of *Speculum Mentis*, in which an attempt was made at re-creating a union after the fragmented forms of human experience since the Renaissance, a union "in which the full development of art, of religion, and of philosophy is possible" (SM,36). With this Collingwood claimed "no more than to be following and working out the tradition founded over a century ago by the great men of the Romantic movement" (SM,38), a claim amply demonstrated by the Hegelian character of *Speculum Mentis*.

AN ILLUSTRATION FROM HISTORICAL THOUGHT (1920-'21)

Collingwood shows a less speculative and more prosaic form of criticism of realism in a short piece under the title "An Illustration from Historical Thought", dated 1920 or 1921 and attached to a "draft of opening chapter of a 'Prolegomena to Logic'". We have seen how according to *Speculum Mentis* history is implicitly based on realism, which leads to its breakdown[9]. It is interesting to note in this connection that while in the earlier "Illustration from Historical Thought" the conclusion is also drawn that realism breaks down, this time history is not linked with it. On the contrary, a problem in historical epistemology is even used by Collingwood as a demonstration of the untenability of the realist position.

The problem is the following. If we take the interpretations of a historical period as given by different historians, for instance Grote and Mommsen, we get different accounts: "Grote would lament over incidents over which Mommsen would exult. That is, they would give a different account of the *values* of the history". But – and Collingwood considers this an important point – "their account of the *facts* would also conflict. Each would present a picture differing not merely in its colours but also in its forms from that of the other. Their sympathies or ideals are not superimposed on their knowledge, the two things (historical fact and value) do not reside on different planes ... but the sympathies and ideals act upon the knowledge and distort it"[10].

An objectivist, Collingwood says, may react in two ways to this fact: 1) He may say that the different accounts are due to the different selections made by Grote and Mommsen with regard to the same object: "Grote will be sharp-sighted exactly where Mommsen is blind, and blind where Mommsen sees most clearly. But the object contains everything that each sees – it contains both the merits and the defects of each form of constitution"[11]. 2) He may say that the different accounts are the result of different sympathies, which act on knowledge and thereby become not knowledge, but error. With real knowledge these sympathies should be avoided.

With the first alternative "reality is presented as infinitely big and various" and is seen as "a kind of stuff out of which the practical or willing mind sculpts its objects"[12]. This view is untenable and breaks down, Collingwood contends, because the different selections contradict each other: " ... the whole point of a conflict of views consists in the fact that each not only asserts itself but denies the other". The first alternative of objectivism, however, does not take this circumstance into account, since "[it] has attempted to put all thinking on a level – to say 'every thought is equally true of reality, equally a selection from reality', thereby eliminating error ...". This view implies that "[a] ll theories are true, so it doesn't matter which we hold"[13].

The second alternative takes the opposite line. Both the accounts of Grote and Mommsen are seen as defective, because of their sympathies. Since all our thinking is affected by subjective influences, it is implied that truth is unattainable.

"The result", Collingwood concludes, "is a *coincidentia oppositorum*. In 1) everything is true: in 2) nothing is true. In 1) Reality includes all subjective views, which are mere selections from its own infinite bulk: in 2) it excludes them all, and they are aberrations from it. In 1) it is a goal always attained, in 2) never attained at all. Now these opposites coincide. For the truth at which thought aims is the whole truth, the truth which includes all reality in its view and excludes all alternative views"[14].

On each interpretation, however, this is impossible. The reason must be found, Collingwood argues, in the fact that both of them make a distinction between thought and action "arguing ... that action vitiates thought ... and therefore truth is defined as 'that which thought would attain if there were not action' and reality as 'that which action prevents thought from cognising'". Collingwood concludes from this that "[r] eality is thus defined as the unknowable, and realism breaks down. For this is the invariable symptom of the break-down of realism, that it defines reality as the unknowable"[15].

With both objectivist alternatives, "reality is sundered from knowledge: true knowledge is the unattained and unattainable". Collingwood ends with the contention that "the distinction between value as something peculiarly subjective and fact as objective must go. That stands to reason as soon as we have seen that differences of opinion about fact cut just as deep as those about value"[16].

Though the relation between fact and value is not further worked out by Collingwood, it is clear that he has – using an illustration from historical thought, well grounded objections against the realist point of view. What is puzzling, though, is the fact that in subsequent writings he still (or again) adheres to a realist position with regard to history. According to *Speculum Mentis*, for instance, "the historical consciousness asserts concrete fact" (SM,208), "The object of history is fact as such" (SM,211), "An historian must state the facts as they happened" (SM,216)[17]. In "The Nature and Aims of a Philosophy of History" we find a similar realistic theory of knowledge: Collingwood speaks there about a world of facts "independent of the [sic] being known"[18] and about the object of history as "the fact in all its actuality"[19].

These statements are in opposition to the view developed in "An Illustration from Historical Thought". It is unclear if the latter is to be considered an isolated phenomenon in Collingwood's main development or a precursor of the anti-realist position he was to develop after 1925 with regard to history. It may also be seen as only one part of his "flank attack" on realism and not as the result of a systematic treatment of the special characteristics of history, as Collingwood was to undertake at a later date.

SOME PERPLEXITIES ABOUT TIME (1925)

On 4 July 1925 Collingwood wrote a first draft of "Some Perplexities about Time", a paper whose final version was read on 15 February 1926 at a meeting of the Aristotelian Society[20]. The first draft differs quite considerably from the final version and is especially interesting because in it we find Collingwood arguing about the ontological status of the past, with all its consequences for the possibility of historical knowledge. He objects to the realist notion of an objective existing past. On the other hand, however, he still sticks to the realist idea, that in order to be known an object must exist, the consequence being that we cannot have knowledge of the past. Collingwood gives the following analysis of the past and our possible knowledge of it:

> It is, we say, a fact that the earth existed before there was life on it; a fact that there were men alive while I was unborn; these facts are known, and they are wholly independent of my memory. That is true; but we must still ask, what are these facts? They are past conditions of things, not present ones; in the sense in which this evening exists, they do not exist at all. We believe that they did exist, because their consequences exist ... but they themselves are as unreal as anything can possibly be; they are utterly disqualified as objects of knowledge in the eyes of anyone who holds that the known must be the real. History, like memory, is a subject on which realists have every ground for reticence; for not only do they, like other people, think that what is known must be real, but they cling to the common-sense dogma that in history and memory

we possess knowledge[21].

Collingwood's conclusion is that the past is, like the future, "wholly unreal": "they are 'ideal' in the sense that they exist solely as objects for certain acts of mind that are not, or at any rate not in the realistic sense, knowledge: their *esse* is respectively to be expected and to be remembered, or, since history is not memory, to be contemplated by a process of historical thinking into whose nature we shall not now further inquire, contenting ourselves with saying that it also cannot be knowledge. What is real, then, is the present, and that alone"[22].

We have seen before[23] that in the published version of "Some Perplexities about Time" Collingwood comes to the same conclusion, with the addition that we disentangle the past "out of the present in which it actually exists, transformed, and re-transforming it in thought into what it was"[24]. In the same article it is also maintained that "[w]hat we know must ... really exist. And if that is so we cannot really know either the past or the future"[25].

4.3. "PRELIMINARY DISCUSSION" (1927)

From 1926 on Collingwood lectured on philosophy of history[26], and his lecture notes of 1926, written out discursively in January, are among the manuscripts. In 1927 he added to these lectures in a "preliminary discussion" a piece entitled "The Idea of a Philosophy of Something, and, in particular, a Philosophy of History". In 1928 he added to it the remark: "Written in Rome[27], by fits and starts, April 1927. I haven't read it since, but from my recollection of the frame or frames of mind in which it was composed I suspect it of being chaotic and practically valueless. Die, April 1928". Notwithstanding this critical remark it is a highly valuable piece, because it clarifies what Collingwood meant by philosophy of history and therefore places his subsequent writings on the subject in a proper context.

The influence of Hegel on Collingwood's thought is often emphasized, this judgment usually being based on his *Speculum Mentis*. The lectures of 1926 and 1928 prove however that he was not less influenced by Kant: their main characteristic can even be seen as a "transcendental analytic".

In his "preliminary discussion" Collingwood considers the concepts of philosophy, history and philosophy of history. The distinguishing feature of philosophy is that it deals with the universal and necessary (Kant's criteria for "pure concepts of understanding" or categories). In contrast to this, scientific thought can only claim an empirical universality. Philosophy is concerned with all facts, and its concepts must be seen as different from the arbitrary, hypothetical, or classificatory concepts of science: "A philosophical concept is universal in the sense that it arises necessarily whenever anybody thinks about a subject"[28]. "The view which I am putting forward",

Collingwood continues a few lines further on, "is that the concepts which compose the body of philosophy are transcendentals"[29]. So philosophy has nothing to do with empirical concepts, but only with transcendental ones; and Collingwood mentions in this connection thought, action, art, science and history (it is striking that religion is left out here, while in *Speculum Mentis* it played an essential role). Philosophy of history is thus "the exposition of the transcendental concept of history, the study of history as a universal and necessary form of mental activity"[30]. Besides this transcendental concept of history there exists an empirical one: history as what historians do. A similar distinction holds for art. But each empirical concept – for example, that of art – also has its transcendentals: " ... every work of art (that is, every operation of the mind *qua* work of art) must display a number of different characteristics which are the transcendentals or categories of art... Thus the relation between the particular work of art and art in general is parallel to that between a particular philosophy such as the philosophy of history and philosophy in general"[31].

Although Collingwood does not explicitly mention history in this connection, the implication is that the same analysis will hold. (In "The Philosophy of History" (1930) he says that "historical knowledge ... has to be built up by each historian for himself, using the universal and necessary principles of historical thought to interpret the data which the past has left behind it"[32]). So "history" can refer both to a transcendental aspect of the mind's activity and to an empirical concept (the business of historians), with certain transcendental characteristics of its own. Both can be the object of a philosophical analysis and they are of course related to each other. Collingwood gives the following clarification of them:

> Clearly, history is an empirical conception if it means that activity which distinguishes persons called historians from others called scientists, trombone-players, or ophthalmic surgeons. History in this sense, as an empirical concept, means the investigation of certain arbitrarily defined problems known as historical problems. Consider for instance what is involved in the fact that a book 350 pages long may be called "History of England". It implies either that everything which has ever happened in England can be discussed in 350 pages, which is absurd, or that everything known to the author about what has happened in England can be stated in 350 pages, which is equally absurd, or else that there are certain quite arbitrary conventions as to what ought and what ought not to be included under that title. That this is the case, everyone knows; and everyone knows that the conventions change, and that whereas once the names and dates of kings and battles were considered to form the main bulk of the History of England, that position is nowadays accorded to a description of social and economic conditions. Thus if you consult professional historians on the question what ought to be contained in books of history, you will

> find that they give various answers which, just because they are merely empirical, cannot be reduced to agreement, precisely as artists will differ about the proper subjects for artistic representation ...
> If on the other hand history means the acquisition or possession of historical knowledge, and not merely the retailing of certain parts of it to others, it must be a transcendental conception. For the object of this knowledge is not the history of England or the history of this or that particular empirical thing, but history as such, whatever history there is, everything historically knowable; and this is a perfectly universal conception. Moreover it is a necessary conception, in the sense that it is implied as a condition in all mental activity...
> Thus history is a transcendental conception, like art and science, when regarded as a pure form of activity; though it becomes, like them, an empirical conception when it is arbitrarily restricted to certain specialised embodiments of that form. If anyone says 'that isn't history, because there isn't a book about it in the historical section of this library, or because a professor of history would not bother to lecture about it, or because it never occurred to the people concerned to call it history', he is using a perfectly legitimate criterion to exclude it from history in the empirical sense, but he is not even attempting to deny that it is history in the transcendental sense: that is to say, that it contains those characteristics which, in a more conspicuous degree or form, confer the name of history upon the things generally so designated. For the empirical concept is nothing but the *prima facie* application of the transcendental concept[33].

The lectures of 1926 – and also of 1928 – deal extensively with the transcendentals of the empirical concept of history, that is, the study of the past as practised by historians. Herein lies their unique value, not only because this is the first time such an analysis was offered by an English philosopher (except for Bradley's *The Presuppositions of Critical History* of 1874), but also because it is a fresh starting-point after the blind alley explored by his German predecessors, with all their ontological and metaphysical implications. Collingwood describes this new concept of philosophy of history as follows:

> The philosophy of history, so understood, means bringing to light the principles used in historical thinking, and criticising them; its function is to criticise and regulate these principles, with the object of making history truer and historically better. It thus arises by an absolute necessity out of the practice of historical thinking, and the historian can evade the necessity of engaging in the philosophy of history only so long as he can evade entangling himself in the problems of methodology; that is, the problems of how he ought to handle historical materials and what kind of result

> he ought to aim at attaining ... The philosophy of history, so understood, is the methodology of history. Arising spontaneously in an unsystematic form out of actual historical work, it cannot ever be expressed in the form of a completed doctrine; it must consist of topics raised and discussed in the shape given them by the peculiar circumstances in which they arise, and the natural method of treating it is by isolated and self-contained discussions[34].

Concerning possible themes for the new study Collingwood makes a distinction between ought-questions and can-questions. His catalogue of the different questions to be asked provides an excellent survey of the special problems in which the science of history is involved and therefore deserves to be quoted fully. Beginning with the ought-questions he gives the following examples:

> Ought history to pay special attention to any one side of human life, such as (according to Marx) economics, or (according to the present Regius Professor) politics? Is it possible, or desirable, to write separate histories of art, of religion, of warfare, of constitutional law, and the like; or do these things, by being separated from their historical environment, become unintelligible in their development? Is the ideal of history a single universal history, a history of the world, or a number of separate histories, and if the latter, how ought they to be divided up? Is it possible to produce good history by portioning out different parts of the subject to different authors after the Cambridge fashion, and if not, why not? Ought history to aim at biographical form, at presenting the reader with individual portraits, or ought it to suppress the biographical element and describe movements whose magnitude transcends the individual? Ought it to admit an imaginary element, a conjecture as to what may have happened when evidence fails as to what did happen; or ought it to state nothing but what, on the available evidence, is certain? Ought the historian to write with an eye to his own times, and to see the past in the light of the present, as Grote saw Athenian democracy in the light of nineteenth-century radicalism, or ought he to leave behind as profane all interest in the present when he enters the temple of Clio? Ought the historian to pass moral judgments on his characters? Ought he to take sides in the conflicts whose history he narrates? Ought he to ascribe their issue to necessity, or to chance, or to the agency of human wills? Questions like these are concerned with the historian's duty in matters where, at least to all appearances, a choice is open to him; but there are others, no less urgent, which begin not with the word *ought* but with the word *can*. Thus, can history exist in the absence of written records? Can there be a history, in the proper sense of the word, concerning the Bronze Age, for example? Can the historian determine why things happened, or only what it was that happened? Can he appreciate

> the motives of his characters, or do their actions necessarily remain for him mere opaque facts? Within what limits, if at all, can the historian go behind his sources and criticise and correct them? and if at all, on what principles?[35].

These questions are interrelated, according to Collingwood, and they all "revolve round one central question, the question of the fundamental nature, meaning, purpose and value of history: the question: what *is* history?"[36]. This question is specified as follows: "is it a genuine form of knowledge, or is it an illusion? can it really make good its claim to be a mental discipline and an approach to reality, or is it a confused mass of heterogeneous and half-developed tendencies of thought? If it is a genuine form of knowledge, what place has it in knowledge as a whole, and how is it related to other forms?"[37]. The reason that Collingwood considers an answer to the question "what is history?" so important is that in his opinion a properly thought-out concept of history is an indispensable condition for the solution of the various methodological questions mentioned before.

The second point Collingwood mentions with regard to the methodological questions is the fact "that these various questions bring us face to face with problems from every department of philosophy". An example he gives is the problem of human freedom and necessity. He concludes his analysis as follows:

> There are thus three aspects of the philosophy of history. First, as a complex of particular methodological problems, growing immediately out of historical thinking. Secondly, as the attempt to answer the question, what is history? Thirdly, as identical with philosophy in general. Now clearly, these three aspects are in no sense three distinct departments of the subject. They are bound up together in such a way that neither can exist without the others. The first is the *matter* of the philosophy of history; the second and third together make up its *form*. The matter is a mere plurality of particular philosophical problems, in themselves chaotic, shapeless, capable of enumeration to infinity; the form is a unity which brings unity into this matter by relating its parts to one another in the light of a whole which is the form itself[38].

It is only when we know what history is, that we may see a rational answer for the various methodological questions, Collingwood contends. On the other hand, "it is only in this concrete experience of historical work and its difficulties that I can be said to know what history is at all. Take away the matter, and the form becomes an empty and worthless formula. The form makes the matter intelligible, the matter makes the form actual"[39]. Collingwood was unique in providing both - as an historian the matter, and as a philosopher the form. It is this quality which makes his philosophy of history so valuable, as is amply demonstrated by his lectures on the subject.

4.4. LECTURES ON THE PHILOSOPHY OF HISTORY (1926)

Collingwood wrote his lectures from 9 to 13 January and they were delivered in Hilary Term (January-March) 1926. At the very outset he gives the framework:

> The purpose of these lectures is to raise and, as far as I can do so, to answer certain questions relating to the study of history and to the object, called history, which in that study we investigate. The fundamental question is, what are we doing when we study history? and this raises three allied questions: 1. What are we doing it *for*? in other words, how does this study fit into our general view of the aims and purposes of human life? 2. What is the best way of doing it? in other words, what are the principles of method by which historical study is or ought to be guided? 3. What are we doing it *to*? in other words, what is the true nature of the thing which we call the past, which historical thought takes as its object?

"[M]y only object in thinking out the notions which I shall lay before you", Collingwood tells his audience, "has been to settle accounts with myself as to why I study history and how I can do it more intelligently". He begins with the last question - the nature of the past, turning then to the second and ending with the first. The lectures are systematically built up and divided into four chapters: a) Introduction: general idea of history, b) The sources of history, c) The interpretation of sources, d) Narrative.

a) GENERAL IDEA OF HISTORY
This part is a short one and largely based on the ideas developed in "Some Perplexities about Time". Collingwood resists again the habit to "spatialise" time with its corollary to see the past and future as "real" and stretched as it were behind and before us. Against this he contends that " [t]he present alone is actual; the past and the future are ideal and nothing but ideal"[40]. Since he argues that what is known must have a real existence, neither memory nor history, according to Collingwood, can provide us knowledge of the past as such. In this connection he also points to the difference between memory and history: the first is subjective and immediate, while the second is objective and mediate. The immediacy of memory means "that we neither have, nor can have, nor can even want, any guarantee or ground for it except itself"[41]. Memory is also confined to the subjective and personal past, whereas the objectivity of history means "that its concern is not with my own personal past but with the past in general, the past depersonalised, the past simply as fact"[42]. Against memory the mediacy of history means that its statements "are always made on grounds which the historian can state when challenged"[43].

b) THE SOURCES OF HISTORY

Statements about the past can only be gained in an inferential way, that is, they must be grounded on sources. These sources must be interpreted and Collingwood calls this the "formal" element of history, while the sources are its "material" element, using Kant's wording: "The 'receptivity' of the historian towards his sources is counterbalanced by his 'spontaneity' in respect of the principles by which he interprets them"[44].

With regard to the treatment of sources Collingwood makes a distinction between dogmatic and critical history. The first treats sources as authorities, which are passively accepted (Collingwood would call this later "scissors-and-paste history"). This attitude necessarily breaks down, however, when it is realized that the various authorities contradict each other, so that a choice has to be made between them.

c) THE INTERPRETATION OF SOURCES

In the first instance this awareness may lead to scepticism, but "[t]he way out of this ... is found when it is realised that sources are not authorities but only sources; that the historian's attitude towards them must consist neither in acceptance nor in rejection, but in interpretation"[45]. This will eventually lead to critical history, a form of history where the historian is, as it were, his own master. At this point the emphasis is upon the interpretation of the sources, and authorities as such have vanished. Such interpretation "requires us to ask not only 'what did this writer intend to convey when he used these words?' which is a question of merely linguistic interpretation, but 'what is the historical truth that lies behind the meaning he intended to convey?' which is a question of historical interpretation in the proper sense, and assumes that the truth of which we are in search was not possessed, ready-made, by the writer whom we are studying, or at any rate not intended by him to be conveyed to us in the words he is using. In short, we are now trying to get behind our authorities, which is exactly what, in the dogmatic stage of historical thought, we said could never be done"[46].

The historian must put his source in the witness box and cross-question it. Collingwood considers a necessary element in this operation to be the critical study of what past historians have written about a certain historical problem. He calls this history of the second degree, or history of history, which cannot be separated from history of the first degree, that is, the study of a given event in the past: "History of the second degree is an absolutely necessary element in history of the first degree; no historical problem about any past event can be settled until we have settled the problem of the history of its history"[47].

This argument does not sound very convincing and one may really doubt the a priori or "transcendental" character of history of the second degree. The justifications Collingwood gives for it are not satisfying either: "In the first place, no problem of the first degree can be solved without a preliminary review of the history of thought on the subject, which enables the inquirer consciously to insert himself in his proper place in the succession of inquirers; in the second place, it fertilises and revivifies the achieved solution of every problem to look back at past attempts to solve it, and without such revivification the solution hardens into a mere formula repeated, parrot-like, without intelligence"[48]. These arguments are more of an empirical than a priori nature, which is even more obvious when Collingwood refers to "the uniform and indispensable practice of all historians"[49].

"The interpretation of sources must proceed according to principles", Collingwood asserts. "It is not enough to interpret them according to the dictates of intuition, to deal with individual cases as if each was unique and unlike any other"[50]. He rejects the idea that these principles have an empirical origin. It is obvious, he contends, "that we accept the principle not because we have seen an example of it but because the principle itself proves acceptable; and that it possesses a certainty far more complete than the certainty that attaches to the fact which, we fancied, guaranteed it. The function of the instance now seems to be, rather, to reveal to us the principles which we implicitly accept, not to introduce to us principles to which till now we were strangers ... [O] ur principles of interpretation have their origin, not in the facts as we observe them, but in the thought which we bring to bear upon them"[51].

So the principles of history should be justified a priori, that is, "made into objects of critical study and discussion by a scientific methodology of history"[52]. Collingwood makes a distinction here between "empirical" methodology and "general or pure" methodology. The first is exemplified by auxiliary sciences such as palaeography, numismatics, diplomatics and epigraphy, which have been developed in order to interpret different kinds of evidence. These sciences combine a theoretical and practical side, the first one "consisting of general propositions concerning such things as the period of history at which this or that moulding or piece of ornament was used" and the second "consisting of general recommendations as to the search for the special kind of evidence in question"[53].

General or pure methodology Collingwood sees as much more important: "This is concerned with problems of method which are never absent from any piece of historical thinking"[54]. He cites the problem of the argument from silence as an example: can we conclude that a certain event did not happen if our sources tell us nothing about it? "On the one side, it may be argued that we cannot, because our sources do not exhaust the whole of the events in their period, and any number of things may have happened about which they say nothing. But on the other side, it may be argued that all historians always do rely on the argument from silence when they accept a narrative based on a certain source because they have no other sources and therefore cannot check the one which they possess"[55].

According to Collingwood the argument from silence is in principle indefensible, but in practice used by every historian. He complains that the general or pure methodology which deals with "the perfectly general question of the principles on which evidence must be interpreted"[56] is nevertheless almost wholly neglected by historians. "They live in this respect from hand to mouth", he observes sarcastically, "and on the rare occasions when they start thinking about the subject they are apt to conclude that all historical thought is logically indefensible, though they sometimes add a saving clause to the effect that they personally can interpret evidence pretty well because they have a mysterious intuitive *flair* for the truth ...[T]he ordinary historian can give no account of the processes by which he extracts narrative from sources; all he can say is that he succeeds in doing it somehow, that something, which he may call instinct in order to mark the fact that he does not know its real name, guides him in deciding what evidence is sound and in what direction it points"[57]. This is unacceptable for Collingwood, because it implies obscurantist and subjective interpretative principles. What is needed, he concludes, is "a general logic of historical thought, and this must be a philosophical as opposed to an empirical science, and must establish *a priori* the pure principles on which all historical thinking is to proceed"[58].

d) NARRATIVE

In his last chapter Collingwood comes back to the concept of the ideality of the past. This is a difficult part of the lectures and not always very clear: he seems to be wrestling with the problem. It is therefore not accidental that this is the problem with which his lectures of 1928 begin and that only then, starting afresh, does he come to a provisional solution. In his 1926 lectures he emphasizes that the past is in no sense whatever actual; this explains why we cannot know the past as it actually happened (Ranke's well-known dictum that he only wanted to say "wie es eigentlich gewesen" is rejected accordingly). Both past and future are ideal elements in the present and since "[t]here is, properly speaking, only one knowable object, namely the actual - that which now exists", the corollary is that " [w]e cannot know the future, because it is not there to be known; we cannot know the past, because it is not there to be known"[59]. Collingwood continues his argument as follows:

> Does this, then, prove that history is an illusion and that to pursue it is folly? No; because the past has no actual existence, it is an ideal element in the present, and can therefore be studied in the same general way and to the same extent to which any abstraction may be studied. The present is the past transformed. In knowing the present, we are knowing that into which the past has changed. The past has become the present, and therefore if we ask where the past is to be found in living and concrete actuality, the answer is, in the present. But whereas the past exists actually as the present, it exists ideally as the past - as what it was before it

turned into the present[60].

A little further on he continues:

> Our knowledge, so called, of the past, is therefore not knowledge of the past as of an actual object, and therefore not true knowledge; it is only the reconstruction of an ideal object in the interest of knowing the present. The purpose of history is to enable us to know (and therefore to act relatively to) the present; that is the truth contained in the pragmatic view of history. But the knowledge of the past must not be misconceived as knowledge of one object, the past, which when achieved serves as means to the knowledge of another object, the present. That is the error of the pragmatic view. The past and the present are not two objects: the past is an element in the present, and in studying the past we are actually coming to know the present, not coming to know something else which will lead us on to know or to manipulate the present. This principle, the ideality of the past, explains both why we cannot and why we need not know the past as it actually happened. We cannot, because there is nothing to know; nothing exists to be studied; there are no past facts except so far as we reconstruct them in historical thought. And we need not, because the purpose of history is to grasp the present, and therefore any past fact which has left no visible traces on the present is not, need not be, and cannot be a real problem to historical thought[61].

Collingwood discusses in this final chapter various subjects, such as the pragmatic theory of history, moral judgments, periodization, the history of history again, universal versus particular history and history as theodicy, developing a speculative theory of Hegelian character of "die Weltgeschichte als Weltgericht". Every now and then the principle of the ideality of the past is used to support his viewpoints. Moral judgments on past events, for instance, are meaningless: "there is nothing to be done about them; the dead must be left to bury their dead and to praise their virtues and lament their loss"[62]. Because the past is not a reservoir of facts with certain characteristics of themselves, we are also justified in making distinctions of our own in how the past is segmented into periods[63]. And the dilemma of a realist view of the past with regard to universal history, since the past in its totality can never be grasped, is also solved by Collingwood with the theory of the ideality of the past: when we start from the principle that we always see the past from a certain point of view, so-called universal histories are feasible and may even be reconciled with particular ones[64].

Towards the end Collingwood refers again to the transcendental concept of history. Having dealt with the relation between universal history and monographs, he says:

> ...[T]his may suggest that history is mainly the concern of

professional persons called historians. This is the last suggestion that I should wish to make. History is nothing but the attempt to understand the present by analysing it into its logical components of necessity, or the past, and possibility, or the future; and this is an attempt that is made by everybody and at all times. Nobody ever attempts to do a job of plumbing or to ride a motor bicycle without historically reconstructing the preconditions of the situation with which he is faced, and there is no difference in principle, only a difference in degree, between the historical thinking done by a bricklayer in the exercise of his craft and that done by a Gibbon or a Grote. The problem is the same, the categories of thought involved are the same, and the solution is the same. History is one of the necessary and transcendental modes of mind's activity, and the common property of all minds[65].

4.5. OUTLINES OF A PHILOSOPHY OF HISTORY (1928)

These outlines, written in April in Die (France), are a complete revision of the lectures of 1926 and were meant to be delivered in Trinity Term (April-June) of the same year. They are arranged under the four headings of Quality, Quantity, Relation and Modality "out of compliment to the Kantian critiques". Under "Quality" "the question is raised whether history is real, and if so in what sense". Under "Quantity" Collingwood deals with the question of universality versus particularity: "the question whether history is properly conceived as a single universal world-history, or a plurality of particular histories". In the chapter headed "Relation", "the question raised is that of the inner structure of historical fact", and finally under "Modality", "the question of the certainty or logical status of history is dealt with"[66].

In an "Introductory Lecture" Collingwood begins again by describing art, and this time also religion, as "a universal and necessary form of human activity", and for this reason deserving the attention of philosophy. Similarly history is described as "a necessary form of human experience". The term "philosophy of history" will be used, Collingwood says, as "the idea of a philosophical science of historical thought" and further on he declares: "Our business, then, may be defined as that of discovering how historians always and necessarily think". We are warned, however, that this analysis of history *a parte subjecti* cannot be separated from history *a parte objecti*, the historical facts:" ... if historical method is adequate to the study of its proper object, as it must be if it is really historical method, then it follows that in studying the necessary and universal features of historical method we are studying the necessary and universal features of historical fact, its proper object". In his Preface to the lectures both aspects of history are linked with the doctrine of the ideality of the past, "for that doctrine lays it down that historical fact, as known to the historian, is essentially relative to the thought that knows it".

Collingwood ends his Preface by asserting that "[h] istorical thought and its object are seen to be inseparable, the latter having only an ideal existence in and for the former; and therefore a methodological theory of the necessary forms of historical thought is also a metaphysical theory of the necessary forms of historical fact". This conception is not dogmatically stated by Collingwood, but worked out in his first chapter.

I QUALITY

The problem dealt with by Collingwood is formulated at the outset as follows: "History *a parte objecti*, the object of historical thought, is of course in some sense real, for if it were not, there would be no sense in which historical judgments could be true, or indeed false. But in what sense are historical facts (using that term to denote the objects of historical thought) real?"[67].

Collingwood begins by giving the realistic view of reality. This theory equates reality with existence, that is, the reality of actual things, or with actual occurrences of events. Historical facts, however, are events that, having already occurred, do not occur at the moment. They are therefore not real, but ideal. By the word ideality Collingwood intends to signify "the quality of being an object of thought without having actuality", while actuality implies "simultaneity with the thought in question"[68]. It is, according to Collingwood, important to note in this connection the difference between things and events, the latter being the object of history: "A thing (e.g. the Matterhorn) may be both ideal and actual: the Matterhorn as I remember it ten years ago is ideal, the Matterhorn as I see it now is actual: but the mountain as it was then and the mountain as it is now are the same mountain. But an object of historical thought cannot have this double reality"[69].

To illustrate this point Collingwood takes an example from the history of music, possibly choosing it for its ambiguity. For one could say that music is both ideal (as past music) and actual (as present music). No event, however, in musical history can fall into both these categories at once and this Collingwood considers its essential difference from an object like the Matterhorn. This illustrates according to him "another and an important point":

> No historian of music deserves the name unless he has studied for himself the old music whose growth and development he is trying to describe. He must have listened to Bach and Mozart, Palestrina and Lasso, and possess personal acquaintance with their works. This means that he must have been present at actual performances of these works, either physically or in imagination; and in the latter case the imaginative power is acquired only by actually hearing similar things performed - e.g. a man who had never heard an orchestra of the Beethoven period could not read a symphony of Beethoven in score with any chance of obtaining a good imaginative hearing of it. We may therefore boldly say that the

> *sine qua non* of writing the history of past music is to have this past music *re-enacted in the present*[70].

In the margin Collingwood wrote an important addition, which declares the theory to be applicable to all history:

> Similarly, to write the history of a battle, we must re-think the thoughts which determined its various tactical phases: we must see the ground of the battlefield as the opposing commanders saw it, and draw from the topography the conclusions that they drew: and so forth.

After this marginal addition Collingwood continues:

> The past event, ideal though it is, must be actual *in the historian's re-enactment* of it. In this sense, and this sense only, the ideality of the object of history is compatible with actuality and indeed inseparable from actuality...
>
> The historical event is this actual and ideal at once: but not at all in the same way in which the Matterhorn is actual and ideal at once. The Matterhorn, because it is a physical thing, not an event, persists in time and may therefore be at once perceived and remembered. But the object of historical thought is an event, and does not persist. Its very permanence, so far as it has permanence, consists in its complete nonexistence: death once dead, there's no more dying then; the event, once over and done with, can be re-enacted in the historian's mind anywhere and any time because it nowhere and at no time can actually recur. Its actuality is only another name for its ideality: regarded as itself, it is purely and only ideal: regarded as the object of *this* act of historical thought it is actual in so far as the act of thought is actual.
>
> This re-enacting of history in the historian's mind is the opposite or complementary aspect of the ideality of history. Because the historical fact is ideal, it has an actuality of its own, an actuality of a peculiar kind: it is *actualised* by the activity of the thought for which it has its ideal being. The object of history, then, while having no existence at all apart from thought, and being so far ideal, is actualised by the thought that thinks it[71].

In a revised version of this passage, attached to the original page, Collingwood restates the latter part as follows:

> The historian, then, re-enacts the past in his mind: but in this re-enactment it does not become a present or an actuality. The actuality is the actual thought of the historian that re-enacts it. The only sense in which the object of historical thought is actual, is that it is actually thought about. But this does not confer any kind of actuality upon *it*, taken in itself. It remains wholly ideal[72].

Collingwood's answer to the question how the historian can re-enact the past is:

> ...[T]he historian may re-enact a past event if that event is itself a thought. When Archimedes discovered the idea of specific gravity he performed an act of thought which we can without difficulty repeat: he was drawing certain conclusions from certain data, and we can draw the same conclusions from the same data. Not only *can* we do this but if we are to write the history of Hellenistic science we *must* do it, and must do it knowing that we are repeating Archimedes's thought in our own mind. Similarly, if we are to narrate the history of a battle, we must see for ourselves the tactical problem that the victorious commander saw, and see the solution as he saw it. If we are to narrate the history of a constitutional reform, we must see what the facts were that the reformer had before him, and how his way of dealing with the facts seemed to meet the necessities as he felt them to exist...
>
> Not only is the history of thought possible, but, if thought is understood in its widest sense, it is the only thing of which there can be history. Nothing but thought can be treated by the historian with that intimacy without which history is not history; for nothing but thought can be re-enacted in this way in the historian's mind. The birth of solar systems, the origins of life on our planet, the early course of geological history - all these are not strictly historical studies because the historian can never really get inside them, actualise them in his mind: they are science, not history, because, however much they may take the form of narrative, they are generalised narratives, accounts of how things must have happened in any world, not accounts of how things actually happened in this world. They are hypotheses, which, however probable, do not even approximate to the status of documented history.
>
> *All history, then, is the history of thought*, where thought is used in the widest sense and includes all the conscious activities of the human spirit. These activities, as events in time, pass away and cease to be. The historian re-creates them in his own mind; he does not merely repeat them, as a later scientist may re-invent the inventions of an earlier: he re-enacts them consciously, knowing that this is what he is doing and thus conferring upon this re-enactment the quality of a specific activity of the mind[73].

In answer to a possible objector who holds that no kind of re-enactment is possible because nothing can happen twice, Collingwood retorts:

> ...[H]e would himself not hesitate to speak of dining twice in the same inn, or bathing twice in the same river, or reading twice out of the same book, or hearing the same symphony twice.
>
> ...[T]o re-enact the past in the present is to re-enact it in a

> context which gives it a new quality. This context is the negation of the past itself...
> For Dante, the *Commedia* was his whole world. For me, the *Commedia* is at most half my world, the other half being all those things in me which prevent me from literally becoming Dante...
> I thus genuinely re-enact Dante's medievalism... but I re-enact it in a context (namely the rest of my mental outfit and equipment) which gives it a new quality, the quality of being *one element* within a whole of thought that goes beyond it, instead of being a whole of thought outside which there is nothing. This quality of being an element within my experience, an element checked and balanced by others and so contributing to the equilibrium of the whole, is the ideality of history. The whole is actual and only actual[74].

What conclusions may be drawn from this important chapter, Collingwood's subsequent view on history in fact being based on it? I think the following may be said about it:
1) It is clear that his theory must be seen in the context of his notion of the ideality of the past, and that the latter, in its turn, is a reaction against the realistic theory of knowledge, which, according to Collingwood, makes knowledge of the past impossible. That the past is ideal and only the present real was already stated, as we have seen, both in "Some Perplexities about Time" and the lectures of 1926. The important difference, though, between these and the lectures of 1928 is that in the former Collingwood draws the conclusion that we cannot have knowledge of the past, because knowledge is confined to the actual present. His re-enactment doctrine now provides a solution for the problem how we *can* have knowledge of the past, though it is obviously not an object in the present. For in the re-enactment of past thought the past is actualised in re-thinking it. This can only be done with thought, the corollary being that we can only have knowledge of the past as thought, that is, that all history is the history of thought. This also demonstrates how history *a parte subjecti* and *a parte objecti* are inseparably linked to each other.

So, while in 1926 Collingwood still adhered to the realistic view that knowledge of the past is impossible, he says now about it: "... [F]rom the point of view of an ordinary realistic theory of knowledge, history is impossible. A theory which regards knowledge as 'apprehension' of an independent object is reasonable if perception is taken as the only legitimate example of knowledge ... it has no shadow of plausibility in the case of history ... The battle of Marathon was an event which ceased happening some 2400 years ago; there is nothing there to apprehend; in the realistic sense of the term object, there is no object whatever for the historian to know. And therefore, since without object there can be no knowledge, history as a form of knowledge is, realistically speaking, an absurdity ... In opposition to all realism, then, any philosophy of history must assert the ideality, as opposed to the reality, of historical fact. It asserts that the past as past has no existence whatever, consisting as it does of occurrences no longer occurring, events that have finished happening: and it holds that these events can be historically known not by anything in the least analogous to perception, observation, or any process or act intelligibly describable as 'apprehension', but by their re-enactment in the mind of the historian"[75]. One could remark in this connection, though, that while Collingwood definitely puts aside the realistic theory of knowledge with regard to its view of grasping an independent and existing object, he still adheres to it in a certain sense, in so far as it asserts that only the actual is knowable, since through the re-enactment of past thought this thought is actualised and Collingwood considers this a necessary condition for its being known.

2) The last observation leads us to the question of the status of re-enactment. It is useful to remember that the context in which Collingwood develops his theory is to provide "a logic of historical method", which means a transcendental analysis of its characteristics. Accordingly the Kantian criteria of universality and necessity are applicable to them. That Collingwood considers re-enactment to be such a universal and necessary characteristic of history is made clear, when he speaks several times of re-enactment as a "must" for history[76], that it "is an essential element in all history"[77], or – to refer to the last quoted passage - that past events can be historically known "but by their re-enactment in the mind of the historian". So the re-enactment doctrine, as developed in the lectures of 1928, is not a proposal for an historical methodology, a sort of methodological device for arriving at reliable knowledge of the past. It is a response to the question how historical knowledge is possible, not to the different question how we can arrive at it.

3) The thought of the past is not an object observed from the outside, but must be actually re-thought by the historian. This activity "differs *toto caelo* from the imitativeness which may induce a man or a beast to do what others do because these others are observed to be doing it", Collingwood asserts. "For the historian does not observe others to be doing the things which he does over again. Until he has done them over again he does not know what they are"[78]. Nor are the re-thought thoughts to be seen as copies of the ones of the past: "The re-enactment of the past in the present *is the past itself* so far as that is knowable to the historian. We understand what Newton thought by thinking - not *copies* of thoughts – a silly and meaningless phrase - but his thoughts themselves over again. When we have done that, we know what Newton thought, not mediately, but immediately"[79]. This past thought, however, is but an element in the total present thought of the historian (in his *Autobiography* Collingwood speaks of its "incapsulation" and of the difference between the "primary" and "secondary" series (Aut,112-114)). As Collingwood expresses this aspect: "[The historian] must enter into [a past frame of mind], reconstitute it with his own mind, and at the same time objectify this very reconstitution, so as to

prevent it from mastering his mind and running away with him"[80].

4) The identity of a past thought with the one re-thought by the historian must be seen as conceptual, as the example of "dining twice in the same inn, or bathing twice in the same river", makes clear. Collingwood's conception is in fact connected with the problem of our knowledge of other minds, as is demonstrated by his answer to a possible objector: "Is the binomial theorem as known to him ... the same theorem that Newton invented, or not? If he says yes, he has admitted all we want. If he says no, we can easily convict him of self-contradiction: for he is assuming that in our mutual discourse we have ideas in common, and this is inconsistent with his thesis"[81]. And Collingwood ends his first chapter by saying: "A person who failed to realise that thoughts are not private property might say that it is not Newton's thought that I understand, but only my own. That would be silly because whatever subjective idealism may pretend, thought is always and everywhere *de jure* common property, and is *de facto* common property wherever people at large have the intelligence to think in common"[82].

5) Past feelings or emotions as such cannot be re-enacted by the historian. Giving the example of a new discovery Collingwood says: "... [T]here is a peculiar quality in the experience of discovery or invention, a peculiar feeling of being the first human being to penetrate into the presence of this particular truth, which the historian can never recapture just because it attaches to discovery as such ... Surely everybody knows that the peculiar thrill with which the victorious commander watches the collapse of an enemy's defence is a thrill which the historian cannot recapture. No one thinks that the historian of Hellenistic science ought to leap out of his bath and run about the town naked when he comes to Archimedes in writing his history"[83]. Appealing in this connection to Hume's distinction between impressions and ideas, the first ones being livelier and more vivid than the second, Collingwood asserts: "We may, and must, recognise that the historian is unable to share the emotional heat with which the characters in his narrative did the things narrated of them, and that *his* emotional heat attaches only to feats of historical research, historical discoveries made and historical perplexities removed"[84]. It is with reference to this assertion that Collingwood then makes the distinction between the different contexts of the thoughts in the past and the ones re-thought in the present.

II QUANTITY

"The question here to be considered", Collingwood says, "is, what is the scope of historical thought? Practically, this is equivalent to the question, what is the right or best form of historical composition? Theoretically, it amounts to this: - what are the limits of historical knowledge?"[85].

The simplest form of historical composition, Collingwood remarks, is the memoir or contemporary history, as for instance exemplified by

Thucydides. For in this case the subject does not have to be chosen by the historian, but it is, as it were, forced upon him. With the widening of interest, however – the potential subjects becoming in principle limitless – the problem arises which subject or aspect the historian has to choose as his field of study. On what principles is this to be done? Because the past in its totality can never be grasped, only a special aspect of it will be studied in a monograph. Since the need is felt to place these specialized studies in a wider context, attempts were made from the eighteenth century on, to compose universal histories.

These histories, however, had a bad reputation among historians. As Collingwood says, they have "fallen into the hands of two classes of persons: the dishonest and the ignorant"[86]. On the other hand, detailed monographical studies are open to the charge of futility. Collingwood then develops an ingenious argument in which he finds a way out of this dilemma by demonstrating not only that both approaches are based in their extremes on a wrong epistemological view of history, but also that they may be vindicated when one starts from the right view. The wrong view on which they are based is to see historical facts as exhaustible in principle. Beginning with universal history, Collingwood comments:

> The essence of the error was the thinking of history as a kind of pattern, a complete body of fact, with articulations of its own and a structure of its own, which the historian had simply to discover. This involved denying the ideality of history: for if history is ideal, it cannot be a single self-contained body of fact awaiting discovery, it must be a growing and changing body of thoughts, decomposed and recomposed by every new generation of historical workers, and the exhaustibility of historical fact, which is implied in the idea of universal history, is an illusion.
>
> The same illusion lies at the root of historical particularism. The devotee of the historical monograph aims at collecting all the evidence there is, and interpreting it completely, so as to give a final account of some point of detail. Here again, we meet with the notion of exhaustibility. 'All the evidence there is' implies that upon any point there is a finite quantity of evidence which is capable of being exhaustively handled in a monograph. But this is simply untrue. A given writer, or a given generation, possesses only a finite quantity of evidence on a given subject; but another writer, or a later generation, succeeds in tapping new sources of information; and where is the process to end? It cannot ever be ended until historical research is ended. Therefore the reasons which our historian gave for confining himself to minute details are bad reasons. They amounted to this: that such details admit of rigorous and scientific handling, which the 'larger' questions do not. But we now see that, precisely as the ignorant man thinks that the larger questions can be definitively settled, and is thereby merely showing his own ignorance, so, when the professional historian thinks that minuter questions can be definitively settled, he too is

> betraying, not ignorance of what has been done, but ignorance concerning the possibilities of future discovery. The idea of the evidence concerning this or that point as a given finite whole is just as false as the idea of history at large as a given finite whole. In both cases the ideality of history is denied[87].

Though their theoretical starting-point is wrongly conceived, Collingwood vindicates the writing both of monographs and universal history. On the former he comments: "...[T]here is no such thing as a large or small question; any question that any historian actually and effectively studies is just large enough to fill his mind, and no larger. The monograph-writer is thus justified by the fact that, because historical fact is ideal and not actual, there are no historical problems except those which historical thought raises; and if I devote my life to the monetary policy of Honorius, the monetary policy of Honorius is for me the whole of history"[88]. "But by the same principle the writer of a universal history is equally justified", Collingwood continues. "He is justified by his very failings. What proves him right is what we thought had proved him wrong -namely the fact that, after all, his universal history is not universal, not complete, but a mere selection of facts arranged to illustrate or prove some particular point. For this makes him a monograph-writer, and removes the sting from that appellation. All that is wrong with his book is, now, its title: it was called 'A History of the World'; it ought to have been called 'The Oppression of the Proletariate in the last Twenty-five Centuries', or 'The Growth of the Modern Conception of Liberty', or the like"[89].

History of the second degree, or the history of history, is again dealt with by Collingwood. Compared with his treatment of the subject in the 1926 lectures this is an improvement in the sense that its a priori nature is made more clear. For this time it is related to the problem solving activity of the historian: "... [A]nyone who is anxious to solve a particular historical problem must find out where he stands, and what his problem exactly is, by looking into the history of the problem itself: that is, into the history of research concerning the subject"[90]. "A theory framed without reference to previous theories", Collingwood asserts, "denies itself the help that may be got from seeing the points that have been already emphasized, and it runs the risk - which in practice is more than a mere risk, it is a practical certainty - of advocating views which have already been conclusively disproved"[91]. The notion of the history of history is in fact placed by Collingwood within the context of his "logic of question and answer" (Aut,29-43), when he uses the example of a monograph on the Peasants' Revolt: "I am not merely asking in a quite vague and general way 'what was the Peasants' Revolt?'", he says in this connection, "I am asking for answers to certain definite and specific questions about it; and these are the questions which have been raised by previous inquiry. Now, unless I am careful to go over this previous inquiry in my mind - to re-enact it, or narrate its history - I shall not clearly see what the problem before me is and how it arose. And in that case I am not likely to be successful in trying to answer it. The presupposition of answering a question is that one should know what the question is that is being asked; and this means finding out how it came to be asked"[92]. The history of history therefore is called by Collingwood, " ... a permanent and indispensable element in history itself. It is the historian's consciousness of how he has arrived at the particular problem which confronts him"[93].

III RELATION

In this chapter Collingwood deals with the inner structure of historical monographs. A monograph necessarily shows a unity and a plurality in its composition. The first is its subject and the second the many events which are seen as parts of it. The subject – the theme or point of view – is constitutive for the parts and the latter must be seen with reference to the first. Collingwood describes their relation in the following way:

> ...[T]he monograph as a whole is a sum of parts, each part being so designed as to make its proper contribution to the whole, and the whole being simply the organised system of parts. For instance, we should describe the battle of Trafalgar in different ways according as we were composing a treatise on naval tactics, on the Napoleonic Wars, on the life of Nelson, or on the influence of sea-power on history. Or again, we might be simply composing a monograph on the battle of Trafalgar, which would demand a different treatment again. Thus the whole must precede the part, in this sense, that the part must be thought out in relation to the whole. The converse is not true. The whole is not thought out in relation to the part. The whole simply is the mutual organisation of the parts. For instance, an history of the Napoleonic Wars contains nothing except accounts of the various operations which collectively go by that name. *The whole, then, is a regulative scheme dictating the details of the work:* apart from the details, it is a mere abstraction, or, at most, a name for someone's intention of writing an historical work, or the bare fact that someone has done so[94].

The point of view from which a certain event is described is regarded by Collingwood as decisive for the value or even the truth of a certain monograph. Giving the battle of Trafalgar as an example he contends that "every historian will recognise that the significance of this battle from the standpoint of the Napoleonic Wars is not the same thing as its significance from the standpoint of the biography of Nelson, and that a perfectly accurate account of it from the latter standpoint would be valueless, or indeed misleading and therefore inaccurate, from the former". "Granted the ideality of history", he adds, "this is intelligible enough: for on that theory, the truth about an event is relative to the point of view from which one approaches it, and an account of an event written from a wrong point of view is therefore not merely irrelevant but false, for the giving of it amounts to claiming that it is relevant, and this misleads the reader and makes him seem to see connexions where there are none"[95].

The parts of a monograph must be put in sequence, that is, one must see "the earlier phases as preparing the way for the later, and the later as explaining the true meaning of the earlier"[96]. Development, Collingwood warns us, "is an ideal process, not an actual process: it consists in something's becoming more and more intelligible"[97]. The same is the case with the idea of progress.

Collingwood also refers to the principle of the ideality of the past with regard to the concept of historical fact: "Because historical fact is ideal, those parts or aspects of it which we are not studying do not exist; what exists is the abstract possibility that we might have been studying them. This abstract possibility is the only kind of reality that attaches to chronological schemes and abstracts of history in general. These things are enumerations – very incomplete enumerations- of the various ways in which we might employ ourselves in historical thought. They resemble guide-books regarded as lists of possible excursions; but they do not resemble them regarded as descriptions of actual places"[98].

In a most interesting passage Collingwood describes how the idea of a process must be conceived in an historical narrative. Since all history is the history of thought he rejects the idea of causal relations, giving the following argument:

> A thought can never be either an effect or a cause; but thoughts may form a sequence of conditioned and conditioning elements. For instance, in a game of chess, it is because White has moved in a particular way that Black replies with a particular move: and this again determines the next move of White. But this determination is not causal. What happens is that White's move places Black in a certain situation, and in this situation there is only one move by which Black can avoid defeat: in order to avoid defeat, he therefore chooses to make that move, and this again creates a new situation for White. It is only because each is a free and intelligent agent that he acts as he does; what is said to determine his act only creates a situation in which he exercises his freedom and intelligence[99].

According to Collingwood this is "the nature of historical sequence. Every event, so far as that event is an expression of human thought, is a conscious reaction to a situation, not the effect of a cause"[100]. The resemblance of this description of causality to the theory of "rational explanation" of actions as developed by Dray is remarkable. There is a resemblance also to what Collingwood himself wrote later about "causation in history" in *An Essay on Metaphysics*[101], the only difference being that he speaks there about "cause" in a special sense, while in the present lectures he refuses to use this word at all. The "new" element Collingwood brings into his example of the chess-players is the importance of rules:

> ...[T] he only reason why a given situation leads to a given action is that the agent is guided by certain principles: in the case of chess, the rules of the game. Apart from these rules, his reaction to his opponent's move would have no meaning and would be unintelligible: but if you know the rules of the game and know that he wants to win, you can see why he moved as he did; unless indeed his move was due to an oversight, in which case the best you can do is to understand what he meant, but failed, to achieve.
> The principles here referred to are different from the causal laws of natural science in that they do not operate except consciously. It is only because the player knows the rules of the game that the rules of the game explain his moves. Consequently these principles cease to operate when people cease to think of them; and therefore they are themselves historical phenomena"[102].

IV MODALITY

In the last chapter "the certainty of history, the nature of the grounds on which it rests, and its status as genuine knowledge"[103] is dealt with. Its general theme is a vindication of the science of history against realist and empiricist contentions. The latter either deny the historian's claim to have knowledge of the past or "assert that the object of historical thought is not the past at all, but a trace or residue of the past in the present"[104]. Collingwood agrees with the objection of the realists that the historian cannot have knowledge of past events as they actually happened: "All he can do is to interpret the evidence at his command"[105]. His dependence on evidence means that "an overwhelming majority of past events must remain permanently unknown and unknowable", and, Collingwood contends, "even in the most favourable cases, one's ignorance is infinite, and one's historical knowledge consists only of a few atoms lost in the void of endless space"[106]. "It is necessary to reflect carefully on this point", Collingwood adds, "because we are apt to think that we know 'all about' something, that is to say, possess a complete knowledge of it, when we know all *that is known* about it; we mistake the coincidence between our information and the extant information for a coincidence between our information and the object. Once this confusion is cleared up, no historian would hesitate to say that, even in the period he knows best, there are infinities of things he does not know for every one that he does"[107]. Collingwood therefore has no difficulty with the conclusion that if history "means knowledge of the past in its actuality and completeness" it is an illusion.

The historian is bound by his evidence; this evidence, however, must be not only recognisable as such and interpreted, but also "relevant to a question, pointing towards an answer": "until a question has been asked, there is no evidence for it ... [E] ven the most complete and striking evidence, is convincing and indeed significant only to one who approaches it with the right question in his mind"[108]. Through this interrelation between question and answer, Collingwood maintains, "the question as to the certainty of history appears in a new light":

> The historian cannot have certain knowledge of what the past was in its actuality and completeness; but neither has he uncertain knowledge of this, or even conjecture or imagination of it. The past in its actuality and completeness is nothing to him; and, as it has finished happening, it is nothing in itself; so his ignorance of it is no loss. The only knowledge that the historian claims is knowledge of the answer which the evidence in his possession gives to the question he is asking. And the question itself is relative to the evidence, as the evidence is to the question: for, just as nothing is evidence unless it gives an answer to a question which somebody asks, so nothing is a genuine question unless it is asked in the belief that evidence for its answer will be forthcoming. A question which we have no materials for answering is not a genuine question; such a question is never asked by the historian, unless inadvertently; and his inability to answer it, if anyone asks it of him, is a sign, not of his incompetence, but precisely of his competence: it is a sign that he knows his business.
> The certainty of history, then, is the certainty that the evidence in our possession points to one particular answer to the question we ask of it[109].

As in the 1926 lectures Collingwood makes a distinction between dogmatic and critical history, the former called by him "an elementary or nursery stage of historical education"[110]. It is in this context that he uses for the first time the concept "scissors and paste", well known from *The Idea of History*: "The historian's sources are, from this point of view, 'authorities', that is to say, places where he finds his statements ready-made; his equipment consists simply of a retentive memory, and his methods of work are comprised in scissors and paste"[111].

In critical history authorities become sources, Collingwood repeats from his 1926 lectures, and he criticizes here the distinction between written and unwritten sources. He also emphasizes again the necessity for the technical equipment of the auxiliary sciences, called by him now "archaeology". These sciences are abstract and classificatory and a *sine qua non* of critical history. "They are not themselves history", Collingwood says; "they are only methods of dealing with the sources of history; but without them history cannot pass beyond the dogmatic or nursery stage. They form, as it were, the bones of all historical thinking. History itself must be flexible, but it must have rigid bones, unless it is to lose all power of independent locomotion and become a parasite"[112]. "Pure" methodology is equated this time by Collingwood with philosophy of history: "the science which lays down universal canons of method for dealing with all kinds of sources and constructing any kind of narrative about any subject... a science dealing with the universal and necessary characteristics of all historical thinking whatever, and differentiating history from other forms of thought"[113].

Collingwood ends his lectures with a strong plea for history, rejecting equally strongly the pretensions of a philosophy which denies it a rational respectability: "A thing like art or science or history does not ask for justification at the hands of philosophy. It is capable of justifying itself. The fact that numbers of people have worked at it for a long time, building up between them a coherent system of thoughts by means of methods devised and elaborated for the special purposes of their pursuit, is itself the proof of its rationality"[114].

A philosopher studying historical thought should not do it, however, only from the outside, Collingwood contends: he will find history to be a rational and necessary form of thought, "but he does not find in it the same necessities or logical connexions which the historian finds. Therefore he thinks of the historian as, at best, somewhat illogically logical and irrationally rational"[115]. Collingwood's solution, then, obviously inspired by his own experience, is the following:

> This difficulty is only removed when the philosopher studies history from the inside: that is, when the philosopher and the historian are the same person and when this person's philosophical and historical work react on one another. In this case the philosopher is sure that the historian's historical thought is rational, because he is himself the historian, and he is merely assuring himself of the rationality of his own thought. It is no mere act of faith, but an examination of conscience, that makes him accept historical thought as a reasonable pursuit for a sane man. But conversely, the historian is able to depend for some things upon the philosopher. The philosopher is concerned, in his theory of historical knowledge, to think out certain questions concerning the limits, validity, and purpose of history: and the historian is able to bring his historical research into conformity with the results of this enquiry.
> Thus a double result will follow. The philosopher's philosophy will become more trustworthy because of his personal and intimate experience of the subject about which he is theorising; and the historian's history will become more rational because it is being brought into increasing conformity with the philosophical idea of itself. History supplies philosophy with data, and philosophy supplies history with methods[116].

4.6. COLLINGWOOD'S DEVELOPMENT

In an often quoted passage in his *Autobiography* Collingwood says: "My life's work hitherto, as seen from my fiftieth year, has been in the main an attempt to bring about a *rapprochement* between philosophy and history" (Aut,77). What he meant by this *rapprochement* is in the first instance not very clear and accordingly it has been variously interpreted. The lectures of 1928, however, provide an answer to this question. Further on in the *Autobiography*, speaking about his conception of history advancing another step forward, Collingwood says:

> This step was taken, or rather registered, in 1928, when I spent a vacation at Le Martouret, that pleasant country-house near Die, sitting under the plane-trees on the terrace and writing down as shortly as I could the lessons of my last nine years' work in historical research and reflection upon it. It is difficult to believe that so obvious a point was reached so slowly; but the evidence of my manuscripts is clear (Aut,107).

It is of course very fortunate that the evidence of this particular manuscript is now at our disposal. With regard to the relation between philosophy and history it is worth seeing what he wrote about it "on the spot" in the Preface of his lectures (he calls it an essay) of 1928. It is remarkable how this corresponds to his description in the passage in the *Autobiography*:

> In its present form it [i.e. the essay] is certain to mislead a reader, because its argument appears to rest on a single point – the ideality of history – and to be developed deductively from that. The reader who wants to destroy the argument will therefore naturally concentrate his attention on the ideality of history and try to undermine that proposition, thinking that when it falls the whole argument will fall with it. But he will be mistaken. The various points made in the course of the argument are in point of fact observations made in the course of historical studies pursued with a special eye to problems of method. Not one of them has been reached deductively from the conception of the ideality of history. On the contrary, the idea of considering them in the light of that conception only occurred to the writer very late in the day, after most of them had been long familiar to him as the fruits of experience in historical research. Therefore, when they are set out as they are here, in the form of a single chain of argument, the reader is asked to remember that the position of each link in the chain is guaranteed not simply by its relation to the first link but by cross-bearings from experience of historical inquiry.

Of the main interpreters of the development of Collingwood's thought, neither Knox, nor Donagan nor Mink take seriously the description Collingwood himself gave of it in his *Autobiography*, although for different reasons[117]. In the light of Collingwood's lectures on philosophy of history of 1926, and especially those of 1928, I would contend, however, that Collingwood's own description of his development is after all in one aspect trustworthy, namely the important one concerning his view of the relation between philosophy and history. The *rapprochement* between them was realized in his philosophy of history conceived as the "pure" methodology of history, as is especially made clear in the lectures of 1928.

There is another piece of evidence for this conclusion. In 1927 Collingwood was appointed "University Lecturer in Philosophy and Roman History". In a Report of 1932 to the "Faculty of Literae Humaniores"[118] about his activities he says:

> By appointing me Lecturer in Philosophy and Roman History, I understand the University to mean, not only that I am to study and teach these two subjects, but also that I am to study and teach them in their mutual connexions: i.e. in philosophy, to investigate the philosophy of history, and, in history, not to neglect the methods and logic of historical work, and to emphasize the relation between history and its sources.

He then gives a list of five projects, the last being

> A study of the philosophical problems arising out of history: especially a) logical and epistemological problems connected with the question 'how is historical knowledge possible?', b) metaphysical problems concerned with the nature and reality of the objects of historical thought.

Collingwood finally describes how his projects have developed and says of the last one:

> This I regard as my chief work, involving the whole of my philosophical and historical studies in their mutual connexions. I am of opinion that there is important work to be done here, and that it cannot be done except by a trained and practising historian who is also in constant work as a philosopher. This opinion has been strengthened by much reading in the last four years, and by gradually reaching, in that time, a provisional solution of most of the chief problems.

Collingwood then refers to his publications in this field, especially *The Philosophy of History* (1930), and says about the latter that it is "in effect the synopsis of a complete treatise, but I do not intend to begin writing such a treatise until I have done several years' work on various aspects of the subject".

So here we have Collingwood making up his mind, in January 1932, about the relation between philosophy and history. He has come by now to "a provisional solution of most of the chief problems" and in the next years he will turn to other subjects: the problem of method in philosophy, resulting in *An Essay on Philosophical Method*, cosmology and metaphysics, while in 1935 the first part of *Roman Britain and the English Settlements* was written[119].

4.7. LECTURES ON THE PHILOSOPHY OF HISTORY: 1929-'32

From 1929 till 1932 Collingwood wrote lectures on philosophy of history to be delivered in Trinity Term[120]. Among the manuscripts are partial notes for the lectures of the years 1929, 1931 and 1932. The lectures of 1931 could not be given, because in April of that year he fell ill, and he had not yet recovered by June[121]. Similarly the lectures of 1932 were not given, because Collingwood had taken leave again for ill-health. The lectures from 1929 have a different plan from those of 1926 and 1928 in that Collingwood deals as well with the history of the idea of history, as he was to do in his lectures of 1936, on which *The Idea of History* is based. In our treatment of them we will only consider those aspects not treated in the latter book.

LECTURES OF 1929

Only parts of the lectures (from the second on) are left. They begin with a treatment of Kant's *Idee zu einer allgemeinen Geschichte in weltbürgerlicher Absicht* (1784). Coming to Hegel Collingwood criticizes his philosophy of history not for using a priori conceptions as such, but for the way they are conceived, especially his artificially forcing of historical facts into a logical framework. "Hegel realised that history is logical", Collingwood contends, "but he then made the mistake of jumping to the conclusion that it was logic". Collingwood asserts particularly that his division of history into three periods, in which one is free, some are free and all are free, "corresponding to the three kinds of quantity in judgments", cannot be convincing[122]. Besides this, Hegel himself knew, according to Collingwood, that the moral freedom of man is a modern idea and that the Greeks did not possess it[123]. On the one hand Collingwood values Hegel's conception of the difference between the various periods of history, but he criticizes him for not noticing their individuality: "... Hegel, trying as he does to state the difference between different historical periods in terms of logical formulae, shows that he is unaware of the individuality, the concreteness and uniqueness, of these historical periods. He is treating individuals as if they were universals. He is dealing with unique historical events as if they were mathematical or physical abstractions. He is applying to history the methods of thought proper to science"[124].

After Hegel, historical studies began to flourish, but there developed a sharp division between the historians and the philosophers of that time, Collingwood contends. Though by no means all Hegelians, the latter "were almost equally unanimous in accepting the fundamental error of Hegel's philosophy of history. This fundamental error was the principle of dividing history into periods and characterising these periods by means of abstract concepts: in other words, it was the fallacy of substituting the scientific universal for the historical individual"[125]. Collingwood mentions here Saint-Simon, Comte, Marx and Spencer. All suffer from offering certain abstract distinctions as the key to a concrete historical distinction. Of Comte's division of history into theological, metaphysical and positive ages Collingwood says for instance: "Deeper historical knowledge would have shown Comte that even what he called the theological age had its own metaphysics and its own science; and that what he called the positive age–the age of modern science – had its own metaphysics and its own religion, and did not consist purely of positive natural science"[126]. Similarly with Marx: "...[H] is antithesis between a present age of capitalism and a future age of socialism is vitiated by the fact that capitalism and socialism are co-existent tendencies in the economic organisation of one single age. If capitalism should disappear, socialism could not survive it, for socialism is essentially an opposition to capitalism; what would survive would be a quite new type of economic organisation"[127].

People like Taine and Buckle tried to elevate history to the status of a science by contending that the historian in the strict sense should occupy himself with the discovery of facts, while the sociologist or "scientific" historian would discover their causes and explain them scientifically. Collingwood considers this programme "a grave danger to historical research, bound up with grave logical errors"[128]. For in the first place the sociologist will tamper with the historical facts, trying to adapt them to his theories. This is inevitable, according to Collingwood, "because the sociologist, claiming to be a scientist in possession of universal and necessary laws, could not help claiming to know what the historian ought to find before he had found it", and consequently "[t] he result could only be a tendency to lower the standard of historical accuracy and to undermine the conscience of historians"[129].

Secondly, Collingwood maintains, the separation of facts from causes is a logical error:

> Even in the natural sciences, we do not first observe facts and then discover their causes. The cause is part of the observed fact. Thus we examine a case of a disease, and the observations which we make include the fact that a certain micro-organism is present, which we call the cause of the disease. It is not one man's work to compile clinical accounts of a disease and another's to inquire into its cause. Even when something of that kind is possible, it is necessary that hypotheses as to the cause should be tested by the most accurate observations and experiments; and thus the last word lies with the person whose business it is to observe the facts. But all this was ignored by the men who proposed to elevate history to the rank of a science. They had too vague an idea of what

> science really was. They thought that a scientist was a man who sits in an arm-chair and elicits startling generalisations out of facts collected for him by assistants and bottle-washers. They did not know how completely a real scientist is steeped in actual experimental work, how close he keeps to observed facts, how quickly a generalisation sends him back to his laboratory in search of empirical tests[130].

Since the end of the nineteenth century, philosophers like Windelband, Dilthey, Simmel and Rickert had claimed that history was an autonomous form of thought not dependent on science. Collingwood criticizes them, however, for not going further: "How it is possible for the historian to apprehend historical fact in its individuality, they cannot say. They can only fall back on a vague and uncritical conception of psychological intuition which somehow gives the historian an inner conviction that he has apprehended the truth". "Their own philosophy of historical knowledge", Collingwood observes, "was precisely the type which Kant condemned as dogmatic – that is, it claimed an immediate intuition of its object without explaining how such an intuition was possible or what kind of object it must be in order to be so intuited"[131].

A real advance in the theory of historical knowledge was made by Croce, Collingwood continues. While the Germans had only said what history was not, showing that it was not science, Croce said what it was, namely art. This theory, developed by Croce in 1893, could not distinguish, however, between the real and unreal in history. It is in his *Logic* that Croce made an improvement in his theory of history, according to Collingwood, by stating that in every judgment both the universal and individual are implied. Collingwood's treatment of Croce closely corresponds with that in *The Idea of History* (IH,190-204). He ends the part of the lectures that has survived, as follows:

> With this theory [i.e. the one of Croce], the philosophical analysis of historical thought has brought itself abreast of the modern advances in historical technique, and we have at last a theory which does justice to the facts in their main principles. The task of successors – and no one has yet made any important addition or modification to Croce's theory – is to think out that theory still more clearly, with a still closer grasp on the actualities of historical method. That is the task towards which these lectures are in the first instance a contribution. This can be best done by concentrating on certain central problems that beset the theory of historical knowledge, and trying to solve them. I have selected four such problems[132].

Collingwood then shortly mentions the contents of the four chapters of his lectures of 1928[133]. So the part of the 1929 lectures which has survived was apparently meant as an introduction.

LECTURES OF 1931

What is left of the 1931 lectures, entitled "The Origin and Growth of the Idea of a Philosophy of History", is again only a fragment. Collingwood this time begins by stating that the traditional theory of knowledge has arisen from the seventeenth century out of the problems of natural science. Since the nineteenth century, however, there has developed a critical and methodical history, the growth of which, Collingwood contends, "is an event no less characteristic of the nineteenth century than that of natural science is characteristic of the seventeenth; and just as the seventeenth century was confronted by the need for a new theory of scientific thought, so our own age is confronted with the need for a new theory of history"[134].

Treating the idea of a philosophy of something, Collingwood asserts that a philosophy of, for instance, logic, ethics, art or religion studies both a subject and an object. If we leave out the first, he remarks in a note, philosophy becomes a kind of science and if we leave out the second it becomes psychology[135]. Collingwood calls it the paradox of history that it is knowledge of the past, "i.e. of that which, because it no longer exists, is not 'there' to be known – not 'given', not 'here', not 'anywhere' – completely faded and vanished into the gulf of nothingness which is past time". "[H]istory lives by solving this paradox in practice", Collingwood says, "the philosophy of history lives by solving it in theory". Philosophy of history, then, "is the philosophical reflexion on historical thinking"[136].

Collingwood sees that there is a similar reflection in the case of science: because science and history "are forms of *organised and critical thought*, the corresponding philosophies have a peculiarly important part to play. They are impossible without *method*, and the method must be *thought out* and consciously adapted to its ends, i.e. the philosophical problem (what *are* the ends of science and history and how can these ends be attained) is a presupposition of the properly scientific or historical problem"[137]. There is even a parallelism between science and history, Collingwood continues, in that both depend on a paradox: "Science is knowledge of that which is not *here and now*, but universal – how *can* we know this? History is knowledge of the past – how can we know *this*? In both cases knowledge has to reach out to an object which prima facie seems beyond its grasp – prima facie knowledge is of the given, the here and now, that which exists"[138]. Unfortunately the manuscript breaks off here. Subsequent pages then deal very shortly with historical thought in antiquity and the Middle Ages, and more extensively with Bacon, historical scepticism, Voltaire, Herder, Kant, Hegel, Schiller, Fichte and Schelling.

LECTURES OF 1932

Of these lectures only the first one is left. Collingwood begins by stating three senses of the phrase philosophy of history, which he argues are incorrect. As in *The Idea of History* (IH,1), he mentions the conceptions of Voltaire, Hegel and the notion of discovering the laws governing the course of history.

In dealing with the question of what the correct sense of the term philosophy of history is, Collingwood begins – as in his "Preliminary Discussion" of 1927 – with the question of what philosophy is. His analysis partly differs, though, from the one offered in the "Preliminary Discussion". Contrasting philosophy with science, he first mentions their difference in method. Scientific thought proceeds from hypotheses and draws consequences from them. Philosophy, on the other hand, is not satisfied with hypotheses or assumed starting points, but sets itself the task of "*criticising* its own assumptions and going behind them". "Hence the two seem to move as it were in opposite directions", Collingwood observes: "scientists think forwards, moving from one point to the next, advancing to new knowledge and building up elaborate structures of thought, whereas philosophers think backwards, undermining their own edifices and trying always to reach something deeper down and more fundamental"[139]. Corresponding to this difference of method there is the difference of object, that while science takes the reality of its object "on trust" philosophy is always in search of ultimate reality. Another difference to be discerned is that in science the universals or concepts only concern a certain limited field of reality, while a philosopher tries to determine the concept of being. "In this sense it has been said", Collingwood says, "that all the concepts of philosophy are *definitions of the absolute*, i.e. they all apply to reality as a whole and define its absolutely universal and absolutely necessary characteristics"[140].

Given this analysis, it is not yet clear why there should be a philosophy of history, especially when philosophy has reality as a whole as its object. One object, however, may have characteristics of many kinds, studied by different sciences. And, argues Collingwood, while "[i]n the same way the various philosophical sciences all study a single object – reality as such – ... there is no reason why they should not differ from each other in a way corresponding to distinctions between the various attributes possessed by this object"[141]. Now, if history is a branch of knowledge, the philosophy of history is a branch of the theory of knowledge. If history were adequately dealt with already in the traditional theory of knowledge there would be no reason to deal specifically with it. However, Collingwood denies that this is the case, repeating his argument of the 1931 lectures: " ...[T]he theory of knowledge, in the traditional sense of that phrase, consists of a body of thought, mostly put together in the 17th and 18th centuries, concerned with problems arising out of reflexion on natural science ... Historical knowledge, the knowledge of a world of human acts in the past, raises a rather different set of problems, and with these our current theories of knowledge are not able easily to deal; the tendency is to slur them over or solve them falsely by reducing them to cases of scientific knowledge". This has the consequence not only "that the special problems of historical knowledge are left unsolved, but, what is more serious, that a false picture is given of knowledge as a whole: the scientific elements in it are over-emphasized, the historical elements neglected"[142].

There are thus, Collingwood sums up, two reasons for paying special attention to the philosophy of history: 1) It studies the special problems of historical knowledge, divided into two questions: "the logic and methodology of historical thinking" and "the metaphysics of historical fact", that is, the attempt "to answer the question 'what really is it about which the historian is thinking?'" 2) "... [B]y raising these problems the philosophy of history may hope to assist philosophy as a whole, by dealing with subjects hitherto to some extent neglected"[143].

With regard to the latter Collingwood refers again – as in the 1931 lectures – to the similarity between the epistemological problems of science and history in that both deal with what cannot be apprehended here and now. For "[s]cience becomes science by going beyond the objects of experience and setting before itself an ideal object, an object of thought which can never be an object of sense ... But in this way history resembles it; the historian's object is not what he sees happening now, but what he thinks to have happened in the past ... Thus, if the ideality of science raises the question 'how is science possible?' the ideality of history must raise a corresponding question there"[144].

However, Collingwood considers the difference between science and history as important as their resemblance: "The object of science is a universal, timeless and unchanging, a Platonic form or in modern language an eternal law of nature. The object of history is an individual fact or event, having a situation in time. The answer to the question how we can know timeless universals cannot also be an answer to the question how we can know past facts"[145].

It is interesting to note the difference between the argument Collingwood develops here concerning the need for a philosophy of history and the earlier one in his "Preliminary Discussion" of 1927. In the latter the necessity for a philosophy of history was argued from the conception of history as a universal and necessary form of human experience, while in the 1932 lecture Collingwood's argument is based on the peculiar epistemological characteristics of the science of history. Its universal importance this time is related not so much to human experience in general, as to philosophy in general. This does not imply, however, that Collingwood did not hold the first view any longer. It must be seen, rather, as a reflection of his being especially occupied in the years before 1932 with the methodological questions of history.

4.8. *REALITY AS HISTORY* (1935)

This manuscript, written in December 1935, was not meant as notes for lectures, but is – in Collingwood's own words – "[a] n experimental essay designed to test how far the thesis can be maintained that all reality is history and all knowledge historical knowledge". It is important for the elaborate analysis Collingwood makes in it of the ideas of historical understanding and human nature, and the distinction between historical and natural process.

The problem from which he starts is put by Collingwood as follows. The Greeks distinguished between a world of appearance, as perceived and experienced by us, and reality. The first is a constant flux of events, while the latter is permanent, the world of fixed forms. Only of the latter can we have real knowledge, the flux of events being outside its reach. In modern science a similar distinction is made between the flux of things, in itself not intelligible, and a permanency which is. This time the latter is not found in a world of forms outside the flux but in it, namely in the permanent relations between its elements. Thus a network of classifications is built up in the form of laws or uniformities. On such a view, Collingwood says, "[t] he individual as such ... is unknowable; we experience it, but that is all; to the intellect, it is inscrutable"[146]. All that counts is its general characteristics: "The practical test by which you can see whether you understand a thing in this sense of the word is to try replacing it with something else like it and see whether you understand that. Do you understand why your clock keeps time? Try thinking of it with a different pendulum of the same length. If you realize that it is not the pendulum but the length of the pendulum that counts, you understand the clock's keeping time"[147].

Collingwood rejects the conceptions of both the Greeks and modern science with regard to knowledge and reality. He wants to develop the thesis that reality is to be found in the flux itself: not in its recurring uniformities, "but in the actual sequence of the elements which make it up". And he adds that "[t] his is the principle of history, in the wider sense of that word; where history means process in time"[148]. If we want to understand something, we have to know how it has become what it is, that is, we have to discover its past and this can only be done by historical thinking. By reducing individual things respectively to fixed forms and certain regularities, Collingwood maintains, "[b] oth Greek thought and scientific thought ... methodically evade the real question, the question how *this* came to be what it is. The only kind of thinking that does not evade the question is historical thinking"[149].

Collingwood does not agree with the distinction made by German philosophers between "Geisteswissenschaften" and "Naturwissenschaften", the first being the sciences of mind and historical, and the latter sciences of nature which has no history. He is realistic enough to see that this distinction will not be recognised by natural scientists, because their thought is based on the belief that their way of thinking is the only correct one. If scientific thought is left unchallenged in its own field, Collingwood asserts, its prestige will react on historical thought in a negative way, twisting it into pseudoscientific forms. The scientist studying nature usually ignores its historical aspect. That this, however, may also be unsatisfactory is demonstrated by Collingwood with the following example:

> The dandelion-head whose seeds I now watch a sparrow eating is as individual and unique a thing as the French Revolution. The sparrow is this sparrow, not any sparrow. Its appetite for the seed I now see it eating is, no doubt, an example of a kind of appetite common in sparrows; but if I cannot be content to say that the French Revolution happened because oppressed populations rebel against rulers too weak to control them, I cannot be content to say that this sparrow eats this seed because sparrows like dandelion-seeds. In both cases, the ground of my discontent is the same: it is, that the general rule, just because it explains every case of the kind indifferently, does not explain this case in its concrete actuality, but only those features of it in which it resembles the rest. In short, if I am content with a scientific explanation of a natural fact, the reason is that I am content to think of it not as the unique fact which it is, but merely as an example of a certain kind of fact[150].

"It often happens that our interest in things falls thus short of their concrete actuality, and would be satisfied with the substitution of another thing not conspicuously unlike them", Collingwood continues. The interest in the sharpness of his razor-blade is given as example, but also the capacity of a particular man to dig his garden. Scientific understanding is enough here and the latter example demonstrates that "[t]he scientific attitude is thus no less natural, no less adequate, when we think about human beings and their actions than when we think about anything else"[151].

When Collingwood contends that it is untrue to say "that there can be no science of mind but only of nature", and that "[t]here is ... no such division as was suggested by the theorists of Geisteswissenschaft, between mind as the proper object of historical thought and nature as that of scientific"[152], this is contradicted by his view as expressed in "Human Nature and Human History", which was written only three months later. For in the latter essay he argues that "whereas the right way of investigating nature is by the methods called scientific, the right way of investigating mind is by the methods of history" (IH,209). The only way in which these views might be reconciled would be to point to the fact that though there can be, according to Collingwood, a science of mind, this would not be the "right way" of investigating it, and though the right way of investigating mind is by historical thought, mind is not its "proper object", because nature can be its object as well. This interpretation, however, seems so artificial as to be hardly convincing. It is refuted, moreover, when Collingwood in his draft of "Human Nature and Human History" flatly declares that "whereas science studies nature, history studies mind"[153].

With regard to the argument presented in "Reality as History" it is clear, however, that according to Collingwood scientific thinking cannot take historical aspects into account. As yet Collingwood has not made clear, though, what the special characteristics of historical thought are. In order to deal with this question he first states the position of scientific thought:

> In science, understanding a thing is simply thinking of it as an example of a certain general law: explaining it is showing it as an example of such a law. According to the theory of knowledge on which science is based, this is more than the way in which science explains or understands (which would permit the possibility of other ways of doing it); it is what explaining or understanding is; nothing that is not subsumption under a general law can be rightly called explaining or understanding at all. From the point of view of a theory such as this, history, it is clear, cannot explain or understand things. All it can do is to ascertain that certain events have happened in a certain order. It narrates these facts, that is, it exhibits them in their succession; but exhibiting a succession is one thing, exhibiting its necessity is another; and to explain a thing is to exhibit its necessity.[154]

The statement of this view is striking indeed as an anticipation of that of Hempel as developed in his well-known article "The Function of General Laws of History"[155], which appeared seven years later and was to start a lively discussion on explanation in history. It is of course most interesting to note Collingwood's argument against this conception. It is again surprising how much he anticipates with it the view developed by Dray in his *Laws and Explanation in History*, the most outspoken attack on the scientific position, or, as it is called by him, "the covering law model"[156].

The argument of scientific explanation "begs the question at issue", Collingwood argues. "It contains two statements as to what explanation is: first, that it is to exhibit the necessity of something, secondly, that it is to subsume it under a general law. The question is whether there is any way of exhibiting a thing's necessity except by subsuming it under a general law. If there is, scientific explanation is not the only possible kind of explanation"[157]. According to Collingwood, it is possible to exhibit the necessity of a sequence, without subsuming it under a general law, though of course not in every case. He gives the following example:

> If it is said: 'on September 25 Great Britain went off the gold standard; on September 26 there was an earthquake in Japan', that statement exhibits a succession, but not its necessity. If it is said:

> 'on September 25 Great Britain went off the gold standard; on September 26 we changed what was left of our francs back into pounds and recovered the whole cost of our holiday in France', not merely a succession, but the necessity of that succession, is exhibited: the second event could not have happened unless the first had happened already. The function of history is to make statements of this second kind; in so far as the historian merely exhibits successions without thus exhibiting their necessity, he has failed in his proper task; instead of being an historian he is being a mere annalist. No doubt there is much that goes by the name of history which in reality is just annals; but that does not affect the question what history is[158].

Though the sense of this argument is more or less clear, it is largely spoilt by the careless way it is expressed. For, in the first place, by putting the necessity of the sequence of the events in his example on the same level as the fact that "the second event could not have happened unless the first had happened already" Collingwood confuses a sufficient condition with a necessary one. The fact that Great Britain went off the gold standard is certainly not a sufficient condition for my changing my money. This fact, however, is a necessary condition for the latter, not with regard to my changing my money as such – since that could have been done irrespective of Great Britain going off the gold standard – but with regard to recovering the whole cost of my holiday. It would be reasonable to argue, though, that the relation between the two events is stronger than that the first simply is a necessary condition of the second. It is obvious that this is exactly Collingwood's point, though it is expressed by him in an unfortunate way in speaking of a necessary succession. The example given by Collingwood in fact shows how much to the point is Dray's conception of "the rationale of actions"[159] in order to express the special relation between the two events.

A defender of the scientific type of explanation would retort, Collingwood says, that "[t]he reason why we understand the connexion between Great Britain's going off the gold standard and your recovering the cost of your travels in France is that we know certain economic laws, and see that your recovery of the cost of your journey is an instance of these laws"[160]. Collingwood refuses to accept this argument and contends that historical understanding "is simply seeing how events happened, and to do this is to see why they happened. Our thought follows the movement of the events themselves, and in so doing finds them intelligible"[161]. It is not the historian who adds something to the mere sequences of the annalist, but the annalist who takes something away from those of the historian, Collingwood argues: "The flow of events in itself is intelligible; the annalist makes it unintelligible by dropping bits out of it and making it discontinuous"[162]. In a crucial passage he then makes clear how the concept of understanding should be understood:

> The difference, then, between exhibiting a succession and exhibiting its necessity, or making it intelligible, is that a succession so exhibited as to be unintelligible is a succession discontinuously exhibited. What makes a succession intelligible is its continuity. And to understand, in the most general sense, is simply to see continuities. Scientific understanding is one way of doing this: it is seeing general types of continuity, the continuity between anything of one general kind with something of another general kind. Historical understanding is another way of doing it: seeing the continuity of this individual thing with this other individual thing[163].

Other examples of historical understanding as seeing the continuity between individual things which are mentioned by Collingwood are the understanding of a piece of music, a poem or a novel. "The principle of historical understanding in the widest sense ... as a form of understanding distinct from scientific", Collingwood concludes, "is that the flux of things in itself and as it actually flows is intelligible"[164].

However, Collingwood goes even further when he adds: "So far from its being true that the single sequence in itself is unintelligible, and only becomes intelligible when ... subsumed, that in fact the opposite is true: unless the sequence in itself, as a single and unrepeated sequence of events, were already intelligible, it could not be made intelligible by showing it to be an example of a general law"[165]. Though Collingwood is certainly right in contending that "what is intrinsically unintelligible does not become any more intelligible for being repeated", his conclusion that "scientific understanding would not be understanding at all unless there were also, and first, what I am here calling historical understanding"[166] is more doubtful, at least in the form as stated by him. For he gives the example of taking opium being followed by sleep. In a case like this one can only speak of a law of nature expressing a true universal proposition, he says, if there is a necessary connection "in the flux of events as it actually happens"[167]. The latter he considers of primary importance, "[f]or if we can really see no connexion between this taking of opium and this going to sleep, how are we entitled to believe that another antecedent of the same kind will lead to a like consequent?[168]. The aspect neglected here by Collingwood, however, is the role played by scientific theories in our understanding of the relation between taking opium and going to sleep. His account is a rather simple inductive one and in this respect he is right in saying that unless we follow his example "[t]he very basis of induction is destroyed"[169]. The already weak position of the inductive approach, however, is certainly not strengthened by Collingwood's one. The problematic aspect of it is the fact that he gives no indication of the nature of the necessity of the relation between the individual events of taking opium and going to sleep, this being the more pressing as there is here no question of a possible rational necessity.

A point Collingwood quite rightly makes with regard to historical explanation is that an explanation can only be given in terms of something else which itself is unexplained. If this is objected to, it amounts to saying that unless we know everything we know nothing. The same principal would apply, however, not only to historical knowledge, but to knowledge in general: "For it is true not only of history, but of every field of knowledge, that it is inexhaustible; not merely in the sense that, however far we explore it, some part remains unexplored, but in the sense that the very advance of knowledge brings us into contact with new problems which had not arisen before ... The point made in this criticism, therefore, has no special relevance to the case of historical knowledge"[170].

Having demonstrated that the flux of historical events can be made intelligible Collingwood comes to the question how the historical process must be conceived. He speaks here only of human history – under the heading "Human Nature and Human History" – and leaves the question whether one can speak of the history of nature in the same sense for a later part of the essay.

With regard to process, Collingwood says, it is usually thought that there must be a permanent substance that underpins it, because otherwise there would be nothing left in the changing reality. Accordingly there must be a distinction between "what a thing is and what it does or what happens to it"[171]. Collingwood strongly rejects this idea in its application to human history, with its corresponding conception of a fixed human nature. He regards the latter as a "metaphysical error". What is ordinarily called human nature, he avers, should be resolved into history.

Collingwood illustrates his view in the first instance by analysing the concept of human character. A man's character must not be seen as a fixed and unchangeable entity "from which his actions flow automatically", he emphasizes. On the contrary, it is modified by his actions: "Because he has a certain character, certain choices are open to him which to a man of different character would not be open[172]. When he has acted in a determinate way on such an occasion, the fresh action leaves (as it were) a deposit in his character, develops it in this way or that: so that, when he comes to his next action, his character has been, however slightly, modified"[173]. It is interesting to note that Collingwood not only speaks of a man's actions bringing about the development of his character, but also "the things that happen to him". So his character is, Collingwood asserts, "partly something which he has made, and for which he is responsible, partly something which has been made in him by force of circumstances"[174]. One could wish that this crucial distinction had been elaborated, because it is obviously highly relevant for the much discussed role of objective conditions in history as well.

Character being shaped by a combination of (free) actions and the force of circumstances, Collingwood makes it clear that "a man's character is something constructed or built up by his history; what character he has, depends on what history he has had"[175]. "His past and his character are the same thing", he says further on. "His character is the name we give to his past as existing here and now, in so far as it determines his present action. As past, it is dead and does not exist at all"[176]. What exists here and now is only the present complex of actions, but this may be analysed into two factors: "the indeterminate activity or will, and the determinant which fixes the will into this particular action. This determinant is what we call his character, and we are now in a position to identify this with his past ..."[177]. So we have the situation not only of the past being built up of a determinate part (the force of circumstances) and an indeterminate part, but also, in the case of present action, of the past or character being the determinate part and the will the indeterminate. Again, however, the nature of the relation between the determinate and indeterminate parts in present action is not clarified by Collingwood.

So a man's character is nothing else than the historically conditioned part of his present or, as Collingwood also calls it, "history itself now living in the shape of fact"[178]. What is the difference between this view and one which reduces this aspect to a permanent and fixed "nature", Collingwood asks. "If in either case it is unalterable, what point is there, other than purely academic, in replacing one view by the other?"[179]. His answer is as follows:

> The practical and very unacademic difference is this: that what history has produced is mere fact – unalterable, as fact must be, but nothing more than fact – whereas 'nature' is something more than fact; it is compulsion. If I do actions of a certain kind because, as a matter of historical fact, I have acquired the habit of doing them, it is an unalterable fact that I acquired that habit: but it does not follow that, in the further course of my history, the habit cannot be modified or broken. On the contrary, since the habit is a mere fact, it falls away, like any other fact, into the past, unless it is constantly renewed by fresh action[180].

The example Collingwood then gives is the difference between seeing the habit of smoking as one which may be overcome by future actions and considering it "natural" with the implication that it cannot be overcome and that one will become a "drug addict". "Thus", Collingwood concludes, "if the habitual 'sets' or recurrent patterns of actions in a given man or people are historically produced, the gates of the future are open; if natural, they are shut"[181].

So Collingwood is far from denying the role of certain determining conditions; his point only concerns the kind of influence we should attach to them. It is interesting that he makes the following observation: "I do not deny that the phrase [human nature] may be loosely used as a collective name for those sets or patterns of human activity which we regard, at any given moment, as permanent, and accept as things beyond our power to change"[182]. What he strongly objects to is a definition of human nature as "a substantial and unchanging block of characteristics"[183], which alternatively is called by him "a mythical entity"[184], "a relic of pre-Darwinian biology"[185] or "a relic of ancient metaphysics"[186]. His position is summed up as follows:

> In denying that there is such a thing as human nature, I am thus not denying the reality of what goes by that name: I am denying the implications of the name and asserting that what is so called is the historic past of mankind as conditioning, and in turn modified by, mankind's present activity. Metaphysically, this contention implies that the reality of man is an historical reality, to be resolved without residue into terms of historical process. There is not, first, a substantial and changeless entity called human nature, and then a series of historical activities and changes somehow performed and undergone by this changeless substance: the substance is nothing else than the activity itself, determining itself and developing itself in time[187].

In the last part of his essay Collingwood deals with the concept of process in the world of Nature[188], concentrating on the question whether natural processes can be equated with those of human history. He begins by qualifying Hegel's dictum that Nature has no history, since Darwin has made it clear that the organic world at least is historical. But even with regard to the inorganic world we cannot avoid the conclusion, according to Collingwood, that it undergoes in its temporal changes an historical process. Otherwise life could not have begun at a certain time in the history of the earth.

Another similarity between inorganic matter and organic life is that, according to the modern conception, both are made up of what they do. Even in the case of inorganic matter, its past is not irrelevant to it. Collingwood then expresses the qualification, however, that "its historicity is not complete":

> There seems no reason to believe that an atom of iron, if it came into existence through the loss of electrons by a heavier atom, would carry that past with it in the sense of acting differently from an atom of iron otherwise produced. If it does not in fact behave differently, we must express this by saying that atoms, and other pieces of inorganic matter, do not conserve in their present being the whole of their past: or, from a historian's point of view, that their present being, even to an ideally perfect historian, would not afford evidence of their whole past. This again might be put by saying that whereas in the full blown historicity of mind there is only one possible past which is a *vera causa* of any given present, in the case of matter there are for any given present a number of possible pasts, or in other words that different causes may produce

> the same effect[189].

Modern physics and the philosophical cosmologies of Alexander and Whitehead based on it, which state that just as human "nature" resolves into history, so the "nature" of inorganic matter resolves into process, have resulted, according to Collingwood, in "a metaphysical achievement of the first importance"[190]. This consists roughly, he says, "in the resolution of existing into happening". But, he warns us, "we must not jump to the conclusion that because all is process therefore all is history". "The difference ... between history and a process which is not strictly historical" is indicated by Collingwood by the difference between identifying "what a man is with what he has done (and what has happened to him)" and identifying "what a piece of matter is with what it does (and what happens to it)"[191]. He elaborates this point in the following way:

> It is not possible to resolve what a man is into what he does. To say that his character is only a collective name for the ways in which he behaves is to say too little. His character is not the way in which he acts, it is something which leads him to act in that way; it is a principle of continuity in his actions, a power which he draws from the past, which enables him to act thus in the present...Thus when we say that a man acts thus because he is this kind of man, his being this kind of man (his character) is his having acted in certain ways in the past. History is this gathering-up of the whole past into the present, as determining that novelty which the present, by thus being itself, creates[192].

However, "Nature's process is not mere change", Collingwood asserts: "It is a creative process". He even calls it progressive, like history, in that "the past survives in the process as the foundation of the present"[193]. "But, even here", he observes, "there is a difference between Natural process and historical process":

> The sense in which the past survives is different. Suppose a number of scattered atoms unite to form a molecule, which if you like may be a molecule having chemical properties that never existed before: a new creation in the sense of being not only individually but specifically new. Here the past is preserved in the present only in the sense that the atoms, which existed in the past, still exist, individually and specifically the same, in the present. But their past scatteredness does not still exist. The process has this element of changefulness about it, that the past scatteredness perished altogether. The new molecule, once formal, forgets the past, is no longer possessed of that past as the substance of its present being, and is only what it now does, not what it has done[194].

Unlike matter, however, "mind not only possesses its own specious present, it possesses its own past as well, in the shape of character or substance"[195]. Collingwood thinks it necessary for this reason to make a distinction between natural and historical development. "Hegel is so far right", he concludes. "He is even right in speaking of the *Ohnmacht der Natur*, as a way of describing the fact that Nature differs from Mind in being thus devoid of a certain power: the power to conserve its own past alive in the present"[196].

4.9. NOTES ON THE HISTORY OF HISTORIOGRAPHY AND PHILOSOPHY OF HISTORY (1936)

From 1932 till 1936 Collingwood did not lecture on philosophy of history and his manuscripts show that, besides *An Essay on Philosophical Method*, he worked mainly on cosmology and metaphysics. In 1935 the work on philosophy of history was resumed, with his Inaugural lecture and the essay "Reality as History". This was the prelude to an intensive focus on the subject in the following year. In January 1936 he read a paper "Can Historians be Impartial?"[197] and in Hilary and Trinity Terms 1936 (January-March and April – June) he lectured again on philosophy of history[198]. It is these lectures on which the main part of *The Idea of History* is based. Unfortunately only some scattered pages of them are left.

The notes which are considered in this section contain among other things some annotations made in preparation for the lectures, particularly on Fueter's standard work *Geschichte der neueren Historiographie* (1911), and Meyer's *Zur Theorie und Methodik der Geschichte* (1910), but also on Hegel and Spengler. Beside these there are interesting fragments on human nature, the status of past events, re-thinking, historical importance, selection and certitude. We will deal here only with the first three subjects.

Collingwood's annotations on Fueter's history of historiography demonstrate that he had studied the work carefully. They concern the period of the Enlightment and Romanticism, from Mabillon to Ranke (pp.329-485).

He then comes to the subject of "Human Nature and Human History", to which the date "9 March 1936" is added. This may be considered a continuation of his treatment of the subject in "Reality as History" and a preparation for his lecture on "Human Nature and Human History", which he gave to the British Academy on 20 May 1936 and which is reprinted in *The Idea of History* (IH,205-231). A first draft of this lecture, which is among the manuscripts, was written in March of the same year.

Collingwood again states that "*what is falsely called human nature is really human history*" and adds that "[t]he fundamental theses of such a view would be something of this kind":

> 1) *Human nature is mind.* We are not talking about bodily nature: only of mental (with the proviso that mind always means embodied mind).
> 2) *Mind is pure act.* Mind *is* not anything apart from what it *does.* The so-called powers or faculties (δυνάμεις) of mind are really activities (ἐνέργειαι). Activity does not a) *exhibit* or *reveal* the nature of mind, or b) *develop* or *explicate* its unrealised potentialities: it *is* mind.
> 3) *The pure act posits itself and its own presupposition at once.* The past belongs to the present, not the present to the past. Whereas in nature the present is the caused effect of the past, in mind the past is the analysed content of the present. Thus what the mind is and what it does are its past and present respectively.
> 4) *Past time therefore is the schema of mind's selfknowledge.* It *can* know itself only *sub specie praeteritorum.* To know oneself is simply to know one's past and vice versa. The philosophy or science of the human mind thus = history[199].

Collingwood then remarks that his "present difficulty" is that he "can make good the above theses without difficulty in the case when human nature = rationality", for instance "in the extreme case" where one studies the history of philosophy, as Hegel showed. "But how far down the scale can you go?" he asks. "Hegel thought you could go down as far as objective mind, but not as far down as subjective mind", Collingwood observes and then continues: "If so, subjective mind is nonhistorical (though its *esse* is certainly *fieri*: but not *historicè fieri*). Psychology (in the objective sense) has no history. In this part of his being, man is simply an animal. His instincts, that is, are not historically conditioned. Is this so? I think it probably is. Subjective mind might be said to = unconscious mind, and historicity = consciousness = selfconsciousness. This must be worked out"[200].

Rationality, historicity and selfconsciousness are put on a par by Collingwood, but the problem for him is how far down the scale one can go with them. His solution will be that one must see them as matters of degree. In his published essay "Human Nature and Human History" he says:

> The idea that man, apart from his self-conscious historical life, is different from the rest of creation in being a rational animal is a mere superstition. It is only by fits and starts, in a flickering and dubious manner, that human beings are rational at all. In quality, as well as in amount, their rationality is a matter of degree: some are oftener rational than others, some rational in a more intense way. But a flickering and dubious rationality can certainly not be denied to animals other than men. Their minds may be inferior in range and power to those of the lowest savages, but by the same standards the lowest savages are inferior to civilized men, and those whom we call civilized differ among themselves hardly less.

> There are even among non-human animals the beginnings of historical life: for example, among cats, which do not wash by instinct but are taught by their mothers. Such rudiments of education are something not essentially different from an historic culture.
> Historicity, too, is a matter of degree. The historicity of very primitive societies is not easily distinguishable from the merely instinctive life of societies in which rationality is at vanishing-point. When the occasions on which thinking is done, and the kinds of things about which it is done, become more frequent and more essential to the life of society, the historic inheritance of thought, preserved by historical knowledge of what has been thought before, becomes more considerable, and with its development the development of a specifically rational life begins (IH,227).

In the draft of the same essay one may find this as well, but it contains also another passage which, though not included in the published version, is relevant here:

> ...[T]hought is not a kind of gold-dust, occurring in greater or smaller quantities, or not at all, in the river-silt of sensitive and emotional organic life. It is more like a principle of crystallisation organizing the elements of that life itself. When it is somewhat highly developed, we can recognize it without difficulty. When we trace it further and further back into its more primitive forms, we find it harder and harder to say whether what we are dealing with is thought at all, and not mere instinct. There is no point in its development at which we can say: 'here it comes into existence, visibly itself: below this it is absent'. Perhaps rationality, in some very primitive shape, is as widespread as life itself, at least among animals with brains, or even those with a nervous system but no brain... This at least is clear, that any formula in which we try to define the minimum that we mean by thinking must be altogether arbitrary, and will define only a certain stage in its development. So far as this development is a development of mind, it is already an historical process, and once more we are driven back on the conclusion that the knowledge of mind is historical knowledge[201].

Coming back later in his notes on the subject of human nature and human history Collingwood deals with the rational and irrational aspects of mind and the corresponding sciences of history and psychology. Expressed in a concise way it gives a clear framework of the argument of the published version of "Human Nature and Human History":

> 1) All processes are not historical processes. 2) That alone has historical processes which is rational. 3) That which is irrational has a natural process. 4) To have an historical process = having a

> history = being historical. 5) To have a natural process = having a nature = being natural. 6) That which is historical therefore has no nature, and to speak of its nature is already to falsify it by placing it on the wrong side of this frontier. 7) But here, as elsewhere, there is overlap of classes. *Man* occupies an ambiguous position. He stands with one foot in nature and one in history. 8) I am not here distinguishing body and mind. Man *qua* mere body (matter) is a subject of physics and chemistry and these take no account of man as such: the carbon in the human body is not human carbon but carbon merely. 9) The distinction is within the human mind (i.e. man *qua* mind). It is the distinction between τὸ ἄλογον (brute-mind in man) and τὸ λόγον ἔχον . 10) τὸ ἄλογον includes senses, instincts, impulses and in general the subject-matter of psychology. 11) τὸ λόγον ἔχον includes the intellect, will and their synonyms and implicates: 'intellectual and imaginative faculties' (Hume). 12) The idea of naturalistic psychology is well-founded, but it runs into error if it is identified with a science of mind... 13) The idea of a special science of the human understanding (Locke) or of human nature (Hume) is well-founded, but the right name of this science is history. Locke and Hume went wrong by modelling their new science on natural science[202].

In a last point Collingwood then contends: "The conception of a philosophical science is thus a very dangerous one: by using the phrase, we may be committing ourselves unawares to thinking of such a science as a special kind of naturalistic science, and this will produce strange paradoxes, *either* falsifying the conception of spirit *or* reducing the phrase philosophical science to a case of contradictio in adjecto"[203].

Under the heading "Historical Events as Eternal Objects" Collingwood deals in his notes with the question of the status of the past. In reaction to Whitehead's concept of "eternal objects", which are conceived as embodied, Collingwood contends that in historical knowledge we are concerned with disembodied forms of an individual nature. Giving the example of the Norman Conquest he says:

> The form in question, a structure-pattern called the Norman Conquest, is disembodied because the matter in which it was once embodied has perished, i.e. there is no longer anything organized in that way. It is individual because there never could be anything *else* organized in that way. It is eternal because a) subjectively it remains as a possible object of knowledge b) objectively it remains as a presupposition of the present state of historical affairs – the effects of the Norman Conquest, as we say, are permanent, never wiped out[204].

The relation then, between the "objective" past and our knowledge of it, is clarified by Collingwood as follows:

> The objectivity of historical fact is this: that *there was* such a fact. Historical fact has its objectivity precisely in being past. To be past here means to be past in the historical sense of the word past. The *mere* past is that which merely was; the historical past is that which not only was, but remains historically knowable, which it does only because it remains: remains not in its actuality (as form embodied in matter) but in its ideality (as pure form)[205].

"History is creative", he adds, "in the sense that it brings into being that which, once brought into being, is eternal". The use of the word "eternal" may be in this context surprising and even confusing, but what Collingwood is referring to in a rather unusual way is in fact nothing else than his idea of a living past, which again is related to the common-sense notion of historical tradition[206]. This is made clear, when he says:

> We can know the Norman Conquest because, being its heirs, we have it in our own minds (in our actual political consciousness) as an integral element. Its eternity is therefore nothing but a grandiloquent (and inaccurate) way of stating its survival as an effective force down to the present. On this showing the eternity of historical fact is only the continuity of historical tradition: the continued embodiment of the past in the present[207].

This does not mean, however, Collingwood warns, that "the *esse* of the past is to be historically known. Tradition here does not mean conscious knowledge of the past". It is only when we analyse our present political consciousness, that "we discover it to have been formed by such past experiences"[208].

Collingwood then comes to the concept of history as "the re-enacting of the past in the present". "In so far as all history is the history of thought", he says, "this must be so: for one can only apprehend a thought by thinking it, and apprehend a past thought by re-thinking it"[209]. Since his Die-manuscript of 1928 Collingwood had not dealt explicitly with this conception and this is possibly the reason that he makes the remark that "the formula needs a good deal of clearing up"[210]. Immediately following, he then offers a clarification. The way this is done is extremely important, because it is to be found nowhere else in his writings, either published or unpublished. For this reason his argument will be fully quoted:

> It may be said: the word thought is equivocal. It may mean το νοεῖν or το νοούμενον (νόησις)(νόημα)[i.e. the act of thinking or the object (content) of thought]. Now, you can re-think a νόημα ; for two acts of thought may have the same object. But you can't

> re-enact a νόησις, for the new νόησις is different from the old. If history means thinking thoughts (νοήματα) that have been thought already, it is only a name for any thinking that isn't a discovery, isn't original ... Obviously we must qualify. History means not re-thinking what has been thought before, but thinking of yourself as re-thinking it. The learner of mathematics is an historian of mathematics also just so far as he thinks: I am here thinking what Pythagoras thought before me.
>
> But whereas the thought (νόημα) in *this* case is a *pure* νόημα, in the case of the Norman Conquest it is a νόησις-νόημα. I mean: Pythagoras was thinking about the triangle: William the Conqueror was thinking about a political situation, and this situation was not merely *apprehended* but *constituted* by certain acts of thinking. For Pythagoras, knowing how other people thought about the triangle was no necessary part of thinking about the triangle; for William, knowing what other people thought about the political situation was not only an integral part of thinking about the situation, it was *the essential* part of it. There *was* no political situation distinct from what people thought about it.
>
> Thus there are not *two* but *four* senses of thought: a) the act (νόησις) b) the object (νόημα) c) a peculiar kind of act whose object is an act (νοήσεως νόησις)[i.e. an act of thinking about an act of thinking] d) a peculiar kind of object which is itself an act (νόησις νοούμενη) [i.e. an act of thinking being thought]. What we are dealing with in history is this third sense. History is νοήσεως νόησις [21].

This passage is important not only because Collingwood makes a distinction between different senses of the concept of thought, but especially because he clarifies the context in which the conception of re-thinking past thoughts must be seen. The distinction he makes here between re-thinking Pythagoras' thought and the thought of William the Conqueror is not made in "History as Re-enactment of Past Experience" in *The Idea of History* (IH,282-302). It is illuminating for a better understanding of the special characteristics of the concept of re-thinking past thoughts in the case of "normal" history, as the example of William the Conqueror apparently is meant to illustrate. But what about re-thinking the thoughts of Pythagoras in the context of a history of mathematics? one might ask. Collingwood himself also speaks of an historian of mathematics, so apparently not all history is νοήσεως νόησις.

What Collingwood has not stated clearly either is that history, as a peculiar kind of act whose object is an act (c) (νοήσεως νόησις), cannot be separated from a peculiar kind of object which is itself an act (d) (νόησις νοουμένη). For not only is the act of thinking by the historian peculiar, but also his object, because this is an act of thinking as well. So, completely formulated it would run: History is a peculiar kind of thinking whose object is of a peculiar kind, because it is itself an act of thinking (νοήσεως νοουμένης νόησις).

After the passage quoted Collingwood continues:

> Why do I say that c) is a peculiar kind of act? Someone will say that it differs in no way from a) : it is just thought over again, its peculiarity being only in its object.
> I say this because it is the peculiar character of this act that it does not *contemplate* its object. It is not θεωρία . It alters the situation which it apprehends. The mind of William is a factor in the situation and it is this because William understands the other minds which are also factors. The thought of Harold, his view of the situation, is for Harold not a *factor* in the situation but the *whole* situation. When William knows what Harold is thinking, that thought (νόησις) is for William a *factor* and no more. It is an *object* to him: one of the objects about which he thinks: at the same time it is a *factorial way in which he thinks,* for he 'enters into' Harold's thought, thinks as Harold thinks, but thinks other things as well. Thus νοήσεως νόησις is not merely a peculiar *case* of νόησις , it is a peculiar *kind* of νόησις , where not the object merely but the act's relation to its object (and therefore the character of the act) is peculiar: for the act in this special case absorbs the object into itself, makes it a factor in itself. But conversely, *any* νοήσεως νόησις goes beyond its object. The *mere* re-thinking of a νόησις is the transcending of that νόησις and the reduction of it to the status of a factor. This is a crucial point in all theory of history. It has e.g. consequences in the history of philosophy, where the philosophical understanding of someone else's philosophy is already the transcendence of that philosophy and its reduction to the status of a past or absorbed moment in one's own thinking[212].

The confusing aspect of this passage is that Collingwood gives as an example of historical thinking the thoughts of William the Conqueror, while we would expect these thoughts to be used as example of an object of the thought of an historian. On the other hand, it makes clear that the conception of re-thinking past thoughts must not be taken too literally in that re-thinking may also refer to present thoughts: as a matter of fact it is concerned with the question of our knowledge of "other minds". So what the historian in fact must do in a case like William the Conqueror is to re-think the thoughts the historical agent re-thinks. But, to make it even more complicated, while the thoughts the historical agent re-thinks are but a factor in his total thought – going in his thought beyond its object and transcending it – for the historian in his turn the total thought of the historical agent which he re-thinks is for *him* but a factor in *his* total thought, transcending it, or – as Collingwood puts it in his *Autobiography* – being "incapsulated" in his own thought (Aut,114).

It is also confusing that in saying that this principle "is a crucial point in all theory of history" Collingwood gives the history of philosophy as example. For understanding the thought of a philosopher is more like understanding the thought of Pythagoras than that of William the Conqueror, since for a philosopher it is no necessary part of his thinking about a subject to know how other people think about it, as is the case with William the Conqueror's thinking about the situation. So the thought of a philosopher should also be interpreted as a pure νόημα, as that of Pythagoras. But in his re-thinking of the thoughts of a philosopher or mathematician the historian also transcends them and reduces them "to the status of a past or absorbed moment in [his] own thinking". The difference, then, between a case like understanding the thoughts of William the Conqueror and the cases of a philosopher or mathematician would be that in the former there is a transcendence at two levels (by William the Conqueror and the historian), while in the latter there is only a transcendence of thought by the historian.

4.10. NOTES ON HISTORIOGRAPHY (1938-'39)

On his trip to the East Indies from October 1938 to March 1939 Collingwood wrote some notes under the heading "Notes on Historiography". They are more concerned, however, with various topics in philosophy of history than with historiography. Apparently they were meant as preparation for *The Principles of History* and the notes include a scheme for this book[213]. At the end of the notes we find under the heading "Introduction to Book I" a passage which corresponds with the beginning of the chapter on "Historical Evidence" in *The Idea of History* (IH,249-250), though not always exactly in the same words. It begins: "'History' said Bury, 'is a science, no less and no more'. Let us grant that it is no less", and ends: "If he [the historian] is studying the Hundred Years' War or the Revolution of 1688, this is not for him a preliminary stage in an inquiry whose ultimate object is to ..."[214]. The sentence breaks off here with the remark "Copied out and continued elsewhere, 10-2-'39". This continuation was realized in his actual writing of *The Principles of History*, because on the same day we find in his diary: "Began writing *Principles of History*"[215].

We have seen that the manuscript of *The Principles of History* was used by Knox – though not completely – in his edition of *The Idea of History*, but that the manuscript itself has been lost[216]. However, Knox included some quotations in his preface to *The Idea of History* which were obviously taken from the non-edited parts of *The Principles of History*. These quotations were used by him to demonstrate that Collingwood became in his later years a complete sceptic and historicist. For instance, Collingwood is quoted as saying "philosophy as a separate discipline is liquidated by being converted into history" (IH,X) and that "history is the only kind of knowledge" (IH,XII).

We will not deal here with all the topics treated by Collingwood in his notes – they are mostly short and isolated observations – but only some passages which show relativistic and historicist tendencies. Because Knox had not seen the notes[217] it is interesting to compare them with the quotations given in the preface of *The Idea of History* and to see them in a broader context.

Under the head "That History is the only kind of knowledge" Collingwood says:

> I have already shown that metaphysics is what I have called an historical science, i.e. that the problems of metaphysics are without exception historical problems. It is easy to show ... that this is true of every so-called philosophical science. Thus philosophy as a separate discipline is liquidated by being converted into history. To deny this proposition is legitimate only so long as the person who denies it has a false view of history, which he feels it necessary to amend by *adding* to it what he has left out (and *correcting* the errors he has put in) by calling a conception of *Philosophy* into existence to redress the balance of his conception of History[218].

On logic he argues the following:

> *Logic an historical science.* The aim of logic is to expound the principles of valid thought. It is idly fancied that validity in thought is at all times one and the same, no matter how people are at various times actually in the habit of thinking; and that in consequence the truths which it is logic's business to discover are eternal truths. But all that any logician has ever done, or tried to do, is to expound the principles of what in his own day passed for valid thought among those whom he regarded as reputable thinkers. This enterprise is strictly historical. It is a study in what is called contemporary history = history of the recent past in a society which the historian regards as his own society. (Thus people like Mill, Jevons etc. in their 'inductive logics' are giving as good an account as they can of the conditions under which 'modern scientists' regard a piece of 'scientific thinking' as valid, or consider a 'scientific theory' well-founded). Often this kind of historical study is accompanied by a good deal of confusion or error as to whether the principles of 'contemporary' thought have been recognized at other times also, and if not, what follows. Actually, the kind of thinking which was investigated by the 19th century inductive logicians hardly existed before about the 16th century. Logic as 'theory of scientific method' is in effect, at any given time, a fragment of a history of scientific method[219].

Finally Collingwood maintains on ethics:

> *Ethics as an historical science.* a) Ethics as an account of the principles of action depends for its content on the structure of the

moral world of which it tries to give an account. Thus ancient Greek ideas of conduct are different from Christian ideas and consequently Aristotle's ethics (say) differ widely from any XVII – XX ethics, without this implying error on either side. Any ethical theory is an attempt to state what kind of a life is considered worth aiming at, and the question always arises – by whom? b) There are departmental ethical sciences like politics, economics. These, at any given time and place, describe the political and economic principles accepted at that time and place. For economics, this has been seen by the Marxists, and it has been admitted by J.M. Keynes, with the odd result that he has tried to construct a 'general' economic theory, stating the supposedly permanent general principles of which any 'special' economic theory, like Adam Smith's, is a special case. This of course is illusory. c) Even the distinction between logic and ethics is an historical one and no more. As we inherit it from the Greeks, it certainly has no permanent validity: the Indians or the Chinese do not make the distinction between thought and conduct in any such way as that which we presuppose when we make it[220].

4.11. FOLKLORE (1936-'37)

Collingwood made elaborate studies in the field of folklore and anthropology, concentrating on fairy-tales and their background. His notes and writings on these subjects date most probably from 1936-'37[221] and there is no evidence from the manuscripts that he had worked on them before. The chapter on "Art as Magic" in *The Principles of Art*[222] is based on these studies, but presents only one element of them and it does not show the context in which they were developed. Collingwood's interest in the subject of folklore has not previously been known, nor has the fact that he was well-read in anthropological literature. The manuscripts are also of great interest for other reasons. In the first place, they show how Collingwood "put into practice" his ideas about history in the field of the social sciences, demonstrating how the methods used in these sciences had prevented the understanding of so-called primitive societies. Interesting as well is the criticism he develops in this connection – in the context of his treatment of the concept of magic – of some fundamental characteristics of European civilization.

From the point of view of history, however, what is perhaps most important is the reason Collingwood gives for his involvement in the subject of fairy-tales. At the very start he says about it:

> Everything men have made can be used as evidence for their history; but in order so to use it, we must find out how to interpret it. For a long time, it seemed impossible to use anything effectively as historical evidence, except written documents

> attesting the occurrence of certain events. During the nineteenth century, archaeologists learnt to use very ancient implements as evidence for periods of history which have left no written memorials. The result has been a vast enlargement of historical knowledge. Similar enlargements will perhaps be made in the future. The purpose of this book is to consider the possibility of one such enlargement, by suggesting how fairy-tales may be used as historical evidence[223].

"What I am here proposing", Collingwood says in another place, "is in effect a new kind of archaeology. Like all archaeology, it aims at reconstructing the past of mankind from the evidence of things he has made which are still with us. But whereas archaeology in the ordinary sense studies the fragments of his industry and manufacture, this new kind studies fragments of his customs and beliefs handed down in traditional stories"[224]. Or as he also puts it: "A theme contained in such a story is a fragment of ancient custom or belief, very much as a stone implement is a fragment of ancient technical skill. In each case the historian uses the fragment by reconstructing in his mind the life and thought of the people who have left him this sample of their work"[225].

At the end of the last passage Collingwood arrives at the question how fairy-tales should be interpreted. In order to use fairy-tales at all as historical evidence, Collingwood elsewhere observes, it is necessary to accept the principle "that the themes found in fairy-tales are organically connected with the customs and beliefs of the people who originated them", but also "that customs and beliefs are things which have a history". If one starts with the psychological conception of fairy-tales derived from "the unconscious" as a common characteristic of human mind, Collingwood warns, "they clearly cannot be used as historical evidence"[226].

For his "new enterprise"[227] of using fairy-tales as historical evidence the first thing to do is to establish its proper methods. As regards its form a fairy-tale is a traditional story, Collingwood asserts, and its content is essentially magical[228]. Following his own principle of recapitulating the "history of history" Collingwood begins by describing how subjects such as mythology, fairy-tales and magic were studied in the past. He distinguishes three methods of approach: the philological (1810-'70, Grimm and Max Müller), the functional (1870-1910, Tylor and Frazer) and the psychological (1900-'20, Freud and Jung). It is Collingwood's contention that all three approaches are fundamentally defective for the reason that they are in essence naturalistic: "Each of them treats its subject-matter as something to be contemplated from without, something external to the thinker, something that is not himself but something else"[229]. A "primitive mind" as the essential characteristic of the "savage" is constructed as an outward phenomenon, completely separated from the "civilized" mind that is studying it. It is classified, according to naturalistic methods, as mythopoeic insanity (Müller), folly (Frazer), or neurosis (Freud)[230].

Not all past studies, however, are given a similar assessment by Collingwood. Tylor's *Primitive Culture* (1871) for instance he values for stating that at the stage of "savagery" a culture already displays rational characteristics and not only folly or depravity. Collingwood even calls this "a kind of Copernican revolution ... in the anthropologist's conception of his own relation to his materials"[231]. Though he is critical of Tylor's comparative approach and his view of magic as "the technique of animism", Collingwood values the argument in Tylor's theory of mythical "survivals" in modern civilization coming from an earlier cultural phase – the argument that myths were to be seen in close relation with the customs of the people. This led to the functionalist school, with Frazer's *The Golden Bough* (1890) as its high-water mark. According to Collingwood, "[t]his method of approach has proved extraordinarily fruitful. It has been found possible to trace a close connexion, in a large number of cases, between the rituals and the fairy-tales of this or that people"[232]. What he criticizes Frazer for, however, is the spirit in which he approaches his subject. For in the preface to his work it is described "as a record of the long tragedy of human folly and suffering". "Such words show", Collingwood comments, "that he approaches his subject-matter as a thing external to himself and the civilization which he feels as his own: without any attempt to work himself into the spirit of it and to re-create in his own mind the experiences whose outward expression he is studying"[233].

Freud's *Totem and Taboo* (1913) is in fact criticized for the same reason, since his experience of neurotics was projected by him into the life of savages ("'The Savage' – who is he?" Collingwood writes in his notes on *Totem and Taboo*[234]). "Does Freud mean us to believe", Collingwood asks, "that he can psycho-analyse the savage on the strength of a general impression gained from reading books about him? ... The real Freud, the endlessly resourceful psychologist of the consulting-room, whom we all revere as a man and a scientist, is hardly recognizable in this arm-chair student of savages in the abstract"[235]. Finally Jung's *The Psychology of the Unconsciousness* (1916) is qualified by Collingwood as "a pseudo-history in which the writer's desires and fears are projected on the blank screen of a past which to him is absolutely unknown"[236]. Jung's interpretation of myths as dreams of a people's childhood is a particular target for Collingwood's criticism.

All the theories treated by Collingwood suffer, according to him, from a naturalistic approach by separating the subject from the object they study. This is justified when the object is not man, as in physics. "But in historical knowledge", Collingwood contends, "where the object is the subject's own past, this opposition has a peculiar emotional significance"[237], and he continues:

> The savage is not outside us; he is inside us. Conceiving ourselves as rational and civilized people, which is what we want to be, we

> are aware within ourselves of savage and irrational elements, parts of ourselves which we would willingly disown. Hating these things in ourselves, and hating ourselves for harbouring them, we throw them in fancy away from us into the picture we form of other people. Thus we create the mythical figure of the savage, no actual historical person but an allegorical symbol of everything which we fear and dislike, attributing to him all the desires in ourselves which we condemn as beastly and all the thoughts which we despise as irrational. This abstract idea of the 'savage' of 'Primitive Man', however, is not merely an object of our hatred and contempt. It has also a kind of horrid fascination. Because these beastly desires and irrational thoughts are actually present in ourselves, we wish to indulge them, though our ideal of ourselves as reasonable and civilized men will not allow us to do so; our idea of the savage is therefore a wish-fulfilment fantasy in which these repressed wishes are given licence for imaginary gratification[238].

For this reason Collingwood considers the fundamental difficulty of anthropological study to be not so much intellectual as emotional: "If we could contemplate the life and mind of the savage without horror, we should not find them hard to understand"[239].

As part of the study of man there is only one right approach for anthropology, according to Collingwood, and that is the historical one, which does not separate the subject from its object:

> The reason why anthropology is an important study for civilized men is not, as might have been thought in the heyday of imperialism, because civilized men have to rule over savages and must learn, therefore, to understand them. It is because the civilized man contains a savage within him, in the special sense in which any historical present contains within itself its own past, and must therefore study this savage – not savages in the abstract, but the savage that he himself in this sense is – for the same reason for which all history is studied, namely to make possible a rational human life in the present day. The problem of anthropology is a special case of the problem of self-knowledge; and history is the only way in which man can know himself[240].

Collingwood emphasizes that anthropology is an historical science, "where by calling it historical as opposed to naturalistic I mean that its true method is thus to get inside its object or re-create its object inside itself". What the anthropologist studies is man; "and to man his fellow-man is never a mere external object, something to be observed and described, but something to be sympathised with, to be studied by penetrating into his thoughts and re-enacting those thoughts for oneself"[241].

The "historical method" Collingwood wants to see based on two rules, which he calls the maxim of Spinoza and the maxim of Butler.

The maxim of Spinoza is "neither to condemn nor to deride the feelings and actions of men, but to understand them"[242]. For this reason Collingwood would like to get rid of terms like "savages", "primitives", "natives" and even "magic" and "taboo", because of their prejudicial connotations, or to use them only as descriptive terms. The maxim of Butler is "that every thing is what it is, and not another thing"[243]. This maxim recognizes facts as facts and not as instances of generalizations to be arrived at in a naturalistic way. The methods of approach dealt with by Collingwood offend against both maxims. With Spinoza's maxim this is obvious: in Collingwood's eyes the so-called savages are derided in a fundamental way. This, however, makes historical knowledge impossible, because " ... history demands, or rather brings about, a peculiar intimacy in the relation between knower and known. The historian can only understand a thought by thinking it over again for himself. If there is any type of thinking which for any reason he is unable to do for himself, he cannot thus re-think it and cannot understand it historically ... He has therefore to stop pretending that the subject-matter of his study is the life of societies inferior to his own; he has to face the fact that the distinction between savage and civilized man is a fiction designed to flatter his vanity"[244]. Similarly they offend against Butler's maxim in that they generalize instead of recognizing facts as facts, showing a tendency " ... to slur over details, emphasize similarities at the expense of diversities, and thus reduce to a spurious uniformity things which may well be essentially different". This is all against history, which "does not work in terms of classes, types, and generalities. Its business is with the individual. Hence, when naturalistic methods are applied to an historical subject-matter, the necessary ignoring of detail becomes a neglect of essentials"[245].

In an analysis of magic Collingwood offers an "application" of the historical method. Magic, he contends, must be seen as an expression of a certain complex of emotions arising out of a certain type of situation[246]. It is not a pseudo-science as for instance interpreted by Frazer. In order to understand magic, then, we must look to magical practices in our own civilization:"... since we can understand what goes on in the savage's mind only in so far as we can experience the same thing in our own, we must find our clue in emotions to whose reality we can testify in our own persons"[247]. In order to understand magic among primitives it is therefore first necessary to look for and understand certain emotions among ourselves, on which magical practices may be based. An example Collingwood gives is the emotional feeling we have towards things we have made or which have acquired a similar status by use. If they are lost or damaged in one way or another we feel hurt. Another person can also express his hatred against me by damaging such a thing on purpose in order to express his emotions or to hurt me accordingly.

In this connection Collingwood remarks: "I have heard a philosopher confess a desire to dance upon a book whose doctrines he disapproved of"[248]. It becomes understandable in this way that in certain societies one is careful not to leave one's hair or nail-cuttings to be acted upon by possible enemies.

Our civilization, however, has suppressed magic by "the deliberate cultivation of a thick-skinned or insensitive attitude towards emotion itself[249]:

> After a long and hideous experiment in suppressing it by force, by burning witches, we came to see that burning witches meant believing in them, and that their victims' belief in them, what I have called emotional vulnerability, was the source of their power. So we changed our own attitude towards them: replaced persecution by ridicule, and gradually developed a whole system of education and social life based on the principle that magic was not a crime but a folly, whose success depended on a like folly in its victims. This hard-headed or thick-skinned or rationalistic attitude towards life, which our civilization invented in the seventeenth century, worked out in the eighteenth, and applied to all aspects of human affairs in the nineteenth, is the dominant factor in modern civilization. The best single-word name for it is utilitarianism. Our civilization prides itself on being sensible, rational, businesslike; and all these are names for the same characteristic, namely the habit of justifying every act, every custom, every institution by showing its utility[250].

"To live within the scheme of modern European-American civilization", Collingwood comments, "involves doing a certain violence to one's emotional nature, treating emotion as a thing that must be repressed, a hostile force within us whose outbreaks are feared as destructive of civilized life"[251]. Our horror of savages Collingwood interprets therefore as in fact a horror of something that is repressed in us, but symbolized by the savage, namely emotion. "[F]or magic", he argues, "which sums up all that we dislike in savage life, is beginning to reveal itself as the systematic and organized expression of emotion"[252]. Because of their "utilitarian obsession " anthropologists also develop a false view on magic, regarding it as a pseudo-science instead of an emotional expression.

Magical practices still have, however, in some sense a utilitarian aspect, Collingwood contends, "in the fact that they resolve emotional conflicts in the agent and so readjust him to the practical life for which these conflicts render him unfitted"[253]. There arises then, Collingwood argues, "a new problem... about our own civilization. We pride ourselves on always acting from utilitarian motives on scientific theories; but that very pride should warn us that this belief about ourselves may perhaps be unjustified. We may be conceiving our own civilization not as it actually is, but as, with our utilitarian obsession, we should like it to be. We think that our rationalism has done away with magic because that is what we want to think; but is it true?"[254].

This question is important for two reasons, according to Collingwood: "... we should not live under illusions about the nature of our own civilization" and our rationalistic conception of man prevents us from correctly interpreting the savage's life[255]. For this reason Collingwood gives a description of certain magical practices among ourselves, such as washing hands and wearing certain clothes. The most interesting part, however, is the one where he comes to a critical analysis of certain magical aspects of our modern technical culture. "Tools protect us against the recalcitrance and hostility of inanimate things", Collingwood contends, "in the same kind of way in which clothes protect us against the hostility and criticism of our fellow-men. The housewife values her Hoover not only for utilitarian reasons, but also because it gives her this feeling of protection; like a familiar spirit, it saves her from exposing her personal forces in the struggle against dirt"[256]. It is an experience he compares with "the glory of notching the table-edge with your first knife or touching eighty in your new car". Collingwood then continues:

> In the second case, it is the motive of many things in modern life: the habit of turning on the wireless, whatever kind of noise comes out of it; the delight in tearing over the face of the land in fast cars, no matter where you get to; the desire to scrap one's old machines and instal bigger and better ones, irrespective of whether one can afford the cost or sell the output ... But since with our utilitarian obsession, we cannot rationalize this delight except by calling it a desire to increase production ... the glory of machinery for its own sake is apt to become a demon, darkly working in the unconscious recesses of our minds and breeding overproduction and bankruptcy and slaughter on the highway and the battlefield. It has these ill effects because we will not face it. We repress it as something that will not fit into our utilitarian picture of ourselves. So we drive it into the unconscious and it becomes a thing of evil[257].

The "sheer sense of glory" with the kindred impulse to conceal it behind "a mask of sham utilitarianism" has never been so strong as it is today, Collingwood contends: "If this is the case, the reason may be that we lack that self-confidence which is fostered by doing things for ourselves, but is to some extent weakened by every new tool that we invent. As we come to rely more and more on our machines, the glory of using them becomes a drug whose magic conceals from us the fact that behind this façade of machinery, we ourselves are sorry examples of the human kind"[258]. The same experience one may get from the power over human beings. Since men are harder to control than machines, Collingwood maintains, " ... the feebler souls among ourselves forget their self-dissatisfaction in the cult of machinery, the stronger do it by becoming kings of business, political bosses, or dictators". "But here again", he says, "as in the desire for clothes and tools, the impulse in itself is universal and healthy. It is only when it is disowned by a world of obsessed utilitarians that it becomes a madness"[259].

So, according to Collingwood, our utilitarian and rationalistic civilization is in fact charged with many irrational aspects, particularly in the way we deal with our tools and fellow-men. Complementary to this assessment, however, Collingwood emphasizes that the magical attitude of the so-called primitive "does not imply failure to understand the mechanical principles on which the tool works"[260]. Or, as he puts it elsewhere: "What we call primitive man does not lack science. He understands enough of mechanics, of chemistry, of biology, to plough and fish and throw spears, to light fires and bake pots, and to sow seed and breed animals"[261]. In developing the view that our civilization is full of irrational elements – though not recognized – and primitive cultures are full of rational elements – though in their turn not recognized – Collingwood provides a common basis for human culture. We have seen that he considers this a necessary precondition for our understanding other people and that, in his view, it is precisely the lack of this which has prevented our understanding of them in the past.

Collingwood's shrewd observations with regard to the way in which in our civilization emotions are surreptitiously experienced have not lost their topicality. On the contrary, the subsequent technological development and the rise of the so-called consumer society with the kindred mentality have demonstrated their rightness in a way he could not possibly have dreamt of.

In his analysis of "primitive" and "civilized" mentalities Collingwood not only provides the theoretical basis for understanding the former, but by confronting them he uses at the same time the theme – well-known from the eighteenth century – of providing with "primitive" man a critical mirror for the "civilized" version. In the first instance the exact nature of this mirror is not altogether clear. The following passage, however, in which Collingwood describes the emotional relation of the savage to his tools, is illuminating: "The better a savage understands his plough, the more keenly he feels the joy of using it. The ritual which we find savages performing in connexion with their use of all tools ... is their way of expressing this feeling of glory. Like all joy, it demands expression: and since he does not suffer from the utility-complex, the savage sees no reason why he should not express it, although the expression has no utilitarian value. He dances and sings his joy. As he does not worship machinery for its own sake, but because he needs it, the joy does not wear off. It renews itself with every using of the tool"[262]. In contrast, the relation of "civilized" man to his tools would read in the last two sentences as follows: "As he does worship machinery for its own sake, not because he needs it, the (surreptitious) joy does wear off. It may renew itself only with another tool". The difference, then, between both attitudes would amount to the fact that the "savage" openly rejoices in the expression of his emotions, while "civilized" man does it secretly, projecting his emotions into impersonal machinery, with the corollary that the former consolidates his personality and the latter loses it. Collingwood's sympathy for the former attitude and his plea that we should pay more attention to it may be seen as being in accordance with his philosophical position of always stressing the interpretative activity of man as against the alleged "objective" conditions. The following passage not only makes this contention clear, but also points to what Collingwood considers the consequence of not realizing its truth:

> ... [A]ll these ways of working with tools are specialized ways of working without them. Man digs with a plough or a hoe or stick because he has first dug with his bare hands. He hunts with weapons because he once hunted unarmed. He judges the sowing time by the stars because he once judged it by guesswork. And however far he goes on the road of mechanization, he must still have a strong enough faith in his immediate activity to assure him that he can handle the tools he has invented and make them do what he wants... Behind all his array of scientific tools, man is still the same naked animal. If he needed valour and wisdom to face nature unarmed, he needs them all the more if he is to use these tools without destroying himself.
> Civilized man, when he takes it upon himself to despise the savage, forgets this. He mistakes the superiority of his tools for a superiority in himself. He forgets that unless his immediate action, his power to make his tools do what he wants them, has advanced in proportion to the advance of his tools themselves, he is sacrificing all the ends of life in his elaboration of means to procure them. Even now, he seems to be cowering in the midst of his machinery, helplessly looking on while it works of itself, like the sorcerer's apprentice who raised the Devil[263].

Though starting with fairy-tales, Collingwood did not in fact get very far with his study of this subject. He thought it necessary first to clear the ground methodologically, showing that the traditional "naturalistic" approach of anthropological studies is a blind alley and that the "historical" approach is a *sine qua non* of *all* studies of man. His demonstration of the irrational aspects of our civilization and his pointing to the rational aspects of "primitive" culture must be seen as preliminary in the same way. His analysis is really a prolegomenon to anthropology. As such it is in essence a transcendental analysis of the subject, though empirical studies are included in it; and when he speaks in this connection of a methodology this must be interpreted as "pure" methodology, as explained by him in the lectures of 1926 and 1928.

4.12. METAPHYSICS AND COSMOLOGY (1933-'34)

In a "prefatory note" to *The Idea of Nature*[264] Knox says:

> When his *Essay on Philosophical Method* was passing through the press in 1933 Collingwood remarked to a friend that, having propounded a theory of philosophical method, he was now proceeding to apply it to a problem which had never been solved, namely, to the Philosophy of Nature. From August 1933 to September 1934 he was working intensively at this subject, studying the history of both natural science and cosmological speculation, and elaborating a cosmology of his own.

Knox revealed at a later date that he himself was the friend to whom Collingwood had made his remark, saying: "After Collingwood had published his essay on *Method*, he said to me that, having issued a Method, he was now applying it to 'a problem that has never been solved', namely the philosophy of nature, and he was writing his answer in a series of 'Woolworth notebooks' which I saw, but which were destroyed later"[265]. Fortunately Knox's belief proved to be wrong, because the manuscript – in the form of five red-covered notebooks – has survived. Collingwood began working on it in August 1933 and the last notes are dated 8 May 1934. In Michaelmas Term 1934 and 1935, Hilary Term 1937 and Michaelmas Term 1939 Collingwood lectured on "Nature and Mind" and in Hilary Term 1940 on "The Idea of Nature in Modern Science"[266]. Knox mentions that Collingwood drastically revised the lectures in September 1939 and made a beginning on rewriting them in book form for publication[267]. This is confirmed by the fact that the manuscript of 1933-'34 indeed has a completely different framework from *The Idea of Nature*.

When Collingwood began writing on the subject he did not have a clear idea what the result would be and it does not seem that he ever intended to lecture on it. For he begins as follows:

> If these notebooks fall into the hands of anyone but myself, let him take notice that they are intended as receptacles for purely experimental, disjointed and desultory notes and attempts to 'think on paper' at problems whose solutions are not in my mind when I begin to write on them. Much of what I have written here is mere groping down blind alleys. At best, these notebooks can be regarded as quarries containing, together with much rubbish, materials out of which I hope to build a metaphysical treatise[268].

On the cover of the notebooks, however, is written "Περι Φυσεως" and the subject he discusses is called by Collingwood "cosmology", while the first page of the first two notebooks is again headed "Notes towards a Metaphysic". In the beginning of his first notebook Collingwood added at a later date the remark that "[f]or reasons which I partly stated in my Balliol lecture on Bradley (Jan. 1934) I regard cosmology as the main subject at present demanding attention from serious philosophers"[269]. The lecture Collingwood refers to was given by him on 17 January 1934 as the second of two lectures entitled "The Nature of Metaphysical Study", which are among the manuscripts. Since his "Notes towards a Metaphysic" were conceived by Collingwood as the possible starting-point for a metaphysical treatise it is interesting first to have a look at his conception of metaphysics, as elaborated in the lectures of January 1934, especially as they were given while he was working on his notes on the subject.

It is worth mentioning that the lectures of 1934 are also of great interest with regard to the development of Collingwood's views on metaphysics. In his *Autobiography* and *An Essay on Metaphysics* Collingwood developed the theory that metaphysics is a purely historical science, laying down the absolute presuppositions of a certain age, group of persons or even individual[270] (he concentrated in fact on the first). In his *Autobiography* it was implied that he had already supported this view for a long time. In Knox's opinion, however, "[t]his is hardly credible". "I have documentary evidence", he says, "that in 1936 he still believed in the possibility of metaphysics as a separate study, distinct altogether from history, a study of 'the One, the True, and the Good'". And he concludes from this that "... his philosophical standpoint radically changed between 1936 and 1938 ..." (IH,X-XI).

Collingwood's lectures of 1934 are now interesting not only because they show an anticipation of his later view of metaphysics as an historical science, but also because they demonstrate a connection between this view and that of metaphysics as a separate study. In his first lecture, given on 15 January 1934, Collingwood develops the latter view. Metaphysics is called by him there "the keep or central stronghold of the castle of philosophy, or the culminating peak of that mountain"[271]. It "is not concerned with any special department of being", he contends, "but with being as such, being in general, or pure being"[272]. The idea of metaphysical inquiry is seen by him as "an inquiry always concerned with the same fundamental problems: being, nothing, becoming, and of course others necessarily arising out of these and their interrelation"[273]. He then makes a crucial move in his argument, saying:

> Now, this by itself would suggest that the metaphysician always begins again at scratch, having learnt nothing and forgotten nothing since his last incarnation; that the problem is always the same, never advanced or altered in any way by the labour of those who have gone before; that, in a word, there is no such thing as a present situation with which the metaphysician is called upon in a special sense to deal.
>
> This implication would not at all express what I take to be the truth. Every science goes through a process of historical development in which, although the fundamental or general problem remains unaltered, the particular form in which this problem presents itself changes from time to time; and the general problem never arises in its pure or abstract form, but always in the particular or concrete form determined by the present state of

> knowledge or, in other words, by the development of thought hitherto. This is a universal rule governing all forms of human thought; there can be no reason why metaphysics should be an exception. And indeed the attempt which I have made to formulate the fundamental problems of metaphysics is an attempt which could have been made, in exactly that way, only at the present stage in the history of the world[274].

In the second lecture Collingwood deals with the question "What are the problems with which metaphysics has specially to concern itself in European countries towards the middle of the twentieth century?"[275], making the proviso, however: "I shall only indicate certain features of them"[276]. He then begins with a discussion of the rise of natural science in the seventeenth century. It was based, he says, on two principles, namely, "that nature works according to fixed and definite laws" and "that things in nature are really measurable and that whatever is not measurable is not real"[277]. Collingwood makes the following comment:

> These two principles are the assumptions on which 17th century science rested, and if that science was to be regarded as real knowledge of the real world these two assumptions must be true. But obviously physical science could not prove their truth; it could only begin to use its own methods when they had been assumed. Their truth was a matter for investigation by metaphysics. Consequently seventeenth-century metaphysics, from Descartes to Locke, took this as one of its main tasks, to prove the truth of these two assumptions[278].

According to Collingwood physics and metaphysics "were engaged on a co-operative task" and they "fitted each other as a glove fits the hand": "The physicist selects certain aspects of nature for study and ignores others; the metaphysician shows that the ones he selects are real, the ones he rejects only apparent ... There is a perfectly happy and harmonious symbiosis of physics and metaphysics: the physicist is assured by the metaphysician that the world really is what in his work he assumes it to be, the metaphysician is assured by the physicist that his *a priori* theories are vindicated by every appeal to the facts"[279]. With Berkeley this co-operation was broken, Collingwood contends, because it was stated by him that all reality was but appearance. This idea was further developed by Hume and Kant. Reality was accordingly sought in the mind instead of in nature. In this way a gulf was created between metaphysics concentrating on mind, and natural science following the lines laid down in the seventeenth century.

Coming to the present time, Collingwood then takes Bradley's *Appearance and Reality* (1893) as his starting point. Bradley has always been misunderstood, he maintains. For he is usually interpreted as an idealist, stating that everything is mere appearance. According to

Collingwood this is not correct, since it was Bradley's opinion, he contends, that there are no mere appearances and that "[e] very appearance belongs to and qualifies reality ... [A] ppearances are not something other than reality; they are reality itself, appearing"[280]. He therefore reaches the paradoxical claim that Bradley is "the father of modern Realism"[281]. The revolt of people like Cook Wilson or Moore against Bradley, Collingwood argues, is in fact not directed against Bradley's ideas, but against the phenomenalist philosophy he himself also criticized. "This, then, is the position in which we stand in metaphysics", Collingwood concludes this part of his lecture, adding to it: "After three centuries of attempting in vain to separate appearance from reality, Bradley has shown that the attempt must be given up... How does this new Bradleian metaphysics stand towards the scientific movements of the same age?"[282].

Collingwood then gives two examples of modern science, where the distinction between appearance and reality has disappeared as well: the relativistic physics of Einstein repudiating the distinction between apparent and real motion and Freud's theory showing the relation between abnormal and normal psychology. "Here again", Collingwood concludes, "there is a kind of pre-established harmony between the new metaphysics and the new psychology. And these facts, I suggest, give an orientation and an impulse to the work of the modern metaphysician"[283]. After this he continues:

> His special task, I conceive, is this: to begin from Bradley, with the principle that all appearances belong to or qualify reality, and with that principle in mind to approach the physics of Einstein and the psychology of Freud in the same spirit of free co-operation in which Descartes approached the physics of Galileo, or Locke that of Newton... It is for the metaphysician of today to work out a general conception of reality into which these findings of modern science will dovetail naturally, to the mutual comfort and confirmation of metaphysics and science. This is an opportunity granted to metaphysics three hundred years ago, when the foundations of modern science and modern philosophy were being laid; now for the first time it is offered again, at an intellectual crisis in which are being laid the foundations of a new age ... [I] t is this desire that has led me to put before you frankly my hopes for the future of metaphysics and my idea of how we ought to conceive the special nature and function of metaphysical inquiry at the present time[284].

Though we will not go deeper into Collingwood's conception of metaphysics, I think some observations may be made with regard to the lectures of 1934. It is obvious that the view Collingwood develops in his second lecture corresponds to his later theory of metaphysics as the science of absolute presuppositions. So, when Knox contends that Collingwood drastically changed his view on metaphysics between 1936 and 1938 this should be placed at least some years earlier. In two aspects, however, Collingwood's view as elaborated in 1934 differs from that in *An Essay on Metaphysics*. In the first place he considered it in his second 1934 lecture the task of metaphysics to prove the truth of the assumptions of natural science. In contrast, however, this is emphatically denied in *An Essay on Metaphysics*[285]. Another difference is that in the latter book it is stated that metaphysics "in fact ... has always been" an historical science[286], while it is implied in the first 1934 lecture that this view itself is the product of an historical development. With regard to the latter aspect, though, Collingwood is not clear. For in his description of "the situation in which the problems of metaphysics stand today"[287], he begins by describing the situation of seventeenth century metaphysics.

Because Collingwood's "Notes towards a Metaphysic" contain no less than 522 pages (around 130.000 words) it is of course not feasible to give even a superficial survey here. We will therefore limit ourselves to some remarks on their starting-point and general framework. "The science which I am discussing", Collingwood says in the beginning, "is *cosmology*. In discussing it, I am following the lead of Alexander, Lloyd Morgan, Smuts and Whitehead"[288]. From the philosophical tradition he then first mentions Plato's *Timaeus* and, he says, "several Aristotelian concepts and treatises have often recurred to my mind". His "comparative ignorance of medieval thought" has prevented him from making great use of St. Thomas, he asserts, but "later, my deepest and most constant debt is to Spinoza, though Leibniz also has given me much". Finally he mentions Hegel's *Naturphilosophie* "containing certain fundamental ideas without which no cosmological theory can get along"[289].

Collingwood's starting-point is the relation between Matter, Life and Mind. How the changes from Matter to Life and from Life to Mind occurred he considers a question for science to deal with. The question, however, how it was possible for them to occur at all is a philosophical one, he contends. With regard to the rise of Life from Matter Collingwood rejects the "two escapes from the problem" provided by the theories of materialism ("the denial that there is any new principle at all – reduction of Life to the level of Matter") and hylozoism ("the denial that it is really new – raising Matter to the level of Life")[290]. Similarly, with regard to the rise of Mind from Life he rejects the arguments that "1)... there is no such thing as mind but only conditioned reflexes 2) that Mind already prevades the whole of Nature". "I would rather adopt", Collingwood continues, "like Alexander, a word like *nisus* and say that matter has a *nisus* towards life and life a *nisus* towards mind: but that I take to be the statement of the problem, not of the solution. What, in itself, *is* this nisus?"[291] This concept of a *nisus* is central to Collingwood's argument, and elsewhere he equates it with ἔρως, the child of πλοῦτος and πενία [292] or the "élan vital"[293].

With regard to the method to be used in his study of the problem Collingwood observes:

> I begin these notes immediately on completion of my Essay on Method, in the hope of finding that my method may help me to clear up these difficulties. I have already found that a grasp on the principles of my method has made it possible for me to accept mutatis mutandis a great deal of philosophical material, contained in the works of contemporary and other writers, which formerly I had rejected for the flaws in it: the method enables me to disentangle the sound from the unsound where hitherto I had been obliged to throw away both together. What I am now trying to do is to treat in this way the evolutionary philosophies (so to call them) like Alexander's and Lloyd Morgan's and Whitehead's, about which I have felt hitherto that there was 'something in them' but that too much was vitiated by methodological flaws[294].

The Scale of Forms Collingwood considers the main principle of his method. Matter, Life and Mind must be seen in this connection as a scale of species of the genus Reality. These species, then, should be both different kinds and degrees of reality, the highest being the most real. Each is distinct from the next, as well as opposed to it. Mattter, Life and Mind Collingwood considers "only a rough preliminary triangulation for a much more detailed survey", of which he says the following: "each should sum up the whole scale to that point and from its own point of view be identical with the genus, so that, from the point of view of a lower term (e.g. Matter) that term *is* the genus (Reality) and the higher terms (Life, Mind) are simply nothing – the names of errors. Whereas from the point of view of a higher term (e.g. Mind) although that term *is* the genus (Mind = Reality) the same term includes the lower terms (Matter, Life) and, instead of denying these, asserts them as implicates of itself. It does deny them in one sense, but only in the sense that it denies their denials, viz. their claim to be the only or highest term in the scale"[295]. In *An Essay on Philosophical Method* nothing was said of the force which drives a lower term in the scale of forms to the next one. It is for this reason that Collingwood introduces his concept of *nisus*. It is, he says, "what Spinoza would call an immanent causality: something in the first term which converts it into the second"[296]. It cannot be equated, though, with Spinoza's *conatus*, he observes, because this is only directed towards the preservation of a thing's being, while *nisus* is conceived by Collingwood as dynamic and teleological. Though he speaks in this connection of development as "the progressive enrichment of something by the emergence into actuality of its potentialities"[297], Collingwood criticizes the so-called "emergent-theory" of evolution, especially Lloyd Morgan's, because it is "a restatement of the problem, not a solution of it"[298].

Nisus is conceived by Collingwood as a cosmic force working from matter to spirit. The developmental series is specified as follows: space-time, matter, life, sentience, consciousness, reason, spirit[299]. The *nisus* towards life "must have been present in the world from the first", Collingwood contends, just as the *nisus* towards mind in life. Mind, he asserts, "as we are here discussing it, means in a scientific sense the perceptive and appetitive function of organisms having a nervous system, and therefore it is evident from the scientific point of view that, being a specialised function of life, it must arise in the course of world-history later than life in general"[300]. Or, as he puts it a little further: "Mind is a specific type of acting (viz. perceptual activity) present in a body which in order to act in that way must have a specific bodily (physical, and proximately physiological) character. So considered, mind may be regarded as an emergent quality of matter when this reaches a certain determinate and very complex organization"[301]. But, Collingwood warns, mind should not be seen as "a mere by-product of a process not teleologically orientated towards it from the beginning ... The evolution of life is from the first directed towards mind: though the specific way in which this nisus finds realisation is contingent and arrived at by *tâtonnement*"[302].

Sentience Collingwood regards as the lowest level of mind. The next level is consciousness. Collingwood rejects the concept of the unconscious, which, he asserts, is used by psychologists to describe the function of sentience and appetite. "Consciousness and self-consciousness are the same", Collingwood contends. In this connection he remarks that the concept of consciousness is misused for sentience when one says that A is conscious of B, meaning with it that A perceives B: "Properly it should be said, not that A is conscious (or aware) of B, but that A is conscious (or aware) of perceiving B"[303]. "Co-perception" Collingwood regards as the essence of consciousness. It is "a power of organising ... various [perceptual] fields into one single whole, which is one single percipient's experience of one singe perceptual world"[304]. While mere sentience is "perception simply of the here-and-now", the act of consciousness "must be regarded as a creative synthesis: the present perceptual field is by this act enriched with a context which is emphatically not 'given', but 'constructed'"[305].

At the level of thought or reason the new elements are the ones of concepts or notions: "The special name for this kind of distinguishing or analysing is *abstracting*; and what we discover by abstracting is the *principles of structure* that are at work in a complex whole"[306]. Reason may be distinguished into theoretical and practical kinds. Of the first Collingwood says that it "does not abolish consciousness or experience, but it appeals to a new world which is not and never can be experienced: a world of ideal objects towards which the world of experience has an asymptotic nisus". It further judges: that is, "it distinguishes truth from falsehood, the real from the unreal"[307]. Practical reason is for Collingwood the same as will. It implies the consciousness of alternatives, the capacity of making choices and the capacity of inference: that is, the construction of plans and policies.

"Metaphysically defined", Collingwood asserts, "[t]hought is ... the reflexion in mind of the distinction between essence and existence. That distinction is the most fundamental feature in the structure of the universe"[308].

The latter contention is elaborated by Collingwood elsewhere, where he makes a distinction between the working of *nisus* with reason and at lower levels. Reason is driven to self-transcendence, while this is not the case at the lower levels, or, as Collingwood calls it "in finite beings": there "the next stage *must* be brought about by a cosmic nisus working through, but not in, the finite nisus of the last"[309]. It is described by Collingwood as follows: "Matter as such, by the operation of its own laws, will never yield anything but matter. Life will never propagate anything but life. Living beings will of themselves go on adapting themselves better and better to their environment, but they will not of themselves evolve minds". The appearance of minds, then, is realized by "the world's creative force ... bending life's development in this direction, in order to create the opportunity for introducing it ..."[310]. "But when we consider the nisus of reason (i.e. its dialectic)", Collingwood says, "as driving it to become more and more reasonable, and as leading it through the stages science-history-philosophy or utility-right-duty, it appears that the nisus overshoots its own mark and leads thought to transcend itself"[311].

This eventually leads to the realm of Spirit, philosophy being between this and reason, leading thought from the latter to the former: "it seems to precipitate thought outside itself, to swallow up its own distinctions in a unity, to posit Spirit as the solution of the problem which Reason states"[312]. As Collingwood puts it elsewhere, the criteria of spirit may be looked for in knowledge "where knowledge arrives at something that can be called ultimate truth" and in action "where will not only transcends the domination of desire but even that of reason (I do x *because* of y) and becomes an end to itself in absolute action. This spiritual action is most clearly envisaged by Kant"[313].

After this very general survey of the framework of Collingwood's "Notes towards a Metaphysic" it should be noted, however, that the value of this manuscript should not only be judged by the cosmological theory elaborated in it. Many might be put off by the sometimes speculative nature of the argument and this would not do justice to the importance the manuscript has for other reasons. For elaborating his argument Collingwood often digresses on important subjects in a way which makes his expositions highly valuable in themselves. They are especially important with regard to his philosophy of mind, but they also concern topics such as the nature of process, causality and the relation between nature and history. Many later views may be traced back to those developed here and others are clarified. In this respect too the "Notes towards a Metaphysic" are important for a better understanding of Collingwood's thought.

Photo taken at the Roman fort at Ravenglass, in september 1927. From right to left: W.G. Collingwood (father of R.G.G.), miss M.C. Fair, R.G. Collingwood.

CHAPTER FIVE

COLLINGWOOD AS AN ARCHAEOLOGIST AND HISTORIAN

5.1. INTRODUCTION

Collingwood made it abundantly clear in his *Autobiography* how closely his theory of history was based on his archaeological and historical practice. In his own view, this was especially the case with regard to the question and answer approach, which was philosophically linked with Cartesian and Baconian principles (Aut, 24-25). We have seen in the second chapter how he also used his archaeological "laboratory of knowledge" to mount a "flank attack" on the realist position (Aut, 28). The importance Collingwood himself attached to his studies in the field is further confirmed by his statement of his activities in the mid-thirties: "It was time to begin arranging and publishing the lessons which all this archaeological and historical work had taught me about the philosophy of history" (Aut, 121).

With this in mind, it is amazing how little attention has been paid by philosophers interested in Collingwood's philosophy of history to his activities in the fields of archaeology and history. Collingwood is certainly not to be blamed for this neglect, because he was not only explicit in his ideas on the subject, but even devoted the largest chapter of his *Autobiography* to "Roman Britain". For a long time it was only Schneider and Donagan who paid any attention to Collingwood's "practical" work as an archaeologist and historian[1]. There are signs, though, that things may have changed. Goldstein who, in the study of his theory of history, has called for consideration of Collingwood's historical work, himself gives some examples[2]. Mink calls attention to "remarkable lacunae both in the understanding of Collingwood's views on history and in the criticism of them" and mentions in this connection, as the first lacuna, that "...critics have made few attempts to range Collingwood's theoretical reflections against his own actual practice as a working historian of Roman Britain..."[3].

In connection with Collingwood's archaeological and historical activities it is first of all necessary to clear away some misunderstandings. The first thing to note is the fact that Collingwood was active in the fields of both archaeology and history. His specialization being Roman Britain it is obvious that these subjects cannot be sharply divided, especially in Collingwood's own practice. The subjects are not, however, the same, and if one mentions only one of them in connection with Collingwood's activities this can be misleading. An obituary by Richmond for instance is entitled only "Appreciation of R.G. Collingwood as an Archaeologist"[4]; Donagan devotes a section to "Collingwood's Practice as an Archaeologist"[5]; while Goldstein speaks of Collingwood merely as an historian[6], and Knox describes him as a "philosopher and historian"[7]. The degree of Collingwood's involvement in archaeology and history seems also to be easily misunderstood. Goldstein for instance asserts: "His own special field was Roman Britain, and for a time he did quite a bit of writing on the subject"[8]. This can only be interpreted as a gross understatement with reference to someone who published on this subject in every year between 1913 and 1939, except in the years 1917 and 1918[9]. Roberts makes it even worse by saying that "R.G. Collingwood was an able and distinguished historian, one of the leading authorities of his day on Roman Britain, before he turned to philosophy and made philosophical thinking his main interest and preoccupation"[10]. In fact Collingwood always worked – from his days at College till near the end of his life – on the three tracks of philosophy, archaeology and history. Thinking particularly of his publications in these fields, one could only make the proviso that he published on the history of Roman Britain from 1920, and that from October 1938 – the date of his departure to the East Indies – he did not work any more on archaeology and history.

As in the case of other aspects of his development, the influence of his father, W.G. Collingwood, who educated his son at home till his thirteenth year, was decisive for Collingwood's interest in archaeology. For W.G. Collingwood, a graduate from Oxford, artist, writer and last secretary to and biographer of Ruskin, was a practising archaeologist as well[11]. Collingwood writes in his *Autobiography* that his father, when excavating Hardknot Castle, took him along as a three weeks old baby in a carpenter's bag and he grew up, he says, "...in a gradually thickening archaeological atmosphere" (Aut, 80). Both father and son concentrated their archaeological activities in and around the Lake District, where they lived. This part, the north-western border area of the Roman empire, is rich in Roman antiquities, the famous Hadrian's Wall being the most impressive. The most important archaeological institution in this area is the "Cumberland and Westmorland Antiquarian and Archaeological Society", founded in 1866. It has a long tradition of organizing excavations and the publication of their results. W.G. Collingwood became a member of the Society in 1887, afterwards from 1900 till 1925 Editor of its Transactions and from 1920 till his death in 1932 President. R.G. Collingwood followed in his footsteps, becoming a member of the Society in 1909, Editor of the Transactions from 1920 till 1934 (so five years together with his father), President from 1932 till 1938 and Honorary Member from 1938 till his death in 1943[12].

In Oxford Collingwood studied with Haverfield (1860-1919), since

1907 Camden Professor of Ancient History, and called by him "the great master of the subject" (Aut, 120). After Haverfield's death "I was left the only man resident in Oxford whom he had trained as a Romano-British specialist", Collingwood says: "and even if my philosophy had not demanded it, I should have thought myself, in piety to him, under an obligation to keep alive the Oxford school of Romano-British studies that he had founded, to pass on the training he had given me, and to make use of the specialist library he had left to the University" (Aut, 120). Of the persons with whom Collingwood co-operated during the rest of his life mention should be made especially of F.G. Simpson (1882-1955), an ex-marine engineer, who from 1906 devoted his life to the excavation of Hadrian's Wall[13]. Of the people of a younger generation, whom Collingwood not only co-operated with but also had a decisive influence upon, the most prominent are I.A. Richmond (1902-1965), since 1956 Professor of the Archaeology of the Roman Empire at Oxford, and E.B. Birley (born 1906), from 1956 Professor of Roman-British History and Archaeology at Durham[14].

In Collingwood's work in the field of archaeology and history three apects may be distinguished. In the first place he did some excavations, especially in the beginning of his career, and wrote an important handbook on the archaeology of Roman Britain[15]. He further showed a continuous interest in the problems concerning Hadrian's Wall and made important contributions to its interpretation. Finally he wrote some books and articles on Roman Britain in general. We have seen that as a philosopher Collingwood became to an increasing degree an isolated figure. This cannot be said, however, of his activities as an archaeologist and historian. On the contrary, his activities in these fields have been highly appreciated up to the present day. That Collingwood himself recognized this is demonstrated by a letter he wrote to de Ruggiero on 4 October 1927 saying: "I find myself writing gloomily. For four months I have been deep in historical studies, and there I find myself among friends and willing collaborators; the return to philosophy means a return to a work in which I become more and more conscious of being an outlaw"[16].

5.2. ARCHAEOLOGY

5.2.1. SCIENTIFIC EXCAVATION

In the chapter on "Roman Britain" Collingwood gives in his *Autobiography* a description of three principles which guided his work in archaeology and history (he does not make a distinction between them). "Most of these principles were", he says, "more or less unconsciously, common ground among historians; but not all of them were generally accepted; or perhaps it would be truer to say that comparatively few were consciously recognized, and of those by no means all were generally regarded as principles by which the historian ought to stand firm through thick and thin" (Aut, 121-122). The first principle is described by Collingwood as follows:

> ...[L] ong practice in excavation had taught me that one condition – indeed the most important condition – of success was that the person responsible for any piece of digging, however small and however large, should know exactly why he was doing it. He must first of all decide what he wants to find out, and then decide what kind of digging will show it to him. This was the central principle of my 'logic of question and answer' as applied to archaeology (Aut, 122).

This principle is contrasted with the "blind" digging that was the practice in the beginnings of archaeology, usually for reasons of curiosity or nostalgia. Collingwood's approach is eloquently described by Toulmin in his introduction to the *Autobiography*: "The primary business of archaeologists and historians is not to dig up art objects for display in museums – gapers at King Tut, please note! – so much as to formulate and answer significant questions about past modes of human life and thought" (Aut, XVII-XVIII)[17].

In this passage Toulmin also refers to Collingwood's second principle, which reads:

> ...[S] ince history proper is the history of thought, there are no mere 'events' in history: what is miscalled an 'event' is really an action, and expresses some thought (intention, purpose) of its agent; the historian's business is therefore to identify this thought. For the archaeologist this means that all objects must be interpreted in terms of purposes. Whenever you find any object you must ask, 'What was it for?' and, arising out of that question, 'Was it good or bad for it? i.e. was the purpose embodied in it successfully embodied in it, or unsuccessfully?' These questions, being historical questions, must be answered not by guesswork but on historical evidence; any one who answers them must be able to show that his answer is the answer which the evidence demands (Aut, 127-128).

Collingwood's third principle was "...that no historical problem should be studied without studying what I called its second-order history; that is, the history of historical thought about it" (Aut,132).

Of these principles the first one is particularly important, because it expresses the Baconian approach of "question and answer", which is even called by Collingwood a revolution (Aut,124). He credits Haverfield for having started it: "Haverfield and his colleagues of the Cumberland Excavation Committee in the eighteen-nineties had been consciously and completely Baconian in their methods. They never dug a trench without knowing exactly what information they were looking for; they knew both that this information was the next thing they needed for the progress of their study, and also that this trench would give it them" (Aut,124). It is interesting to note that Collingwood already in 1919 was of the same opinion, writing about Haverfield after his death: "His reputation with posterity will rest on a few monographs and numerous short papers. But he did much to inaugurate the era of scientific investigation which has replaced, in Romano-British archaeology, the cruder methods of the last generation"[18]. Strangely enough Haverfield himself apparently was not aware of his inauguration of the era of scientific investigation, for in his *Autobiography* Collingwood comments: "Haverfield himself, least philosophical of historians, cared nothing about the principles or the potentialities of the revolution he was leading" (Aut,83).

It is to be doubted, though, if Collingwood is altogether correct in crediting Haverfield with the "scientific revolution" in archaeology. It is more probable that he projected his own ideas and ideals on to his teacher, whom he valued so highly, the more so because he admits that Haverfield himself was not conscious of the novelty of the approach. There is more reason to believe that it was Simpson who worked out the principles of "scientific" archaeology, apparently in the beginning of his career together with J.P. Gibson (1838-1912).

With regard to Hadrian's Wall Haverfield was of the opinion that a wall made of turf – parts of which have survived – was built in the reign of Hadrian (117-138) and the stone one in the reign of Severus (193-211). In a joint paper, published in 1911, Gibson and Simpson proved, however, that the stone wall was of Hadrianic origin as well[19]. Birley says about this crucial period in the Wall's interpretation:

> By 1903, the last season of the Cumberland Excavation Committee's work, Haverfield had become the acknowledged leader in the study of Roman Britain, and his views on the mural problem showed signs of hardening into a dogma of despair, assigning a Turf Wall from sea to sea to Hadrian ... and the Stone Wall to Severus ...; but J.P. Gibson ... remained unsatisfied, and it is fair to say that it was his persistence and his vision which kept the idea of research alive, and found in F.G. Simpson the man through whom it could be revived and reinvigorated.
>
> Gibson lived long enough to see Simpson firmly established as an excavator and as a planner of excavations ...; a detailed diary of Simpson's digging, from 1907 onwards, would be needed to show how complicated a series of interlocking problems led him from one site to another, with more and more threads of research interwoven as the work proceeded. When one reviews what has been done on the Wall in the past half-century, it is not easy to find any branch of the subject in which the key work and the basic thinking and planning were not done by Simpson[20].

The archaeologist Corder gives a similar assessment:

> Between 1909 and 1913 the work carried out on Hadrian's Wall by the late John P. Gibson, F.S.A., and Mr. F. Gerald Simpson showed an advance in the technique of excavation and its record as significant for our studies as anything the General [Pitt Rivers] ever achieved. For the first time in this country excavation was directed to solving specific problems. In place of the very costly clearance of a large site in the hope of discovering what might be there ... each site was carefully selected as part of a planned campaign of research, and the excavation was limited to a definite objective clearly in mind before it was undertaken[21].

Simpson, however, did not easily come to publishing the results of his findings, let alone to a description of the principles he used. It is in this connection that Collingwood, who was closely associated with Simpson's work for many years, appeared as an ideal counterpart. For with his keen interest in methodological issues he was not only aware of the importance of the new approach in archaeology, but also put it in the broader philosophical framework of a "logic of question and answer". After Simpson's death Birley says in an "In Memoriam" about the relation between Simpson and Collingwood: "[T]he two men became firm friends, and for many years Collingwood acted as interpreter of Simpson's work to the world at large, putting his pen and his influence in academic circles freely at the disposal of a man in whom he recognised the qualities of genius and of real humility" and a little further on he remarks on Simpson that "...he was responsible for initiating and directing a long series of further excavations, all marked by the sureness of touch and the economy of effort which the world at large has come to think of as R.G. Collingwood's approach, but which Collingwood himself always regarded as the most striking features of Simpson's work in archaeology"[22]. In his subsequent book on Hadrian's Wall Birley gives a similar assessment:

> R.G. Collingwood was the third of the group, and the one who stood closest to Simpson, in many respects serving as the interpreter to the outside world of the work which Simpson was doing; but he was also the master planner of financial and moral support for that work, particularly in Cumberland, and the diplomat who saw to it that what needed doing was not only done but also published promptly...
>
> Without Collingwood's support and generalship, Simpson might never have been able to continue his geometric progression of research, and there would certainly have been far less widespread recognition of the importance and the validity of the results which Simpson's system of selective excavation was able to achieve. The essential basis was a cumulative analysis of detailed observations, followed by digging carefully planned to produce answers to specific questions at the points where answers could be obtained most economically[23].

Richmond is of the same opinion as Birley, saying about Collingwood that "[h] is attitude to excavation was profoundly influenced by the selective method of excavation, introduced and perfected by F.G. Simpson on Hadrian's Wall: and the first lesson which he drew from these methods, that excavations should be conducted with specific problems in mind upon sites likely to provide an answer, was salutary and useful"[24].

In order to deal with another aspect of Collingwood's view on archaeology we may begin by quoting Richmond again, when he adds, after the passage quoted above, the following criticism of Collingwood's approach: "But Collingwood's corollary, that to pose a problem permitted its answer to be predicted, was a product of the study rather than the field. For there are problems thus soluble, particularly those purely philosophical problems which it was Collingwood's daily task to consider. But to the field-worker excavation, no matter how carefully planned in advance, is always a plunge in the dark ... the expected is always accompanied, and often overshadowed, by the unexpected ..."[25]. Birley, in a letter to the present writer, is of the same opinion: "As an excavator, he [Collingwood] was not very effective, partly (I think) because he tended to find what he had expected to find, even if it was not in fact there". Though I am not in a position either to corroborate or to refute these assessments, I think a few comments can be made upon them. It is in the first place not correct, as Richmond seems to think, that with philosophical problems to pose a problem permits its answer to be predicted. This is certainly not Collingwood's doctrine and we have seen that when dealing with cosmology he even explicitly stated at the beginning that he did not know the answers to the problems he posed[26]. With regard to historical problems as well, Collingwood always emphasized that these could only be solved by appeal to evidence, and the idea that one could deduce an answer from a question he would certainly reject as ridiculous[27]. It is another question, of course, whether Collingwood's archaeological practice was always in line with his theory, and he indeed may have given the impression to fellow-archaeologists that this was not the case. In the subsequent section we will try to demonstrate, however, that at least on occasion this impression may have been unfounded. G. Simpson – daughter of F.G. Simpson – gives a possible explanation of the fact that Collingwood leaves this impression on many interpreters: "Haverfield's insistence on the need to have working hypotheses, always followed by Simpson, was also followed by Collingwood, although his persuasive style has given his readers the impression that everything he wrote was to be accepted without question"[28]. That this impression was rather a matter of style than of principle is confirmed by what Collingwood himself says in "The Nature and Aims of a Philosophy of History": "[An historian] does not perpetually qualify his statement with 'in my opinion', 'probably', 'so far as the available evidence goes', just because a qualification of this kind is assumed as a standing order in all historical thinking", adding to it that the omission of these qualifications easily leads to the misunderstanding" ... that what the historian gives as a probable opinion, based on the available evidence, is a statement of ascertained fact"[29].

Richmond also criticizes Collingwood on a related point: "...[W]hile Collingwood's favourite *dictum*, that 'what you are not looking for, you do not see', is of wide application, an excavator's first duty is to see everything without the blinkers imposed by prearranged concentration of vision. Collingwood's own excavations, however, illustrate that he was by no means alive to these and other pitfalls"[30]. The archaeologist Hogg has recently leveled a similar criticism against Collingwood:

> Any excavation ... is sampling; but there is no agreement either as to the basis on which the samples should be chosen or as to how best to analyse the results obtained. Rescue excavations now provide a fairly unbiassed selection of sources to supplement earlier work, so there is ample material from which hypotheses can be devised. Research samples should be chosen to investigate these. This approach is related to Collingwood's 'Method of Question and Answer' which led notably to great advances in the interpretation of the archaeology of Roman Britain; but in principle the Method was intended to answer questions raised by the excavator, so that he could produce a rational narrative account of his subject, and the excavation report was not regarded as a primary source of raw material for others to study. By contrast it is now almost universally accepted that once a site has been chosen all information revealed should be recorded, whether or not it is relevant to the original question[31].

Here again we must distinguish between Collingwood's theoretical attitude with regard to this question and the possible impression he left on others in his practice. Donagan has made clear that, so far as the first is concerned, Collingwood is quite explicit. He quotes the following passages: "...[T]he past leaves relics of itself ... We preserve these relics, hoping that in the future they may become what now they are not, namely historical evidence" (IH, 203) and "If in the case of one object [an archaeologist] does not understand [what it was for], he has, as an archaeologist, no use for the object; he would throw it away, but that he hopes some one more learned or more resourceful than himself may solve the riddle" (Aut, 108)[32].

It is interesting to note, however, that as a practising archaeologist, too, Collingwood proves to be very sensitive to this issue. Writing on the problem of dating the Roman occupation of Hardknot, which could only be solved by pottery evidence, Collingwood says:

> It is now possible to date a mass of coarse pottery with reasonable accuracy, especially in the second century and the latter part of

> the first: a thing which was impossible till a few years ago.
> I did not know to what extent the Hardknot pottery had been preserved, and whether enough had been kept to make such an analysis as I proposed a possibility...I found that it had all been preserved, even the least promising potsherds having been scrupulously collected, in many cases neatly packed in paper parcels and labelled ...
> Historians owe a debt of the liveliest gratitude to these men, who, long before such evidence could be deciphered, and while most excavators were cheerfully throwing away everything of the kind except a few show-pieces which happened to be more or less whole, spent time and trouble to preserve for our use several thousand fragments of pottery, which in their reports could only figure as 'twenty pounds of coarse pottery found in this site', and so forth. Thanks to their forethought I was able to examine I believe every potsherd yielded by the excavations[33].

Though Collingwood welcomes the forethought of the people who had kept the pottery remains, in spite of their being of no use to them, it could be questioned whether it is really possible for an excavator "to see everything" or to record "all information revealed", as is demanded respectively by Richmond and Hogg. It rather seems only sensible to interpret this recommendation as meaning that one should record all the information with regard to possible future questions as one can see them at the moment. The possibility always remains, I would contend, that we should fail to record information that will be highly relevant in the future, not only because we do not know what questions will be asked but also because we are not aware of possible types of evidence, that will be used by future investigators, and are unknown at the moment.

5.2.2. EXCAVATIONS

Collingwood's first publication was a report of the excavation of a Roman fort at Papcastle in 1912[34]. The excavation was directed by Collingwood himself and four to seven men were employed. As with Collingwood's other writings on archaeological and historical subjects we will not deal extensively with the actual content, but concentrate on some examples illustrating certain methodological principles. In this connection we will especially keep in mind the three principles of question and answer, interpretation in terms of purposes and second-order history.

The results of the excavations at Papcastle were not very striking. The fort apparently had not been an important one and the remains were too fragmentary for Collingwood to allow definite conclusions. The excavation had a limited aim which is clearly stated by Collingwood: "The primary object of the excavation was to fix the limits of the camp"[35]. In his conclusion to the report the limited scope of the excavation is re-emphasized: "So brief an exploration was not intended to be exhaustive, and its results are not final. The whole site has been so thoroughly disturbed in the search for building stone that a complete excavation would perhaps be of little service"[36]. He comes, however, to the positive conclusion that there were two forts at Papcastle, an earlier and later one. It is characteristic that Collingwood ends his report by posing questions to be resolved by future research: "Other problems remain, and new ones have arisen. If excavation should ever be renewed, it might determine the size of the earlier fort by following its north wall, and reveal some of its interior buildings. The principia of the later camp are still to seek, and the whole of its western area is untouched"[37].

At Ambleside, at the northern end of Lake Windermere (Cumbria) there is a site near the shore of the lake where the remains of a Roman fort are clearly visible. The site is open to visitors and at the National Park Centre at Brockhole near Windermere there is not only an exhibition of the finds from the site, but also a hand-made model of the Roman camp. This camp at Ambleside has been excavated by Collingwood between 1913 and 1920 and is undoubtedly his most important achievement in archaeological excavation. The model exhibited at Brockhole was made by Collingwood and is a fine example of his artistic talent. The decision to excavate the Roman fort at Ambleside was taken in 1913. An Exploration Committee of the Cumberland and Westmorland Antiquarian and Archaeological Society, under the chairmanship of Haverfield, appointed Collingwood as director of the excavations[38]. They took place in 1913, 1914, 1915 and 1920, taking altogether 29 weeks of which Collingwood was not present for only four weeks. In his work he was always assisted by several men.

Though Richmond, as we have seen, is rather critical of Collingwood's archaeological practice he says of the excavations at Ambleside: "Collingwood carried out the task with diligence and ability: details were, indeed, revised from year to year, but the ultimate result was sure. In particular, his treatment of the Agricolan fort, buried below the Hadrianic, was an essay in reconstruction neither imitated nor equalled"[39]. Collingwood wrote extensive reports on the excavations. At the end of the first report he gives a fair example of his question and answer approach and his practice of asking about the purpose of the remains. On the shore of the lake he found in the water the remains of a wall, built of cobbles and boulders. "The two problems suggested by this construction are its purpose and its date"[40], Collingwood says. He thinks it possible that the Romans may have built it as some kind of sea-wall, but also discusses another purpose in the following way:

> The alternative view is that the wall was intended for a quay ... The shape of the wall and its shortness tell in favour of this theory.

On the other hand the water is much too shallow at present to admit of even the smallest boats using such a landing. But this may be due to silting-up, caused by westerly and southerly winds banking up the silt brought down by the Brathay and Rothay. A second and more serious objection is the absence of any approach to the wall from the land side. If it was a quay there must have been a solid paving behind it, like that which we found outside the east gate of the fort: whereas nothing of the sort is visible, and it can hardly have disappeared. Thirdly, the harbour, if such it was, is very much exposed to all the most violent winds; and if boats were ever brought here they must have been drawn up on the shore for safety, as is done at the modern Waterhead landing. But this is impossible with a quay, though they may, of course, have been beached elsewhere. Fourthly, the roughness of the building, though suitable in a breakwater, would be out of place in a landing stage. But the Romans may have used boats, and must have landed them either hereabouts or in the river.

It is, however, not certain that the work is Roman at all. We found no positive evidence of any date. But it is difficult to see what other date could be assigned to it. As a breakwater, it can never have been of the least value to anyone who was not living in the fort or within a very few yards of its east gate. The only possibility is that it may have been a boat-landing of another date. If so, it must have been reached by a road; but the 'Borrans Road' is of comparatively modern construction, long subsequent to the establishment of the Waterhead landing, which is the easiest and best landing for Ambleside. If, therefore, this was a mediaeval landing, in spite of the difficulties, stated above, in the way of explaining it as a landing at all, there was no road to it. It can, therefore, only be explained as built in connection with the Roman station and accordingly of Roman date"[41].

Though Collingwood's judgment is only conclusive with regard to the date of the wall and not its purpose he says in his report of the next year that "[t]here seems to be no doubt that this gravel was intended to protect the wall [of the fort] against the force of waves during winter floods and storms"[42].

In the same report Collingwood deals among other things with the location of buildings inside the fort. After mentioning how they were grouped in other forts, he says:

> It seemed possible then by means of a few trial trenches to ascertain (1) the size of these blocks of barracks; (2) the direction in which the barracks lay; (3) the material of which they were built; and (4) whether they showed good stratification or any characteristics which would make it worth while to explore them more fully[43].

"For this purpose we chose the north-east block", Collingwood continues, "and made a few trenches in it near the east gate. These fixed the edges of the *via praetoria* and of the road running behind the east rampart ... We then cut a trench diagonally from the neighbourhood of the east gate across the block in question, emerging on the *via principalis* some 50 feet south of the north gate. This supplied us with material for answering the rest of our questions"[44]. We will not follow Collingwood through the rest of his report. In 1914 he made the important discovery that there must have been two forts at Ambleside: an earlier one made of turf and timber, built during the British governorship of Agricola (78-84) and a stone one, built during the reign of the emperor Hadrian. The latter was used till the second half of the fourth century. In his report of the excavations of 1915 Collingwood then gives a survey of several problems previously unsolved which found a solution through his discovery of the earlier fort below the stone one[45].

Besides giving in his reports a careful account of the excavations, Collingwood also provides a meticulous exposition of a mass of small finds obtained during the excavations. These include pottery of various kinds, coins, metal objects, glass, tools and even a pair of leather shoes. Many of these objects are carefully drawn by Collingwood, showing for instance the different types of pottery. The reports also contain many valuable plans of the site, both general and detailed, which demonstrate Collingwood's excellent draughtsmanship. Since Collingwood's excavations at Ambleside no new one has been carried out inside the fort and his conclusions are still considered valid. In a recent booklet on the fort at Ambleside it is only stated that present-day archaeologists, on pottery evidence, now prefer to place the foundation of the earlier fort a few years later than Collingwood contended[46].

We will not deal with Collingwood's excavations at Bainbridge in 1926, of which Richmond is rather critical[47], but only mention the one he conducted in 1937 – his last one – of King Arthur's Round Table[48]. Richmond's judgment on it is severe: "Still less fortunate was his excursion into pre-history at King Arthur's Round Table, near Penrith. There he had made up his mind in advance what he was to find and found it with fatal precision. When the writer ... came to translate the German report upon a second examination of the site a year later, it was sad and inexorable reading, a very verdict of Nemesis"[49]. The latter remark refers to the report of the distinguished German archaeologist and prehistorian Dr. Bersu, who carried out excavations of the same site in 1939[50]. We are not in a position to comment on the reports of Collingwood and Bersu or to compare them. It seems that Collingwood indeed came to conclusions which were rather weakly based. It is striking in this connection that he ends his report with a conclusion of six pages in which he distinguishes three pre-historic periods of the site while Bersu's conclusion is sober and reserved, saying: "All the facts lead to the conclusion that the earthwork was erected as a sepulchral monument and then left without further attention. Finds to date the work are lacking"[51], and a little further on: "In view of the extensive destruction of the monument we must thank our chances that it was even possible to make any sort of reconstruction of the site in its guise as a prehistoric tomb"[52].

In archaeology Collingwood's accomplishments have not primarily been in the field of actual excavations. Though fellow-archaeologists may be critical of the few he carried out, there is reason to believe that his excavation of Ambleside fort is of a high standard. As I have tried to show, it also provides a good illustration of how certain principles of method were put into practice.

5.2.3. *THE ARCHAEOLOGY OF ROMAN BRITAIN* (1930)

Collingwood's synthesizing mind, always alive to methodological principles, was well able to deal with the more general interpretative aspects of the science of archaeology. His *Archaeology of Roman Britain*, published in 1930, is the result of his studies in this field. It is probably Collingwood's most successful book and its influence in archaeological circles can only be compared with that of *The Idea of History* among philosophers and historians, although it has been far less controversial.

Before dealing with the book itself we should remind ourselves of Collingwood's ideas on the science of archaeology, especially in relation to historical science. We have seen how Collingwood in his lectures of 1926 and 1928 describes archaeology as "empirical methodology", setting it against philosophy of history as "pure methodology"[53]. Empirical methodology is considered a necessary precondition of critical history, that is, history which interprets its sources in a critical way and is not dependent any more on authorities, as is the case with dogmatic history. "... [E]verything depends on the ability of the historian to discover materials that he can use", Collingwood says, "and these will be of the most widely divergent kinds and the principles of their employment infinitely various. This is the *raison d'être* of such sciences as palaeography and diplomatics, epigraphy, numismatics, historical architecture, and all the ramifications of archaeology in its application to various kinds of implements and relics"[54]. These sciences are also described by Collingwood as "means to criticise sources and extract history from them", to which he adds that "... this implies, on the part of the historian, a technical equipment of the kind that is generally called scientific"[55]. Or, as he puts it a little further on: "Critical history classifies its sources into groups, and then subdivides these groups, framing rules for the manipulation of the various subdivisions. Taken as a whole, this technique is an abstract or classificatory science, which has no general name, unless that of archaeology is used for it, and is subdivided into numerous departmental sciences ..."[56].

It is important to note that Collingwood considers archaeology to be abstract, classificatory and aiming at generalizations: features which he alwyas strongly objects to with regard to history. The relation between the two fields of inquiry is clarified by him as follows:

> Classificatory and abstract thought is the negation of history, which is individual and concrete through and through; but the concreteness of history can only be reached through the abstractness of the archaeological sciences.
> Every advance in critical history rests on an advance in the interpretation of evidence, that is, an advance in archaeological science. Every advance in archaeological science consists in the discovery that some class of facts can be made to yield historical knowledge, which has hitherto yielded none. The archaeologist feeling his way towards new advances is constantly asking himself whether this or that detail of script or moulding or pottery can be proved characteristic of a certain date or a certain origin; he collects instances, perhaps thousands of instances, to test the suggestion, and may end by committing himself to the generalisation that this feature has a definite significance[57].

In a note in his 1929 lectures it is put even more clearly: " ... [T]he historical determination of facts is a complex matter, and archaeology, which is a scientific discovery of general laws (empirical descriptive science) is a means to it"[58].

Though Collingwood wrote an important manual on archaeology it is interesting to observe how reserved in fact he was with regard to an endeavour of this kind and that his book should be primarily considered therefore the result of sound archaeological practice. In discussing the archaeological sciences in his 1926 lectures he says about them:

> In part, these sciences can be discovered set forth in textbooks; but only in a very small part. The student who is anxious to learn them must get himself apprenticed to the trade by working in company with skilled exponents; he will find them in museums, in libraries, on the staffs of excavations, and even [sic] in universities. These bodies of skilled historical investigators, handling down by personal instruction and word of mouth a vast amount of knowledge that never finds its way into books, form one of the most interesting features of our civilisation on its intellectual side. It reminds one of the medieval gild system, and it has the same strong points: it ensures, as nothing else can, a high and fairly consistent level of work, and makes it difficult for a totally incompetent or untrained person to undertake a delicate piece of research and impose his valueless results on the public ... [N]o amount of book-learning can make up for this personal instruction and personal experience in the handling of actual objects. The fisherman who found his way home in a fog by smelling the lead, after sounding with it, was hardly more independent of book-learning than the archaeologist

> who rubs his thumb along the edge of a potsherd and says 'they never feel like that much after the reign of Domitian'[59].

The prolific studies in the field of the archaeology of Roman Britain, however, apparently changed Collingwood's opinion in the sense that he thought a manual not only useful but even necessary. In the preface to *The Archaeology of Roman Britain* this view is expressed by him as follows:

> The purpose of this book can be best explained by explaining the situation out of which it has arisen. Romano-British studies have for generations been among the usual interests of educated Englishmen; but it is only lately that they have become a branch of scientific archaeology, demanding for its prosecution a special training and the thorough mastery of a special literature. Year by year, as this movement gathers force, the special literature grows in bulk. For the most part, it consists of monographs on particular sites, or rather, on particular pieces of work carried out at particular sites. They are highly technical, repulsively unintelligible to the general reader, and usually published (for that reason) only in the journals of societies, whence they never emerge into the general book-market except at the death of these societies' members ...
>
> The longer this situation is allowed to last, the worse it inevitably becomes. Owing to the inaccessibility of our technical literature, our amateur antiquaries become more and more amateurish, foreign archaeologists lose touch with our work more and more completely, and even the best archaeologists in this country show signs of not knowing what each other has done. The time seems ripe for some one to make a first attempt to digest the mass of technical detail into a manageable form; and that is the attempt which is represented by this book. It is strictly, as the title of the series implies, a handbook of archaeology, not an history ...
>
> So elementary a book ought to have been easy to write; but it has not proved so, because, on some of the subjects with which I felt obliged to deal, elementary and simple ideas were just what no one seemed previously to have worked out. No one, so far as I know, has reviewed the types of Romano-British forts, villas, or town-houses, as a whole, with attention to their varieties; still less has anyone tried to draw up a list of the easily recognisable and certainly datable forms of coarse pottery or of the varieties of brooch. I have tried to do these things, and I cannot hope to have done them otherwise than rather badly; the wonder is (to adapt a saying of Dr. Johnson's) that it should be necessary to do them at all. I shall be well satisfied if the defects of my treatment stimulate others to do them better[60].

The sixteen chapters of the book deal with the following subjects:

roads, camps, fortresses and forts, signal-stations and light-houses, frontier works, towns, villas, temples, tombs, native settlements, inscriptions, coins, Samian ware or terra sigillata, coarse pottery, brooches and (finally) weapons, tools and utensils. Again, we will not go into the actual content of the book, but only mention some methodological points it raises.

In his first chapter Collingwood discusses not only the general characteristics of Roman roads, but also the ways in which they can be recognized. This is an excellent illustration of his principle of question and answer, and the way he recommends the asking of specific questions in order to determine whether a road is Roman or not thus deserves to be quoted in full:

> Evidence of age is important in trying to distinguish a Roman road from a disused modern road. It may take two forms, documentary and topographical. Documentary evidence of age may come from old maps, or charters, or the like, or from parish boundaries; for Roman roads have often been used as landmarks, and wherever a parish boundary runs along a road, or follows a straight line across country for a mile or so together, it gives rise to the suspicion of a Roman road. Topographical evidence of age depends on the time required by certain changes in the landscape, notably changes in the exact position of water-courses. This cannot be accurately estimated, but accurate estimate is not necessary for the present purpose. A metalled road, for instance, is found crossing a stream, and the stream has eaten into its bank so deeply that the road now ends in mid-air ten feet above the water-level. By its construction, as tested by eye and probing, it is either Roman or modern; the question then is whether the erosion in question could have taken place since (say) 1750. If it could not, the road is Roman. Or again, a choked culvert in a metalled road has resulted in a wash-out, scouring away metal and bottoming alike. Could all this have happened in the last century or two, or is it necessary to allow much more time? In the latter case, the road is Roman. In order to form judgments on points of this kind it is useful to study modern roads which have been abandoned and allowed to fall into decay. When this has been done, it is often possible to distinguish their condition from that of Roman roads similarly constructed in similar country, in which the same processes have gone perhaps fifteen times as far. Where excavation is possible, the character of the metalling may be decisive; and in this case special care should be taken to look for kerbstones, which are normal in Roman work and almost always absent in modern macadam; but much can be done even without excavation by a careful and practised observer. It is also necessary to distinguish Roman roads from the pack-horse roads of the Middle Ages. These are normally much narrower, and are commonly paved with large stones laid transversely so as to give the horses a hold on the road; they are not aligned like Roman

roads, and in lay-out they are more like cross-country footpaths[61].

Collingwood's treatment of the different types of archaeological remains corresponds to his definition of archaeology as a generalizing science. For he constantly tries to give the general characteristics of Roman camps, forts, pottery, brooches etc., in order that they will be interpreted in the right way to be used as evidence[62]. In this connection he mentions for instance camps as forming "a recognisable class of earthworks"[63] and speaks of "[a] generalised description of the main features of a Roman fort"[64]. The generalizations Collingwood develops are often limited to a certain period. Under the head "General Features of Flavian Castella" he thus says: "Flavian forts in Britain show certain constant features. Their ramparts are never of stone, but always of earth (puddled clay, as a rule) or, more rarely, of turf. Their shape is always more or less rectangular; even Ambleside, the least regular in outline, approximates to a rectangle. Their internal buildings, so far as we know, were always of wood, except that their bath-houses, being heated with hypocausts, were necessarily of stone; but these were nearly always outside the ramparts. When they were more than two acres in extent, they always had four gates. Other generalisations are less binding"[65]. Another example is Collingwood's description of "Trajan-Hadrianic Forts on Virgin Sites": "A brand new Trajanic or Hadrianic fort has, as a rule, certain definite characteristics. It is rectangular (oblong or square), though the accuracy of its right angles is often defective. It has a composite rampart of earth and stone . Its gates and corner towers are of stone, and it may have intermediate stone towers as well. Its central buildings are of stone, and its barracks are generally, though not always, of stone likewise. Forts of this pattern may always (in Britain at any rate) be construction provisionally assigned to the reigns of Trajan and Hadrian; and nothing in plan or construction permits us to discriminate between the reigns of these two emperors"[66]. Collingwood then describes three "abnormal Trajan-Hadrianic forts"[67].

Many other examples of generalizing descriptions could be given. Of ditches it is said that "[t]emporary and semi-permanent camps normally have a single ditch; permanent works normally have two, but occasionally three"[68], and the chapter on towns begins with: "Roman provincial towns fall into two classes: a larger pattern, laid out in a plan of streets crossing each other at right angles ...; and a smaller pattern, which has no such formal plan, and consists of a group of houses clustered by the side of a road or at the place where two roads meet, or where a road crosses a river"[69].

Collingwood's typologies are always carefully worked out: his chapter on villas for instance is divided into sections – besides the one on "General Features and Types" – dealing with the winged corridor house, the bipartite and tripartite corridor house, the courtyard house, the basilican house, outbuildings and other types of villa. Since the archaeological remains are so different in character – from roads to brooches – Collingwood's discussion of each kind differs as well. In his treatment of the various types of pottery and brooches, for instance, he is very detailed, giving dozens of examples, illustrated with many drawings by himself. He sees certain types of objects as having their own problems of interpretation; this is the case, for example, with brooches, on which Collingwood comments: "The brooches we are to examine do in a sense grow out of each other, but their growth is a complicated matter. A new type is never produced as it were by parthenogenesis from a single parent; the new modifications arise through reminiscences of one type being combined with imitation of another. In the offspring we can detect the likeness of both parents, and often, if we look close enough, of grandparents, uncles and aunts. Thus the groups which it is the business of this chapter to distinguish are related to one another like a number of families all inhabiting the same district, which intermarry and produce unexpected combinations of facial likeness, armorial bearings, and landed property"[70].

There are also many illustrations to be found of Collingwood's principle of interpreting archaeological remains in terms of their purpose. This is especially the case in the chapter on frontier works. Since we will deal with Collingwood's work on Hadrian's Wall separately in another section, we will not discuss it here. An example relating to this principle could also be given, however, from Collingwood's discussion of the history of the Roman fort in Britain, where he describes a change in "tactical purpose": "The early forts housed a garrison trained to fight in the open; they are therefore supplied with plenty of broad gateways, and the problem of turning their rampart-walk into a defensible fighting platform is not taken at all seriously. As time goes on, walls become higher, gates become fewer and narrower, bastions are added to enfilade attacking parties, and *ballistaria* or gun-platforms are multiplied. The word *castellum*, from its original sense of 'little camp', is moving towards its medieval sense of 'castle'"[71].

We have seen how Collingwood contrasts in his *Autobiography* the scientific approach towards excavation with the one based on feelings of curiosity and nostalgia (Aut,123). It is interesting to note that in *The Archaeology of Roman Britain* there is also a passage where he refers to this difference, in explanation of the fact that so little is known of native settlements: "The number of native sites hitherto excavated is very small in proportion to the number of Roman. This is not because they are fewer; it is because they have been in the past less interesting to antiquaries in search of 'finds'. Sites of this humble quality are attractive only to the excavator who approaches his work from a rigidly scientific point of view, in search, not of collector's specimens or *objects d'art*, but of knowledge. And this attitude is rare"[72]. With regard to native villages Collingwood then makes a little further on a remark, which serves to refute the claim that it was his practice to predict the answers to the problems he posed: "The question how far, and why, these villages were abandoned in the latter part of the Roman period is one which has not yet been seriously discussed; nor is it very much use discussing it until more facts have been collected by excavation"[73].

The Archaeology of Roman Britain had a very positive reception. It was reviewed in no less than eighteen scholarly journals[74] and nearly all of them in eulogizing terms. Richmond says of it: "Mr. Collingwood ... grew up with most of the new facts, and that is part of the secret of his success, the other part being clarity of thought and expression. All who are interested in the Roman world, and, philosophically, in the attitude of mind which the study thereof develops, will read this book with admiration and gratitude. It provides the text-book which the study of the British province has long demanded"[75]. According to Birley "Collingwood's new book supplies the student of Roman Britain with a masterly and authoritative introduction to its archaeology"[76]. Wheeler says of the material discussed by Collingwood that it is "... for the most part, of a kind that is likely to remain valid for an almost indefinite time; with a periodical revision 'Collingwood's Handbook' is destined to serve many generations of students, and to serve them well"[77]. Bosanquet praises the book as "... a compendium of the highest value to all students of the Roman Empire, historians as well as archaeologists"[78], and even calls the last six chapters "miracles of judicious selection"[79], while it is described by Droop as " ... a masterly summary of the confusing mass of detailed evidence ..."[80]. To quote finally another reviewer, Casson called the book " ... a closely reasoned and documented study in which scientific methods of thought and classification are strictly adhered to, and in which the value of evidence is scrupulously tested before acceptance" and his conclusion is that "[o] ne can safely say that there is no aspect of Roman Britain which is not dealt with adequately (for students or excavators) in this volume"[81].

The importance of Collingwood's book is enlarged by the new areas which are explored in it. The classification of brooches, for instance, is called by Wheeler " ... the most original contribution to Romano-British archaeology", representing " ... the first comprehensive attempt to establish a systematic nomenclature for our Roman brooches"[82]. Likewise, in Richmond's judgment, " ... the classification of brooches covers new ground to most experts, and will form the basis for all future study of Romano-British brooches"[83] and he is joined by others in this appraisal[84]. According to Birley the chapter on fortresses and forts " ... contains more matter than a library full of excavation reports, set out with a clarity that few excavation reports attain"[85] and Casson says that Collingwood's " ... account ... of the native villages of Romanised Britain and his classification of Roman forts is a new contribution to knowledge of the first order"[86]. Similar positive judgments are expressed with regard to other subjects treated by Collingwood[87].

In 1969 *The Archaeology of Roman Britain* reappeared in a completely revised edition[88]. It was Richmond who had taken care of the revision, but the work appeared posthumously after the latter's death in 1965. In his preface the editor (D.R. Wilson) says: " ... Collingwood was the first to arrange the greater part of his material in accordance with typological principles, and this typology was largely of his own devising. Not all of Collingwood's ideas have stood the test of time, and much of his text has naturally been outdated by subsequent discoveries. A new edition has been long awaited, and its appearance now, thoroughly revised and enlarged by the late Sir Ian Richmond, must be regarded as an event equal in significance to the original publication"[89].

The impression that Collingwood's book left on a generation of archaeologists is demonstrated as well by the reviews of the revised edition. Thompson, for instance, asserts: "Those of us who are autodidacts in archaeology will perhaps recall how the original 'Collingwood', published in 1930, was the key to the whole magical world of Roman Britain ... [I]t was ... outstanding among Methuen's early series, *Handbooks of Archaeology*. It became a rarity, of course, seldom appearing in booksellers' catalogues and, to judge by its continued absence, purloined by the unprincipled from the shelves of reference libraries"[90]. Wheeler, having reviewed the original edition nearly forty years before, remarks on the new one with an air of nostalgia that "[h]ere and there, where the original text has been retained with verbal alteration, Collingwood's friends will sometimes miss the happy diction of the master"[91]. He also gives the following description from his memory of Collingwood's involvement and state of mind when working on the book:

> I saw much of Robin when he was writing the book. For him, with his essentially broad habit of thought, it seemed at first a trifle off-beat that he should immerse himself in so much museum-like detail at the expense of his more specific interests in the diverse fields of professional metaphysics, the wide movement and meaning of history, and the intriguing puzzledom of epigraphy, to say nothing of his periodical journeyings up and down Roman Britain. I remember wondering whether he was not really fashioning this textbook as a sort of imposed personal discipline. In many ways he was by nature a survivor from that long and distinguished line of amateurs – in the fullest and best sense of the term – who for three centuries or more had sustained the study of British antiquities as an inevitable and engrossing duty of the educated mind. The true end of this line had been marked by the brief but powerful transit of Pitt Rivers ... But Pitt Rivers had died in 1900, and now it was 1930 and time for a receptive mind like Collingwood's to get his details right, along modern lines. In the process, he might help other aspirants into new, often ill-sorted, knowledge; but that admirable motive, I felt sure at the time, was incidental to a primary mission to organize his own thinking[92].

EPIGRAPHY

In a treatment of Collingwood's archaeological work, however short, his activities in the field of epigraphy cannot be left unmentioned. Though he was well versed in the various scientific subdivisions of archaeology Collingwood was an acknowledged expert on Roman epigraphy. It was only in 1965 that the results of his work in this field were posthumously published in *The Roman Inscriptions of Britain*[93], edited by R.P. Wright, who had succeeded Collingwood in the project of editing the Roman inscriptions in Britain.

The background to this enterprise is the following[94]. In 1887 the responsibility for publishing the British inscriptions for the *Corpus Inscriptionum Latinarum* was transferred by the German historian Mommsen from the archaeologist Huebner to Haverfield. It was Collingwood whom Haverfield later attracted as draughtsman of the inscriptions. After the death of his pupil Cheesman in 1915 Haverfield tried in 1919 to give the supervision of the enterprise to the well-known ancient historian and Russian refugee Rostovtzeff, who left, however, for the United States[95]. After Haverfield's death in 1919 Collingwood then decided to go on with the work himself[96]. On Collingwood's activities Wright says: "Between 1921 and 1929 and again in 1936 Collingwood carried through the scrutiny, drawing, and fair-copying of most of the inscriptions on stone and of many of the inscriptions on other materials"[97]. The drawings were made by Collingwood with his own devised method of "contact-drawing"[98]. In 1938 Collingwood selected Wright as Junior Editor and when in 1941 Collingwood gave up his Roman epigraphy the work on the Roman inscriptions was entrusted to him[99]. *The Roman Inscriptions of Britain* is a monumental work with data on more than 2000 inscriptions. Nearly all are illustrated with drawings, the majority of them made by Collingwood. It is striking indeed to find one of the composers of this highly technical study referred to on the title-page as "Sometime Waynflete Professor of Metaphysical Philosophy", but it illustrates well Collingwood's versatility and the range of his interests and activities.

5.2.4. PLANNING OF RESEARCH

Collingwood's main contribution to archaeology was in the field of interpretation. *The Archaeology of Roman Britain* is the best example and we have seen that Collingwood himself saw its aim as " ... a first attempt to digest the mass of technical detail into a manageable form". Collingwood's interpretations of Hadrian's Wall, which will be dealt with later, are another example. On the analogy of what Collingwood elsewhere says about the relation between philosophy and science[100], one could call this the approach of "backward thinking"; that is, it is not aimed at the discovery of new finds, but at the interpretation of finds already made and sometimes even at framing theories from them. This is not to say, of course, that these interpretations of theories would not be relevant for future research. On the contrary, it has always been Collingwood's contention that they are and in the next section some relevant examples will be given. One can say, however, that at the level of interpretation and theory-formation one is not directly aiming at making new discoveries. It is obvious that this approach is especially useful, when new discoveries are increasingly abundant, as was the case with archaeology in Collingwood's time.

It is also obvious, however, that this approach is less useful for fields which are short of relevant data. This was the case, in Collingwood's view, with the prehistory of north-west England, the area where he had grown up and which always kept his archaeological interest. In 1932 Collingwood wrote an important article on this subject[101]. In his 1935 Report as University Lecturer he mentions it "... as one example of the 'prehistoric' studies which I have lately been undertaking in order to arrive at a more adequate conception of the substratum upon which the fabric of Romanized provincial life was erected"[102]. While Collingwood's ultimate aim is to provide a better understanding of certain aspects of Roman Britain, the scarcity of prehistoric data does make his approach here totally different from that taken to the archaeology of Roman Britain. What is primarily needed in this case is the discovery of more relevant data and for this reason Collingwood intends in his article to give directives for future research. In contrast with the interpretative approach and on the analogy mentioned earlier, one could call this an example of "forward thinking". In this connection it is an interesting example of Collingwood's treatment of a specific methodological aspect of archaeology, though it illustrates as well more general methodological principles.

"The aim of this paper is to sum up", Collingwood begins his article, "in the briefest possible outline, the present state of knowledge about the prehistory of our district ... in the hope of providing a basis for discussion and a framework for research to be carried on ..."[103]. The three requisites for further advance of prehistoric studies are, according to Collingwood: "First, to become acquainted with the discoveries already on record, and to think out their implications. Secondly, by new field-work to make fresh discoveries. Thirdly, to master the apparatus of the modern prehistorian, and, in particular to know one's way about the best recent literature"[104]. The first inventories of prehistoric finds in the area had already been made by Collingwood's father. On the next steps to be taken Collingwood says:

> The first step towards a prehistory of the district is to sort out the entries from these inventories and plot them according to their kinds on distribution-maps. This is what I have attempted to do in the present paper. The result of such a task should be a *vindemia prima* (as Bacon called it) giving a first approximation to the truth, and at the same time raising problems of a more or less definite

> and concrete kind which further field-work might solve.
> This preliminary digestion of the material would thus prepare the way for field-work, which ought to depend, not on the haphazard selection of interesting sites, but, partly at any rate, on the question 'what it is that, in the present state of our knowledge, we particularly need to discover; and where can we best discover it?'...
> The third thing is to get up-to-date in one's reading; and with this in view I have compiled a short bibliography to be printed as an appendix[105].

Of these three steps the first and the last could be interpreted as examples of Collingwood's principle always to start from the "history of history" in one's research. The larger part of the article contains a detailed survey of the prehistoric material so far found in the area. Five maps are added, which show the distribution of stone axes, great stone circles, stone axe-hammers, flint arrow-heads, flint knives or daggers, bronze implements, barrows, settlements and forts. At the end of his survey Collingwood provides a "summary of conclusions", described by him as "... a general sketch – very tentative, and presented merely as a working hypothesis to be tested by field-work and excavation ..."[106].

Under the head "suggestions towards a policy of research" Collingwood then gives certain concrete directives, this being in fact the most interesting part of the article. "The work needed for the advancement of knowledge on this subject", Collingwood asserts, "falls under four heads: office-work, field-work, excavation and publication"[107]. The first is divided into three aspects: "By office-work I mean work that can be done by examining and arranging material already to hand. This paper is an example of it. It includes the compiling, from published works or from museum cases, of lists of sites and finds, properly described and classified; the plotting of these sites and finds on distribution-maps, and the drawing of generalised conclusions from this work"[108]. Collingwood's comment on the last aspect is of special interest:

> The third, drawing conclusions or forming theories, is a thing for which few rules can be given; one has to ponder over the catalogue and the maps until some idea presents itself to the mind, and then test the idea as best one can. But there is one rule; never form theories except on the broadest possible basis: that is, a person making a study of one small district had better form no theories at all until he has brought his results into relation with those reached in adjacent districts. Cataloguing and map-making can be done bit by bit; theories can be formed only by putting these bits together into a whole[109].

It is clear that Collingwood is afraid that theories are formed over-hastily and are based on unsound evidence. Excavations, further, should be well planned, and this planning has to be done in the field: " ... work done at the sites themselves by every means short of excavation. This involves the study and planning of known sites and the discovery of new ones"[110]. This field-work could be described as "middle-range theory"; that is, it aims to provide background knowledge relevant for the excavation of certain concrete sites, rather than drawing more general conclusions. There are certain points, according to Collingwood, which demand special attention "under the general heading of field-work":

> First, we are very badly supplied with plans of ancient settlements, forts, circles and barrow-groups. It is most urgent that persons wishing to advance our prehistoric knowledge should set to work and make plans of such things. This is even more important, in the present state of our knowledge, than excavation; for, until we know more than we do about the appearance and general character of our sites, as shown by plans, intelligent excavation is much impeded by lack of preliminary staff-work. Also it is far easier and quicker to plan sites than to excavate them; and if anyone is tempted to plead that he can dig but does not know how to plan, the answer is that a person unable to draw a plan is for that very reason unqualified to dig[111].

In order to avoid the practice of "blind digging" Collingwood also gives some concrete examples of "particular questions which field-work alone can solve": "Are there always barrows close to ancient settlements? If not, can we recognise any particular type of settlement which in general has no barrows near it?" and "Where and how did the people of the great circles and the polished stone axes bury their dead?"[112].

Excavation is called by Collingwood "the ultimate test of all archaeological theories and the source of most archaeological knowledge". Though he asserts that "its necessity needs no emphasis" Collingwood warns that "... there is sometimes danger of forgetting that the scientific value of any excavation depends less on the site than on the excavator. He must be, to some extent, an expert, trained in the general work of excavation and in the special kind of problem presented by the site he is digging". With regard to the first of these qualities he insists that " ... the excavator ... must keep a detailed written record of the work, marking every trench accurately on his plan, and recording the exact place of every find"[113]. And "as examples of the particular problems towards which excavation might be directed forthwith" Collingwood mentions: "(a) What culture or cultures can be found by digging in the barrows of south and west Cumberland? ... (b) What is to be found in the barrows that lie close to ancient villages? ... (c) Where hut-villages lie close to Roman sites or to Roman roads, they are sure to contain Romano-British pottery if they were occupied during that period. Therefore it would be a comparatively easy matter to decide by excavation whether they were occupied then, or not"[114].

Collingwood's recommendation, finally, with regard to publication reads:

> To excavate a site is to destroy it; and every excavator is responsible to his contemporaries and to posterity for destroying what he digs. To dig without a due sense of this responsibility is a crime pardonable only in the vulgarest treasure-seekers; anyone professing any interest in antiquity is morally bound to accept the responsibility, and to discharge it partly by digging with all possible care and after due training and partly by publication. A dig whose results are not published is a dig whose results are lost to the world, which therefore ought never to have been undertaken[115].

This short consideration of Collingwood's article on prehistory clearly shows how his systematic mind worked not only on the interpretative aspects of archaeology, but also on those relating to the planning of its research. Here again not only his main interest, but also the value of his approach lies in the elaboration of particular principles[116].

5.3. HADRIAN'S WALL

5.3.1. INTRODUCTION

Britain was situated at the north-western corner of the Roman Empire. Since it was never completely conquered and no natural boundary was available – as in other parts of the Roman empire with the rivers Rhine, Danube and Euphrates – the Romans decided to build an artificial one. This is the famous Hadrian's Wall, built from around 122 in the reign of Hadrian. Large parts of it are still visible, winding in a beautiful scenery and forming a unity with it. It is certainly one of the most unique and impressive remains of Roman antiquity.

The Wall ran across North-England from coast to coast, from Wallsend in the east to Bowness-on-Solway in the west, being about 73 English miles in length (80 Roman miles). It was part, however, of a more comprehensive frontier-system. Other elements of this system included a V-shaped ditch on the northern side of the Wall of some twenty-seven feet wide and nine feet deep, and the Vallum, a flat bottomed ditch, twenty feet wide and ten feet deep, situated south of the Wall at a varying distance; parallel to it on both sides run mounds twenty feet wide and six feet high. Between the Vallum and the Wall a military road was constructed. Every mile along the Wall there was a milecastle and between each pair of them there were two turrets. Finally there were forts at various distances along the Wall, different in size and position, altogether seventeen in number. During the building of the Wall plans changed; originally it was designed to be ten feet wide; this was later changed to seven and a half feet. The remains of the Wall nowhere reach its original height; it was probably about fifteen feet high with a parapet of five feet on top. Originally only about 2/3 of the Wall (the eastern part) was to be built in stone, the rest of turf. Gradually, however, the latter part was replaced by a construction in stone.

This simplified description does not do justice to the many intricate questions that arise with regard to the Wall and the frontier-system in general of which it forms a part. It should especially be noted that not all its elements were planned from the beginning, the different phases of its construction having been a subject for discussion till the present day. We are not in a position, however, to go into this matter. Here my only aim is to give some essential data which will make Collingwood's involvement in the problems of the Wall understandable.

5.3.2. THE PURPOSE OF THE ROMAN WALL (1921)

In an obscure local journal, *The Vasculum* ("The North Country Quarterly of Science and Local History")[117], Collingwood wrote in 1921 an important article entitled "The Purpose of the Roman Wall"[118]. It is a clear illustration of his second principle of looking for the purposes of archaeological remains, but also of his first of asking the proper questions. The article begins as follows:

> Among the many questions which historians and antiquaries have asked concerning the Roman Wall... there is one that has been curiously seldom raised. What precise object had its builders in view? How exactly did they intend to use it, and how did they in fact use it?
> The question has not been raised because the answer has been taken for granted. It has always been assumed that the Wall was a military work in the fullest sense, a continuous fortification like the wall of a town, designed to repel or at least to check invading armies not, in this case, attacking the outskirts of any mere city but those of a province. The Roman troops have always been imagined lining the top of the Wall and from that strong position, entrenched as it were on the rampart-walk behind the parapet, repelling the attacks of Caledonian armies that attempted to carry the work by breach or escalade[119].

Against this "ordinary view", popularized by Kipling's *Puck of Pook's Hill*, Collingwood contends that all the evidence is against the idea of the Wall's garrison being supplied with artillery; it was only armed "in the ordinary Roman fashion" with spears and sword. A second argument against the conception of the Wall as a defensive barrier is its inconsistency with the general plan of the Wall itself: "The rampart walk cannot have been more than three or four feet broad, and that is a very narrow fighting-front", Collingwood asserts. "There would be barely room for a man to pass behind the actual firing line; no room, if the 'firing' line consisted of men throwing, with the necessary freedom of gesture, a heavy six-foot pilum. It would be practically impossible to reinforce a threatened point, even in the most favourable conditions: wholly impossible to move wounded men. And a few corpses, or a couple of Caledonians who had effected an escalade, would block the walk entirely"[120].

Other arguments against the traditional view, Collingwood mentions, are the facts that the Wall had no provisions for artillery and especially (called by him " a very striking piece of evidence") that bastions were never added to the Wall. By the fourth century the Romans were using bastions for defensive works and added them to existing ones in Britain. This was never done, however, with the Wall[121]. Taking these arguments into consideration Collingwood comes to the conclusion that the Wall cannot have been meant as a defensive work and proceeds therefore "to suggest an alternative explanation".

When conquering a country the Romans adopted the principle, Collingwood argues, of constructing a network of fortified posts. Since the time of Agricola these forts tend to be grouped in what may be called a defensive line, without being connected, however, by a wall. About the latter he then says:

> The continuous wall or fence, or ditch, begins about the same time or a little later. But in its origin it serves a different purpose from the line of forts; for while the latter contain bodies of troops intended to cope with armed enemy forces, the continuous line was at first designed to serve simply as a mark to show where the Roman territory ended. With this primary function was combined the secondary function, not always emphasized by the character of the work, of being an obstacle to smugglers, or robbers, or other undesirables[122].

"All this is abundantly clear", Collingwood remarks, "from the actual character of the German frontiers of the Roman Empire, which are closely analogous to the British. But the British frontier is apparently a later and more highly-developed example of the same type"[123].

The Vallum has been a puzzle to antiquaries, Collingwood says, but his theory not only provides a solution to this problem, but also suggests an alternative interpretation of the purpose of the Wall:

> The puzzle of the Vallum simply disappears when it is suggested that it was not a defensive work but a frontier-mark, a line indelibly impressed upon the earth to show the wandering native where he might not go without accounting for his movements. Here the line-element is emphasized, and the secondary obstacle-element is wholly absent, for even in its first youth the Vallum was no obstacle to anything short of wheeled traffic.
> The non-military character of the Vallum is to-day an accepted fact ... But the abandonment of a military theory of the Vallum

> seems to prepare the way for a similar process of thought in relation to the Wall ... The Wall which took the place of the slightly earlier Vallum was a work of the same pattern, belonging to the same series, as the Vallum itself: the series of frontier-marks whose primary function might or might not be combined with the secondary function of being an obstacle to raiders or smugglers. The Wall took the line of the crags not for tactical reasons but in order to increase the outlook of the sentries; for in essence the entire structure was an elevated sentry-walk, the sentries being supplied from the garrisons of the milecastles and having the turrets as their immediate quarters when on duty[124].

We will not discuss here the question whether Collingwood's interpretation of the Vallum and its relation to the Wall is a correct one. His interpretation of the Wall's purpose is still accepted by present-day archaeologists and historians. Birley, for instance, asserts: "The older antiquaries tended, for the most part, to think of the Wall as a defensive line, built primarily for military purposes ... It was Collingwood, as recently as 1921, who demonstrated that the methods of the Roman army and its equipment were such that the Wall could never have had military significance as a fighting-platform" and a little further on he says that "... ever since the publication of R.G. Collingwood's paper on 'The purpose of the Roman Wall' in 1921, it should have been evident that the Wall itself was in no circumstances expected to serve as a military fortification ..."[125]. In a recent book on Hadrian's Wall the same statement is made: "R.G. Collingwood pointed out long ago that the wall top did not serve as a fighting platform but as this idea constantly re-emerges his arguments bear repeating"[126].

5.3.3. HADRIAN'S WALL: A HISTORY OF THE PROBLEM (1921, 1931)

In the same year that Collingwood's article on the purpose of the Wall was published there appeared another article which illustrates in an exemplary way his third principle in the study of history. As the title of the article indicates – "Hadrian's Wall: A History of the Problem"[127] – the principle referred to is the one of studying the "history of history" or history of the second degree when one studies a historical problem.

Collingwood begins the article by saying: "The theories that have been advanced concerning the Roman Wall in England and its attendant works have been so many, so divergent, and at times so rapid in their succession as almost to justify the favourite taunt of irresponsible criticism, that their sequence is a matter of fashion or caprice rather than of rational development. Such a criticism, whether directed against historical, scientific or philosophical thought, hardly merits refutation"[128]. It is of interest to compare Collingwood's views on past theories of the Wall with his view on the general idea of studying past solutions of historical problems, as expressed seven years later in the 1928 lectures: "The history of history arises when the historian, in trying to solve a particular problem, proceeds by collecting and criticising the solutions which have already been offered. This collection and criticism of previous solutions may be done in two ways: either by treating the various solutions in a disconnected manner, dealing with each separately and discussing them in a haphazard order, or else by treating them historically, showing how each expressed a certain attitude which was itself an historical phenomenon, and established itself by criticising its predecessors"[129]. Or, as he expresses it a little further on discussing the fact that different solutions of a particular problem have been given: "If you are determined to get at the truth, you must begin by trying to reduce these differences to order, and this can only be done by discovering how the various accounts grew out of each other"[130].

It is obvious that Collingwood in his discussion of the history of the problem of Hadrian's Wall is not content to describe the various theories in an incoherent sequence, showing them to have been " a matter of fashion or caprice", but that he rather tries to demonstrate the "rational development" or logical order of them. Especially with the theories of the Wall as they had been developed during the last three centuries he sees a logical continuity: "The object of this essay", Collingwood says, "is ... to tell a plain tale, the story of the process by which, in the three centuries that have elapsed since Camden took it up, the problem of the Wall has been attacked first in one way and then in another till finally, within the last generation, a complete solution seems to have come within the range of possibility"[131].

The article is also worth considering for other reasons than being an example of Collingwood's principle of second order history. For it demonstrates as well his critical evaluation of various types of sources, among others traditional "authorities", and his conception of scientific excavation in accordance with question and answer. Collingwood's survey is built up logically: after a description of the remains of the Wall – or rather, of the frontier-system in general – there are sections on "the ancient authorities", "the native historians", "the period of surface inspection", "the beginnings of excavation" and "the period of scientific excavation".

Of the ancient authorities the most important is, according to Collingwood, Spartian (third century), who is confusing, however, since he first attributes the construction of the Wall to Hadrian, and then in a later writing to Severus. In the next century the Wall is ascribed by Victor to Severus only. Collingwood is critical of this authority, however; for the passage in which Victor refers to the Wall, he says, "...is simply 'lifted', with very slight verbal changes, from Spartian. It is not independent evidence but mere quotation"[132]. The four subsequent writers who mentioned the Wall, all copied Victor or each other, Collingwood contends. He therefore ends by giving the following critical appraisal of the ancient authorities, illustrating his principle of using authorities not as authorities but only as sources, to be dealt with critically:

> English antiquaries in general have been greatly impressed by the cloud of authorities for the story that Severus built a Wall in Britain. 'Here', they argue, 'are no less than six ancient writers all unanimous in ascribing a Wall – or may we not say *the* Wall? – to Severus. Such a mass of testimony greatly outweighs the one unsupported statement of Spartian that a Wall was built by Hadrian: and consequently it must be accepted as the first fixed point in any treatment of the problem that, whoever did not build a Wall in Britain, Severus *did*'.
> This argument ought to be finally disposed of by the mere chronological quotation of the authorities. It must by now be clear to the reader that there are not six mutually corroborative stories ... but one story due to one author, namely Spartian, and repeated by a number of compilers whose repetition adds nothing to its credibility. The testimony of six authors for the Wall of Severus, when valued by weighing instead of counting heads, is precisely equivalent to the single testimony of Spartian for the Wall of Hadrian[133].

Of the later "native historians" – the most important being Gildas and Bede in the early Middle Ages – Collingwood says: "These are no longer primary authorities standing in the tradition of ancient historians, but belong in a sense to the modern period in that they combine a certain archaeological knowledge of the remains with a certain literary knowledge of the ancient writers, and thus produce what can only be called *theories* of the Mural Problem ..."[134]. Bede especially is highly valued: "Here we have the first complete Mural theory, with a reasoned account of the Vallum, the Stone Wall and the Scottish Turf Wall. It is a fine piece of historical work ... Such a level of historical thought was not reached again, in connexion with our problem, for eight and a half centuries"[135].

From the end of the sixteenth century, beginning with Camden, the Wall was examined by surface inspection. The period during which this approach was practised ends with Horsley (1684-1732), of whom Collingwood remarks: "The whole period from Camden to (say) 1800 culminates in the work of Horsley ... For his period Horsley is as indispensable as Gibbon for his; and, bearing in mind the difference between the extent of their fields, Horsley is Gibbon's equal. With him we feel that we have emerged from a tentative and amateurish, a pre-scientific, study of the subject, in which grave oversights and fundamental errors are expected and pardoned, into an age of clear thinking, where problems are faced and evidence mustered in a scientific spirit. The eighteenth century in him, as in his contemporary and neighbour David Hume, reaches maturity"[136]. This positive judgment does not imply, though, that Horsley's theory is correct. On the contrary, Collingwood even comes to the paradoxical judgment that "[a] less clear-headed man would not have been driven to invent so absurd a theory ..."[137]. The cause of Horsley's wrong interpretation is to be found in his limitation of method: "It became clear that the method of surface inspection, combined with uncritical acceptance of the literary authorities, could be pushed not further. The method had reached a point where it only produced absurdities, and further thinking along the same line was useless. So far from removing difficulties, it was multiplying them. If the problem of the Wall was to be solved an absolute break was necessary: a new method must be devised and the problem approached with a fresh eye from a different point of view"[138].

This new method was provided by excavation, first practised by Hodgson (1779 - 1845). In the beginning, however, it also had its limitations. Collingwood mentions in this connection Clayton (1792 - 1890), who bought up land on which the Wall stood in order to excavate: "It was, of course, not what we call scientific digging. That had not yet been invented. It was pioneer work, and inevitably destroyed much evidence which to-day would be valuable: for Clayton's main object was only to clear the chief walls and to collect inscribed stones"[139].

We have seen that Collingwood considers the essence of scientific excavation to be a method which approaches a site with a specific question in mind. In the section on "the period of scientific excavation" Collingwood gives some striking illustrations of this method, particularly with regard to the problem of who should get credit for the construction of the Wall. According to traditional opinion the stone Wall was built under the emperor Severus. The last to express this view was Haverfield, whose theory was that a turf Wall was built under Hadrian and a stone one under Severus. The way this theory was proved wrong is explained by Collingwood as follows.

In his excavation of a turret along the Wall Gibson found in 1891 that there were three floors, one above the other, with a layer of burnt matter and rubbish between them. He concluded from this that the turret had been built and destroyed three times. So there arose now the problem not only of the original construction of the Wall, but also of subsequent destructions and reconstructions. In another fort Gibson also found evidence of three occupations. How these discoveries ultimately proved that Hadrian was the builder of the Wall is described by Collingwood in the following way:

> Gibson, now assisted by Mr. F.G. Simpson, in 1909-10 explored the milecastle at the Poltross Burn, and there made some remarkable discoveries. The familiar three floors were not only identified once more, but it now became possible to date them. The lowest represented an occupation beginning in the first half of the second century and ending disastrously about 180: a disaster obviously to be connected with Dio's story of the British war of 181. The

> second floor ended in another disaster probably soon after 270: and the third lasted down to about 330. These dates were established on quite satisfactory coin-evidence, and proved that the milecastle went back to Hadrian. But the Hadrianic floor-level was found to overlie perpendicularly the foundations of the stone Wall: which showed that the stone Wall forming the north wall of the milecastle could not be the work of Severus – for in that case its foundation-trench would have cut off the edges of the Hadrianic floors – but must itself be Hadrianic. This suggested that the whole stone Wall was Hadrianic too: but it might still be argued that Hadrian's Wall was of turf between the milecastles, and the milecastles themselves of stone.
>
> There was obviously one way of settling this question. For two miles at Birdoswald the turf and stone walls lie apart. If the stone Wall is Severan, and if, of the well-known three floors, the lowest is Hadrianic and the second Severan, there will be buildings on the stone Wall in this sector in which the first or Hadrianic floor is absent and the two later floors alone present. If all three floors are present, that proves that the stone Wall is the work of Hadrian and not of Severus.
>
> In 1911 Gibson and Simpson examined this section of stone Wall and, on digging the three turrets and one milecastle which it contains, found in every one the complete series of three floors, together with a sufficiency of dated material to make it absolutely certain, quite apart from analogy with other sites, that the lowest floor and therefore the stone Wall dates from the first half of the second century.
>
> The Turf Wall theory was thus exploded. It could no longer be maintained that Hadrian had built a turf wall, replaced in the time of Severus by one of stone[140].

After the death of Gibson, Simpson took up the problem of the relative chronology of the Wall, the Vallum and the forts. In his description of Simpson's treatment of this problem Collingwood gives a clear example of the logic of question and answer in practice. For with regard to the question of the chronological relation between the Wall and the forts he asks about the latter: "When they were built, was it planned at the same time ... or was it an afterthought?", after which he continues:

> This question could be answered by digging. If the forts were designed as separate works, they would have ditches round them, which would have to be filled up when the Wall was carried across them. This, Mr. Simpson finds by excavation, is really the case: filled-up ditches underlie the Wall in such a way as to show that the forts were at first isolated works and that the building of the Wall necessitated altering their plan to the extent of filling in their ditches. Then arises the further question: granted that some of the forts were enlarged, as Haverfield has shown they were, was

> this enlargement done before or after the Wall was built? Excavation again answers the question. Both at Birdoswald and at Chesters it is perfectly clear that the forts were enlarged first and the stone Wall built up to them later[141].

When discussing the conception of the history of history in his 1926 lectures Collingwood says: "The real formula will run: 'A thinks it was thus; B thinks it was thus; C thinks it was thus; *and I*, having diligently studied their views and all other evidence, *think it was thus*'. Here the history of history culminates where it ought to culminate, in the present"[142]. Though Collingwood concentrates in his survey of the history of the problem of Hadrian's Wall on methodological aspects, he demonstrates that here the same principle holds. For, after the description of Simpson's theory, he asserts at the end of his article:

> No earlier theory has been tested throughout by the spade. From Bede to Bruce, the theories turned on mere inspection of the visible remains, reinforced to a greater or less extent by dependence upon the ancient historians. As knowledge of the remains increased, the balance between these two sources shifted. It became clear that the ancient authorities were neither sufficient nor wholly reliable; and the possibility gradually came in sight of reconstructing the history from an intensive study of the remains, carried out by digging. This is the only method which has not broken down in the hands of the user, and by this method the results of Haverfield, Gibson and Mr. Simpson have been reached[143].

The question whether Collingwood is correct in all the details of his article cannot be discussed here and is also outside the scope of the present study[144]. Of Collingwood's treatment of the ways in which the problem of Hadrian's Wall had been dealt with it can be asserted with certainty, however, that he was right in valuing highly the new approach of "scientific excavating". He was very much aware of its revolutionary nature, but in order to interpret it as such it must be compared with previous practice. This is that Collingwood has done in such an admirable way in his article.

In 1931 Collingwood wrote an article in which he gave a report of the work done on the Wall during the previous ten years[145]. At the beginning he discusses "the problem of method" in a way which fits in well with the treatment of the same subject in his article of ten years before. Of the time when the earlier article was written Collingwood now comments: "Ten years ago, in spite of the few professional scholars to whom I have referred, there was not even the nucleus of an expert staff; the bulk of the work remained where it had always been, in the hands of the amateur and the local antiquary"[146]. Because of this excavators had to cope with a scarcity of funds and the results were only published in local transactions. Collingwood then develops a most interesting argument: "In spite of these drawbacks, progress of a remarkable kind had been made by 1921. I am not sure that the drawbacks did not, in a sense, facilitate the progress. The problem to be solved was so peculiar in certain of its features that, if its solution had been undertaken by the methods current among professional archaeologists, the discovery of the right method could only have been delayed"[147].

The current method among professional archaeologists was to excavate a site as completely as possible. With the system of Hadrian's Wall, however, this would be completely impossible, Collingwood argues: being 73 miles long and around a quarter of a mile broad it contained no less then sixteen square miles to be excavated. "The excavation of Silchester took twenty years", Collingwood remarks in this connection; "at that rate, to excavate the Wall would have taken two thousand. It would therefore have been useless to apply the ordinary methods of excavation to the Wall"[148]. The forts along the Wall excavated in the traditional way did not yield an immediate advancement of our knowledge of the Wall, Collingwood contends. "That was progressing all the time", he then remarks, "but it was progressing chiefly through amateur work, which enjoyed no publicity and was able to develop, almost under cover of secrecy, a method adequate to its peculiar problem". Collingwood then continues:

> This method consisted in what I may call selective excavation. The whole sixteen square miles of the site were first of all intensively studied on the surface; much of this study was done by Horsley in the eighteenth century, and Hodgson and MacLauchlan and Bruce in the nineteenth. Tentative theories were then put forward to explain the whole complex of works: and finally these were tested by bringing the problems to a focus at particular points where they could be solved, or at any rate advanced, by some quite small piece of excavation, planned and supervised and recorded with the utmost care[149].

Collingwood is not consistent in his argument, however, when he says after the passage quoted above that " [t]he classical instances of this procedure are set forth in Haverfield's reports ..." and that " [i]t is chiefly to Haverfield that we owe this method". For it would certainly not be correct to consider Haverfield an exponent of amateur work[150].

The new method and its relevance for the study of the Wall is described by Collingwood in the following way:

> The method of selective excavation is not the method traditionally employed by professional archaeologists. They have generally begun by thinking of a site as a unit that admits of complete excavation, and their ideal is to excavate it completely. While part of a site remains unexplored, they think that their duty to that

> site remains undone. Therefore, when they look at the digging hitherto done on the Wall, and this they have begun to do oftener in the last ten years, they are offended by its scrappiness, its incompleteness, in a word, its selectiveness. They would prefer, and they have been heard to advise, the complete excavation of some chosen site on the Wall; in order that, in this one instance at least, we should know what the facts in their entirety are.
> To this I would reply: the methods here in question are methods intended to increase our knowledge, not of Housesteads or Birdoswald or Chesterholm, but of the Wall. The Wall is our unit; and the Wall is a hundred times the size of Silchester. The only way in which we can hope to solve the problems of a site is to keep steadily before our minds an idea of the site as a whole, and to direct every detail of our work towards that idea. Where a site is so large as this, the difficulty of seeing it as a single unit is correspondingly great; and it is all the more necessary to insist on the idea, and to reject any proposed method of work that is not based upon it. There may be good reasons for taking some single Wall site and excavating it completely; but these reasons cannot include among themselves the advancement of our knowledge of the Wall; and any such undertaking, in so far as it diverts men and money from the study of the Wall by the true method of selective excavation, must directly impede that study[151].

Though according to Collingwood the pioneer work of the research on the Wall was realized by amateurs he is of the opinion that there are signs that their principles are taken up by "a generation of workers trained in Roman history and in the archaeology of the Roman provinces": "They accept, as any one must who understands the problems of the Wall, the method of selective excavation; in fact, they delight in it, as a method scientifically superior to that of complete excavation; demanding more constructive thought and, in consequence, yielding a richer return of knowledge in proportion to the expenditure of time and money. In number they are small; but the significant fact is not their numbers but their existence, showing as it does that the study of the Wall has now arrived at a point where professional archaeologists are willing to take it up and make it their chief occupation"[152].

5.3.4. HADRIAN'S WALL AND THEORY

It is obvious that when Collingwood emphasizes that what is needed in the study of the Wall is first and foremost the thinking out of a general idea of the Wall to be kept in mind in the actual research, we come to a point where his views on archaeology, history and philosophy of history are closely related to each other. We may see here as well that these different intellectual concerns of Collingwood have a positive and fertilizing influence upon each other.

This fortunate cross-fertilization of Collingwood's various interests and experiences, inspired by his creative and imaginative mind, is well illustrated by his involvement in the research on Hadrian's Wall. The importance of Collingwood's activities with regard to the study of the Wall has been in the field of its general interpretation rather than in actual excavations. In a note in his article of 1931 he remarks: "My own part in the work has been that of an interested spectator, with whom the workers have always been willing generously and without reserve to discuss their hypotheses, their methods, and their results"[153]. In his role of an interested – and, one should add, highly competent – spectator Collingwood was in an ideal position to evaluate the results of the various excavations within the context of a general idea of the Wall, to formulate certain hypotheses with regard to them, and to emphasize the need to excavate in a selective way, with specific questions in mind. It is clear that this approach is not only an example of his logic of question and answer, but that this logic in its turn may be considered the elaboration of his experience in archaeology, a fact which is mentioned by Collingwood himself in his *Autobiography* (Aut, 24-25,30).

In April 1937 Collingwood delivered a lecture on "John Horsley and Hadrian's Wall"[154]. We have seen that in his article of 1921 on Hadrian's Wall Collingwood expressed a high esteem for this eighteenth century antiquary[155]. Working out Horsley's ideas and approach with regard to the study of the Wall it is interesting to see how Collingwood projects in his lecture of 1937 his own ideas on those of Horsley, using them as a standard for his appraisal. Comparing Horsley with contemporary antiquaries Collingwood contends:

> The contrast between Horsley and all these other writers is startling; and the more closely it is studied, the more striking it becomes. The difference, when one analyses it, is not a mere difference of degree – Horsley's being immeasurably the best – but a difference of kind. These other writers were travellers, visiting ancient remains and recording what they happened to see. Horsley studied the Wall in a quite different way: as a field archaeologist, approaching the remains with definite questions in his mind, and returning to them again and again until he could read the answers. If we thus distinguish between antiquarian travel and field archaeology, we must recognize in Horsley's study of the Wall the first piece of field archaeology done in this country, and in many ways the best[156].

The questions to be asked in archaeological research should not be asked of course haphazardly. In order to be meaningful they should be posed from a certain theoretical background and the answers given to them should be evaluated by reference to this background. In this process the theory may have to be revised – a thing which happened often enough with Hadrian's Wall – with the result that new questions will have to be asked etc. The need to have a theory guiding the question and answer activity is mentioned explicitly by Collingwood nowhere else in his work. It is for this reason of importance to see how he discusses the concept of theory in his lecture on Horsley. What Collingwood especially values in Horsley is the fact that he explicitly intended to frame a theory of the Wall and that he did it, in his opinion, in an admirable way. We should keep in mind in this connection, however, the limitations to which Horsley was subject in his time. For he could only make use of two sources of archaeological remains – surface inspection and the texts of ancient writers. Discussing the way these sources were used by Horsley in framing a theory of the Wall Collingwood argues as follows on theories in general:

> A theory, in this connexion, means a way of combining two sets of data so that they fit into each other. We have, first, the remains themselves as we have become acquainted with them through archaeological study; and secondly, the ancient historical texts referring to them. These are the two sets of data. The business of a theory is to bring them into relation, so that what is mentioned in the texts can be identified on the ground, and *vice versa*. But this way of putting it somewhat over-simplifies the facts. You would not expect your texts to mention every detail of the remains; consequently you must be allowed to dot the i's and cross the t's of the ancient writers, to supplement what they tell us by reading into their words implications which are not expressed there. Conversely, you may be sure that some things mentioned in the texts will have left no visible trace in the remains as they now exist before your eyes; consequently you must be allowed to reconstruct these remains in your head, in the hope that as so reconstructed they will tally with the statements and implications of the texts. This is as much as to say that the two sets of data which have to be fitted together are not rigid data, like the pieces of a jigsaw puzzle, but flexible data, like a foot and a shoe, which accommodate themselves to each other if you handle them skilfully. And the consequence is that the difference between a good historical theory and a bad one is not like the difference between assembling a machine rightly and assembling it wrongly, so that in the one case it will work and in the other case it will not, but like the difference between a good fit and a bad fit in shoes. You can force your foot into an ill-fitting shoe, but the two things irk each other; the foot blisters, the shoe goes out of shape. In the same way, bad historical theories can seldom be logically refuted; but we can feel and locate the strains which they impose on the delicate organism of archaeological thought on the one hand or the scholarly interpretation of texts on the other[157].

What corollary may be drawn from this exposition with regard to modern archaeological practice? It is obvious that the latter is completely different from that of Horsley's time. For the ancient writers have lost their importance and the remains are not studied by surface inspection but by systematic excavation. In the case of Horsley's practice Collingwood defines a theory as "a way of combining two sets of data so that they fit into each other". In modern practice, however, there is only one set of data, namely the finds provided by excavation. But the role to be played by theory has not lost its importance. For not only do the various data have to be combined so that they fit into each other, but also it is from the standpoint of theory that the relevant questions have to be asked.

It is not surprising to see Collingwood as a philosopher stressing the importance of theory in archaeology and we have seen that it is precisely this attitude which he praises in Horsley. Though there has been a revolution in archaeology since then, this has not been one concerned with the need for sound theoretical interpretation. With regard to the latter, Collingwood even emphasizes the continuity between Horsley and the present time, regarding the revolution in archaeology only as a technical one. "Excavation, in Horsley's day undiscovered", Collingwood says, "has given the archaeologist a weapon as powerful as the astronomer's telescope or the industrialist's steam-engine. The scientist's business is to theorize on a basis of fact, and when a revolution in technique reveals a new order of facts he has to reorganize his whole world. The old theories are superseded; the old controversies become meaningless; the old problems become unimportant"[158]. After this Collingwood digresses on the idea of a scientific revolution – and this is important enough to be quoted in full:

> In such a landslide, it is difficult for any scientist to keep his head; impossible perhaps, were it not that revolutions of this kind (like all revolutions, when you understand their true history) happen very gradually. The temptation is to think that such a revolution changes everything; whereas in fact it changes only one thing – the materials which in his theorizing the scientist has to use. The ways in which he can use them remain unchanged. A revolution in scientific technique reveals a new order of facts; but, once revealed, they must be reflected on exactly like any others. The starting-point of the work of theorizing has changed, but the nature of that work itself is unaltered.
>
> This tends to be overlooked by people who have recently acquired a new scientific technique. They become engrossed in the use of their new method, and accumulate fact upon fact, forgetting that to the scientist facts are useless except so far as they become a basis for theories. Absorbed in the excitement of hunting new facts, they neglect the discipline of theoretical thinking, and in consequence an advance in scientific technique is often accompanied by a falling-off in the quality of scientific thought.

> Improvements in scientific technique are a mixed blessing, because they delude us into thinking that they make scientific work easier, whereas really they leave it just as hard as ever it was. Repeatedly, in the history of science, you will find that the best theoretical thinking is done just before a revolution. It is, in fact, the high quality of this thinking, the finality of its conclusions on the basis of existing evidence, that necessitates the revolution, by convincing men that nothing more can be done until we have learnt to explore a new region of facts.
> Of this generalization Horsley is an instance. In the history of archaeology there has never, so far as I know, been a better field-worker, and never a thinker more highly gifted or more successful in distilling the entire bulk of the data into a coherent and acceptable theory. On the subject which I am here discussing he said, so far as the evidence available in his day was concerned, the last word. It was not until the archaeological revolution of the nineteenth century, which gave us the technique of excavation, that anyone could find a flaw in his theory[159].

We are not in a position to give an assessment of Collingwood's judgment on Horsley. It is possible that he is idealizing the latter's work in an unacceptable way[160]. Our interest, however, is directed to the way Collingwood expresses his own views in his treatment of Horsley. His final judgment on him is clear enough: " ... none have surpassed Horsley in the sheer scientific quality of his thought"[161]. Collingwood ends his lecture by appealing again to the importance of theory in archaeological research. Mentioning the problem of the Vallum he asserts: "Does the solution await the discovery of still further technical methods, as yet undreamed of? I suspect that it does not; I suspect that what is at fault is not our technique but our logic; and that if our methods were used by a man equal to Horsley in the faculty of clear and coherent thinking, they would yield a solution which, even if not final, would satisfy the twentieth century as his satisfied the eighteenth"[162].

After this digression on theory it is interesting to point out Collingwood's own contributions to the theory of Hadrian's Wall. One of them we have discussed already, namely his interpretation of the purpose of the Wall. We have mentioned the lasting acceptability of his view of the Wall as an elevated sentry-walk rather than a solid defensive barrier.

Thinking out the consequences of this interpretation, Collingwood came to another theory (or, in his own words, "A question answered causes another question to arise" (Aut,129)). It is described in his *Autobiography* as follows:

> If the Wall was a sentry-walk, elevated from the ground and provided (no doubt) with a parapet to protect the sentries against

> sniping, the same sentry-walk must have continued down the Cumberland coast, beyond Bowness-on-Solway, in order to keep watch on vessels moving in the estuary; for it would have been very easy for raiders to sail across and land at any unguarded point between Bowness and St. Bee's Head. But here the sentry-walk need not be elevated, for sniping was not to be feared. There ought, therefore, to be a chain of towers, not connected by a wall but otherwise resembling those on the Wall, stretching down that coast. The question was, did such towers exist? (Aut,129).

This theory was developed by Collingwood for the first time in a paper read in April 1928[163]. He expresses in it the opinion that the view of the Wall as in essence a chain of signal-stations " ... necessarily re-opens the question of the Cumberland coast"[164]. He then begins with a comparison with the system of signal-stations on the East-Coast against the Saxon sea-raidings, saying of them that " [t]hese fourth-century signal-stations are descended by a perfectly regular genealogy from the turrets of the Wall and similar works"[165].

Collingwood also points to the analogous situation along the river Danube, where the emperor Commodus set up a system of signal stations: "Frontier raiding, which ... explicitly accounts for Commodus's works on the Danube, is the true explanation of Hadrian's Wall; and the situation on the Solway is like enough to that on the Danube to suggest that in Cumberland, as in Pannonia, petty raiders might be provided with boats"[166].

"When these facts are borne in mind", Collingwood concludes, "it seems very unlikely that the Hadrianic system of signal-stations from Wallsend to Bowness-on-Solway should stop at Bowness"[167]. Some signal-stations had indeed been discovered before on the Cumberland coast and Collingwood mentions is this connection a certain Robinson who found four of them in 1880[168]. "It was clear enough from these finds that there was, as Mr. Robinson said, 'a system' of these towers", Collingwood says, "But no one except himself seems to have recognised this at the time ... It was nearly half a century later that the subject cropped up again from another quarter. The present writer, trying to think out the conception of the Wall as a chain of signal-stations, realised that it has a logical consequence; namely the existence of other signal-stations along the Cumberland coast. This led to an examination of published materials and so to the discovery of the fine field work already done by Mr. Robinson. It remained to visit the coast and look for the sites of other signal-stations; for one thing was clear, that considerable numbers of them must await discovery"[169].

In July 1928 Collingwood made a short survey of the Cumberland coast for remains in the places where they should be according to his theory and indeed found some. He warns, however, at the end of his report that "... emphasis must be laid on the fact that this paper consists largely of guesswork... To confirm or controvert the conjectures here put forward, therefore, much field-work is required; in the first instance, repeated examination of suspected sites under winter conditions, and then excavation"[170]. Subsequent research has amply proved Collingwood's theory to be correct: a complete system of signal stations and forts was found on the Cumberland coast from Bowness to St. Bee's Head[171]. It is undoubtedly the most concrete result achieved by Collingwood's theoretical approach to Hadrian's Wall and lasting evidence of its fruitfulness.

Another lasting contribution by Collingwood with regard to Hadrian's Wall was his design of a numbering system for sites on the Wall, which has been in use since till the present day[172]. In this system all milecastles, turrets and forts are numbered, beginning from the east.

Collingwood's work on Hadrian's Wall was not confined, however, to scholarly contributions. For he also wrote a popular and most readable *Guide to the Roman Wall*[173] and took care of the ninth edition of *The Handbook to the Roman Wall* by J. Collingwood Bruce[174], whose first edition had appeared in 1863, and which was in fact completely re-written by Collingwood[175].

5.4. HISTORY OF ROMAN BRITAIN

5.4.1. *ROMAN BRITAIN* (1923, 1932)

In the field of the history of Roman Britain Collingwood also made important contributions, especially, as with Hadrian's Wall, with regard to general questions of interpretation. In 1923 his *Roman Britain* was published[176], written for a general public. A second edition, which appeared in 1932[177], was in fact mostly re-written. What about the relation between archaeology and history in Collingwood's work? At the beginning of this chapter it was stated that, though interrelated, these subjects should not be equated. Discussing the fact that in traditional histories of Rome, Britain gets little attention, partly because ancient writers rarely refer to it, Collingwood himself asserts the following concerning their relation: "Granted the poverty of these ancient references, can we hope to know anything at all about Roman Britain sufficiently certain and sufficiently detailed to be worth knowing? We can; but only by using the methods of archaeology. What we know about Roman Britain is not derived, except to a very small extent, from reading old books. It is derived from studying, in a systematic and accurate manner, the remains left by the Romans on British soil, and interpreting them in the light of everything else that we know about Roman history"[178].

The difference, though, between archaeology and history of this kind is not only that in the latter the archaeological data are interpreted in the more general context of Roman and British history. Of more importance is the fact that in a history of Roman Britain different questions are asked than in an interpretation of, for instance, Hadrian's Wall. Though Collingwood does not mention this explicitly it is only natural to see the matter this way, in view of the importance he attaches to the question and answer activity in the study of history and archaeology. The point is also made clear by the specific questions he actually asks in connection with the history of Roman Britain, which indeed are of a special nature.

The problem from which Collingwood starts, and which is also the general theme of his study, is the question how the "Roman period" in British history should be interpreted. According to the traditional view it is to be seen as an isolated period. Or, as it is described by Collingwood: "The Romans came, conquered, and departed, and left no mark except the ruins of their buildings. When the Saxons landed, Britain was once more a country of Celtic tribes living in a state of barbarism and mutual warfare"[179]. From the Roman side as well Britain is usually considered of little importance: "Histories of Rome ... think of Britain as a frontier province, held by the army but untouched, or relatively untouched, by civilization ... They also tend to pass lightly over the whole subject of Britain because of the poverty of references to it in ancient writers"[180].

Roman Britain is divided into the following chapters: "Introduction" (in the 1932 edition changed to "Britons and Romans"), "History of the Conquest and Occupation", "Town and Country LIfe", "Art and Language", "Religion" and "Conclusion" (in the 1932 edition changed to "The End of Roman Britain").

The 1932 edition is mostly re-written and it is interesting, of course, to compare the earlier and later editions. Many changes are certainly due to the growth of knowledge on the subject between these dates. We will not deal, however, with this aspect, but only mention some changes in interpretation.

Some of the latter are striking. In the 1932 edition the questions are not only sometimes posed in a slightly different way, but the answers given are, on the whole, theoretically better worked out and based on a theory of British Romanization which is not to be found in the earlier edition. Generally speaking, one could say that in the later edition there are more references to certain general principles. To give one simple example: on the question whether it is correct to consider the Roman period only as an episode in British history, Collingwood makes the following statement, which is found only in the second edition: "No doubt the Roman occupation would be an isolated episode in English history if there were such things as isolated episodes in history. But there are not"[181].

In the introduction to the 1923 edition Collingwood warns against the danger of anachronistic interpretations of Roman Britain, particularly in the form of comparisons with certain characteristics of modern imperialistic empires. "The Roman, compounded of Celtic and Mediterranean elements, could claim kinship, physical and spiritual, with every one from the Tyne to the Euphrates and from the Sahara to the Rhine", Collingwood asserts. "It is this that makes the Roman Empire a quite different thing from all modern empires. The empires of modern times are rent by a racial cleavage between a governing race and a governed, which are too far apart to unite into a single whole. We have barriers of colour and race and language which were absolutely unknown in the Roman world ... Hence all attempts to understand the Roman Empire by comparison with, say, the British rule in India or the French in Algeria are frustrated by a false analogy ... The Roman Empire was a society of peoples in which intercourse was nowhere checked by barriers such as separate races or even nations in the world of to-day. That can be proved by the three tests of travel, residence, and marriage. In these three ways the Roman Empire was far more cosmopolitan than modern Europe"[182].

The problem of the anachronistic interpretation is formulated differently in the 1932 edition: "Was the Roman occupation of Britain the occupation of a barbarian country by foreign conquerors, who held it and ruled it for three centuries and more without leaving any mark on the natives? In other words: was there, in race, language, and civilization, a gulf between the Romans and the Britons, of the same kind as there is between, say, the British and the natives in the Sudan?"[183]. So in the first edition travel, residence and marriage – examples of certain habitual practices – are mentioned as tests for the cosmopolitan character of the Roman Empire, while in the later edition more fundamental questions are raised concerning race, language and civilization. On the first of the latter questions Collingwood says: "Rome was never the name of a race or stock. It was the name, first and foremost, of a city; and, secondly, of an empire. This empire was never, from its earliest days, homogeneous in race or blood; its homogeneity lay in its law and its civilization"[184]. With regard to language Collingwood says that "... the ancient world was always a polyglot world"[185] and on the question of civilization he remarks: "... [T]he Britons whom the armies of Claudius conquered were by no means savages. The more advanced tribes at least had a very considerable civilization of their own ..."[186].

With regard to the Roman occupation of Britain Collingwood considers as important questions not only its general nature, but also how it came to its end. This second question is the more compelling if it is compared with Roman Gaul. For after the departure of the Romans the latter kept, in contrast with Britain, its Romanized character to a considerable degree. It is in connection with this question that the answers given in the two editons differ in a way which is striking indeed. In the first edition Collingwood maintains:

> [W]hy did Roman Britain not carry on its Roman tradition into the Middle Ages, as Roman Gaul did? In a word, the answer is that Britain had more and deadlier enemies, who succeeded in destroying her civilization. Gaul defeated Attila and absorbed the Franks; her Romanized population weathered the storm, and their

> Latin speech developed quietly and steadily into the dialects of French. Britain was less fortunate. Romanized though she was, she was not so thoroughly Romanized as Gaul: her civilization, it has been well said, 'like a man whose constitution is sound rather than strong, might perish quickly from a violent shock'. The shock was administered by the triple invasions of Saxons, Picts, and Scots, enemies more dangerous, because harder to crush, than Attila himself. And, just when the danger was greatest, a succession of usurpers drained Britain of troops to support their own claims to the throne of the Empire. But for these facts England would to-day be speaking a Latin tongue, though in race she would perhaps be no less and no more Teutonic than she is[187].

In the second edition, however, the same question is discussed as follows:

> The Britons, then, became Romans; Romans in civilization, in speech, in patriotism and sentiment. At the same time, they did not cease to be Britons; their participation in the cosmopolitan life of the Empire was not of such a kind as to swamp or obliterate their original character and peculiarities. The business of this book is to show how this happened, to show in what ways the Britons became Romans and in what ways they remained Britons.
> There is one test by which any answer to this double question must be judged. Roman Gaul was sufficiently Romanized to survive the barbarian invasions, to turn Franks into Frenchmen, and to preserve the Roman tradition and language into the Middle Ages. In Britain, this did not happen. If the fate of Roman Britain had been like that of Roman Gaul, we should now be speaking a Romance language. Why is it that, when Gaul defeated Attila and absorbed the Franks, when the Gaulish cities weathered the storm and developed into the towns of medieval France, the opposite results came about in Britain? We cannot now say 'Because Gaul was Romanized and Britain was not'. We must discover exactly what kind and degree of Romanization came about in Britain; and if we can do that, the ultimate fate of Roman Britain will become intelligible, and the Roman occupation, instead of seeming a mere irrational episode in English history, will reveal a logic of its own, not without significance for the meaning of civilization and the fate of empires[188].

So in the first edition the end of British Romanization is seen as the result of bad fortune on Britain's part. Its fall was caused by outside forces: a violent shock "administered by the triple invasions of Saxons, Picts and Scots", combined with the draining of troops by usurpers. The causal relation between the two events is seen as a strong one, the invasions and draining of troops functioning as a sufficient and even necessary condition: "But for these facts England would to-day be speaking a Latin tongue ..." This interpretation is in sharp contrast to that in the second edition. No emphasis is put here on outside forces; it is certain characteristics inside Roman Britain itself that are regarded as the cause of its decline, particularly the nature of its Romanization. The latter conclusion may be seen as an illustration of Collingwood's mature conception of history with its emphasis on the way "objective conditions" are conceived and reacted to by people rather than on those conditions in themselves.

In contrast also to the earlier edition the second one does not give an immediate answer to the question concerning the end of Roman Britain. After posing the problem, only an indication is given as to where we have to look for an answer. It is interesting to note in this connection that we must discover, according to Collingwood,"... exactly what kind and degree of Romanization came about in Britain". Collingwood interprets this pivotal concept of Romanization as a philosophical concept, as envisaged by him in *An Essay on Philosophical Method*. As we have seen, such concepts cannot be precisely classified or defined, and their species differ in both kind and degree, overlapping on a "scale of forms", and realizing the generic essence in varying degrees[189]. At the end of the chapter on town and country life the concept of Romanization is explicitly discussed by Collingwood. He comes there to the conclusion that "... we cannot be content simply to assert that Britain was Romanized. The civilization which we have found existing in the towns, the villas, and the villages is by no means a pure, or even approximately pure, Roman civilization bodily taken over by the conquered race" [190]. Collingwood then gives the following specification of "a scale of Romanization", describing how its generic essence is embodied in varying degrees:

> What we have found is a mixture of Roman and Celtic elements. In a sense it might be said that the civilization of Roman Britain is neither Roman nor British but Romano-British, a fusion of the two things into a single thing different from either. But this is not a quite satisfactory way of putting it; for it suggests that there was a definite blend of Roman and British elements, producing a civilization that was consistent and homogeneous throughout the fabric of society. The fact is rather that a scale of Romanization can be recognized. At one end of the scale come the upper classes of society and the towns; at the other end, the lower classes and the country. The British aristocracy were quick to adopt Roman fashions, but the Roman fashions which they adopted were rather those of Roman Gaul than those of Rome itself, so that their borrowings are already Romano-Celtic rather than Roman. But this Romano-Celtic civilization gradually becomes less Roman and more Celtic as we move from the largest towns and largest villas to the small towns, the small villas of humbler landowners, and lastly to the villages. Here we encounter a stratum of the population in whose life the Roman element hardly appears at all;

> if we must still call their civilization Romano-Celtic, it is only about five per cent. Roman to ninety-five Celtic[191].

This view of Britain's Romanization also provides an explanation for the later de-Romanization after the invasions. For although Collingwood denies that the un-Romanized peasants made common cause with the invaders, he asserts the following about the latter's actions and their consequences:

> ... [T]hese raiders must necessarily have attacked the wealthier and more Romanized elements in the population, and thus de-Romanized the British countryside by the simple process of sacking the villas. Consequently, from the late fourth century onwards, Britain became less Roman and more purely Celtic, not because the Roman element was composed of foreigners who left Britain at the so-called 'departure of the Romans', but because it was composed of a minority of wealthy Britons of the upper classes, whose wealth and power, indeed to a great extent their very existence, came to an end in the troubles that marked the close of the Roman occupation of Britain[192].

In the first edition no analysis of the concept of Romanization is given like the one discussed above. This concept therefore does not allow a distinction of degrees, which brings Collingwood to an altogether different argument than the one expressed in the later edition: "There was no division between a Romanized upper class and a peasantry or town proletariat that clung stubbornly to its Celtic traditions; the two traditions blended more or less harmoniously in all classes of the people, and all classes derived benefit from the blend"[193]. It is obvious that this view does not allow for the same explanation of the subsequent de-Romanization as in the later edition and consequently a less satisfactory explanation of this process is given.

Roman Britain had a good reception, though it was reviewed in fewer journals than *The Archaeology of Roman Britain*. This time reviews also appeared in Germany and France and Collingwood's reputation abroad was echoed by a reviewer who remarked that "M. Collingwood est assez connu chez nous pour que nous n'ayons pas besoin de préciser que l'exposé est de tout premier ordre"[194], while according to another the book was written "... par un des meilleurs spécialistes de la Bretagne romaine"[195]. Of the English reviewers Myres says in his review of the 1932 edition that the earlier one "... has been the standard introduction to the subject for the general reader ever since"[196] and Hawkes that "... [it] for the last nine years ... has been eagerly and widely read as the one and indispensable small-scale introduction to the subject"[197].

A most interesting review is that by the well-known anthropologist Evans-Pritchard[198], then in the beginning of his career.

Evans-Pritchard highly values the way the process of Britain's Romanization is treated by Collingwood and thinks it a real contribution to anthropological theory. For this reason his assessment of *Roman Britain* is worth quoting at length:

> It is short, clear, vivid and may be read with interest and understanding by those whose knowledge of history is slight. For this reason I commend it to ethnologists whose impressions of the material and the methods of history and archaeology so often seem a dull, grotesque reflection of ethnological theories. Since the archaeologist has to rely mainly on what has been preserved in the ground through many centuries, his data are necessarily concrete and objective, and contrast in these characteristics with much of the loose data of ethnology that are so often a product of introspection followed by projection into savage behaviour.
> Some anthropologists speak about 'culture contacts' as though the only way in which they can be studied is by investigations among primitive peoples. To these people Mr. Collingwood's book may be recommended, for it deals almost entirely with questions of diffusion. The diffusion of Roman culture traits into Celtic culture is mainly in the form of indirect diffusion, it exemplifies diffusion by contiguity as well as by conquest, and it illustrates diffusion of process as against mechanical diffusion of objects. By indirect diffusion I mean that the Romans did not just come from Rome and dump down their culture on the Britons. Nothing so crude happened... Not only was the medium of diffusion, the Roman legions, a product of cultural fusion, but most of the objects of Roman culture which they brought to Britain had been deeply influenced by foreign techniques...
> By diffusion by contiguity as well as diffusion by conquest, I mean that before the conquest of Britain in the reign of Claudius, the Celts of south-east England appear to have already borrowed widely from their semi-Romanized neighbours, the Gauls. By diffusion of process as against mechanical diffusion of objects, I mean the importation of objects which the Celts themselves could manufacture and transform through their indigenous technique. Hence, one can tell those objects which were directly imported from Italy from the same type of object made in Britain.
> Diffusion is never a simple mechanical process, a wholesale taking over of foreign products without changing them. It is not a thrusting of culture traits on a submissive and unresponsive people. It is a dual rather than a unilateral process; the people who take over a trait assimilate it to their own modes of behaviour[199] ... Hence we find imported types and forms of artifacts which nevertheless display the dominant traits of indigenous technique. The history of Roman Britain well illustrates these processes of culture fusion, and it has been excellently brought out by Mr. Collingwood that diffusion spells fusion[200].

5.4.2. *ROMAN BRITAIN AND THE ENGLISH SETTLEMENTS* (1936)

Collingwood ends his 1935 Report as University Lecturer, dated 18 January, with the announcement that he intends to begin writing a book on Roman Britain in the next few months. It would be in his own words, "the fullest account of Roman Britain yet written" and would appear, possibly in 1936, as the first volume of *The Oxford History of England*[201]. Collingwood kept his word, for the book indeed appeared in 1936, his preface being dated 14 January 1936.

As the title *Roman Britain and the English Settlements* indicates, the book not only deals with Roman Britain, but also with the subsequent period of the English settlements. The latter part, covering the period from the beginning of the fifth till the middle of the sixth century, is written by J.N.L. Myres. The two parts were written separately, or, as Collingwood declares in the beginning of his preface: "This volume is not a work of collaboration. It consists of two independent studies of two distinct, though interlocking, subjects"[202].

Roman Britain and the English Settlements is divided into five books. The first four, which are written by Collingwood, are entitled: "Britain before the Roman Conquest", "The Age of Conquest", "Britain under Roman Rule" and "The End of Roman Britain". The third book is the most extensive and also the most interesting one, containing chapters on "The Machinery of Government", "The People", "The Towns", "The Country-Side", "Industry and Commerce", "Art" and "Religion". Collingwood's contribution to *Roman Britain and the English Settlements* is undoubtedly his best known historical work and on the few occasions that philosophers have paid attention to Collingwood's historical practice it is this book that has been referred to[203]. We will not try to give a survey of it, but parts will come up for discussion in the next chapters as illustrations of Collingwood's view on history.

The book was reviewed in no fewer than eighteen scholarly journals, attention being paid mostly to Collingwood's contribution. According to Richmond it " ... is likely to remain for some time the standard book on its subject"[204]. This opinion is endorsed by Wheeler, but the latter also criticizes Collingwood for the – in his opinion – too liberal interpretations in the earlier part dealing with the invasions and conquest of Britain: " ... Mr. Collingwood has adopted a personal and subjective attitude towards History that must either be accepted or rejected by the reader at the outset", Wheeler maintains, "it admits no compromise. He interpolates motives, builds characters, constructs episodes with a liberality or even licence that is great fun, but is liable to shock the pedant. Fact and speculation stand shoulder to shoulder, no documents are cited, the innocent student may know not with what voice his author speaks. Mr. Collingwood's feet are on the mantelpiece, he enjoys himself, and his reader with him"[205]. Many years later the editor of *The Oxford History of England*, Sir George Clark, was to report a similar opinion: "When the late professor's Collingwood's masterly instalment on Roman Britain was published one of the other contributors remarked rather tartly: 'He gives the impression that we know more about Roman Britain than about any subsequent period'"[206].

It is not surprising that Collingwood leaves this impression. As his starting-point is that all history is the history of thought, he sometimes describes actions which are the expression of certain thoughts in a way which gives us the feeling of being eye-witnesses, so to speak. At the same time, he sometimes leaves the impression that there is reason to doubt whether his reconstruction is really based on sound evidence[207]. To give one example: in his treatment of the Claudian invasion Collingwood gives the following description of the engagement of the commanding officer Plautius with the local chiefs:

> Meanwhile, however, the two most active sons of Cunobelinus were mustering their forces and hastening to meet the invaders. Even now, disunion was their undoing. Instead of acting in concert, each of the two brothers independently gathered his own men around him and rushed blindly upon the Roman force. Caratacus, the abler and more vigorous of the two, reached Kent first. It is tempting to conjecture that he took up his position on Caesar's old battlefield at the crossing of the Stour; for Dio's narrative suggests that he fought on ground of his own choosing somewhere in the eastern half of the county, and no better defensive ground is to be had. But Plautius found no difficulty in driving him headlong from his position. He escaped with his life and the remnants of his force along the line of the Watling Street; and Plautius in his pursuit, somewhere along that line, met with Togodumnus and crushed him. The loss of these two engagements made it impossible to hold East Kent, and some part at least of its inhabitants submitted to Plautius. They are described as a section of the Bodunni, who were doubtless one of the four unnamed Kentish tribes mentioned by Caesar[208].

Referring to this passage among others one reviewer comments: "In the account of the Claudian invasion there is so much which is not to be found in the only extant authority, that the reader who is familiar with the text of Dio is inclined to wonder if Professor Collingwood has not rediscovered the lost books of the 'Annals'"[209].

The same reviewer, however, remarks a little further on that "[w]hen he has the not so bald narrative of Tacitus to follow, and when he is interpreting archaeological evidence, Professor Collingwood rides his imagination with a tighter rein"[210]. Wheeler is of the same opinion: "Only when the major written authorities with their personalities begin to fail him, after the end of the first century, does Mr. Collingwood settle down to a detached recension of his evidence. From Chapter viii onwards it were difficult to find anything but admiration for his sober, workmanlike survey, and his fluid prose"[211]. These opinions seem to demonstrate that the strength of Collingwood's historical reconstruction lies more indeed in the interpretation of unwritten sources than written ones. This may be seen as the result of his experience in archaeology; but it is also in line with the special value he attaches to this science for the reconstruction of the history of Roman Britain.

5.4.3. OTHER WRITINGS

Besides the books discussed above Collingwood wrote much else on Roman Britain, both on general aspects and on specific subjects. These ranged from popular writings, meant for a general public, to detailed and scholarly studies. A unique example of the first is a series of six articles on "Rome in Britain", which appeared in *The Home-Reading Magazine*[212]. At the end of them we find "questions for discussion", such as "Compare the conditions and consequences of the Roman conquest of Britain with the conquest by European peoples of a) the American continent, b) India"; "What consequences would have followed if Agricola had been able to complete his proposed conquest of Scotland and Ireland?"[213]; "Could the Romans have devised more effective means of protecting Britain against the barbarians?"[214]. The articles are written with admirable clarity and in a fluid style. The same can be said of Collingwood's popular booklet on Roman Eskdale[215], called by Wheeler "a model of its kind"[216].

Of the more scholarly contributions we should mention Collingwood's discussion in the early twenties with the ancient historian Bury about the date of the end of the Roman presence in Britain[217]. In the thirties he wrote the chapters on Roman Britain in the standard work *The Cambridge Ancient History*[218]. Collingwood's most lasting contribution to the social and economic history of Roman Britain is the part contributed by him on this subject to the third volume of *An Economic Survey of Ancient Rome*, edited by T. Frank[219]. This contains chapters on land and population, public finances, communications, mining and minerals, public buildings and works, money, education and professions, agriculture, industry and commerce. All these aspects are discussed by Collingwood in detail, using not only the evidence of written and unwritten sources but also secondary literature.

The article Collingwood wrote on "Town and Country in Roman Britain"[220] deserves special attention for the way an unconventional historical problem is discussed, namely that of estimating the total population of Roman Britain. Collingwood deals with this problem on the occasion of a new edition of the Ordnance Survey Map of Roman Britain, which gave for the first time a detailed account of the distribution of population. The theoretical background on which Collingwood bases his analysis is provided by two studies by Carr-Saunders on demography[221]. The way Collingwood discusses the problem of coming to an estimation of the population of Roman Britain is not altogether convincing and there are serious flaws in his argument.

Collingwood takes over Carr-Saunders' contention that "[t]he more skilled a race is, the denser is its population, at least as a general rule. Agricultural races are more skilled than hunting races, and have as a rule denser populations; while the more skilled agricultural races have a denser population than the less skilled agricultural races"[222]. Now the population of Roman Britain was, according to Collingwood's estimate, under 9 to the square mile. He reaches this number on the basis of two approaches. The total population of the Roman Empire being estimated at 50 to 70 million and taking into account that the eastern provinces were far more densely populated than the western, Collingwood comes to the conclusion – rather abruptly – that " ... it is difficult to conceive any distribution of the remainder which could allow to Britain more than half a million or at most a million inhabitants"[223]. His second approach is based on Carr-Saunders' estimate of the population of England and Wales in the Middle Ages: "In 1066 this is reckoned at a million and a half; in 1415 at 3 millions; that is, it has doubled itself in 350 years, and in another 350 years it has rather more than doubled itself again, having reached 7 millions in 1760 ... Working backwards from 1066 according to the same formula, we get three-quarters of a million in A.D. 700, after the Anglo-Saxon settlement has taken place. This suggests that the population of Roman Britain, at any rate towards the end of its history, was considerably less than three-quarters of a million"[224].

Having reached a density of population of less than 9 to the square mile for Roman Britain, Collingwood asserts: "Contrast this with 26 in 1066, 52 in 1415, or 99 in 1714, and the extreme thinness of the Romano-British population is at once apparent"[225]. From this he concludes: "The inference is that the Romano-Britons 'practised a primitive form of agriculture', and not only that but stood quite low down in the scale of even primitive agricultural methods"[226]. Other causes of the low density of population were not acceptable to Collingwood. He dismisses for instance the argument that the latter might be due to warfare with the contention that " ... experts are agreed that warfare and massacre have little effect on the density of populations"[227].

The ease with which Collingwood takes generalizations such as these for granted is surprising. The one on the influence of warfare and massacre is contradicted, moreover, by his own description of Boudicca's revolt at the beginning of Britain's occupation by the Romans. For in *Roman Britain* he asserts that in this revolt no fewer than 150.000 lives were lost[228]. In an estimated population of half a million this would amount to a loss of 30%, a number which could only have been recovered after many generations. Another generalization we may question is the one given by Carr-Saunders and taken over by Collingwood. Carr-Saunders points out, Collingwood says, " ... that the determining factor in all populations must be the extent to which the natural power of increase is allowed to have full play, and argues that ultimately the density of every population depends on the available food-supply, which again depends largely on the technical skill of the food-winners"[229]. This generalization certainly does not hold, however, for certain overpopulated areas in Asia at the present day. To give a concrete example, Collingwood mentions as an estimate of Egypt's population in Roman times a number of 7 to 10 million[230]. It cannot be said, however, of the present population of that country – exceeding 40 million – that it rests on important improvements in agricultural techniques. On the contrary, a present-day Egyptian fellah hardly differs in skill and equipment from his Roman predecessor.

One may also question the way Collingwood reaches the number of 9 to the square mile in Roman Britain by extrapolation into the past by means of a demographic "formula" from an estimate of the population in 1066.

We may conclude from these considerations that Collingwood's argument is based on an unsound premiss as to the estimated density of Roman Britain's population and on certain doubtful generalizations. His argument also affects the conclusions which are drawn from it. He contends, for instance, that " ... the towns of Roman Britain had no real basis in the economic system of the country. The country was too thinly populated either to need, or to support, urban life"[231].

The view developed by Collingwood in his article was criticized by Randall[232]. Analysing the number of people who were not active in agriculture and had to be fed, the latter comes to the conclusion that Collingwood not only underestimates the agricultural population of Britain but also the level of their technique. According to Randall Romano-British agricultural practice " ... supported an extensive town life, fed a large garrison, and had a considerable surplus available for export"[233]. Randall does not try to give an estimation of the population of Roman Britain. Wheeler, however, commenting on the views of both Collingwood and Randall, does attempt this[234]. His estimate is that the population was not half a million, as Collingwood contends, but a million and a half[235]. Referring at a later date to his article "Town and Country in Roman Britain" Collingwood asserts, regarding the population of Roman Britain, that he " ... guessed half a million, probably underestimating the number of villagers; Wheeler ... proposed a million and a half; the truth is most likely somewhere between the two figures"[236].

At the end of this article Collingwood mentions individuals to whom he owed among others " ... the general encouragement to pursue a line of inquiry which is rather remote from ordinary topics"[237]. Discussing the article in his *Autobiography* he also contends that "[q]uestions of this kind about Roman Britain had never been asked before" (Aut, 137). We have seen more than once how Collingwood emphasized the need to ask clear questions, both in archaeology and history, and it is this which is likely to be his most valuable and lasting contribution to the study of Roman Britain. The value of his work lies more in the type of questions asked than in the specific answers given to them. This typical philosophical approach could be considered an example of Collingwood's *rapprochement* between philosophy and history.

PART THREE

CHAPTER SIX

THE HISTORICAL OBJECT

6.1. ACTION

In our final exposition of Collingwood's philosophy of history we will make use of his own scheme for *The Principles of History* as a framework: history as a special science, its relation to other sciences, and history and practice[1]. Collingwood's scheme was not very detailed and it can be used therefore only as a general guide in our discussion of his philosophy of history. We will not follow it exactly. As subjects for his first book Collingwood begins with methodological questions and goes on to the historical object, the conception of re-enactment and history as the self-knowledge of mind.

In chapter 6 we shall look at the historical object, in chapter 7 at historical method, while in chapter 8 some controversial issues will be dealt with, among others the conception of re-enactment. History as the self-knowledge of mind will come up for discussion in chapter 10 dealing with the relation between history and practice, while the relation of history to other sciences will be discussed in chapter 9.

We have seen in the third chapter how the conception of the historical object – history *a parte objecti* – is contrasted by Collingwood with the concept of nature[2]. The difference between them is described by him in various ways, but they all revolve around the notion of human action as the specific characteristic of the historical object[3]. Other claims associated with this notion are that historians focus on the inside of events, looking for the purpose of actions; that all history is the history of thought; and that the historical past, by contrast with nature, is not dead but is living in the present.

When we consider the importance of the concept of action in Collingwood's idea of history it is striking how little attention he paid to it explicitly. Scattered throughout his less known writings, however, are some important observations. In *Religion and Philosophy* he remarks that "[i]n any case of action, it is easy to see that some thought must be present" and a little further on: "... [E]very act depends for its conception and execution upon thought. It is not merely that first we think and then we act; the thinking goes on all through the act. And therefore, in general, the conception of any activity as practical alone, and containing no elements of knowing or thinking, is indefensible. Our actions depend on our knowledge" (RPh, 30-31). It is not enough, though, to define an action as being based on knowledge, because it is also essentially characterized by being conscious of its own activity; "if it were not", Collingwood asserts, "it would be not an activity but a mechanism" (RPh, 34). Consciousness and volition are seen by him as two manifestations of mind, always existing side by side and expressed in action. Mind is not a thing distinguishable from its own activities: "the mind *is* what it *does*; it is not a thing that thinks, but a consciousness; not a thing that wills, but an activity" (RPh, 34).

This self-consciousness of mind implies an element of reflection in all activity. Or, as Collingwood puts it in *Speculum Mentis*: "Of everything that a mind in the full sense does, it gives itself an account as it does it; and this account is inseparably bound up with the doing of the thing. Thus every activity is also a theory of itself and, by implication, of activity in general; but not necessarily a true theory" (SM,84). This aspect of thought is later described by Collingwood as "criteriological". "Any piece of thinking, theoretical or practical", he asserts in *An Essay on Metaphysics*, "includes as an integral part of itself the thought of a standard or criterion by reference to which it is judged a successful or unsuccessful piece of thinking. Unlike any kind of bodily or physiological functioning, thought is a self-criticizing activity ... The mind judges itself, though not always justly. Not content with the simple pursuit of its ends, it also pursues the further end of discovering for itself whether it has pursued them successfully" (EM,107-108). The science of mind, then, studies these criteria in a systematic way[4]. This is the special task of philosophy, defined by Collingwood as a "reflective or secondary experience, ... the return of the mind upon itself to study its own primary experience" (SM,255). In *The Idea of History* philosophy is called "... the organized and scientific development of self-consciousness..." (IH,4). In the same work, however, history is also put forward as the science of mind[5]. Since mind is what it does it is only through the study of its past actions that mind knows itself. So the study of mind includes both a criteriological (philosophical) aspect and an historical one (This *rapprochement* between philosophy and history is well illustrated by Collingwood's historical treatment of the (criteriological) idea of history in *The Idea of History*).

The close relation between philosophy and history can also be illustrated by Collingwood's conception of activity in another way. For in the "Preliminary Discussion" of the 1926 lectures he maintains that philosophy only deals with transcendentals, that is, with universal and necessary concepts "applicable to everything that exists"[6]. Action is mentioned as an example "because everything affords a field or opportunity for action"[7]. What he means by a philosophical analysis of action is clarified by Collingwood in "Economics as a Philosophical

Science"[8]. He makes there a distinction between economics as a philosophical and an empirical concept[9]. This distinction is worked out by Collingwood in *An Essay on Philosophical Method* and we have seen how a philosophical concept is described there as not precisely definable and necessarily having a certain vagueness[10]. It is interesting to note here that this view was already put forward by Collingwood in *Religion and Philosophy*. It is made clear there that action should be considered a philosophical concept – an idea which was developed later. For he contends:

> ...[A]ctions cannot strictly be classified at all. What is a lie? Intentional deceit? Then it covers such cases as ambiguous answers, refusals to answer, evasions; or even the mere withholding of information when none has been demanded; and we cannot easily say when such concealment of the truth is intentional. To lay a trap for an opponent in controversy would probably have to be called lying, as well as countless other cases in which we do not use the word. A classification of actions, in short, can only exist so long as we refrain from asking the precise meaning of the terms employed (RPh,206-207).

6.2. COLLINGWOOD'S PHILOSOPHY OF MIND

The previous discussion of Collingwood's conception of mind and activity is fundamentally incomplete and will therefore easily lead to misunderstandings. It should be added, however, that the picture given corresponds rather well to the traditional interpretation of Collingwood. For he is usually said to hold an overtly rationalistic view of mind and human action in general, summarized in his well-known slogan that "all history is the history of thought". We have seen in the third chapter how a host of critics, both historians and philosophers, have directed their wrath against this alleged one-sided view of history, based on a similarly one-sided philosophy of mind.

In reaction to this all too common interpretation we would contend, however, that it rests on a misunderstanding of Collingwood's views. It is easy enough to explain this misunderstanding, because an interpretation based only on *The Idea of History* indeed easily leads to it. Since this work appeared posthumously Collingwood cannot be blamed too much for it and there is reason to believe that his own planned book *The Principles of History* would have been quite different. When we consider other writings by Collingwood we cannot avoid the conclusion that it is impossible to give a proper assessment of Collingwood's philosophy of history based only on *The Idea of History*. For this work is only understandable within the context of a philosophy of mind, which is not explained in it. Only a few have pointed to this fact: Mink has emphasized in this connection the importance of Collingwood's philosophy of mind as elaborated in *The Principles of Art* and *The New Leviathan*[11] and Skagestad has dealt briefly with certain relevant ideas in *Religion and Philosophy*[12]. With the unpublished manuscripts now at our disposal we may add to these sources the manuscript on cosmology, since it also deals extensively with topics in the philosophy of mind[13].

We will not try to give an overall view of Collingwood's philosophy of mind, but will limit ourselves to some essentials. The subject is extremely difficult to come to grips with – not only because it is discussed in various works, which deal mainly with other topics and therefore discuss the subject in different contexts, but also because Collingwood sometimes seems to have changed his views on details or emphasizes different things. The main line of his thought on the subject, however, is, as far as I see it, consistent.

As remarked above, the picture given of Collingwood's ideas on mind and action at the beginning of this chapter is incomplete. It would seem to corroborate the traditional view of Collingwood as interested only in rational appearances to the neglect of the irrational aspects of feelings and emotions. For a correct interpretation of Collingwood's philosophy of mind it is essential to start from his theory of philosophical concepts as worked out in *An Essay on Philosophical Method*. In the last chapter this theory was also referred to in connection with the concept of Romanization[14]. Now, it is important to remember that concepts like consciousness, thinking, mind, freedom or activity are considered by Collingwood philosophical concepts. That means, to repeat again, that they cannot be defined precisely, but that their species exemplify a "scale of forms", dialectically related to each other, in which their generic essence is realized in varying degrees. These concepts should be taken, therefore, in the technical sense elaborated by Collingwood: if they are interpreted in their common-sense meaning this will necessarily lead to misunderstanding. In the history of interpretations of Collingwood this has indeed been the case. Hence, for instance, if Collingwood speaks of thought or consciousness one must always remember that these concepts may refer to different levels, which may not correspond with common-sense notions. It should be noted here that in our discussion of Collingwood's view of mind and action at the beginning of this chapter these concepts were in fact used at the highest level of their generic essence, that is, that of rational action. The idea of a philosophical concept, which is fundamental for Collingwood's philosophy of mind, is well illustrated by the way he discusses the concept of mind in his manuscript on cosmology:

> In the *Essay on Philosophical Method* it has been argued that a philosophical concept is differentiated into specific forms arranged on a scale; at the bottom of the scale comes the most rudimentary or primitive, which hardly exhibits the generic essence of the concept at all; the higher forms exhibit it more and more completely and at the same time in a more and more developed and elaborated way: the essence acquiring new increments as it

> realises itself more fully.
>
> On this principle, the first thing to ask about consciousness is not whether it is essential to mind, but whether it constitutes the minimum essence of mind: the last determination left when mind is stripped of all its attributes, before it vanishes altogether. Now, the answer to this question is simple. Pure sentience is not the same as consciousness, and seems to be something yet more primitive. In framing a philosophical theory of mind, therefore, we must treat sentience and consciousness separately: sentience first, because it seems the more primitive; and thus find the minimum essence (or pure being) of mind in sentience, with consciousness as a further increment[15].

Another important aspect of Collingwood's philosophy of mind is its monistic nature. Despite certain interpretations, which impute to Collingwood a dualistic view of mind, based on a too literal reading of his metaphor of the inside and outside of events, he was completely opposed to this. Discussing the relation between mind and body in the manuscript on cosmology Collingwood speaks of "the mind of *that* body and the body of *that* mind" and asserts that their relation is "... very unlike the Cartesian dualism of two substances: more like Spinoza's conception of the mind as the idea of the body. Mind is a specific type of activity (viz. perceptual activity) present in a body which in order to act in that way must have a specific bodily (physical, and proximately physiological) character"[16]. And in *The New Leviathan* he maintains that "... man's body and man's mind are not two different things. They are one and the same thing, man himself, as known in two different ways" (NL,11). It is far from true that Collingwood conceived mind as a mysterious entity hidden behind observable events, as the inside/outside metaphor has led some to believe. Already in *Religion and Philosophy* he declares that mind is only definable "... in terms of the object of which it ... [is] conscious" (RPh,152). Elsewhere in the same work he maintains: "The mind is specifically that which knows the object; and to call it a 'thing' already suggests conceiving it as an object one of whose qualities is that it knows other objects ... or, still worse, as a machine which turns out a kind of work called thinking, as a typewriter or a dynamo turns out its own peculiar product. The mind seems to be not so much that which thinks as the thinking itself; it is not an active thing so much as an activity. Its *esse* is *cogitare*". Collingwood continues:

> Again, just as the mind is not a self-identical thing persisting whether or no it performs its functions, but rather is those functions; so the consciousness in which it consists is not an abstract power of thought which may be turned to this object or that, as the current from a dynamo may be put to various uses. All consciousness is the consciousness of something definite, the thought of this thing or of that thing; there is no thought in general

> but only particular thoughts about particular things. The *esse* of mind is not *cogitare* simply, but *de hac re cogitare* (RPh,100)[17].

We have seen in the third chapter how Collingwood is generally criticized for making a sharp distinction between the rational and irrational aspects of mind, between thought and feelings or emotions[18]. That, in Collingwood's view, only a history of the first would be possible, is then usually seen as an unacceptable restriction of the historian's interests. Here again, we would contend that this interpretation is based on a misunderstanding. "The life of the mind is whole", Collingwood declares in *Religion and Philosophy*, "without seam, woven from the top throughout; the only sense in which we can separate one attribute from the others is that we may abstract it, that is, have a false theory that is separate; we can never actually employ one faculty alone" (RPh,154). And in the manuscript on cosmology he says: "I cannot admit the Cartesian body-mind dualism on which the realistic contentions are largely based. Nor can I admit the separation of intellection from other mind-functions"[19].

In order to understand Collingwood's views on the relation between thought and feelings or emotions it is necessary to study his philosophy of mind in general. Since this has been done by Mink in an admirably clear way, suffice it to refer to his analysis, which has already been described briefly in the third chapter[20]. The essence of Collingwood's philosophy of mind is that mind is built up of various levels with a practical and theoretical side. The lowest level consists of an undifferentiated sensuous-emotional flux of "pure feeling", while the highest level is the one of will and itellect. These levels are "connected" through intermediate levels of consciousness or thought. On the lowest level of feeling Collingwood contends:

> Feeling appears to arise in us independently of all thinking, in a part of our nature which exists and functions below the level of thought and is unaffected by it. All that we have said about it, and all that anybody can ever say about it, is of course discovered (or mis-discovered) by the activity of thought; but thought seems in this case simply to discover what was there independently of it, almost as if we were thinking about the anatomical structure and functioning of our body, which would no doubt exist and go on whether we thought or not ... [I]t seems that our sensuous-emotional nature, as feeling creatures, is independent of our thinking nature, as rational creatures, and constitutes a level of experience below the level of thought. In calling it lower, I do not mean that it is relatively unimportant in the economy of human life, or that it constitutes a part of our being which we are entitled to despise or belittle. I mean that it has (if I am right in my opinion about it) the character of a foundation upon which the rational part of our nature is built; laid and consolidated, both in the history of living organisms at large and in the history of each

> human individual, before the superstructure of thought was built upon it, and enabling that superstructure to function well by being itself in a healthy condition (PA,163-164).

The place of feeling in his concept of mind is described by Collingwood in *The New Leviathan* as follows: "The essential *constituent* of mind is *consciousness* or thought (practical and theoretical) in its most rudimentary form ... Feeling is an *apanage* of mind. It is an apanage of simple consciousness, namely its proper object, what there is consciousness of ... Man as mind *is* consciousness ... he *has* feeling ..." (NL,18-19).

Consciousness is identified by Collingwood with thought. At its lowest level it means the turning of attention to the direct and immediately given feelings. This is done by the activity of naming them: "To name the feeling awakens ... consciousness of the feeling ... the practical act of naming your feeling is what sets you off being conscious of it" (NL,42). In *The Principles of Art* this level of thought is referred to by the concept of imagination: "...[I] t is the kind of thought which stands closest to sensation or mere feeling. Every further development of thought is based upon it, and deals not with feeling in its crude form but with feeling as thus transformed into imagination" (PA,223).

Though by no means a dualist Collingwood contrasts thinking and feeling in the following ways: 1) Feeling has a kind of simplicity, while thought is bipolar. In the case of the latter Collingwood refers to the characteristic of being "criteriological": "Whenever we think we are more or less conscious of a distinction between thinking well and thinking ill, doing the job of thinking successfully or unsuccessfully. The distinctions between right and wrong, good and bad, true and false, are special cases of this bipolarity" (PA,157). 2) Feelings have a special kind of privacy in contrast with the publicity of thoughts. 3) The upshot of the previous distinctions can be described as saying that "... thoughts can corroborate or contradict each other, but feelings cannot" (PA,158). Another characteristic of feelings as such is that they are limited to the here and now (PA,159; NL, 21) and evanescent (NL,33). The momentary nature of feelings means that they cannot be re-felt as thoughts may be re-thought. This does not mean that they are not involved in man's activities or thoughts. It only means that their original nature cannot be recaptured. For this is dialectically transformed by the various levels of consciousness: on the practical side through appetite and desire into will and on the theoretical side through imagination to the intellect. Collingwood gives the following description of this transformation:

> Feeling is a here-and-now immediately given to consciousness; from which it follows that any characteristics that feeling may have are discoverable by simply reflecting on that consciousness, and any characteristics that a particular feeling may have are

> disvoverable by reflection on that particular feeling as given to theoretical consciousness after being distinguished from the here-and-now in which it oocurs by the act of selective attention (NL,24).

This passage makes clear that the characteristics of a certain level of consciousness can only be ascertained by a higher level through reflection. However, the feeling as such cannot be recaptured by thought: that can only be directly captured by a primary act of consciousness. "There is *nothing to argue about*", Collingwood says. "Have I a headache? Do not weigh pros and cons; do not reason about it; simply consider how you feel" (NL,24). For this reason feelings cannot be remembered, Collingwood asserts: "People who think they remember a feeling are deceived, never having been careful to make the distinction, by the fact that *a proposition about a feeling can be remembered.* You cannot remember the terrible thirst you once endured; but you can remember that you were terribly thirsty" (NL,34). Already in *Religion and Philosophy* Collingwood points out the evanescent nature of feelings: "If we are asked what we mean by the feelings of triumph, sorrow, indignation and so on, we reply as a rule by explaining the kind of occasion which excites them: 'triumph is what you feel when you have succeeded in spite of opposition'. But this is quite a different thing from stating what triumph feels like". Similarly, "[w]e generally define a scent not by its individual nature but by its associations; we state not what sort of smell it is but what it is the smell of". This "definition by circumstances", Collingwood asserts, "... is apt to mislead us seriously in any attempt to describe our feelings. We think we have described the feeling when we have only described the occasions on which it arises" (RPh,188).

Collingwood makes a distinction between feelings and emotions. The term "feeling" he reserves for the lowest sensuous-emotional level, also called by him the psychical level, while emotions appear also at the higher levels:

> I shall ... use the word 'feeling' only with reference to the psychical level of experience, and not as a synonym for emotion generally. This level contains indeed a vast variety of emotions; but only those which are the emotional charges upon sensa. When thought comes into existence... it brings with it new orders of emotions: emotions that can arise only in a thinker, and only because he thinks in certain ways. These emotions we sometimes call feelings; but ... I shall avoid so calling them, in order not to confuse them with the peculiar experiences we enjoy at the psychical level (PA,164).

In making this distinction between feeling and emotion Collingwood does not use these concepts in their common-sense meaning, according to which they are more or less equivalent. This has certainly supported the intellectualistic interpretation of Collingwood's philosophy of mind and the similar interpretation of his views on history. However understandable, it is nevertheless not correct to interpret Collingwood as making an absolute distinction between thought and emotion. On the contrary, in his view emotions contain thought and thought emotions. All levels of consciousness have an "emotional charge": "Take any form of consciousness", Collingwood says in *The New Leviathan*, "however highly developed, it always has an immediate object, and the immediate object always carries an emotional charge" (NL,25). On the other hand emotions contain an element of thought. To give an example from the lowest level: of fear it is said that it "... contains an intellectual element, an element not of propositional thinking ... but of conceptual thinking: the idea of a not-self" (NL,68). The same is asserted of anger (NL,70,71). Similarly in the manuscript on cosmology it is stated that "[a]n emotion which we find by reflexion to persist as part of the furniture of our mind is a thought, or contains thought. Emotion as such is not destroyed by reason; it is clarified, it comes to know itself, it rids itself of many strange errors; but it survives all these changes"[21]. That this view should be considered a permanent element of Collingwood's philosophy of mind is demonstrated by the fact that we find it already expressed in *Religion and Philosophy*: "... [E]motion is not a totally separate function of the mind, independent of thinking and willing; it includes both these at once ... There is no emotion which does not entail the activity of the other so-called faculties of the mind" (RPh,10).

If emotions contain an element of thought, even the highest level of thought has its specific emotions. "... [I]t is a matter of fact", Collingwood maintains in *The Principles of Art*, "that discourse in which a determined attempt is made to state truths retains an element of emotional expresssiveness. No serious writer or speaker ever utters a thought unless he thinks it worth uttering" (PA,264). "'The proposition', understood as a form of words", he says a little further on, "expressing thought and not emotion, and as constituting the unit of scientific discourse, is a fictitious entity" (PA,266). Generally speaking, Collingwood is of the opinion that all activity has an emotional charge: "For every different kind of activity there is a different kind of emotion. For every kind of emotion there is a different kind of expression" (PA,266). Archimedes' cry of "eureka" is given by Collingwood as proof that intellect has its own emotions. That the interweaving of thought and emotion is not limited, however, to such extreme cases is made clear in the following passage:

> There is no need for two separate expressions, one of the thought and the other of the emotion accompanying it. There is only one expression. We may say if we like that a thought is expressed in words and that these same words also express the peculiar emotions proper to it; but these two things are not expressed in the same sense of that word. The expression of a thought in words

> is never a direct or immediate expression. It is mediated through the peculiar emotion which is the emotional charge on the thought. Thus, when one person expounds his thought in words to another, what he is directly and immediately doing is to express to his hearer the peculiar emotion with which he thinks it, and persuade him to think out this emotion for himself, that is, to rediscover for himself a thought which, when he has discovered it, he recognizes as the thought whose peculiar emotional tone the speaker has expressed (PA,267-268).

What about the concept of thought? We have seen that it is identified by Collingwood with consciousness, both having various levels (consciousness, on its side, is identified with self-consciousness[22]). He makes a distinction here between the primary and secondary form of thought. The first is concerned with the relations between the sensa together with their corresponding feelings, while the second is thought about thought (PA,164-167). "The propositions asserted by thought in this secondary form", Collingwood says, "may be indifferently described as affirming relations between one act of thinking and another, or between one thing we think and another" (PA,167). It is only this level of thought which may be called rational, its main distinguishing feature being that it is "criteriological", or, as it is formulated in the manuscript on cosmology, it makes a distinction between essence and existence[23].

Summing up the previous discussion we may conclude that Collingwood had a pronouncedly unitary and monistic view of mind. It is obvious that this is not in line with the way he is traditionally interpreted. Walsh's version could be given as example. Comparing Dilthey with Collingwood he maintains: "...[F] or Dilthey, to say that history was properly concerned with human thoughts would be the same as to say it was concerned with human experiences: the word 'thoughts' would be used generically, much as *cogitatio* is in the philosophy of Descartes. Dilthey would have denied that all history is the history of thought if that were understood to mean the history of thinking proper, considering such a conception altogether too narrow and intellectualistic to fit the facts. But Collingwood, who was certainly familiar with Dilthey's theories, deliberately opted for this narrow view"[24]. In contrast with this interpretation, however, we have seen that the concept of thought was conceived by Collingwood as philosophical: that is, it should be interpreted in its widest sense – indeed, generically[25].

After the passage quoted above Walsh asserts about Collingwood, in contrast to Dilthey, that he interpreted thought only as intellectual operations. Though these were seen against a background of feeling and emotion, the historian would not be concerned with the latter[26]. Though it should be conceded that Collingwood's distinction between feelings and emotions is confusing and not worked out in *The Idea of History*, we have seen that it is not true that he separated thought from emotions. Discussing Collingwood's "dialectic of mind" Mink asserts of Collingwood's view on their relation:

> He has not, like traditional rationalism and the rather extensive survival of rationalism in 'common-sense', opposed reason to emotion as if one could distinguish between ways of acting which because they are rational are not emotional or which because they are emotional are not rational. He has connected the emotions and passions in such a way that each survives in higher levels but is transformed from being merely a way of being conscious of objects into being itself an object of a more inclusive level of consciousness. ... Moreover, as a higher level gives form to a lower, so the lower gives content to the higher. Whether one moves from a lower to a higher level is always a contingent fact, but it is a necessary truth that the higher depends on and cannot exist apart from the lower. Thus each form of rational activity has its specific emotional aspect. The grain of the marble, so to speak, survives in the finished statue, not eliminated but literally 'transformed' or exploited as in the sculptor's use of its definiteness and intractability for his own purposes[27].

In connection with the metaphor used by Mink at the end, it is interesting to note that Collingwood made clear, indeed, that in a work of art expressing certain emotions a reflective element is necessarily present. "This indeed seems to be the special character of art and its peculiar importance in the life of thought", he says in *The Idea of History*. "It is the phase of that life in which the conversion from unreflective to reflective thought actually comes about" (IH,314). A demonstration of this view on art may be found in Collingwood's interpretation in *Roman Britain* of a colossal Gorgon head, which had been part of the decoration of the temple of Sul at Bath. Collingwood comments:

> Some antiquaries have fallen into the trap of thinking that because the Bath Gorgon is fierce and violent in expression, it is therefore the work of a barbarian artist and expresses the uncivilized character of the Roman Briton's mind. That is an elementary mistake. The artistic representation of fear or anger is beyond the power of a terrified or angry man; a passion cannot be expressed until it has been mastered. The Bath sculptor was a man of high education, deeply versed in the technique of his art and coolly skilful in the execution of it[28].

Though Collingwood does not separate thought and emotion in the mind's activities it is possible, of course, that he was intellectualistic in the sense of being primarily interested in the mind's rational aspects, to the neglect of its emotional side. In his discussion of magic, however, we have seen that this is not the case at all[29]. On the contrary, he sharply criticizes there the utilitarian and rationalist character of European civilization. The repression of emotions, which this attitude implies, involves doing violence to man's emotional nature, Collingwood contends. We have seen how he levels his criticism especially against the way emotions are surreptitiously expressed in European culture in the cult of impersonal machinery.

In *The Principles of Art* Collingwood also remarks that we are accustomed "... to attend far more carefully to our sensations than to our emotions". "The habit of 'sterilizing' sensa by ignoring their emotional charge", he continues, "is not equally prevalent among all sorts and conditions of men. It seems to be especially characteristic of adult and 'educated' people in what is called modern European civilization; among them, it is more developed in men than in women, and less in artists than in others" (PA,162).

The value that Collingwood attaches to emotions is also clearly demonstrated by certain passages in "Man Goes Mad", a manuscript he wrote in 1936[30]. "If man outrages his body by refusing to eat, he dies", Collingwood says. "If he outrages his mind by injuring the foundations of his emotional life, he goes mad"[31]. Discussing then the fact that man essentially possesses a civilization, he mentions three dimensions the latter must have in order to be real: complexity, continuity and vitality. To these dimensions correspond three dimensions of mental life, namely intelligence, memory and emotion. "If any of these failed", Collingwood asserts, "civilization would perish"[32]. Of emotion it is said, that if it failed, "... the whole fabric of...[man's] civilization would crumble in his hands to dust and vanity, and he would sink back into the condition of a human brute"[33]. It is this dimension of civilization that is especially emphasized by Collingwood: "Changes in civilization seem, in fact, to have been due often enough to the dying-away of certain emotions"[34]. And a little further on he maintains: "The sanity of man as man depends on the health of his fundamentally human emotions. The sanity of man as civilized depends upon the health in him of the emotions fundamental to his type of civilization"[35].

Reviewing Collingwood's philosophy of mind, but also some examples of his more general ideas on man and civilization, at least this conclusion can be drawn. Any interpretation which credits Collingwood with a one-sided intellectualistic view on mind and man in general is on the wrong track. That the vast majority of his interpreters nevertheless adhere to this version may be understandable in so far as their interpretation is based only on *The Idea of History*. However, Collingwood's other writings, both published and unpublished, make it clear that this interpretation should be reconsidered. This is not to say that Collingwood had well-elaborated ideas on the subjects discussed above. On the contrary, they have to be grasped from the most diverse writings. In this connection it should also be observed that his premature death prevented him from working out certain crucial ideas more fully, especially in their relation with the science of history. This does not preclude us, however, from discerning some of their general characteristics.

6.3. HISTORICAL PROCESS

We have seen in the second chapter that already by about 1920 Collingwood had expressed as his "first principle of a philosophy of history" that "history is concerned not with 'events' but with 'processes'" (Aut,97)[36]. We have also seen how he considers process an important characteristic of the historical object and how in *The Idea of History* he criticizes certain historians and philosophers for not taking it into consideration. In this connection, then, criticism is leveled by him against the "substantialism" of Livy and Tacitus, the "naturalism" of Spengler, the "pigeon-holing" of Toynbee and the interpretation of history as isolated individual facts by Rickert and Dilthey[37].

Apart from a short description in the *Autobiography* (Aut,97-98) Collingwood made no systematic analysis of the concept of process. As with other concepts related to his philosophy of history, its characteristics have to be derived from scattered statements and short descriptions in his writings. So his treatment of the concept of process in the manuscript on cosmology is illuminating. A process of development should not be seen as a causal process, he asserts. A cause, according to Collingwood, is a factor from outside and this contradicts the idea of a process. In a process something is coming to be: that is, it is genetic. "It comes to be through a sequence of phases", Collingwood maintains, "and these are connected in a necessary sequence – it can only come to be in this way, through these phases in this order; but the order is not a chain of causes and effects but a chain of phases related in a different way"[38]. Its features are described, then, as follows:

> The relation may be put by saying that at each stage one phase is turning into the next. The existence of A is not separable from its turning into B, that of B from its turning into C, and so on. And this means A turning into C and so on. This is quite different from causation, where the impact of a bullet does not turn into the death of a man but remains other than it. Now, in the series ABC, by what name are we to call that which undergoes the changes? Is it A all the time, or B all the time or C all the time or something else, x, which takes on the temporary states A, B, C?[39]

The latter viewpoint is rejected by Collingwood. For when a permanent x changes from A into B there is no question of development, but only of change. He makes a distinction between "a change-philosophy in which a substratum x passes from one state to another, and a development-philosophy in which there is no substratum and no states of it, but always something turning into something else"[40]. Thus development has no substratum. This is clear, according to Collingwood, in the case of a growing organism; but, he says, "[i]t is perhaps clearest of all in history where e.g. the development of the jury-system or capitalist production has obviously no substratum at all"[41].

"Rejecting the name x for that which undergoes the changes or rather the development", Collingwood goes on, "what name are we to use? If A is in the act of turning into B, are we to call it A or B?" He then continues:

> Both, but in different senses. I say that it is actually A and virtually B. But which is it really? – That depends on what you mean by really. You may mean actually as opposed to virtually; then it is A. You may mean in reality as opposed to in appearance; this antithesis doesn't help, for I am assuming that there is no error in describing it as A turning into B, so in this sense it really is A turning into B. You may mean essentially, i.e. you may be asking which name expresses the best insight into its nature and explains best why it is what it is actually. Here reality is *opposed* to actuality as its ground or explanation. And in this sense it is really B. We do in fact think that the development of A into B is a process in which A comes to show what from the first it really was, and this implies that a thing's real nature is that which it is going to exhibit when fully developed, even though now it is a nature not actual but somehow latent. This implies that it is not only legitimate to describe A as a nascent B, but necessary so to describe it – better, in fact, than to call it only A, which would imply that there was not in it any nisus towards B[42].

We have seen how Collingwood rejects especially the idea of mind as a substance. "[I]n the life of mind there are no states, there are only processes", he asserts. "Every case of mental 'being', so called, turns out on examination to be a case of mental 'becoming'" (NL,285). Concepts like civilization or barbarity should not be interpreted, therefore, as states of affairs, but as processes (NL,285). Or, to give a concrete example, England did not yet exist in Roman times, Collingwood asserts: it is "the product of a historical process"[43]. At another point he speaks of "... the process by which the field-armies of the late Republic developed into the network of sedentary garrisons that guarded the frontiers of the Empire"[44]. These examples also illustrate the consequences of Collingwood's conception of all history being the history of thought. For his process-view of mind is transferred as a matter of course to historical phenomena and even to social phenomena in general.

A characteristic of processes is that they "... are things which do not begin and end but turn into one another" (Aut,97-98). In his analysis of the concept of civilization in *The New Leviathan* Collingwood describes the process of civilization as an "*asymptotic approximation to the ideal condition of civility*" (NL,284). The same holds, he contends, for the other end of the process: "Just as the community which undergoes civilization never will be purely civil, so it never was purely barbarous and if the process is reversed can never become purely barbarous". He continues: "All mental processes have this *asymptotic* or *approximative* character. A spatio-temporal process from t_1 to t_2 or from p_1 to p_2 really begins at t_1 or p_1 and really ends at t_2 or p_2; it really begins at its 'initial' point and really ends at its 'terminal' point. But a mental process from ignorance to knowledge or from fear to anger or from cowardice to courage never begins simply at the first term, but always at the first term with a mixture of the second; and never ends simply at the second term, but always at the second term qualified by the first" (NL,284). For this reason Collingwood elsewhere maintains that "[t]he true nature of an institution is shown not in its beginnings but in its developments"[45].

His fundamentally dialectical view of the historical process Collingwood considers an essential element of the historical attitude. This view was discovered, according to him, in the late eighteenth-century. "Before that time", Collingwood contends in the manuscript "What 'Civilization' Means", "history was conceived as a series of 'epochs' or 'periods' each having its own proper character. The historian's task was to study one or more periods separately, and then, if his outlook was wide enough, to show how one of them differed in character from what went before and what came after. The processes leading from one period to another eluded his grasp; being telescoped into vanishing-points, all he could say about them was that they had happened"[46]. Gibbon's *Decline and Fall of the Roman Empire* is given as example of the process-view of history. It soon became generally accepted, Collingwood says: "By the time Gibbon's work had become familiar to the general public, the idea of history as describing static conditions existing in the past at this or that period was obsolete. History had become the history of processes. Static periods had vanished ... It had become a commonplace that every period was a period of transition, and that processes of transition were all the historian had to describe"[47]. The influence this view may have on historical practice is demonstrated by Collingwood's discussion of the problem of dating the end of Roman's occupation of Britain. He considers this "a question of secondary importance" and would rather speak of a process of de-Romanization: "To trace the stages of this process, and to estimate the extent to which Romano-British elements survived and were taken up into English life and society, is a task of more historical interest than to fix the date at which Britain ceased to be governed directly from Rome"[48].

Another characteristic of the historical process, as seen by Collingwood, is its rationality – or, what amounts to the same thing, since all history is considered a process, the fact that the past makes sense. We may find this idea in various places among his writings. It is most explicitly dealt with, however, in a place where one would not expect to find it, in the conclusion to his article on Hadrian's Wall of 1931[49]. "It is much that we should have arrived in this way at a truer estimate of the Wall as a well-planned and efficient frontier defence", Collingwood there asserts, "but if that had been the only fruit of ten years' work, it might have been doubted whether the work was worth doing. Perhaps, therefore, it is desirable to explain why so elaborate a tissue of archaeological minutiae has been allowed to engage the writer's attention for ten more years"[50]. He then points out the importance of studying Roman history and of studying it in a scientific way. After this he makes a third point, which is important enough for a quotation in full:

> ...[A]ll science and all history depend, for their very possibility, on the assumption that 'the real is the rational and the rational is the real'. If the facts of nature and of human history are nonsense facts, obeying no law, forming no intelligible whole, connected by no rational relations, then scientific and historical thought are folly, and their ideals of method and of truth are delusions. The real service which natural science does to the human mind lies in the assurance which it gives, by the forward march of its discoveries, that no part of nature can remain finally impenetrable to human understanding. Science achieves this result by taking a special portion of nature and thinking about it until the object becomes, as it were, incandescent in the flame of thought, and is revealed as wholly intelligible. If there is any residue of unintelligibility left over at the end of the process, when science has done all it can, the whole process is in vain. In the same way, history can only demonstrate its own right to exist by demonstrating the rationality of its subject-matter; and this it must do, not by showing that certain points or tracts, scattered here and there in the abyss of time, shine with the light of rationality, but by showing that any tangle of human facts, patiently unravelled, makes sense. If the real is rational, it is possible, by intense and methodical thought, to see the mass of facts accumulated by the blind or half-blind industry of generations of archaeologists, as a luminous whole, out of which rises the truth. The more tedious the detail, the more apparently irrelevant the facts are to each other and to the whole, the more important it is to show that here, and not only in the visible symmetry of classical Greece or the intellectual glory of the Renaissance, reason still reigns; there is still a thread, if one can find it; there is not chaos, but order and intelligibility. It is only by the determined attempt to make sense of a collection of historical data which at first seem nonsense, that we can discover whether we are right, as historians, to assume that the real is rational and the rational real, or whether the story of human affairs is 'a tale told by an idiot, full of sound and fury, signifying nothing'[51].

Wrapped up in Hegelian phraseology this passage shows a realistic element in Collingwood's view on history. Statements like "any tangle of human facts, patiently unravelled, makes sense", or "there is ... a thread, if one can find it; there is not chaos, but order and intelligibility", while on the one side not eliminating rational efforts by historians, on the other hand do not imply that the rationality of the past is only imposed upon it. On the contrary, as the first part of Hegel's dictum implies ("the real is the rational") the past is seen as having a rationality of its own. It is interesting to note in this connection that the passage quoted above is from 1931, that is, after Collingwood's shift from a realist to an idealist view of the past following 1925.

Dealing with Croce's philosophy of history we find Collingwood already in 1921 in agreement on the same Hegelian principle: "The truth which Croce wishes to express is the same which Hegel concealed beneath his famous phrase, 'the real is the rational'. What happens, happens for a good reason, and it is the business of history to trace the reason and state it"[52]. That the Hegelian view of the rationality of the past is a permanent element in Collingwood's idea of history is demonstrated by the fact that he also explicitly refers to it in his lectures of 1926, 1929 and 1936. In the first an Hegelian vision is given in its most speculative form, interpreting history as a theodicy[53]. In the 1929 lectures Collingwood asserts on Hegel: "When he said that the world is ruled by reason, he was certainly right. When he said that a plan can be detected in history, he was no less right. When he said that history was the realisation of freedom, the progressive discovery by man of his own rational nature, he was, once more, right. But he interpreted all these thoughts hastily and superficially...Hegel realised that history is logical; but he then made the mistake of jumping to the conclusion that it was logic"[54]. In *The Idea of History* the main features of Hegel's view on history are again endorsed by Collingwood. One of them is that "... since all history is the history of thought and exhibits the self-development of reason, the historical process is at bottom a logical process" (IH,117). That Collingwood indeed considers this to be a crucial aspect of history is made clear a few pages further on, when he asserts that "[n]ineteenth-century historiography did not abandon Hegel's belief that history is rational – to do that would have been to abandon history itself..."(IH,122).

Collingwood does not confine himself to a general endorsement of the vague Hegelian principle of the rationality of history. This idea is worked out by him in various places in several ways. We have seen, for instance, how in "Reality as History" Collingwood rejects both the vision of the Greeks and of modern science on the flux of events[55]. For in their view this flux is not real and has to be reduced either to the permanency of fixed forms or to laws or uniformities. Against this Collingwood develops the thesis that reality may be found in the flux itself, calling this "the principle of history"[56]. The way this flux of events is to be conceived in history is explained by Collingwood in his manuscript on cosmology. There he makes the distinction between a logical order of concepts and a temporal order of events. This idea is dealt with briefly and rather problematicly, especially because both orders are seen as irreversible and even necessary[57]. (It should be added that Collingwood himself qualifies his notion "full of difficulties"). Since it illustrates well Collingwood's conception of the rationality of history it is worth seeing how it is applied by him:

> What I want to suggest here is that history is the coincidence of logical with temporal order. I mean that the successive events of history form an order which, so far as it is genuinely historical (not all chronological sequences of events in human life are so), is a logical order as well as a temporal one. If it is temporal but not logical, the sequence is not historical but merely chronological – it is what Croce calls annals, or a mere series of events.
> History begins when we see these events as leading by necessary connexions one to another: and not only that – for history demands more than that – but as the γένεσις of something, the history *of* something which is coming to be in this temporal process. Now, in a mere temporal process, necessary though it is, nothing comes to be; there is only change, not development. What imparts to an historical process its character of development is that the phases of this process are the phases in the self-development of a concept – e.g. that parliamentary government is coming into existence, which can only happen if the concept parliamentary government is articulated into elements or moments which a) are capable of arrangement in logical order, the first being what historians call the germ of it b) are capable of being brought into temporal existence *in that order*.
> Thus history is the development of a concept in a process that is at once logical and temporal[58]

According to this analysis we may thus discern three forms of the succession of events: 1) Only temporal and not logical sequence of events = chronological. 2) Only temporal (necessary) process = change. 3) Temporal and logical process = self-development of a concept or the genesis of something in history. The difference between the second and third form is not altogether clear, particularly since in the second one the temporal process is seen as necessary, but not as logical, the difference, however, not being apparent[59]. Of the second form it is said that history begins there; but only the third is said to be "genuinely historical". This may be seen as an illustration of the different exemplifications on the "scale of forms" of the philosophical concept of history.

In the lectures of 1926 and 1928 there are some passages which may clarify Collingwood's notion of the rationality of the historical process. Discussing the idea of dividing history into different periods he asserts: "So long ... as we think of the past at all, we must think of it as possessing that kind of determinate structure which consists in a sequence of more or less clearly-defined periods having characteristics of their own and each possessing precisely those characteristics which would necessitate their turning into the next, and so on. We must, that is to say, find in history a pattern or scheme which makes it a self-contained and logically-articulated whole"[60]. Similarly, after declaring in his 1928 lectures that each historical event must be seen in relation to a certain point of view, Collingwood maintains: "The various parts of a treatise, however, are not only related to the whole: they are related to each other. Primarily, they are related chronologically: they state a temporal sequence and therefore constitute a narrative. But the relation between them is very far from being merely chronological. They constitute not merely a sequence but a process. Each part leads to the one which follows and rests on the one which precedes"[61]. Collingwood refuses, however, to speak of a causal relation. He develops, then, the idea of a "rational necessity", using the example of two chess-players[62].

In the 1928 lectures Collingwood also gives a clear illustration of the combination of the temporal and logical elements within the historical process, when he discusses the concept of development:

> What appears chronologically as a sequence must appear as a simultaneous whole in the historian's thought... He must feel the earlier phases as preparing the way for the later, and the later as explaining the true meaning of the earlier. He must, in a word, see the inner structure of his subject as a development.
> This conception of development, or progress, defines a necessary character of every historical period, where period means a particular subject of historical study... Development is only possible where there is unity: there must be one thing that develops, and when it changes into something that is not recognisably the same, it cannot any longer be said to be developing. Development also implies a plurality of phases within the process; and it further implies that the process brings out by degrees some characteristic of the one thing which at first was not clear. Development is an ideal process, not an actual process: it consists in something's becoming more and more intelligible[63].

With the last mentioned idea Collingwood refers to what he was to call in the passage quoted above from the manuscript on cosmology "the self-development of a concept".

Since history is considered by Collingwood as in essence a rational process, it is not surprising to see it described also as a plot or even a drama. By this concept he refers to the idea that an historian should see the events of the past in their logical or rational order and not confine himself to a mere temporal order. Already in "Ruskin's Philosophy" Collingwood maintains: "In the hands of a logically-minded person, history becomes a mere succession of events, fact following fact with little or no internal cohesion. To a historically-minded person, on the contrary, history is a drama, the unfolding of a plot in which each situation leads necessarily to the next" (RuPh,19)[64]. In "The Nature and Aims of a Philosophy of History" Collingwood comes back to this conception. Though history has a plan, he contends, it is not determined beforehand: "The plan which is revealed in history is a plan which does not pre-exist to its own revelation; history is a drama, but an extemporised drama, co-operatively extemporised by its own performers. This is a view of history which I, for one, am prepared to defend. To deny it would involve asserting that history consists of an indefinite series of atomic events, each wholly devoid of connexion with those which happen before and after" (NAPhH, 36-37). A little further on Collingwood maintains: "There is in history a necessary relation between one event and another; and the more closely one studies any period of history the more clearly one sees it as a whole whose parts mutually condition one another, the antecedents being necessary if the consequents are to exist, and the consequents necessary if the antecedents are to be understood. The period thus reveals itself to the historian as a drama in the sense of an organised and coherent whole of events" (NAPhH,37). It is Collingwood's contention that to look for a plot is the task of "normal" history and not of the philosophy of history: "If it is the historian's work to discover the details, it cannot be anybody's work but the historian's to discover the interconnexion of the details" (NAPhH,38). Though it should be conceded that Collingwood's speaking of historical plots may easily lead to the misunderstanding that he refers to certain mysterious entities or kindred activities on the part of the historian, it should likewise be emphasized that this is not what is meant by it. For the need to look for plots in history Collingwood considers "... true on whatever scale we are working" and he mentions in this connection the plot of the Norman Conquest or of the French Revolution (NAPhH,38)[65].

That Collingwood with his concept of an historical plot indeed refers to a necessary element in all historical studies is made clear by his argument in his paper "A Philosophy of Progress (1929). After mentioning the growth of historical science in the eighteenth century he asserts: "And when people began to reap the fruit of these historical studies, when the picture of history as a whole began to take shape before their eyes, a startling discovery burst upon them: *it made sense*. It had a plot. It revealed itself as something coherent, significant, intelligible. It was a genuine discovery. It was true, and it remains true, that history lacks plot or significance only when it is told by an idiot" (PhP,111-112). Having discussed before some aspects of the history of architecture he contends: "Told by an idiot, the history of architecture would become a meaningless succession of whims and fashions. Told by a competent person, it has a plot; the various changes which it records are rational changes. The same is true of any other piece of history" (PhP,112).

Collingwood's idea of the rationality of the past is one of the most fundamental principles of his philosophy of history. Following Collingwood's own admonition in his *Autobiography*[66] one could observe, however, that the true meaning of this principle can only be grasped if one is conscious of the problem it is intended to solve and the answers it implicitly rejects.

The study of the past, one could say, has to face two basic problems: the past consists of innumerable "facts" and displays a constant flux. These characteristics are of course valid for all reality. Science, however, has come to grips with them by reducing individual facts to certain types or generalizations and the flux to the permanency of laws or uniformities. If it does not always succeed in this, it could nevertheless be asserted that this is at least its aim. However valuable this approach may be in science, Collingwood stubbornly resists it with regard to the study of the past, or, to put it more correctly, with regard to human studies in general (the latter aspect will be discussed later). In his opinion fundamental violence is done to the reality of the past by reducing its individuality and its character of fluidity. The question, however, is whether science has a monopoly on making things which are individual and in flux understandable, or, to put it differently, making them rational. This idea again, is rejected by Collingwood. Since he is convinced that all reality is rational, he is strongly of the opinion that the world of human affairs, and therefore the human past, is rational as well. It is the task of historians, therefore, to find this rationality, to make it explicit where it is only implicit and to make the past understandable in this way.

There is one influential rival solution to the problem of making past human affairs understandable, which is explicitly rejected by Collingwood. This is the one of imitating the scientific approach, which is called by him "positivism". This positivism can take the form of reducing individual events to fixed types, reducing the flux of events to laws or uniformities, or of seeing individual events disconnectedly.

Collingwood's fundamentally rationalist view of the past has the corollary that its irrational aspects are not recognized. He is indeed of the opinion that when certain aspects or periods of the past are called irrational it only demonstrates the historian's inability to understand them, that is, to see their rationality. To give an example from "The Theory of Historical Cycles" (1927), with regard to the fact that in each historical study there are certain gaps, Collingwood contends: "These gaps will appear as breaches in continuity, periods in which the historian loses track of the development. Necessarily, therefore, the history of these gaps will appear an irrational history, a history of muddle and failure and misdirected energies, the history, in a word, of a Dark Age" (THC,87). "Each period with which we are tolerably acquainted", he says a little further on, "each period which we understand well enough to appreciate the problems and motives of its agents, stands out as something luminous, intelligible, rational, and therefore admirable[67]. But each period is an island of light in a sea of darkness" (THC,88). The lack of understanding of the Enlightment for the Middle Ages is of course notorious. It is also significant for Collingwood's position that the latter period was qualified by the former as "irrational". Collingwood refers to this example when he asserts in *The Idea of History*: "Certain historians, sometimes whole generations of historians, find in certain periods of history nothing intelligible, and call them dark ages; but such phrases tell us nothing about those ages themselves, though they tell us a great deal about the persons who use them..." (IH,218-219). Similarly, he states about an eighteenth century writer on architecture: "His condemnation of the Middle Ages as a period of decadence was simply a confession that his own mediaeval studies – or rather, those of his entire generation – were in an unfinished and unsatisfactory condition" (THC,88).

What cannot be explained by science is also usually called irrational. "That real life is illogical every one admits", Collingwood says in *Speculum Mentis*, "but that is the fault not of life but of logic, of abstract thinking. The scientist wants actual fact to behave as if it were a mere example of some abstract law; but it is never simply this, and the elements he has deliberately ignored upset all his calculations. He then calls the fact irrational, or contingent, meaning unintelligible to him because too solid and hard to be forced into his moulds, too heavy for his scales, too full of its own concrete logic to listen to his abstractions" (SM,226-227). Collingwood then gives the example of certain conventions, such as fashion in dress, which "... are lumped together by abstract thought as irrational". "[B]ut to the eye of concrete historical thought", he continues, "they reveal themselves as informed by the most delightfully subtle intelligence, as inevitable as the plot of a drama and as little capable of scientific or abstract analysis: differing from the rationality of law only in being implicit. Paris dressmakers could not tell you why they alter a certain fashion in a certain way this autumn, or if they did the reason would be a wrong one; but there is a reason, and it can be traced if the problem is approached from an historical point of view" (SM,227).

A very striking example of Collingwood's contention that the real is rational in all human affairs is of course his attitude towards "primitive" cultures. As we have seen in the discussion of the manuscripts on folklore, it is Collingwood's strong belief and indeed the starting-point of his studies on primitive man that the latter's culture makes sense and is rational. According to Collingwood, European science has simply not been able to understand primitive cultures and has therefore called them irrational. He is of the opinion, furthermore, that the historical approach is the only remedy. From this viewpoint, then, he looks for the implicit rationality present in primitive culture and speaks for instance of magic as "... a system of beliefs and practices with definite laws of its own" and of "[t]he laws of the fairy-tale country"[68].

If the irrational is identified by Collingwood with the not understood, he does the same with the contingent or accidental. These concepts should likewise not be applied to history, according to Collingwood. A few quotations may illustrate his view and particularly the (ir)relevance of such concepts for history: "Contingent ... is only a synonym for unexplained" (RPh,50); "For history there is no contingent; no fact is turned away from the historian's door" (SM,209); "...[T]o the historian historical processes are not accidental, because his business is to understand them, and calling an event accidental means that it is not capable of being understood" (EM,289); "What is irrational means what my principles or explanation do not explain. An irrational element in the self is called 'caprice'; one in the not-self is called 'accident'"(NL,110). In his notes made during his journey to the East Indies Collingwood defines an accident as "an event which happens to a person without his intending that it shall happen to him". After this he continues: "If a second person intends that it shall happen, it is still an accident relatively to the first person; but relatively to them both together it is not an accident"[69]. Collingwood goes on to remark that "[i]f history has to do with events, then of course accident plays an enormous part in history"[70]. Since history, however, has to do with human actions and not with events, this observation implies that accident does not play a part in history[71].

The regarding of factors as contingent is rejected by Collingwood as bad historical practice, lacking the idea of historical process. In *The Idea of History* he gives as an example Pascal's remark "... that if Cleopatra's nose had been longer the whole history of the world would have been different". Collingwood calls this view "typical ... of a bankruptcy of historical method which in despair of genuine explanation acquiesces in the most trivial causes for the vastest effects". For him it shows an "... inability to discover genuine historical causes..." (IH,80-81). Collingwood likewise repudiates Bury's view on the role of contingency in history, a product, according to Collingwood, of his positivism: "Contingency just means unintelligibility; and the contingency of history means the 'role of the individual' – which is not at all a bad definition of history – seen through the spectacles of the positivistic dogma that nothing is intelligible except generalizations"[72].

Directly related to Collingwood's view of history as a rational process is his rejection of the concept of revolution in history. This rejection is based on two considerations: the concept of revolution denies the continuity of the historical process and it is used in order to highlight an allegedly contingent and surprising factor, which is only an expression, however, of the fact that certain events are not really understood. "To stop being surprised when the course of history waggles, and to think of it as waggling all the time", Collingwood says in *The New Leviathan*. "...[T]he word 'revolution' has fallen out of use among historians much as the word 'chance' has fallen out of use among physicists... Each is a pseudo-scientific term whereby one's own ignorance of why an event happens is offered as an answer to the question why it happens" (NL,201)[73]. The reason why Collingwood is critical of concepts like revolution is further explained by him in his manuscript "What 'Civilization' Means", when he describes the transition in historical thought in the eighteenth century, beginning with the situation at its start in the following way:

> The historian's inability to grasp any process by which one state of human affairs turned into another was expressed in various alternative ways. What the historian found unintelligible he might hand over to the theologian by calling it a miracle, a divine judgment, or the like. He might hand it over to the natural scientist by calling it a catastrophe. Or he might half-humanize the idea of miracle by calling it a revolution: where 'revolution' means a purely human act replacing the static conditions of one historical period by those of another, but a human act not intelligible to the historian as the conditions of the periods which it divides are intelligible to him: a human act of a quasi-miraculous or mysterious kind.
>
> The historians of the late eighteenth century made it their business to overcome this dualism between the intelligible static conditions of a single 'period', and the mysterious dynamic events leading from one period to another, by focussing their attention on these dynamic events themselves[74].

With regard to Collingwood's view of the historical object his conception of history as a rational process is fundamental. Taking into account the vast literature on Collingwood's philosophy of history it is surprising that hardly any interpreters have noticed this important aspect of his thought. To my knowledge only Mink has called attention to it[75]. He is certainly right in qualifying it as a "recessive" element in *The Idea of History*, but we have seen how it is not only to be found in Collingwood's unpublished manuscripts, but in his other published writings as well. In my discussion of Collingwood's view of the historical process an attempt has been made to give a survey of his ideas on the subject. It is also clear, however, that these ideas form the basis of his better known views on history. One could mention in this connection those on the autonomy of the science of history, explanation and understanding, and also the re-enactment doctrine.

Summing up Collingwood's position one could say that its starting-point is the conviction that the past makes sense, an idea derived from the Hegelian principle that the real is rational. This implies that the past can be made understandable, that is, made explicitly rational by the efforts of historical science. This can only successfully be done, however, if it is recognized that the flux of past events can be shown to have a rationality in itself. In this connection any attempt to reduce the complex and fluid character of the past in a positivist fashion is strongly rejected by Collingwood.

The discussion of Collingwood's view on history as a rational process also shows how his ideas may directly influence historical practice. To give some examples: historical concepts must be seen as processes and not as fixed types; attention should be focussed on the characteristics of processes and not on their beginnings or ends; contingent or accidental events should be avoided in historical explanations: and instead of paying special attention to "surprising" events like revolutions, the continuity of the historical process should be emphasized. The consequences of Collingwood's views on the historical object for historical practice also demonstrate that history *a parte objecti* cannot be sharply distinguished from history *a parte subjecti*. But, as we have seen before, this was also Collingwood's contention[76].

CHAPTER SEVEN

HISTORICAL METHOD

7.1. HISTORY AS A SCIENCE

One of the best known aspects of Collingwood's thought is his ardent defence of the autonomy of history. Strangely enough, though, his views on the methodological aspects of history have not received the attention one would expect – and, one should add, they deserve. The reason for this must be found in the fact that the attention has mainly been focussed on Collingwood's re-enactment doctrine and that this again has generally been given a methodological interpretation. Besides the fact that, in our contention, this interpretation is not correct, it has the corollary that the real methodological questions have been neglected. Donagan was the first to point this out. Commenting on the "received" methodological interpretation of the re-enactment doctrine he maintains: "Misleadingly expressed non-methodological passages were taken to propound an eccentric methodology, and genuine methodological passages were misread or discounted. The crucial passages in the *The Historical Imagination* and 'Historical Evidence' were seldom used: their customary fate was to be either admired and passed over, or mined for glittering things to be displayed to advantage in imported settings"[1]. Recently there are signs, however, of a growing interest in Collingwood's views on questions of historical methodology[2]. These questions are not only discussed by Collingwood, though, in "The Historical Imagination" and "Historical Evidence" (IH,231-282), mentioned by Donagan. They are also dealt with by him in "The Nature and Aims of a Philosophy of History", "The Philosophy of History", and especially in "The Limits of Historical Knowledge". The latter articles, however, have hardly been noticed by Collingwood's interpreters.

In his *Autobiography* Collingwood refers several times to his study of historical method (Aut,85,112,133). In his 1932 Report to the Faculty, too, Collingwood asserts that his aim in the study and teaching of history was – after declaring that he wants to do it in mutual connection with philosophy – "... not to neglect the methods and logic of historical work, and to emphasize the relation between history and its sources". That he saw this aim in close relation to his archaeological and historical practice is obvious when he describes a little further on his third project as "A large-scale work on Hadrian's Wall, with special reference to the problems of archaeological method there arising"[3]. This relation is also explicitly made clear by Collingwood in his *Autobiography*, when he contends: "Obscure provinces, like Roman Britain, always rather appeal to me. Their obscurity is a challenge; you have to invent new methods for studying them, and then you will probably find that the cause of their obscurity is some defect in the methods hitherto used. When these defects have been removed, it will be possible to revise the generally accepted opinions about other, more familiar, subjects, and to correct the errors with which those opinions are perhaps infected" (Aut,86)[4]. Discussing the importance of studying Hadrian's Wall a similar statement is made in his article of 1931 on the Wall:

> ...[I] f history is to be pursued at all – and that is not a question that can be raised here – it must be so pursued as to win respect by the solidity of its logical structure. It is not so much useless as mischievous to practise history in so slipshod a manner that any alleged fact may be slightly wrong. By a combination of slight errors, the general character and significance of the whole are certain to be distorted; and the general sense of an historical narrative may easily be changed, by a distortion of this kind, into the very opposite of the truth. This danger, which increasingly besets history in proportion as history deals with more interesting and important subjects, can only be averted by the discovery and application of the most rigidly scientific methods: methods not borrowed from other sciences, but worked out with strict reference to the special problems and characteristics of history itself. A single highly complex problem, like that of Hadrian's Wall, just because of its richness in apparently pointless and fruitless minutiae of evidence and interpretation, offers a perfect field for that experimental work without which no scientific method can be devised[5].

What are the special problems and characteristics of history, according to Collingwood? In the first instance they are defined by him by contrast with what they are not. For Collingwood history is a third way on the road to knowledge, additional to science and mathematics. In his draft of "Human Nature and Human History" he begins by comparing history with science. In the seventeenth century, he says, "[i] t was evident that physical science had discovered its own proper object and (what comes to the same thing) the proper methods of investigating that object. The question then arose, whether these methods were universally applicable to the problems of knowledge in general, or (what comes to the same thing) whether all reality had the characteristics which render the physical world capable of being investigated by those methods"[6]. At first this question was answered affirmatively, according to Collingwood. For in the seventeenth and eighteenth century human mind was studied on the analogy of the principles of natural science with a view to elaborating a "science of human nature"[7]. "Since then", Collingwood asserts, "the orderly classification of the sciences has been gravely disturbed by the rise of history, which, like physics in the 17th century, has discovered its proper object and proper method: its object, the human past in its entirety; its method, the reconstruction of this past from documents written and unwritten, critically dissected and interpreted"[8].

The close relation between history *a parte objecti* and history *a parte subjecti* is demonstrated by the fact that Collingwood speaks of the proper object and proper method of history in the same breath (In the passage quoted above he even observes about the proper object and method of physical science that it "comes to the same thing"). A little further on Collingwood contends that "...the right way of investigating mind is by the methods of history"[9]. The proper object of history, then, may be defined both as the human past and the human mind. One could observe in this connection that in mentioning the human mind reference is made to the more general (philosophical) concept of history, while the human past is the object of the more specific (empirical) concept of history, this being the business of historians.

The special character of history is discussed, among other things, by Collingwood in his lectures of 1936. "Ever since the Greeks", he asserts, "people have been accustomed to recognize two kinds of knowledge, the empirical knowledge of what is given in perception and the a priori knowledge of eternal and necessary truths: αἴσθησις and νόησις, truths of fact and truths of reason, intuition and conception, matters of fact and relations between ideas, and so on"[10]. The knowledge of the past, however, falls under neither head: "It does not proceed by way of perception", Collingwood contends. "The past is not present, and cannot be empirically perceived. Historical knowledge is not immediate; its object is not given, it must be arrived at by reasoning"[11]. It differs, though, from the reasoning of science in that its object "is not something eternal and indifferent to space and time", but "... has a perfectly definite location in space and time ..."[12]. Collingwood concludes therefore that "[n]either perception nor thought, as traditionally conceived, can give us historical knowledge"[13]. The latter is defined as "...a reasoned knowledge of something not immediately given"[14]. In *The Idea of History* the same argument is developed by Collingwood (IH,233-234,249-252). His conclusion reads there: "History, then, is a science, but a science of a special kind. It is a science whose business is to study events not accessible to our observation, and to study these events inferentially, arguing to them from something else which is accessible to our observation, and which the historian calls 'evidence' for the events in which he is interested" (IH,251-252).

This definition of history, however, is the product of a long development of the idea of history itself. In the past history was seen otherwise, and, Collingwood contends, as it still is by the "common-sense" theory of history. "According to this theory", Collingwood maintains, "the essential things in history are memory and authority. If an event or a state of things is to be historically known, first of all some one must be acquainted with it; then he must remember it; then he must state his recollection of it in terms intelligible to another; and finally that other must accept the statement as true. History is thus the believing some one else when he says that he remembers something. The believer is the historian; the person believed is called his authority" (IH,234-235). One could observe that according to this theory history is reduced to the acquaintance or empirical model of knowledge, although it is a weak example of it. Collingwood is strongly opposed to this notion of history. In the draft of "The Historical Imagination" he clearly shows how it necessarily breaks down[15]. In the first place, an historian has no means to protect himself against omissions, concealments, distortions or lies on the part of his authorities. Further he cannot transcribe everything his authorities tell him: he has to make selections. This, however, must be done according to certain criteria of importance, which are not given by the authorities. The need for an historian to judge his authorities is even more apparent when they contradict each other or when events are reported which the historian thinks could not possibly have happened. This will lead to "critical history", whose main feature is that authorities are no longer considered authorities but sources, to be dealt critically with by the historian.

In the 1926 lectures history based on the ready-made narratives of authorities is called by Collingwood "dogmatic". Since an historian has no reasons for choosing between his authorities this view will lead to scepticism, according to Collingwood. The only way out is neither to accept nor to reject authorities, but to interpret them. This phase of historical thought is also called "critical" history[16]. There is a difference, though, between Collingwood's analysis in the 1926 lectures and his draft of "The Historical Imagination". While in the first a distinction is made only between dogmatic and critical history, in the second a third form of history is added, superseding critical history. This form is termed "constructive history" and defined by Collingwood as follows:

> Its essence is that it takes the statements made by its authorities, after these have been subjected to criticism, not as so many pieces of fact to be fitted into a mosaic, but as so many clues from which the required fact is to be reached inferentially. The finished product, the historical narrative or description, is here not a collection of statements each directly drawn from an authority and merely selected and fitted together by the historian, but something which the historian has arrived at by thinking out the implications of what his authorities have told him. It is not something he has

been told, it is something which he has thought out for himself[17].

In the final text of "The Historical Imagination" constructive history is also dealt with by Collingwood as a higher form of history, in contrast with the common-sense theory and critical history. It is described by him as "...interpolating, between the statements borrowed from our authorities, other statements implied by them" (IH,240).

The kind of history which depends on the testimony of authorities, which is called by Collingwood in 1926 "dogmatic history", and in 1935 the "common-sense theory", is called "scissors-and-paste history" in "Historical Evidence", written in 1939. This concept has become a well-known slogan, and after "re-enactment of past thought" probably Collingwood's most cited expression. Collingwood is very critical of scissors-and-paste history and though he says that people who know little about history still think this is what history is like, he qualifies it as "not really history at all" (IH,257).

Two movements have undermined this pseudo-history, Collingwood contends: the systematic examination of authorities by critical history and the use of non-literary sources. Critical history, however, which started in the seventeenth century, is not enough, according to Collingwood. Its only aim is to find out whether a certain statement is true. This means, that its value is confined to a possible *nihil obstat*, with which historical science cannot be satisfied (IH,259,261). Critical history is considered by Collingwood not more than an advanced stage of scissors-and-paste history. The next stage, which has to supersede it, is the one at which a source is not analysed for its truth, but for its meaning. This viewpoint is described by Collingwood as follows: "...[T]he important question about any statement contained in a source is not whether it is true or false, but what it means. And to ask what it means is to step right outside the world of scissors-and-paste history into a world where history is not written by copying out the testimony of the best sources, but by coming to your own conclusions" (IH,260). It was Vico, according to Collingwood, who developed for the first time this important view of historical science. As he puts it in his lectures of 1931: "Ever since Vico, the great problem of historical method had been this: – how do we know, as historians, what our authorities do not tell us?"[18]. In order to know this, the historian has to ask specific questions. As we will see later in this chapter Collingwood considers this the main feature of what he calls "scientific history".

7.2. EVIDENCE

We have seen that according to Collingwood our knowledge of the past is inferential. That is, we cannot have direct knowledge of it. What it has to be mediated through is called evidence. This concept is considered by Collingwood characteristic for history and is to be found in this form neither in mathematics, nor in philosophy, nor in natural science. "What evidence is there for the binomial theorem?", he asks, "None; the question is meaningless. What evidence is there for Plato's theory of Ideas? Everything is evidence for it, if you believe it; everything evidence against it, if you disbelieve it. In other words, the conception of evidence does not enter into the process of thought by which it is defended or assailed. What evidence is there for or against the inheritance of acquired characteristics? None; what might loosely be called *evidence for* it would be properly described as *well-attested cases of* it. The experiments which corroborate or overthrow a biological theory are not sources or documents, precisely because, if they are impunged, they can be repeated, done over again. You cannot 'repeat' Herodotus, or write him over again, if you doubt something that he says; that is what shows him to be, in the strict sense of the word, evidence" (NAPhH,92).

In "The Limits of Historical Knowledge" the use of evidence is also emphasized and considered all important for history: "All history is the fruit of a more or less critical and scientific interpretation of evidence" (LHK,91); "...[T]here is no way of knowing what view is 'correct', except by finding what the evidence, critically interpreted, proves. A view defined as 'correct, but not supported by the evidence', is a view by definition unknowable, incapable of being the goal of the historian's search" (LHK,98); "... [H]istorical thinking means nothing else than interpreting all the available evidence with the maximum degree of critical skill. It does not mean discovering what really happened, if 'what really happened' is anything other than 'what the evidence indicates'" (LHK,99). We have seen how Collingwood identifies in the same article the science of history with a game played with evidence according to certain rules[19]. Elsewhere Collingwood has made similar statements emphasizing the role of evidence: "...[N]o historian is entitled to draw cheques in his own favour on evidence that he does not possess... He must argue from the evidence he has, or stop arguing" (Aut,139); "History has its own rubric, namely 'the evidence at our disposal obliges us to conclude that' such and such an event happened" (EM,56)[20]; "The process by which [an historian] arrives at [statements] ... is a process of making up his mind what the evidence in his possession proves" (EM,235).

Collingwood also speaks of sources instead of evidence. The difference between the concepts is not altogether clear. One could perhaps say that a source is more concrete and refers to something which provides "evidence for" something else[21]. Though Collingwood emphasizes that the historian is bound by his evidence, it is far from true that he is at its mercy. On the contrary, Collingwood makes it clear that it is the historian who "creates" evidence by using it as such[22]. Instead of evidence being an authority, it is the autonomous thought of the historian which qualifies certain things in the present as evidence for certain things in the past. It is not the historian who trims himself to his sources, but the sources which are trimmed to the historian. Using a Kantian phrase Collingwood speaks here of "a

Copernican revolution in the theory of history" (IH,236). According to Collingwood there are no limits to what may be considered evidence; it is up to the historian to decide: "...the evidence concerning a particular problem consists of everything which historical research has found, or shall find, to be relevant to it"[23]. Or, as he puts it in *The Idea of History*: "In scientific history anything is evidence which is used as evidence, and no one can know what is going to be useful as evidence until he has had occasion to use it" (IH,280).

The way evidence is used by historians being the starting-point, the question arises how this is done or should be done. In other words, what are the criteria an historian uses in his interpretation of evidence? We come here to a crucial point in Collingwood's views with regard to the methodology of history and his philosophy of history in general. For he considers this question to be the heart of the science of history. Evidence, he contends, should not be interpreted haphazardly, but according to strict principles. These principles cannot be set up empirically, but should be justified a priori: we have seen how Collingwood distinguishes in the 1926 lectures between an "empirical" methodology of the auxiliary sciences and a "general or pure" methodology, which deals with the general characteristics of history. In the 1928 lectures the latter is identified with philosophy of history[24].

An important element in Collingwood's views on the methodology of history is the definite ideas he has on the sources to be used by historians. Traditionally historical sources are considered to consist of written documents. It is obvious that this type of source fits in well with scissors-and-paste history. It is for this reason not accidental that one of the factors which had helped to undermine scissors-and-paste history had been the use of non-literary sources. In *The Idea of History* the latter is explicitly mentioned in this connection (IH,258). Collingwood's emphasis on the use of non-literary sources is obviously inspired by his archaeological practice and his experience in the study of Roman Britain, for which literary sources are of little use[25]. On the other hand, "obscure provinces" like these apparently also had an appeal on him because of their methodological challenge[26]. Since Collingwood is of the opinion that literary sources should never be regarded as authorities, but only as "neutral" evidence to be dealt with critically by the historian, he considers the distinction between literary and non-literary sources methodologically vacuous. For this reason he is also opposed to the traditional distinction between history and prehistory:

> From the point of view of this distinction, history is coterminous with written sources, and prehistory with the lack of such sources. It is thought that a reasonably complete and accurate narrative can only be constructed where we possess written documents out of which to construct it, and that where we have none we can only put together a loosely-constructed assemblage of vague and ill-founded guesses. This is wholly untrue; written sources have no

> such monopoly of trustworthiness or of informativeness as is here implied, and there are very few types of problem which cannot be solved on the strength of unwritten evidence... Strictly speaking, all history is prehistory, since all historical sources are mere matter, and none are ready-made history; all require to be converted into history by the thought of the historian. And on the other hand, no history is mere prehistory, because no source or group of sources is so recalcitrant to interpretation as the sources of prehistory are thought to be[27].

The latter contention receives further elaboration when Collingwood speaks elsewhere of "new methods of exploration" in archaeology. They gave the archaeologist, he asserts, "...an array of well-ordered information which enabled him for the first time to convert prehistory into a branch of history: that is, to take the archaeological data of a given region and to read off from these data, as from any other kind of document, the history of that region"[28].

Collingwood has always pointed out the importance of archaeological evidence. Commenting, for instance, on an article by Bury on the question of the end of Roman's occupation of Britain[29], his criticism of the latter's approach is that "...there is one source of evidence which he has almost entirely ignored, namely, Romano-British archaeology"[30]. Collingwood has not confined himself just to a passing glance at the non-literary sources, but has made a thorough study of them. We have seen that *The Archaeology of Roman Britain* is in fact a manual on what is called by him "empirical methodology", interpreting various types of non-literary sources. Examples have been given: for instance, how to determine whether a road was Roman, and the special problems involved in the interpretation of brooches[31]. An excellent example of Collingwood's treatment of a certain type of evidence is his discussion of "coins as archaeological evidence" in the chapter on coins in *The Archaeology of Roman Britain.*

"The principle which governs ...[the] use of coins is in itself simple", Collingwood maintains, "but its application is complicated by various subordinate principles and points of detail. It is this: a place inhabited for an appreciable time by people using coins almost always contains coins which they have accidentally lost or purposely buried there; these coins were in circulation during the period of inhabitation, and therefore give a clue to the date and length of that period"[32]. In the interpretation of coins as evidence various principles are involved. On lost coins Collingwood observes:

> ...[A]ny addition to our coin-list not merely adds a new fact to those already known, but also throws new light on these other facts. If a hundred coins are found on a site, each one tells us more than if only ten had been found. This is partly because all evidence has a negative as well as a positive aspect. If a site yields two coins separated by a gap of a hundred years, the gap is a

> mere gap in our knowledge; but if it yields two hundred coins, forming two groups separated by a gap of a hundred years, the gap represents not a lack of evidence, but the evidence of a lack – evidence that coins were not reaching the site during that century, and that therefore it was perhaps not occupied[33].

In order to use coins for purposes of dating one must have a general idea of their "life", that is, the interval between minting and loss, Collingwood asserts. The degree of wear can give an indication, but cannot generally be trusted. With regard to British finds, though, experience allows certain generalizations to be made, according to Collingwood, and some of them are cited by him[34]. Another matter discussed by Collingwood is the "rate of travel" of coins, that is, the question "[h]ow long did coins take to come into currency at a given place, that is, to travel from the mints into (for instance) a remote corner of a distant province?"[35]. Collingwood finally deals with finds of hoards of coins. He makes a distinction between "a floating cash hoard", which "...will generally consist, for the most part, of recent coins, and will include a diminishing proportion of older pieces" and "a savings hoard", which "...will generally contain a more evenly-spaced selection"[36]. As a rule hoards are savings, Collingwood contends. From a find certain conclusions may be drawn:

> Where there is no well-established banking system, savings must always be hoarded, short of investing them; and there is no doubt that at any given moment an appreciable percentage of the existing Roman currency was being hoarded in this form. Certain economic conditions would, from time to time, accentuate this tendency. When the value of money is falling sharply, users find that it will not buy as much as they are accustomed to think it should; they therefore allow it to hang on their hands, and thus hoards everywhere tend to accumulate. This happened in the depreciation of the third century, which reached its climax in the financial disaster of Gallienus's reign. The result of these events is to be seen in the great numbers of hoards assignable to about the years 260-280 and containing vast quantities of debased silver and silver-washed copper[37].

Normally a hoard of savings is ultimately spent. The reasons that it remains unspent are, according to Collingwood, that it either becomes worthless, or is lost. On the latter possibility he contends:

> The loss of a hoard means that the person who deposited it did not live to reclaim it either by his own hand or by that of his heirs; and that would be likeliest to happen if he met with a sudden death. When a countryside is disturbed by warfare or raiding, the number of hoards does not increase – for these conditions do not cause people in general to have more savings – but the proportion of

hoards that are lost increases with every increase in the intensity of the disturbance[38].

"If, then", Collingwood concludes, "the archaeologist knows of an unusually large number of hoards dated to a certain period, he may regard this as evidence either of economic crisis or of political unrest. If it is a widespread phenomenon, it probably points to grave depreciation in the currency; if a local, to destruction of life and property through war"[39].

Collingwood had a keen eye for different types of evidence which might be used. Though he could not work out, of course, all their principles for himself, as he had done with coins and inscriptions, he was well aware of relevant developments. Discussing, for instance, the survival of the British population during the invasion by the Anglo-Saxons he mentions "the evidence of place-names"[40]. Other examples are the way corn grows and its colour, as indications of a site ("...it has become a regular weapon of archaeological study"[41]), or the use of air-photography to determine a site[42].

It is Collingwood's contention that everything may be used as historical evidence. The great value he attaches to evidence is made clear when he asserts that "...in fact the advancement of historical knowledge and the improvement of historical method comes about mainly through discovering the evidential value of certain kinds of perceived fact which hitherto historians have thought worthless to them"[43]. He adds to it that "[t]he discovery of new kinds of evidence may have two functions: it may help to answer old questions, or it may solve new problems about which, if only because there was no evidence bearing upon them, historians have not previously thought"[44]. We have seen how Collingwood's study of folklore was explicitly set up to consider the possibility of using fairy-tales as historical evidence, and how he saw this as "a new kind of archaeology"[45]. He did not make clear, though, exactly what he considered its function. One could say, perhaps, that it might help both to answer old questions, especially with regard to the more obscure periods of the past, and also to solve new problems.

As stated above, Collingwood's views on the methodology of history have not been at the center of his interpreters' interests. Recently, however, they have been exposed to some criticism. Both Couse and Coady contend that Collingwood unjustifiably rejects the function of testimony in history. They are of the opinion, moreover, that Collingwood himself, in his example of the detective story, worked out under the heading "Who killed John Doe" in *The Idea of History* (IH,266-274), makes use of certain testimonies[46]. This criticism neglects, however, Collingwood's point with regard to testimonies. Collingwood does not claim that in an investigation like the one in his detective story no testimony should be trusted. For instance, there might be no reason for the Detective-Inspector to disbelieve in the first instance the testimonies of the parlour-maid or the constable. The point is that this is always done on the investigator's own responsibility. He will start his investigation, of course, with an hypothesis concerning certain persons who might be suspected. Their statements, however – and this is what is emphasized by Collingwood – are not treated as statements. That is, what interests the Detective-Inspector is not their content, but the fact that they are made and the conclusions that may be drawn from them (IH,275). In connection with Collingwood's example of the detective story it should also be kept in mind that its function is to illustrate the questioning activity of scientific history (IH,273-274). It certainly does not claim to give a complete survey of all aspects of historical methodology, as both Couse and Coady seem to imply. Most of their criticism for this reason is beside the mark.

It would also be a gross misunderstanding to think that Collingwood disregards all literary evidence. This is maintained, however, by Coady when he asserts that "...surely historical evidence includes the transmitted testimony of previous generations", adding that Collingwood would disagree with this[47]. But there is ample evidence in Collingwood's historical work that he did take notice of literary evidence. In his comprehensive survey of the economic conditions of Roman Britain, for instance, he refers again and again to all the literary evidence available[48]. At one point he even explicitly mentions "literary, epigraphic, and archaeological" evidence[49]. Collingwood's argument, however, is that testimonies should never be treated as authorities, but as sources. These sources should always be critically dealt with, that is, interpreted by the autonomous judgment of the historian. This can be done by comparing various testimonies, as has been done by Collingwood with regard to the contention that the Wall was built by Severus[50]. It can also be done with independent archaeological evidence.

According to Couse, Collingwood's "...radical opposition to critical history ...demands special consideration as a challenge to present-day historical practice"[51]. The impression given by this contention – that Collingwood was not interested in a critical interpretation of historical sources – is not correct. We have seen, on the contrary, how he even did pioneer work in this field in *The Archaeology of Roman Britain*. The aspect of "critical history" which is criticized by Collingwood is the fact that it is only interested in the trustworthiness of statements. "...[T]he important question about any statement contained in a source", Collingwood maintains against this view, "is not whether it is true or false, but what it means" (IH,260). This does not imply, though, that the truth of a statement can be disregarded. It only says that for scientific history this is not its most important aspect. This is made clear when Collingwood contends: "No very profound knowledge of coins and inscriptions is needed in order to realize that the assertions they make are by no means uniformly trustworthy, and indeed are to be judged more as propaganda than as statements of fact. Yet this gives them an historical value of their own; for propaganda, too, has its history" (IH,260). This statement does not mean that a source need not be dealt with critically. On the contrary, in order to write a history of propaganda certain sources first have to be interpreted as propaganda. The difference, then, between critical and scientific history must be seen as a difference in approach or attitude: the first is aimed only at establishing the truth or falsity of a statement, while the latter takes a further step and asks for the meaning of a statement. A false statement may contain from this point of view most valuable historical information.

Having given this short survey of Collingwood's views on historical methodology, both in theory and practice, it is obvious that certain contentions are completely unjustified. This is the case, for instance, with Coady's when he asserts that "Collingwood's picture of historical knowledge (for all his talk of proof, evidence, etc.) is dangerously close to fantasy or fiction"[52], or that "[w]hat Collingwood seems most often to be trying to do is to make the historian's data depend wholly upon the historian's theorizing imagination..."[53]. But what to say of Elton's opinion in this connection? Discussing Gardiner's interpretation of Collingwood he even goes so far as to assert: "What is so patently missing here is practical experience of the effect of historical evidence (the relics of the past) on the investigating mind. Not without reason did Collingwood do his historical work in a period exceptionally devoid of evidence"[54]. Another sad experience is to be confronted with Fain's assertion that Collingwood "...took the heroic path of attempting to dismiss historical evidence altogether as awkward impedimenta in the search for historical truth"[55] and that he supposed "...that there is a simple method of dispensing with the evidence, a method that brings one to grips with historical truth directly"[56]

Summing up Collingwood's contribution to historical methodology I would contend that it is of four kinds. In the first place he emphasizes the importance of non-written sources. It is obvious that this conception stands in close relation to his archaeological and historical experience. This experience on the other hand, enabled Collingwood to develop well-reasoned arguments for his view. Collingwood further emphasizes that literary and non-literary sources should be treated in the same way. In contrast with the traditional view he considers in this connection the treatment of non-literary sources as exemplary. Its main feature is that one has to look for the meaning of sources, with the corollary that what is at issue with literary documents is not their content as such, but the information that may be inferred from them. Collingwood is further of the opinion that historical science might be improved by the use of new types of sources. Here he points out especially the importance of the rise of archaeology. His study of folklore also demonstrates how much he was interested in the search for new types of historical evidence. Finally, Collingwood stresses the need that evidence, in order to be used in a satisfactory way, should be interpreted according to sound principles. With respect to archaeology he himself made important contributions in this regard. We have also seen that in his treatment of folklore as a possible historical source he considers it necessary first to work out the principles of its interpretation[57].

It is no exaggeration to assert that Collingwood's contribution to the theory of historical methodology is an almost completely neglected aspect of his thought on history. It is interesting to notice in this connection that the anthropologist McCall in his study *Africa in Time-Perspective: A Discussion of Historical Reconstruction from Unwritten Sources*[58] mentions Collingwood's *The Idea of History*, together with Bloch's *The Historian's Craft*, as the books "...which I found most useful for my purpose ..."[59]. And referring to Collingwood's example of the detective story he asserts that "[t]hat is the point of view adopted here"[60]. Though never fully worked out, Collingwood's ideas on historical methodology are indeed still relevant with regard to the theoretical background of new developments in historical science. They may even have a stimulating influence on them.

7.3. QUESTION AND ANSWER

"Knowing involves asking questions and answering them", Collingwood asserts in *The New Leviathan* (NL,74). Historical knowledge is no exception. On the contrary, we have seen how Collingwood describes the need to start from precise questions as the primary of three principles "...by which the historian ought to stand firm through thick and thin" (Aut, 122)[61]. Likewise, we have seen how this principle was derived, according to Collingwood, from the "laboratory of knowledge" of his archaeological practice and used in a "flank attack" upon the realists[62]. In the *Autobiography*, however, where this view is expounded, Collingwood also speaks of another flank attack upon the realists, though he says that this "ultimately converged" with the first (Aut,26). Collingwood refers here to the principle that one should always try to reconstruct as exactly as possible what it is that a philosopher really said or meant[63]. This could only be done successfully if one was conscious of the problem a certain theory tried to solve, that is, to see the latter as the answer to a certain question. Both the principle of asking precise questions in historical inquiry and that of reconstructing the question and answer complex of a philosophical theory belong to the "logic of question and answer" developed by Collingwood[64].

This logic has some puzzling aspects, which have been questioned[65]. The most evident are Collingwood's contention that two propositions do not contradict each other unless they are answers to the same question (Aut,33) and that a passage from a philosopher contains both the evidence of what the problem was and its solution (Aut,70). I will not go into these difficulties, however, but confine myself to the relevance of Collingwood's "logic of question and answer" for historical methodology.

Mink has given a most interesting interpretation of Collingwood's theory. He is of the opinion that his newly developed logic, conceived as an alternative to "traditional" logic, is indeed tangled with difficulties. But he contends that it can be given a meaningful interpretation if it is not seen as a logic at all, but as a *theory of inquiry*[66]. Instead of a "logic of question and answer" it might more properly be called a "*dialectic* of question and answer"[67]. In calling it a dialectic, reference is made to the fact that the activity of question and answer should be seen as a process. This process, then, has the characteristic of being prospectively open but retrospectively determinate, Mink maintains[68]. That is, what the logic of question and answer may refer to can be both a looking forward and a looking backward[69]. "...[I] t is possible from any point to trace a unique series *backward*", Mink asserts, "but there is no point for which there is only one series *forward*... Yet just here, in the asymmetry of the series (like that of any dialectical series), may lie the clue. History is like this too: we can retrace, as it were, the logic of a random conversation, seeing in retrospect by what relevant associations it found its way from subject to subject, although we never could have predicted at any stage where it would be one subject ahead. *The Logic of Question and Answer does not tell us how to conduct the conversation but how to reconstruct it*"[70].

I think this interpretation of Collingwood's logic of question and answer is not only basically correct, but also proves it to be highly relevant for history. What Mink in the above quoted passage refers to is hermeneutics, that is, the theory of historical interpretation. He also explicitly identifies the logic of question and answer with this conception[71]. We have seen, though, that he interprets it as a theory of inquiry as well. In my view, however, within the conception of the logic of question and answer a distinction should be made between hermeneutics and a theory of inquiry. The reason for this is that the conception is relevant both for the historical object and the activity of the historian. That is, it says on the one side that anything from the past should be interpreted within the context of a question and answer complex; on the other hand, that the activity of the historian himself is part of the process of question and answer. So the logic of question and answer is both relevant for, and may therefore refer to, history *a parte objecti* and history *a parte subjecti*. The first, then, is the aspect of hermeneutics: backward looking and related to the retrospectively determinate, while the second is the aspect of the theory of inquiry: forward looking and prospectively open. Though Mink points out the fact that "[h] istorical inquiry is doubly dialectical, both in its procedure and in its subject-matter"[72], he not only makes no terminological distinction between them as proposed here, but also does not work out the aspect of historical inquiry. I would contend, however, that it is important to make a distinction between the two aspects of the logic of question and answer: that of historical understanding (hermeneutics) and that of historical inquiry (historical methodology in the broad sense). In the present section only the latter aspect will be dealt with (historical understanding will come up for discussion in chapter 8.7).

Discussing the use of historical evidence in his 1928 lectures Collingwood added at a later date with a blue pencil the following remark: "we must *crossquestion* the evidence (cf. Bacon) – not merely *listen* to it – This destroys the conception of *authorities* and leads to that of *sources*"[73]. This idea is worked out in the draft of Collingwood's Inaugural. Under the heading "Critical history: questioning witnesses" he maintains:

> The importance of critical history for the theory of historical knowledge lies in the fact that the critical historian ...has explicitly abandoned a merely receptive attitude towards what his authorities tell him, and thus no longer regards them in the proper sense of the word as authorities. They are now treated not as authorities but as witnesses; and just as natural science begins when the scientist, in Bacon's famous metaphor, puts Nature to the question, tortures her in order to force from her an answer to his questions, so critical history begins when the historian puts his authorities in the witness-box and extorts from them information which in their original and freely-given statements they have withheld, either because they knew it but did not wish to give it, or because they did not know it and therefore could not give it[74].

In the final text, reprinted in *The Idea of History*, we find the same wording. After discussing the autonomy of historical thought by way of selection and historical construction, Collingwood refers to Bacon's metaphor of putting Nature to the question: "[t]he historian's autonomy is here manifested in its extremest form ..." (IH,237). In "Historical Evidence" Collingwood comes back to it, having given his example of the detective story "Who killed John Doe?". This time, however, the Baconian approach is explicitly identified with scientific history. "...[I]n scientific history, or history proper", Collingwood declares, "the Baconian revolution has been accomplished". Though the scissors-and-paste historian and the scientific historian may read the same books, he says, the latter "...reads them in an entirely different spirit; in fact, a Baconian spirit. The scissors-and-paste historian reads them in a simply receptive spirit, to find out what they said. The scientific historian reads them with a question in his mind, having taken the initiative by deciding for himself what he wants to find out from them"(IH,269).

Collingwood's reference to scientific history implies that history is a science. It is, however, as we have seen before, a science of a special kind in his view[75]. In the discussion of Collingwood's ideas on the science of history his views on its special character are usually emphasized. Collingwood is also of the opinion, however, that history is a *science* and it is this aspect that needs attention as well. It is in this connection all important, of course, to know what Collingwood means by science. In his 1936 lectures, edited in *The Idea of History*, he asserts that history "...generically... belongs to what we call the sciences: that is, the forms of thought whereby we ask questions and try to answer them" (IH,9). A little further on he observes that "[s]cience is finding things out: and in that sense history is a science" (IH,9). Discussing Bury's dictum that "history is a science; no less, and no more" Collingwood declares three years later, in "Historical Evidence": "...the word 'science' means any organized body of knowledge. If that is what the word means Bury is so far incontestably right, that history is a science, nothing less" (IH,249). The restriction of the meaning of "science" to "natural science" is regarded by him as "a slang usage". Collingwood does the same in *An Essay on Metaphysics* and asserts there: "The word 'science', in its original sense, ...means a body of systematic or orderly thinking about a determinate subject-matter. This is the sense and the only sense in which I shall use it" (EM,4).

Elsewhere Collingwood is more specific about the characteristics of science. In his lecture on "The Nature of Metaphysical Study", given in 1934, he mentions as one of the two principles which "... form the charter of modern scientific thought" Bacon's view "...that the scientist must not passively observe nature but actively question her, that is, approach her with a ready-made question in his mind and torture her until she gave the answer"[76]. In the beginning of *The Idea of History* "the establishment of modern natural science" is also identified with "the new methodologies of Descartes and Bacon" (IH,6). Further on in the same book Collingwood similarly contends with regard to the questioning activity that "Descartes, one of the three great masters of the Logic of Questioning (the other two being Socrates and Bacon), insisted upon this as a cardinal point in scientific method..." (IH,273).

The interesting conclusion to be drawn from this contention is that, when Collingwood speaks of scientific history or the scientific revolution in history, what he refers to is science in a sense also applicable to natural science. The essence of this idea is the asking of specific questions in a systematic way. Hence the scientific revolution in history, accomplished at the end of the 18th and the 19th century, runs parallel to the one in natural science in the 17th century: while in the latter "Nature is put to the question", in the former the same is done with regard to evidence. The implication of this contention is, surprising as it may appear, that Collingwood in fact endorses a unitarian view of science as far as its questioning activity is concerned. Though he considers this aspect of primary importance, both in natural science and in history, it does not preclude the possibility that there are also important differences between the two sciences. We have seen that Collingwood indeed argues that this is the case[77].

With regard to science instead of questions to be asked

Collingwood also speaks of problems to be solved. This viewpoint, one may observe, is surprisingly analogous to Popper's, when he defines science as in essence a problem solving activity[78]. Like Popper, Collingwood also emphasizes that we always start from a specific question or problem in an inquiry. "The beginning of historical research", he says, "is ... not the collection or contemplation of crude facts as yet uninterpreted, but the asking of a question which sets one off looking for facts which may help one to answer it. All historical research is focussed in this way upon some particular question or problem which defines its subject" (PhH,137)[79]. No solution of a problem is final, Collingwood contends:

> The historian's work is never finished; every historical subject, like the course of historical events itself, is open at the end, and however hard you work at it the end always remains open. People who are said to 'make history' solve the problems they find confronting them, but create others to be solved, if not by themselves, by their survivors. People who write it, if they write it well, solve problems too; but every problem solved gives rise to a new problem (EM,65).

The similarity with Popper's view is evident, for instance, when he says: "We may start from some problem P_1 ... which is submitted to critical discussion in the light of evidence, if available, with the result that new problems, P_2, arise"[80]. It is also interesting to note that Collingwood speaks of the problem solving activity both in connection with the historical object (people who "make history") and the activity of the historian. Popper expands the activity of problem solving into a principle of all life, to be summarized in his slogan that "from the amoeba to Einstein is just one step"[81]. Though it is not clear that Collingwood would agree with this contention – his life-principle of *nisus* might be seen, however, as expressing a similar idea – he certainly considers the problem solving activity as a characteristic of all human life. It is in this connection significant that when discussing his logic of question and answer the first example concerns the questions to be asked when a car will not go (Aut,32). Again, this viewpoint runs parallel to Popper's description of scientific knowledge as "common-sense knowledge writ large"[82]. Finally, to illustrate his view on scientific archaeology and history Collingwood refers to Acton's injunction to "study problems, not periods" (Aut,125; IH,281). Popper does the same in his argument that natural science and history are similar in their problem solving activity. He writes: "Does not a modern historian like Lord Acton tell us to study problems, not periods; that is, to begin with a problem?"[83]

An historical problem is usually posed in the form of one or more questions. In order to be meaningful, however, these must meet certain requirements. Three of these are dealt with by Collingwood: the questions must "arise", their relation to evidence must be evident and they must be put in the right way.

1) In the development of his conception of absolute presuppositions Collingwood gives in *An Essay on Metaphysics* the following definition after asserting that every question involves a presupposition: "*To say that a question 'does not arise' is the ordinary English way of saying that it involves a presupposition which is not in fact being made*". After this he continues: "A question that 'does not arise' is thus a nonsense question: not intrinsically nonsensical, but nonsensical in relation to its context, and specifically to its presuppositions. A person who asks another a question which 'does not arise' is talking nonsense and inviting the other to talk nonsense in the same vein" (EM,26). Similarly, Collingwood says in his *Autobiography*: "Each question and each answer in a given complex ha[s] to be relevant or appropriate, ha[s] to 'belong' both to the whole and to the place it occupie[s] in the whole. Each question ha[s] to 'arise'; there must be that about it whose absence we condemn when we refuse to answer a question on the ground that it 'doesn't arise'" (Aut,37).

The same principle holds for historical research. "A problem only exists for the historian", Collingwood maintains in "The Nature and Aims of a Philosophy of History", "in so far as something in his experience has raised it; and in the case of any *bona fide* problem – as distinguished from the pseudo-problems which may be raised verbally out of idleness but are not actually raised by historical thinking in the course of its development – the way in which it arises must of necessity, to an intelligent mind, convey some hint of the direction in which evidence for its solution is to be sought" (NAPhH,52-53). An example of a pseudo-problem given by Collingwood is: "what ... was the name of the first Roman citizen who died a natural death in the year 1 A.D.?" (NAPhH,53). Collingwood refers in the above quoted passage to the requirement of a sensible question that it is related to possible evidence. Elsewhere he gives a similar example in order to criticize the realist position that anything in the past may be the object of historical thought. From this viewpoint, he says, "... the question what Julius Caesar had for breakfast the day he overcame the Nervii is as genuinely historical a problem as the question whether he proposed to become king of Rome" (LHK,101).

There is, however, also another aspect to the claim that a question must arise. After asserting that historical research starts with asking a question Collingwood says in "The Philosophy of History":

> ...[T]he question must be asked with some reasonable expectation of being able to answer it, and to answer it by genuinely historical thinking; otherwise it leads nowhere, it is at best idle 'wondering', not the focus of a piece of historical work. We express this by saying that a question does or does not 'arise'. To say that a question arises, is to say that it has a logical connexion with our previous thoughts, that we have a reason for asking it and are not

moved by mere capricious curiosity (PhH,137).

Though it is not explicity mentioned by Collingwood one could interpret the passage as meaning that a question must have a logical connection with previous thought in the sense that it must fit in the context of previous discussions. It would refer, then, to Collingwood's principle of second-order history, that is, that the posing of a historical problem should be done against the background of past historical thought about it. We have seen how Collingwood refers to this principle in his 1928 lectures[84], taking the study of the Peasants Revolt as an example[85].

Not only vague and general questions should be avoided, but also matters of detail, which are questioned arbitrarily. Questions like "what had Julius Caesar for breakfast on a certain day" or "what is the name of the first Roman citizen who died in 1 A.D." are not only senseless because of lack of evidence, but also because they are not put within the context of the history of historical thought on a certain problem. Hence they are not real problems at all, but only whimsical questions. They simply do not arise. To give another example offered by Collingwood: Suppose, he says, that someone who has made a special study of the battle of Waterloo was asked "... the name of the hundredth man to be put out of action by musketry fire. He will not be able to answer; but the question is, will he be disconcerted by his inability, or not? He will not; he will think it a silly question, and will be rather annoyed at your asking it instead of taking the opportunity to discuss all the interesting problems concerning the battle on which he has something to say"[86].

2) As has already been indicated, it is also necessary for an historical question to be related to evidence. On the other hand, it is emphasized by Collingwood that one can only sensibly speak of evidence if this in turn stands in close relation to a question. "...[E]vidence means facts relevant to a question, pointing towards an answer", Collingwood declares in his 1928 lectures. "Every historian knows that evidence, even the most complete and striking evidence, is convincing and indeed significant only to one who approaches it with the right question in his mind"[87]. On the historical question, however, he asserts a little further on, that it "... is relative to the evidence, as the evidence is to the question: for, just as nothing is evidence unless it gives an answer to a question which somebody asks, so nothing is a genuine question unless it is asked in the belief that evidence for its answer will be forthcoming. A question which we have no materials for answering is not a genuine question; such a question is never asked by the historian, unless inadvertently"[88]. Elsewhere we find similar statements indicating how closely the relation between question and evidence was conceived by Collingwood. In "The Limits of Historical Knowledge" it is said: "The historian does not first think of a problem and then search for evidence bearing on it; it is his possession of evidence bearing on a problem that alone makes the problem a real one" (LHK,102). In "The Philosophy of History" the other side is emphasized:

"The historian cannot first collect data and then interpret them. It is only when he has a problem in his mind that he can begin to search for data bearing on it" (PhH,137). Similarly in *The Idea of History* Collingwood contends: "...[The historian] has already in his mind a preliminary and tentative idea of the evidence he will be able to use. Not a definite idea about potential evidence, but an indefinite idea about actual evidence. To ask questions which you see no prospect of answering is the fundamental sin in science... Question and evidence, in history, are correlative. Anything is evidence which enables you to answer your question – the question you are asking now". On the other hand, Collingwood asserts a little further on: "You can't collect your evidence before you begin thinking... because thinking means asking questions...and nothing is evidence except in relation to some definite question" (IH,281).

Having discussed the two poles of question and evidence in historical research, one might ask which role could be assigned to the "historical fact", a concept which is so often to be found in historical studies. Collingwood's opinion about this is quite clear: in history no facts are studied as ready-made data, but only problems. Not everything, however, can be questioned in an historical study. The latter always includes certain questions which are considered to be provisionally answered in a satisfactory way. For an historian, then, "[t]hese answers are his data", Collingwood asserts, "and if they have been arrived at by other people these other people are his authorities. The datum is a datum relatively to the present inquiry, because within the limits of this inquiry the question whether our belief in it is justified is a question that we are not reopening". But, Collingwood warns, "[i]t is not an absolute datum, because this question has been an open one at some time in the past, and always may be reopened in the future. To the historical spirit in its universality all questions are open questions; but as it operates here and now in the person of this individual historian, many questions are of necessity judged and for the time being closed"[89].

Collingwood's objection to the view of history as the knowledge of past facts is tersely formulated in his criticism of the German historian Meyer: "I think he does not sufficiently realize the principle of the *actuality of the problem*. He thinks of history as knowing the facts, whereas it is really answering our own questions about the facts. The questions, as well as the answers, arise out of our own view of the facts"[90].

3) Historical questions should not only "arise" and stand in relation to evidence: they should also be put in a way that makes a successful answer possible. In this connection Collingwood's conception of "the principle of the limited objective", as described in *The New Leviathan*, is of primary importance. He calls this principle "the most fundamental difference between the modern sciences and the sciences of ancient Greece" (NL,253). The Greeks, Collingwood contends, tried to find the "essence" of things by asking general and vague questions like "What is

Man", or "What is Nature"[91]. In contrast with this Collingwood admonishes: "Limit your objective. Take time seriously. Aim at interpreting not, as the Greeks did, any and every fact in the natural world, but only those which you think need be interpreted, or can be interpreted...; now, choose where to begin your attack. Select the problems that call for immediate attention. Resolve to let the rest wait" (NL,254). Earlier in the same book Collingwood says: "The technique of knowing proper, or what is called scientific method, depends on replacing questions which, being vague or confused, are unanswerable, by real questions, or questions which have a precise answer" (NL,74).

In his archaeological practice Collingwood strictly adhered to this principle: it is in fact the essence of his method of "selective excavation" and we have seen how he considered this the main feature of "scientific archaeology". What he means by this approach is well illustrated, when he asserts, after saying that tentative theories had been put forward in connection with Hadrian's Wall: "...[F]inally these were tested by bringing the problems to a focus at particular points where they could be solved, or at any rate advanced, by some quite small piece of excavation, planned and supervised and recorded with the utmost care"[92]. A question like "Let us see what we can find out about this site", not in line with the principle of selective excavation, is repudiated by Collingwood as "...only a vague portmanteau-phrase covering a multitude of possible questions but not precisely expressing any of them" (Aut,122).

Beside the argument that questions should be couched in specific terms, matters of principle may also be involved in asking them. What is at issue here are the (mostly implicit) presuppositions of the questions asked. It is obvious that if these presuppositions are incorrect or wrongly conceived the answers to the questions based upon them will be unsatisfactory as well. By making these presuppositions explicit and questioning them, certain problems are not only clarified, but their solution may be looked for in another direction or even directly found. An excellent example is given by Collingwood in his discussion of the problem of the "Celtic Revival" in art after the Romans left Britain. As we have seen in the second chapter this problem was solved by him "not by discovering fresh evidence, but by reconsidering questions of principle" (Aut,145). The principle concerned was to consider the Celtic element as an unconverted residue incapsulated in the historical process[93]. In *Roman Britain and the English Settlements* the Celtic revival is described through the idea of the continuity of a cultural tradition[94]. That in the explanation of an event like this certain matters of principle are involved is made clear when Collingwood warns against any attempt to explain it biologically: "Biological conceptions like that of race throw no light, but only darkness, upon historical problems, breeding error and superstition where what we want is fact"[95].

Discussing the problem of the date of the end of Roman Britain

Collingwood also turns to its presuppositions. "Without deciding which date is the right one" he proposes, "we can still raise the question, 'What was the nature of the event whose date is in dispute?'"[96]. The end of Roman Britain should be seen, Collingwood then asserts, as "...a departure not of Roman officials but of Roman civilization: it was... 'the un-Romanization of Roman Britian'. When and by what stages did this event take place, and how far did it go? That is the problem now before us"[97]. This is a clear example of how Collingwood, by reconsidering a matter of principle, shifts a certain problem in another direction: in this case with the result that instead of the end of Roman Britain, the concept of Romanization itself is discussed[98].

In his review of Rostovtseff's well-known book *The Social and Economic History of the Roman Empire*[99] Collingwood also questions the way a certain problem is posed, when he argues: "On ... the problem of the causes for the decay of ancient civilization, the book closes. The terms of the problem are keenly and penetratingly felt, and the criticism of many current solutions is conclusive. But if, as it seems to be, the problem is insoluble, ought we not to ask whether it is a genuine problem? Was the so-called collapse of ancient civilization really a collapse at all?"[100]. As the question already suggests, Collingwood is of the opinion that this definition is incorrect. It is only when seen from the standpoint of the old, he asserts, that the change is seen as catastrophic[101].

A question may also need to be specified. Thus Collingwood asserts in a discussion of the status of London: "If the question is asked: where was the capital of Roman Britain? no answer can be given unless we distinguish between four possible senses of the word capital: religious, civil, financial, and military" [102]. Collingwood's suggestion then is, that "... before the Boudiccan revolt, Colchester was the civil and political capital, London the financial; and that in consequence of the revolt London became capital in all three senses"[103] (the change of the religious capital from Colchester to London is described earlier).

Another example, finally, of Collingwood's questioning the presuppositions of a question, thereby shifting the problem in another direction, is the following statement of his: "...[T]he right way of putting the question why Claudius invaded Britain is to ask, not why it was done, but why it was done then and not earlier. The conquest was merely the execution, at the right moment, of a policy long accepted"[104].

The examples given of Collingwood's critical discussion of the presuppositions of certain questions are primarily concerned with the more sophisticated aspects of historical interpretation. They must nevertheless be seen within the context of Collingwood's "logic of question and answer". They are indeed the more refined elaboration of this "logic", which, as has been indicated above, can better be defined as a theory of inquiry.

There is no reason to disbelieve Collingwood's contention, that his archaeological practice served as a "laboratory of knowledge" for the working-out of certain aspects of historical methodology. On the contrary, in the chapter on Collingwood's work in archaeology and history we have seen how the questioning activity played a crucial role in his research. "The reason I am talking so much about archaeology", says Collingwood in his *Autobiography*, "is that in archaeology the issue raised by the project of a Baconian revolution is unmistakable" (Aut,133). That is, in archaeology the questioning activity is necessarily of pivotal importance. For sources are dumb and have to be interpreted by asking specific questions. How to extract information from them is highly relevant for historical methodology in general. To have worked this out is an important contribution made by Collingwood to historical methodology, which has been insufficiently appreciated until now.

7.4. INTUITION

As said before, the reason that Collingwood's ideas on the methodological aspects of the science of history have been neglected is primarily the fact that his re-enactment doctrine has been interpreted methodologically. The main form of this version is the one which imputes to Collingwood an intuitionistic theory of historical knowledge[105]. As we have seen in the third chapter, this interpretation is endorsed by a large group of writers on Collingwood, all of them critical. This criticism, then, takes the form either that intuition is subjective, or that it is non-inferential, or that it does not take into account general knowledge[106].

However, the view that Collingwood endorsed an intuitionistic theory of historical knowledge is not correct. This can be demonstrated with regard to all three features that critics have ascribed to this theory. That the third criticism is misplaced is made clear, when Collingwood asserts in his 1926 lectures:

> The interpretation of sources must proceed according to principles. It is not enough to interpret them according to the dictates of intuition, to deal with individual cases as if each was unique and unlike any other. People sometimes advocate this happy-go-lucky or intuitive method of dealing with the problems presented by moral conduct, art, science, or even philosophy under the name of dealing with every case on its merits, and support their contention by a polemic against casuistry and the tyranny of abstract rules. And certainly abstract rules are bad masters. It does not follow that they are not good servants. And it is sometimes forgotten that to deal with a case on its merits is impossible unless it has merits, that is to say unless it has recognisable points of contact with other cases whose merits are of the same general kind. It is doubtless true that every case is unique; but uniqueness does not

> exclude points of identity with other unique cases; and a denial of the genuineness of universals is at least no less disastrous than a denial of the uniqueness of their particulars. In point of fact, no one would dream of trying to interpret an historical document except in the light of general principles[107].

Further on in the same lecture Collingwood criticizes historians for being under the impression that they "...have a mysterious intuitive *flair* for the truth..."[108] and emphasizes the need to use certain principles "...unless [the historian] will condescend to the ignominy of seriously claiming that he has a direct intuitive perception of the difference between a true and a false narrative..."[109].

We have seen that Collingwood considers history an inferential and reasoned form of knowledge[110]. That he rejects the notion that historical knowledge can be obtained by non-inferential intuition, so often imputed to him, is obvious from his criticism of German theorists of history, who did endorse this idea. Discussing in his 1929 lectures the views of Lazarus and Steinthal that "...science depends on logical abstraction and general concepts, history [on] psychological intuition and concrete *Vorstellungen*", Collingwood contends: "This distinction had not served as a basis for a theory of historical method, because the so-called psychological intuition could only be a personal and private opinion, incapable of being rationally defended or criticised"[111]. Further on Collingwood refers again to the "... vague and uncritical conception of psychological intuition..."[112]. After saying, then, that Croce took an important new step by comparing history with art he asserts:

> The progress here achieved may be measured by reflecting that the intuition to which the Germans had appealed was a quite vague and undefined thing, a thing which, so far as they could show, had no rules and no methods, no distinction between truth and falsehood or better and worse. If they were asked how this intuition could ever convince us of its truth, the best reply they could give was that historians did actually experience such a conviction. But that did not prove that on their theory historians *ought* to have experienced it. Nor did it prove that this conviction was in any sense a reasonable one. In short, the German analysis of historical thinking failed to show that historical thinking worked according to any kind of critical principles, and left it exposed to the charge of being a mere matter of taste or caprice[113].

This passage makes clear that Collingwood was also conscious of the subjectivist aspect of the intuitive theory and was indeed opposed precisely at its non-rational nature. In the 1929 lectures Collingwood refers at another place again to the German conception of history. After saying that Croce made in his *Logic* an improvement on his earlier theory, he asserts: "History is now no longer the mere irrational intuition with which the nineteenth-century Germans had identified it; it is no longer simply the aesthetic intuition with which Croce had identified it in his first essay. It is not intuition but judgment"[114]. In the 1936 lectures the same topic is dealt with by Collingwood, again rejecting the German intuitionistic approach[115].

In a paper on Spengler, read in 1929 at the London School of Economics, Collingwood comes back to the German conception of history, because he considers Spengler a typical exponent of it. After philosophers had tried to impose on history the methods of natural science, he argues, they tried to account for the fact that historians "...succeeded in constructing scientific and intellectually respectable accounts of individual facts"[116]. The way this was done by them, however, is again criticized by Collingwood. Since it throws an interesting light on his own conception of the science of history it is worthwhile reproducing Collingwood's argument:

> A science of the individual, they said, is a contradiction in terms; scientific knowledge must be knowledge of the universal: of the individual there can only be intuition, a kind of immediate vision unaccompanied by any demonstration or proof and confirmed only by a mysterious inner conviction: something like the poet's insight, and utterly removed from the cautious reasoned inferences of the scientist.
>
> This theory of historical knowledge as a mysterious intuition, based on a private and personal conviction of truth instead of a reasoned demonstration, was advanced by a number of German writers in the 80's and 90's of the last century. I need hardly say that it is no theory at all; it is merely the consciousness of the need for a theory. Its only merit is that it recognizes that there is a difference – a vital and fundamental difference – between history and science, but it cannot grasp the difference, and, in attempting to state it comes to the disastrous conclusion that history is a matter of intuition in the sense of mere personal belief, devoid of all logic, all argument, and all rationality.
>
> This new irrationalistic theory of historical knowledge is, for good and evil, the mainspring of Spengler's philosophy of history[117].

How negative Collingwood's opinion was of this theory of historical knowledge is made clear when we find in his "Notes on the History of Historiography and Philosophy of History" of 1936 the following observation on Spengler: "'intuitive and depictive through and through' – i.e. he claims that this is idiographic or historical, not nomothetic or naturwissenschaftlich. This reveals the confusion which, implicit in the German movement, bred Spengler's monster"[118].

The passages discussed above definitely prove how incorrect the interpretations are which impute to Collingwood an intuitionistic theory of historical knowledge. The idea that Collingwood would adhere to the view that this knowledge could be acquired by a sort of mysterious activity is simply ridiculous. To be fair it is necessary, of course, to remind ourselves that the passages discussed here are all taken from the manuscripts, and that these have not been available before. On the other hand, one could retort that Collingwood nowhere in his published writings asserts that historical knowledge is acquired by intuition. On the contrary, as we have tried to demonstrate before, he makes it clear in these writings that historical knowledge is rational, inferential and based on evidence. For this reason there are indeed some interpreters who have correctly rejected the idea that Collingwood held an intuitionistic view of historical science[119]. It is interesting to note in this connection that among Collingwood's published writings there are also a few passages which explicity reject intuition as a way of acquiring (historical) knowledge. In his article "Can the New Idealism Dispense with Mysticism", for instance, he asserts:

> Now it is notorious that all idealism since Kant has maintained that ultimate truth is to be reached, if at all, only by hard thinking, by the critical development of rational theory, and not by any kind of intellectual intuition...
> Now if it is true that ultimate truths are to be reached by the path of intuition, and not by the 'labor of the notion', then certainly all idealism is futile. So is all scientific and historical thinking. And the only thing left for the person who wants to get at the truth is to return like Nebuchadnezzar to the level of the instinctive animals and *s'installer dans le mouvement*, instead of trying to raise himself above it in order to understand it[120].

In a short passage in "The Philosophy of History" the idea of intuitive historical knowledge is also implicitly rejected. Collingwood asserts: "...Lazarus and Steinthal ... tried to explain history as 'intuitive' knowledge; but that destroyed its reasoned, scientific, inferential character" (PhH,134). This character of historical knowledge is well-worked out by Collingwood, both in his published and unpublished writings. Unfortunately this fact has been obscured by the intuitionistic interpretation of Collingwood's idea of the science of history.

CHAPTER EIGHT

SOME CONTROVERSIAL ISSUES

8.1. PAST AND PRESENT

History being present knowledge of the past the nature of the relation between past and present is of fundamental importance. Collingwood had definite ideas on this relation. It was, in his own words, the subject of his "first principle of a philosophy of history", developed by about 1920. This principle was, as Collingwood says in his *Autobiography*, "...that the past which an historian studies is not a dead past, but a past which in some sense is still living in the present" (Aut,97). This idea indeed plays an important role in Collingwood's philosophy of history and in order to understand the latter properly it is therefore worth discussing it separately.

Collingwood's view that the historical past is not dead, but living in the present, has come up for discussion several times before in this study. As we have seen in the third chapter he considers this doctrine to express one of the ways in which historical processes differ from natural ones[1]. The way the past lives on in an historical process is described by Collingwood in his *Autobiography*, while in the same book an illustration of this is given with his analysis of the revival of Celtic art[2]. We have also seen that Collingwood criticizes philosophers of history such as Rickert, Simmel, Spengler and Toynbee for not seeing the past as living on in the present, but as dead[3].

Though the idea of a living past was of an early date and was apparently seen by Collingwood from the beginning as essential to a proper understanding of the historical process, it got another dimension with the doctrine of the re-enactment of past thought, which was developed by Collingwood for the first time in his 1928 lectures. The close relation of the latter conception to the idea of a living past is made clear, for instance, in Collingwood's discussion of Oakeshott in *The Idea of History*: "...Oakeshott supposes that there is no third alternative to the disjunction that the past is either a dead past or not past at all but simply present", Collingwood asserts. "The third alternative is that it should be a living past, a past which, because it was thought and not mere natural event, can be re-enacted in the present and in that re-enactment known as past" (IH,158).

The conception of the re-enactment of past thought, however, was related to the idea of a living past only at a later stage in Collingwood's thought[4]. We have seen how the problem of the status of the past was constantly in Collingwood's mind as his thought on history developed[5]. After having held a realist view of the past, Collingwood worked out, for the first time, in "Some Perplexities About Time" the idea of the ideality of the past[6]. This idea was then further elaborated in the 1926 lectures. Since we can only know what actually exists and the past is ideal, Collingwood asserts in these lectures that our supposed knowledge of the past cannot truly be so called[7]. His final view is developed in the 1928 lectures. Though he keeps to the idea of the ideality of the past, he now maintains that the past may be actualised by the present thought of the historian. This actualisation can only be realised, however, in the case of past *thought*[8]. In this way we are able to have real knowledge of the past, which was doubted until then by Collingwood. The idea that we can only know the past through the re-enactment of its thought is accompanied by the idea that all history is the history of thought. The conception of a living past fits in well, of course, with this theory. So, in Collingwood's view, the three features of history are: 1) it is a process of thought, 2) it lives on in the present, 3) it can be known through re-thinking past thoughts. These aspects cannot be separated and they even presuppose each other. This is made clear, when Collingwood asserts that "... the historical past...is a living past, kept alive by the act of historical thinking itself" (IH,226). This means that the past not only *can* be known by re-thinking past thoughts, but *should* be known in this way in order that the historical thought-process lives on in the present[9]. For this reason Collingwood can also assert that "... the historian is an integral element in the process of history itself, reviving in himself the experiences of which he achieves historical knowledge" (IH,164), or that in history "... the process of historical thought is homogeneous with the process of history itself, both being processes of thought" (IH,190).

The idea that the human past is dead is strongly opposed by Collingwood. In his repudiation of the realist view that the historian tries to know the "real" past, he slips, however, at one place into this contention. For in "The Limits of Historical Knowledge" Collingwood maintains: "An event that has finished happening is just nothing at all. It has no existence of any kind whatever. The past is simply non-existent; and every historian feels this in his dealings with it...What the historian wants is a real present. He wants a real world around him...; and he wants to be able to see this world as the living successor of an unreal, a dead and perished, past. He wants to reconstruct in his mind the process by which *his* world...has come to be what it is. This process is not now going on" (LHK,101). This contention is certainly not in line with Collingwood's final view, as described above. It is interesting in this connection that he explicitly comes back to it in his notes on historiography of 1939, saying:

> People are always talking as if history were the same thing as transience or timefulness: 'this has a merely historical interest' means 'this interests only people who are interested in things that have passed away'. I myself have insisted that history is the story of 'events that have finished happening', and must be careful about this in future. But I have also insisted that all history is the history of thought, and that the historian knows a past thought by re-enacting it in the present. As so re-enactable, it is not something that has finished happening. Is it a *condition* of the past's being historically knowable, that it should be still something actual, not something dead? Or is this a *consequence* of its being historically known? Neither: it is the same thing as its being historically known. There are (of course) no 'conditions' of a thing's being knowable – that is the error of realism[10].

Collingwood's conception of the relation between past and present is also of importance with regard to his philosophy of mind. As we have seen in the fourth chapter he identifies a man's character with his past, the latter being the determining factor of present actions[11]. In the same vein Collingwood asserts elsewhere that "...the past is the substantial being of the present: to know the past is to know not *how* the present came to be what it is but *what* it is"[12]. Having a past is a necessary aspect of mind. The relation of man's present to his past is essentially different from the one of nature. When pointing out this fact Collingwood refers to the idea of history as a philosophical or transcendental concept, that is, that history should be seen as a universal and necessary aspect of human mind[13]. How fundamental this idea is, may be illustrated by a passage in the manuscript on cosmology, when Collingwood discusses the concept of consciousness:

> Consciousness lies not merely in passing through any special sequence or cycle of perceptions, but in passing through any sequence of them, no matter what, and holding them together in a present act of mind for which the past *as such*, as well as the present, is an object. This may be called a survival or revival of the past, if we like, but we must remember what a very curious thing it is: not a survival of something in its effects, or a revival in something of the same kind, but its survival *in itself, alongside of* its effects[14].

Collingwood then illustrates this view with the following example: '[M] y having eaten something unwholesome in the past, if I am a merely sentient but unconscious animal, is present to me now only in the shape of its effect, viz. feeling unwell; whereas in consciousness the same thing is present twice over, both in its effects (feeling unwell) and in itself (remembering the act of eating)"[15].

Summing up we may contend that Collingwood's idea of a living past is an important element both in his re-enactment doctrine and his philosophy of mind. We have seen that the same is the case with his process-view of history, with the corollary that historical continuity is emphasized. Besides these, the idea of a living past is also important in connection with Collingwood's opinion on the use of historical study. The latter aspects, however, will be considered later.

8.2. HISTORY AS THE RE-ENACTMENT OF PAST THOUGHT

Collingwood's conception of the re-enactment of past thought has already come up for discussion at various places in this study. In the third chapter it was dealt with in the general survey of *The Idea of History* (3.2), the section on the re-enactment doctrine itself (3.3.5) and the section on explanation and understanding (3.3.6), while in the fourth chapter aspects of it were discussed in the treatment of the 1928 lectures (4.5) and the "Notes on the History of Historiography and Philosophy of History" of 1936 (4.9).

In this section the re-enactment doctrine will be discussed in a more systematic way. Within this doctrine various aspects may be distinguished, which will be dealt with separately: its status, the concept of thought it implies and the notion of re-thinking. Finally some examples will be given from Collingwood's work in archaeology and history.

Status of the Re-enactment Doctrine

As has been argued already in the fourth chapter, Collingwood's re-enactment doctrine should be seen in the context of a transcendental analysis of the universal and necessary characteristics of the science of history[16]. It is useful to remind ourselves that in his 1928 lectures, in which the doctrine was worked out for the first time, Collingwood describes his task as "... discovering how historians always and necessarily think"[17]. Against a large group of interpreters, such as Toynbee, Renier, Gardiner, Mandelbaum, Flenley, Buchdahl, K.M. Martin, Roberts, Grant, Popper and Cebik[18], I would therefore contend that Collingwood's conception of the re-enactment of past thought should not be conceived as a theory of historical method. As was pointed out for the first time by Donagan it should be seen as a theory describing what is logically implied by historical knowledge[19]. Though with different shadings, the same view is endorsed, as we have seen, by Dray, Toulmin and R.Martin[20].

I would contend, however, that a simple rejection of the methodological interpretation of the re-enactment doctrine fails to give a complete picture of this much discussed theory. For there surely are certain methodological aspects involved. It is useful in this connection first to consider a passage of *The Idea of History* in which Collingwood makes it clear that his analysis of history should not be seen as methodological. Discussing the fact that historians are confined to the study of human affairs he contends:

> The question must ... be raised, why do historians habitually identify history with the history of human affairs? In order to answer this question, it is not enough to consider the characteristics of historical method as it actually exists, for the question at issue is whether, as it actually exists, it covers the whole field which properly belongs to it. We must ask what is the general nature of the problems which this method is designed to solve. When we have done so, it will appear that the special problem of the historian is one which does not arise in the case of natural science (IH,213).

After this passage Collingwood describes the historian as investigating human actions, the latter being the unity of the outside and inside of events[21].

It is interesting to compare the above quoted passage, which is from "Human Nature and Human History", with Collingwood's first draft, since the latter demonstrates even more clearly what he is up to. "The question... is", he observes, "whether this customary restriction of history to the history of human affairs is a merely empirical limitation, due to an incomplete realization of what historical method, duly widened, can do, or whether it represents the just recognition of a barrier beyond which historical thought cannot pass without turning into something else"[22].

What Collingwood here refers to is his distinction between history as an empirical concept and as a transcendental or philosophical one[23]. It is clear that he is interested in the latter, when he says: "... it is necessary to consider not the characteristics of historical method (for the question at issue concerns a supposed limitation in that method as hitherto practised) but the terms of the problem which this method is designed to solve"[24].

In the draft of "Human Nature and Human History" Collingwood also contends that an historian is directed towards the "inner side" of human actions, that is, towards "processes of thought". That he considers their re-thinking a necessary aspect of all historical knowledge is made clear when he maintains, in a passage which is not to be found in *The Idea of History*: "The historian's true business is to detect these processes of thought, and re-think them somehow in his own mind. This is not only *the essential feature of all historical knowledge*, but it marks it off decisively from every kind of natural science"[25] (italics mine). A scientist, Collingwood asserts, will never look for the inner sides of events. However, he then makes an observation which implies a link with the methodological issue: "This constitutes the fundamental divergence between the principles of historical and scientific thinking, *and from this divergence of principle flow the differences in their methods*" (italics mine). Collingwood does not enlarge on these differences, but only remarks: "... I intend here to assume that scientists and historians respectively know their own business, and are using methods genuinely appropriate to that which they are investigating"[26].

Likewise, in a review of *Philosophy and History: Essays presented to Ernst Cassirer*[27], Collingwood links the principle of re-thinking past thought with the methodological question. This time, however, there is an illustration. Having given a description of the notion of re-thinking, he contends:

> ...[A] past whose thought the historian is unable thus to make his own, whether through lack of evidence or through defect in his own mental powers, inability to sympathize with it, is a past at once dead and unknowable. This doctrine has a practical bearing on historical method. It implies that in order to understand a certain past event or state of society the historian must not only have sufficient documents at his disposal; he must also be, or make himself, the right kind of man: a man capable of entering into the minds of the persons whose history he is studying[28].

As we have seen before, this requirement is the central topic of Collingwood's manuscripts on folklore. For it was precisely the incapability of the traditional studies of primitive man to enter into primitive mind, which had been, according to Collingwood, the main impediment to its understanding.

The primary methodological bearing of the re-enactment doctrine, however, is stated by Collingwood in his *Autobiography*. In chapter 10 of this book he gives an analysis – in the form of three "propositions" – of the most important characteristics of history[29] (it is an elaboration of the conclusions to which he had come in his 1928 lectures). The second proposition states the doctrine of the re-enactment of past thought (Aut, 112). Then in chapter 11 a description is given of certain "principles of historical thinking", which are of a methodological nature[30]. According to the second principle, an historian should study a human action as the expression of some thought: that is, "the historian's business is ... to identify this thought" (Aut,128). In Collingwood's view this thought can only be identified by re-thinking it, but in the chapter at issue (on "Roman Britain") it is expressed methodologically in the statement that "[f]or the archaeologist this means that all objects must be interpreted in terms of purposes" (Aut,128).

Though the re-enactment doctrine has certain methodological corollaries, it would be incorrect to conclude from this that its interpretation as basically a theory of the universal and necessary characteristics of history is thereby impaired. We have seen how Collingwood considers human action the specific historical object and how he sees thought as an intrinsic element of it. The function of the re-enactment doctrine in this connection is, that it provides a solution to the problem how knowledge of past action (thought) is possible. The way this was worked out by Collingwood in his 1928 lectures clearly demonstrates that it was regarded as a transcendental analysis, the approach even being, in fact, Kantian in nature. This does not mean, though, that Collingwood's theory is not relevant to questions of historical methodology. On the contrary, this is rather implied by the re-enactment doctrine being conceived as stating certain necessary conditions of historical knowledge. If historical knowledge is wrongly conceived this will have the consequence that a wrong methodology will be practised. This is the practical aspect of the re-enactment doctrine and of Collingwood's transcendental analysis of history in general. It is pointed out in his 1928 lectures, when he propounds his conception of philosophy of history as the "pure methodology" of history, "dealing with the universal and necessary characteristics of all historical thinking whatever"[31]. For he contends:

> This science is practical, or methodological in the sense of providing guidance in the pursuit of historical knowledge, in that it studies what history everywhere and always is, and therefore what history everywhere and always ought to be. It is easy to object that on this showing, history always is what it ought to be, and therefore the philosophy of history can have no practical value. This would be true, were it not that people who refrain from pursuing philosophical inquiries are generally more or less at the mercy of philosophical fallacies[32].

After this Collingwood gives some examples of what he considers fallacies:

> ...[L]ogical thinkers have distorted history in various directions. They have advocated historical materialism; they have destroyed the continuity of history by asserting fantastic distinctions between the savage and civilised minds; they have tried to reduce history to a science by suppressing all that makes it history; they have invented the doctrine of historical cycles; they have asserted a mechanical law of progress; they have denied progress altogether; they have committed a hundred fallacies of the same kind, each involving an error in the philosophy of history and each in consequence falsifying the whole structure of their historical thought[33].

At a later date Collingwood added the following observation to the above quoted passage: "To avoid these consequences of bad philosophy there is no way except by finding a better philosophy: in this sense the philosophy of history, as we have tried to expound it here, acts as a practical guide to the logical problems of historical thought". The re-enactment doctrine, then, should be conceived as part of this program. For this reason certain theories and their corresponding methodologies are criticized by Collingwood for not taking into account the implications of this doctrine. Positivism and the traditional studies of primitive man, both confining themselves to the outside of human actions, could be given as examples.

Concept of Thought

If the re-enactment doctrine is to be plausible it is first of all necessary to ascertain if a thought can be re-thought by another person. In Collingwood's view this is surely the case, his position being based both on his philosophy of mind and on his notion of the concept of thought. In the beginning of chapter 6.2 it was shown how Collingwood had already developed in *Religion and Philosophy* a monistic concept of mind, according to which mind is defined in terms of the object of which it is conscious[34]. This is not only the basis of personal identity, he points out in the same book, but also of communication between two minds. "A mind is self-identical ... if it thinks and wills the same things constantly", Collingwood asserts; "it is identical with another, it if thinks and wills the same things as that other" (RPh,116). He even goes so far, in this connection, as to state that if two persons think of the same table, they not only share the same thoughts, but even actually have one mind for this moment (RPh,101). (The latter contention was not endorsed by Collingwood in his later elaboration of the re-enactment doctrine in the 1928 lectures. For he maintains there that though the same thoughts may be thought by different people, they are thought within different contexts and hence different minds. In the *Autobiography* this idea is expressed by the notion of the "incapsulation" of past thoughts in present ones).

In my short survey of *The Idea of History* some characteristics of Collingwood's concept of thought have already been pointed out[35]. As we have seen, it is fundamental for this concept that, though thought is part of mind's experience, it is not conceived as wholly entangled in it: that is, the same thought can be part of different experiences. This is true not only for the thoughts of one person, but also for the thoughts of other persons, both in present and past. With regard to the first aspect Collingwood argues:

> Suppose that, after thinking 'the angles are equal' for five seconds, the thinker allows his attention to wander for three more; and then, returning to the same subject, again thinks 'the angles are equal'. Have we here two acts of thought and not one, because a time elapsed between them? Clearly not; there is one single act, this time not merely sustained, but revived after an interval. For there is no difference in this case that was not already present in the other. When an act is sustained over five seconds, the activity in the fifth second is just as much separated by a lapse of time from that in the first, as when the intervening seconds are occupied by an activity of a different kind or (if that be possible) by none (IH,286).

Collingwood concludes from this argument that nothing prevents a thought of one mind being revived by another mind: "Granted that the same act can happen twice in different contexts within the complex of my own activities, why should it not happen twice in two different complexes?" (IH,288). This view is the basis of Collingwood's contention that thought, besides being part of private experiences, belongs fundamentally to the public world as well[36]. We may find it already in *Religion and Philosophy*, when Collingwood asserts: "...[T]he statement 'my knowledge is my knowledge' must not be so interpreted as to exclude the complementary statement that my knowledge may also be yours" (RPh, 103). This implies not only the possibility of communication with and knowledge of present minds, but also knowledge of past ones. "The spirit of truth is not circumscribed by the limits of space and time", Collingwood also contends in *Religion and Philosophy*, "If a real community of life is possible between two men who share each other's outward presence and inward thoughts, it is possible no less between two who have never met; between the ancient poet and his modern reader, or the dead scientist and the living man who continues his work" (RPh, 160). (Again, this passage clearly demonstrates that Collingwood's re-enactment doctrine is based on a philosophy of mind which is indeed of an early date).

Collingwood's view of thought as being essentially public, besides being part of the private life of mind, is not only crucial for his re-enactment doctrine, but also difficult to grasp. For this reason many interpreters have been led astray, coming to an altogether incorrect view of the doctrine, interpreting it psychologically, mostly in accordance with the intuitive theory. In our discussion of the re-enactment doctrine as developed in the 1928 lectures it was stated that the identity of a past thought with the one re-thought by the historian should be conceived as a conceptual one[37]. This idea is clarified by Collingwood in his manuscript on cosmology:

> Concepts determine facts as their formal cause, as the essence of which they are existence. Now, essence is one where existence is manifold; one form is therefore capable of embodiment in a plurality of instances. Therefore, whereas my experience can only be mine, and nobody else's, the concepts exemplified in it may be exemplified in other experiences. No two people can have the same toothache, but they may both have toothache. Thus concepts provide a common ground on which diverse experiences can meet. Any world of thought is a public world, accessible not indeed to every mind in common, but accessible in common to any two minds which enjoy similar experiences. It is because they have similar experiences that they can share the same thoughts, and it is through sharing the same thoughts that they can know their experiences to be similar ... Experience is nothing but the existence-term of that dyad whose essence-term is thought; consequently what is unified in thought must be dispersed in experience[38].

Though this passage only refers to the sharing of present thoughts, we have seen that in Collingwood's view the sharing of present and past thoughts makes no difference. It gives us, therefore, a better understanding of the re-enactment doctrine as well[39].

Re-thinking

In our discussion of the "Notes on the History of Historiography and Philosophy of History" of 1936 we have seen how Collingwood makes a distinction between various meanings of the word "thought", of which the one between the thought-content (*noema*) and the act of thought (*noesis*) is the most important[40]. Though Collingwood at that place primarily refers to other aspects of the re-enactment doctrine, it is made clear by him elsewhere that a thought-content can only be known by actually re-thinking it. In *The Idea of History*, for instance, he maintains:

> Historical knowledge is the knowledge of what mind has done in the past, and at the same time it is the redoing of this, the perpetuation of past acts in the present. Its object is therefore not a mere object, something outside the mind which knows it; it is an activity of thought, which can be known only in so far as the knowing mind re-enacts it and knows itself as so doing. To the historian, the activities whose history he is studying are not spectacles to be watched, but experiences to be lived through in his own mind; they are objective, or known to him, only because they are also subjective, or activities of his own (IH, 218).

Elsewhere in the same book it is similarly stated that an act of thinking can never be studied as "a ready-made object": "It has to be studied as it actually exists, that is to say, as an act" (IH, 292). Likewise, in his *Autobiography* Collingwood maintains: "...[T]he historian must be able to think over again for himself the thought whose expression he is trying to interpret. If for any reason he is such a kind of man that he cannot do this, he had better leave that problem alone. The important point here is that the historian of a certain thought must think for himself that very same thought, not another like it" (Aut, 111)[41]. It is interesting to note, again, that the same view is already held by Collingwood in *Religion and Philosophy*. For distinguishing the "two senses in the word knowing", he maintains:

> There is, first, knowledge in the sense of what I know, the object; and secondly, there is the activity of knowing, the effort which is involved as much in knowing as in anything else. Knowledge as a possession – the things we know – may be common to different minds, but, it may be said, knowledge in the sense of the activity of knowing is peculiar to the individual mind. It may perhaps be replied that since knowledge is admittedly an activity, an effort of

> the will, there is no difference between thinking and willing to think. And if two minds are identical in thinking the same thing, they are equally and for the same reason identical in willing to think the same thing. All knowing is the act of knowing, and therefore whatever is true of thinking *sans phrase* is true of the act or volition of thinking (RPh, 102).

This view of Collingwood is undoubtedly a most crucial element of his re-enactment doctrine. Its rejection, therefore, necessarily implies a rejection of the doctrine itself. One could also say that most criticisms of it could be traced to an unwillingness to accept the view that no distinction should be made between the thought-content and the act of thinking. This unwillingness is seldom expressed explicitly. Walsh, however, is a case to the contrary, when he rejects outright the idea that two acts of thinking can be the same, while he thinks that two thoughts may have identical contents. "Misled like so many others by the fatal word 'know'", he observes, "Collingwood has put forward an impossible solution for a difficulty which is perhaps not real at all"[42]. Walsh refers here to his own interpretation of the re-enactment doctrine as aimed at providing a criterion for truth. In our discussion of the doctrine, however, we have not found any indication that Collingwood had this in mind. We have also seen that Collingwood was not misled by the different meanings of the words "know" or "thinking". On the contrary, as has been shown, he was sharply conscious of their equivocality. He was not misled, but developed certain views, which, of course, as with all views, may not be found acceptable by everyone. It is these views which are at issue when the re-enactment doctrine is criticized. Taking into account the many criticisms leveled against that doctrine it is surprising how rarely this has been realized.

Only Popper, in fact, has launched a fundamental attack on the re-enactment doctrine by criticizing its basic presuppositions. As has been shown in the third chapter, this was in the context of his theory of the three worlds of physical states, mental states and intelligibles[43]. The re-enactment doctrine, then, Popper interprets as being aimed at the world of mental processes (acts of thinking). Against this he puts his own theory of historical understanding as being exclusively aimed at the third world of thought-contents. As has been pointed out by Skagestad, Popper's interpretation of Collingwood is fundamentally incorrect[44]. For Collingwood's point is precisely that in historical understanding no distinction can be made between acts of thinking and thought-contents, or, in Popper's theory, between the second and third world. Popper's psychological and subjectivist interpretation of Collingwood is therefore beside the point, since for Collingwood thoughts belong as much to the objective third world as they do in Popper's view.

The question at issue – and this has also been made clear by Skagestad – is not one of a second world theory of historical interpretation versus a third world one, but between two theories within the context of the third world[45]. According to Popper a distinction should be made within the third world between the object-level of the thoughts of the historical agent and the meta-level of the thoughts of the historian, who tries to understand them. In Collingwood's view, however, such a distinction between object-level and meta-level is impossible within the realm of thought. We have seen that in his opinion a thought can never be a mere object, but must actually be re-thought in order to be known. This is the core of the re-enactment doctrine and the point of disagreement between Popper and Collingwood. I agree with Skagestad that Collingwood's position is a strong one and more convincing than Popper's. A thought is indeed a very special "object" to think about and cannot be put on the same level as other possible objects of thought. It should be added, though, that this holds only for thoughts as studied by historians and not as studied by psychologists. Collingwood also made this distinction, asserting that, in the former case, thoughts cannot be abstracted from their content, as is done by the psychologist, who sees them as mere events[46]. It is this fact that makes their re-thinking necessary[47].

A matter not discussed by Skagestad is Collingwood's idea that a past thought when re-thought by the historian is "incapsulated" in his present thought. A historian is always conscious of the fact that he re-thinks a past thought. He does this in the wider context of his present thoughts and the latter transcend the re-thought thoughts of the past. Hence there is no question of a complete identification of the historian's mind with the mind he studies. This makes it possible for the historian to criticize the thoughts he studies. Collingwood's notion of the transcendence of past thought by the present thought of the historian should be distinguished, however, from Popper's theory of object- and meta-levels. It is in fact Collingwood's own solution – within the context of the "third world" – to the problem of historical understanding[48].

Examples of Re-thinking

We have seen that re-enactment may be methodologically translated as the principle of looking for the purposes of archaeological remains and the intentions of human actions. This principle, however, is not *derived* from the re-enactment doctrine. The latter must rather be seen, as I argued, as a description of certain necessary and universal characteristics of history, a doctrine which was developed at a later date. This is demonstrated by the fact that the principle of seeking purposes was put into practice by Collingwood long before the re-enactment doctrine was worked out. In chapter 5 some examples were given, of which Collingwood's article "The Purpose of the Roman Wall" of 1921 is the most obvious. In what follows some other examples of various types will be considered.

In the popular *Home-Reading Magazine*, in 1925 Collingwood wrote some articles on Roman Britain[49]. Though the re-enactment of past thought was not yet developed as a doctrine we find a most striking example of its main idea in the following passage:

> The practical study of Roman Britain is best begun by studying Roman roads. Most readers need not go far from home to see one, and there is no better way of thinking oneself back into the Roman point of view than to look up on the map a well attested piece of Roman road and follow it for a few miles across country. Get a Roman road, or, for that matter, any road, under your feet, and you enter into the spirit of the men who made it; you see the country through their eyes; you get into your bones a feeling – obscure, perhaps, but powerful and unmistakable – of what they meant to do with the country and how they meant to do it[50].

A little further on Collingwood continues:

> When the reader has accustomed himself to tracing Roman roads in the field, let him look at their course in detail on the best maps... over a wide area. Let him think out the relation of the road to the modelling of the land as shown by the contours, seeing how often the straight lengths of road are laid out from hill-top to hill-top, and how seldom the Roman builders shrank from steep gradients; seeing where, and for what reasons, they abandoned the method of straight lines and built their roads in a series of curves; and then let him turn to the 'Antonine Itineraries', the official Roman road-book of the second century A.D. ... and follow out all these itineraries, so far as he can, in detail, asking himself why the Romans regarded that particular route as important enough to form part of a stereotyped scheme of communications[51].

In *Roman Eskdale* (1929) Collingwood indicates why Hardknot Castle (between Ambleside and Ravenglass in north-west England) was built. He does it by reconstructing the argument of the Romans, based on geographical and military considerations:

> The reason for its foundation is not hard to guess. Agricola, when he built the road from Lancaster to the Irish Sea, placed a fort at Kendal, another at Ambleside, and another at Ravenglass. Now the distance from Ambleside to Ravenglass was about 20 miles, over two high mountain passes; and Ravenglass may have been thought dangerously exposed to raids from the sea, especially as there was no other Roman fort on the Cumberland coast, so far as we know, and the Irish may already have begun to develop the piratical habits for which they became famous two or three centuries later. In these circumstances it was a reasonable move to give up the fort at Ambleside – excavation there suggests that the early fort was not occupied very long – and replace it with a fort at Hardknot, within easy reach of Ravenglass[52].

One of the things the re-enactment doctrine has been criticized for is its being relevant only to understanding the actions of individual agents. The passages quoted above demonstrate, however, that in practice Collingwood claimed to explain also collective human phenomena. When Collingwood discusses the meaning of Roman roads, or the foundation of Hardknot Castle, it is only presupposed that they are the products of human actions; the question whether these should be seen as individual or collective is undetermined. For the idea of re-thinking past thoughts this is enough. It would be surprising if it were otherwise. For in archaeology – and in history mainly being based on it, like Roman Britain – individuals are seldom referred to. In sharp contrast to the view that the idea of re-thinking past thoughts would only be applicable to the thoughts of individuals, we shall see from other examples as well that reference is mostly made to the thoughts of anonymous persons.

After his article of 1921 Collingwood returned frequently to the question of the purpose and meaning of Hadrian's Wall, both with regard to its general idea and to certain questions of detail. One example will be enough to demonstrate how the thoughts embodied in the Wall were reconstructed by Collingwood:

> The Wall is by far the strongest and most elaborate of all the Roman frontier works, but it differs from the ditches, palisades, and walls of the German, Roumanian or Saharan frontiers in degree, not in kind. All these were primarily lines of demarcation, meant to show plainly where the Roman Empire began and ended. That is clearly the intention of the Vallum. Secondarily, they might develop into obstacles intended to check smuggling and raiding and to lighten the task of the patrols whose duty was to watch the frontier. That is the purpose of the Wall. The Wall was not meant as a fortification in the strict sense of the word – no Roman officer can have intended to line its top with soldiers and so defend it against Caledonian armies trying to carry it by storm – because Roman soldiers were not armed for that kind of work; their training and equipment were designed for a highly-specialised kind of infantry tactics in the open. The Wall checked cattle-lifting, kidnapping, and petty disturbances generally, and gave sentries a good look-out and safety against snipers; when an army appeared before it, we may be sure the north gates of the forts were thrown open and the enemy tackled hand to hand[53].

We have seen that one of the characteristics of thought, according to Collingwood, is that it is critical[54]. For an historian re-thinking past thoughts this means that not only his own thinking as an historian is judged, but also the thought that he re-thinks and which is incapsulated in his present thought[55]. An illustration of this principle is Collingwood's assertion regarding Hadrian's Wall that "[w]hatever its defects as a military work, as a police work it was well planned and no doubt perfectly efficient"[56].

In spite of what has been said above, of course, Collingwood also discusses the thoughts of individuals. The following passage not only shows how the thoughts of Caesar are re-thought, but also how they are appraised by the historian:

> From a military point of view Caesar was entitled to regard his invasion of Britain as a success. The chief problem of strategy had been the discovery and destruction of the fortresses belonging to the British tribes against which he was fighting; and this had been done. The chief problem of tactics had been how to deal with the British charioteers, and this had been solved not only by the discovery that they were helpless against a legion in battle-formation, but by the further discovery that a sufficient body of Gaulish horse, properly supported by infantry, could break them up and rout them by a well-timed charge. Moreover, in spite of Caesar's failure to discover a safe harbour, he had learnt how his ships could be protected against the weather; he also knew that wheat in very considerable quantities could be had in the country; he was therefore right in thinking that it would be possible for a Roman army to winter there. Britain was very far from being conquered; but the first steps towards conquering it had been taken[57].

The actions of groups may also be described in terms of their thoughts. Regarding the revolting tribe of the Maeatae Collingwood says:

> ...[W]hen Albinus withdrew the garrisons from northern England they realized that the time had come to avenge the wrongs of their forefathers. They swept in, no one resisting them, and deliberately treated every Roman building they could find in the same manner, arguing that without their fortifications the Romans could do nothing. If they did not destroy the walls of towns, it was because their inhabitants could beat them off. They were not prepared for siege-work, and could destroy only where no opposition was offered[58].

Another revolt is described thus:

> The Britons had less reason than ever to be satisfied. The emperor whom they had created was neglecting their interests and leaving them to shift for themselves; accordingly they revolted against him, expelled his governors, and sent protestations of loyalty to Honorius while arming themselves for their own defence[59].

The question how far assertions about motives and intentions like these are justified is not at issue here. One is certainly inclined to doubt that they are always based on sound evidence: for instance when Collingwood says: "Furious at this breach of faith, terrified at the prospect of endless future oppression, and burning to avenge the insult to their royal house and nobility, the Iceni rose at Boudicca's call"[60].

That completely anonymous thoughts present in works of art may also be re-thought is made clear, when Collingwood asserts, with regard to the monument of the Bewcastle Cross:

> ...[O]ur first and chief duty to it is read in it the thoughts which those who carved it expressed in their carvings: to understand how its interlaced patterns, with their subtle design and intense feeling, express here as elsewhere in early English and Celtic art a dark and brooding consciousness of eternity[61].

From these examples it may safely be concluded that no aspect of history is excluded by the re-enactment doctrine. This only underlines Collingwood's claim that it reflects certain universal and necessary characteristics of historical science.

8.3. CORPORATE MIND

We have seen that although Collingwood considers all history the history of thought this does not imply that all history is reduced to the thoughts of individual agents. The question then arises whether Collingwood refers to something like a corporate mind. He speaks indeed of the English[62] and Chinese[63] mind, or the Roman and Greek spirit[64]. Apart from expressions like these Collingwood's historical writings show that one cannot simply endorse Donagan's view, that he "was a methodological individualist, in the strongest sense of that disputable term"[65]. Sometimes, however, Collingwood seems to be conscious of the difficulties of the concept of corporate mind. Speaking of the "spirit" of Roman-Britain he says that this word "may seem vague"[66], and when he mentions in *The Idea of History*, "the corporate mind ... of a community or an age", he adds "whatever exactly that phrase means" (IH,219).

In other places Collingwood explicitly rejects the reification of abstract concepts. In "The Devil" (1916) he calls "society" "a fictitious entity", asserting that "'Society' consists of Tom, Dick, and Harry"[67]; in "Some Perplexities about Time" (1926) his answer to the question "What is the State, in itself, quite apart from its members?" is: "Nothing: and that answer is the right answer to all the questions which people ask about the State in abstraction from the persons whose political activities and passivities make them a State"[68]; in "Political Action" (1928) he maintains: "The agent is always a human being ... We speak of a society, but the society is not anything except the people in it. Its actions are their actions"[69]; while in "A Philosophy of Progress" (1929) it is stated that "... society ... is not a mythical superhuman being but just individuals themselves in their mutual relations" (PhP,119).

The concept of society is worked out by Collingwood in *The New Leviathan*, where he distinguishes two senses, called by him "society" and society. The difference between them is described as follows: "[E]ach of them has a *suum cuique* ... in each of them the members have a share in something that is divided among them; but in a society proper the establishment and maintenance of the *suum cuique* is effected by *their joint activity as free agents*" (NL,136). Further on he speaks of a society as a community, whose members share a "social consciousness" (NL,139). He elaborates on this notion as follows:

> A man engaged in a joint enterprise has a general idea of the enterprise as a whole and a special idea of the part in it allotted to himself. Unless he has both these ideas he has no social consciousness, and without social consciousness there is no society... But they are not equally precise. Of the enterprise as a whole he has only a 'general' (relatively vague or indeterminate) idea; of his own share he has a 'special' (relatively precise or determinate) idea... He has to know the nature of his own share accurately enough to do it. Beyond this his knowledge of the enterprise as a whole need only be very vague. He must know that there is a whole; but he need not know what it is, except that it is the whole to which his own share belongs (NL,149).

This passage makes clear that Collingwood's position cannot be described as that of a methodological individualist, since certain holistic ideas play a role in actions of individuals. That he is not a holist either, however, in any disputable sense, is made clear after the passage quoted above, when he asserts that "a society is nothing over and above its members" (NL,149)[70].

The historical dimension of "corporate" mind is provided for by the concept of tradition. This concept is clarified by Collingwood in his discussion of the notion of *genesis* in the manuscript on cosmology. The latter notion is described by him as "... a series in which the earlier continue ... with accumulation or enrichment of the existent by the sum of its own past". After this he continues: "For mind in general, this accumulation is called experience; for consciousness, it is called memory; for a social unity, it is called tradition; for knowledge, it is called history"[71].

It is obvious that the concept of tradition is central in Collingwood's study of folklore. That folklore not only cannot be reduced to actions of individuals, but that, in Collingwood's view, this irreducibility is closely bound up with its peculiar value, is shown by the following passage:

> The conception expressed by the term folklore arose out of ... [the] Romantic doctrine. Folklore meant something not invented by original and individual thinkers, like *Paradise Lost*, or the *Essay on Human Understanding*, or *Tom Jones*, or the *Jupiter* Symphony, but something handed down from mouth to mouth among people whose uneducated condition made them incapable of original

> creation. Yet this very fact gave it a peculiar value and importance. For, in the first place, because it was invented by no individual, it expressed a corporate experience in which the eccentricities and errors of individual thought were cancelled out: it thus achieved a profundity and universality which no individual thinker, whatever his genius, could emulate. And, in the second place, because it was not the work of individual creators, but was the heirloom of tradition, it came down to us unaltered from the remotest antiquity, from the very infancy of the human race[72].

Since Collingwood's starting-point is the use of folklore as historical evidence, this passage plainly demonstrates that Collingwood's view of history did not exclude non-individualistic aspects of human conduct. On the contrary, it makes clear, that it was even especially interested in them[73]. We have seen from the discussion of the manuscripts on folklore that the way this study was conceived by Collingwood is in line with his general view on history. For, in his opinion, corporate activities as well should be studied from their inside, as expressions of a social consciousness, and not from the outside, as mere phenomena.

8.4. "UNCONSCIOUS" ACTION

In the discussion of Collingwood's philosophy of mind we have seen that consciousness is considered an essential constituent of mind, but also that its concept is philosophical: that is, consciousness exists at different levels[74]. According to this view a thought or feeling can exist preconsciously and can be made explicit by a higher level of consciousness. "Any form of consciousness, practical or theoretical, call it C_x", Collingwood maintains in *The New Leviathan*, "exists in what Freud calls a preconscious condition unless and until it has been reflected upon by the operation of a form C_{x+1}" (NL,38). A little further on he gives the following example: "...[A] man is cold. He may be cold 'preconsciously' ...; not that he represses the feeling of cold, but just that he 'hasn't noticed it'" (NL,41).

So actions may be determined, in Collingwood's view, by "unconscious" thought[75]. This is the case, for instance, with tradition, on which Collingwood asserts:

> The continuity of a school is a conscious continuity: it depends on one person's teaching another, explicitly, what to do and how to do it. The continuity of a cultural tradition is unconscious: those who live in it need not be explicitly aware of its existence. The continuity of tradition is the continuity of the force by which past experiences affect the future; and this force does not depend on the conscious memory of those experiences. In the life of a people, a great experience in the past affects the way in which the generation that has had it teaches its children to look at the

future, even though they never knew what that experience was[76].

That a tradition can be made explicit, however, by a higher level of consciousness is made clear by Collingwood elsewhere, when he contends:

> Tradition ... does not mean conscious knowledge of the past. Our present political consciousness has been formed by such events (experiences) as the norman conquest, but not on condition of our consciously remembering them. On the contrary: this consciousness is a given fact, it is the way in which we now think politically. It is only when we ask *why* we thus think that, analysing this consciousness, we discover it to have been formed by such past experiences[77].

This making explicit of an unconscious tradition can be accomplished not only by those involved in it, but also by an historian. Collingwood himself has given an example of this with his analysis of the tradition of Celtic art[78].

That Collingwood is of the opinion that human actions in general are often not guided by articulate thought is shown by a passage in *The Idea of History*, where he asserts: "The extent to which people act with a clear idea of their ends, knowing what effects they are aiming at, is easily exaggerated. Most human action is tentative, experimental, directed not by a knowledge of what it will lead to but rather by a desire to know what will come of it" (IH,42). This does not preclude, though, that the rationality implicit in actions like these may be retrospectively discerned. In the discussion of Collingwood's view of the historical process we have seen how he considered its rationality a matter of principle[79]. It is not necessary, therefore, that the historical agent(s) consciously had a certain thought for a thought of the past to be reconstructed by the historian. The distinction made by Walsh between what a person has *in mind* and what he has *before his mind*, the former also being a possible object for the historian, is certainly clarifying[80]. It should be added, though, that it is not always easy to determine whether certain thoughts were held consciously or not. Collingwood was aware of this difficulty. For discussing concepts like protestantism, as employed by the German historian Fueter, he observes: "...[T]he question whether the people who serve such ideas serve them *consciously* is still obscure - and fundamental"[81].

An historian, however, can also give a rational reconstruction of aspects of the past, which cannot have been even implicitly in the minds of the historical agents concerned. This fact is pointed out by Collingwood in his 1931 lectures, when he discusses Hegel's concept of "the Cunning of Reason" and the kindred view of Kant. "Kant and Hegel are right in thinking this a principle of great importance in historical research", Collingwood contends; "it is foolish to credit e.g. Edward I with any idea of what the English Parliament was going to become. The principle really amounts to this: that the actors on the stage of history lay the foundations of the future but do not know what the future will be"[82].

Another example where the past as reconstructed by the historian could not possibly have been known by the historical agents involved is afforded by economic history. Collingwood asserts the following:

> The historian who sketches the economic history of the Roman Empire depicts a state of things which no contemporary ever saw as a whole; and this whole is not built up in the historian's mind out of parts, each separately seen and reported by a contemporary; because the whole, which is the object of the historian's thought, is not the sum of these parts but the system of relations uniting them, and it is because he grasps this system of relations that he is able to reject certain contemporary statements of alleged fact as inaccurate or misleading, and to interpolate inferences of his own concerning matters on which his sources are silent[83].

In the draft of his Inaugural lecture Collingwood comes back to this aspect of historical knowledge. He calls it "very important for the theory of history ... that the historian can ... recover facts which until he ascertained them no one ever knew at all" and elaborates as follows:

> For example, the Gaulish potters of the 2nd century A.D. supplied most of the western Roman empire with their products; and at that time there was probably no one engaged in the trade who knew, as every student knows today, how widely its products were distributed over Europe. The historian of today who sketches in broad outline the economic condition of the Roman Empire is doing something which certainly no Roman was able to do[84].

Collingwood rejects the idea "... that all the modern historian can do is to put together a mosaic of facts each one of which was known to contemporaries, and that no one fact in his mosaic can have been unknown to them"[85]. For the historian interpolates between his data, he contends. So Collingwood's theory of history explicitly allows us to have not only knowledge of "unconscious" aspects of past behaviour, but also of aspects that could not have been known by the historical agents themselves. The latter is quite normal, of course, in modern historical practice, especially in social and economic history. Though many examples of it could also be given from Collingwood's own historical practice, he did not discuss the matter in his published theoretical writings on history.

In the unpublished manuscripts the question of having knowledge of aspects of the past that could not have been known to contemporaries is only dealt with obliquely. In fact the passages quoted above are the only ones bearing on the point. It is surprising that Collingwood did not say more, especially since he calls the idea that actors in history do not know the consequences of their deeds "a principle of great importance in historical research", while he regards the possibility in economic history of recovering facts unknown to contemporaries "very important for the theory of history". The question arises how far this view corresponds with Collingwood's view of history as discussed up to now. The object is still human actions in the past, but there is the important difference that in this case there is no question of studying them from the inside, that is, looking for their purposes or intentions. As Collingwood puts it in the case of economic history, a "system of relations" is constructed, interpolations being made by theoretical thinking. This can only be done, however, by studying the "facts" from the outside. How far this may be reconciled with the re-enactment doctrine remains a puzzling question[86].

A basic likeness that remains, however, is that a rational and coherent picture of the past is still given. Or, as Collingwood puts it in the draft of his Inaugural: "The entire aim of the historian is so to compose his picture that in any part of it, whether that part is a narrative of events, or an analysis of social and economic conditions, or a sketch of some historical character, all the elements hang together in such a way that each of necessity leads on to or arises out of the rest"[87].

Though the aspects of historical science discussed above were not satisfactorily worked out by Collingwood, one conclusion may be drawn: his awareness not only of their special character, but also of their importance demonstrates that his view of history was not as one-sided as it is generally believed to be.

8.5. CAUSALITY AND OBJECTIVE CONDITIONS

In chapter 3.3.3 we have seen how Collingwood has been criticized in various ways for not taking into account the objective circumstances of natural environment, placing history, as it were, in a vacuum. In this connection it is highly paradoxical to see Collingwood praised in a review of *Roman Britain and the English Settlements* for doing exactly the opposite: "The geographical discussion by Collingwood will be most welcome to those who feel that historians often neglect too much the physical conditions of society"[88]. I think there is more reason to endorse the view expressed in the review than that of Collingwood's critics. One only needs to call to mind the fact that the first chapter of *Roman Britain and the English Settlements*, entitled "The Stage of History", explicitly deals with the geographical conditions of Britain and their relevance for its history.

Collingwood's view on the role of objective conditions has already been shortly discussed in chapter 3.3.3. Its essence is, that such conditions never determine human actions in themselves, but only through the way they are conceived by men. It is obvious that the question of the role of objective conditions in history cannot be separated from that of causality. Collingwood developed explicit views on the causal concept, which have not been discussed until now. I shall deal with these first, making it possible to place the role of objective conditions in its proper context.

Collingwood was originally of the opinion that the concept of cause could not be applied to human actions. "...[A]ction is precisely that which is not caused", he asserts in "The Devil"; "the will of a person acting determines itself and is not determined by anything outside itself ...[T]he Law of Causation ... cannot be applied to the activity of the will without explicitly falsifying the whole nature of that activity. An act of the will is its own cause and its own explanation; to seek its explanation in something else is to treat it not as an act but as a mechanical event"[89]. In the 1928 lectures the same view is held. As we have seen from the example of a game of chess, Collingwood maintains that a thought can neither be an effect nor a cause. Since all history is the history of thought, this implies that the concept of cause is not applicable to history[90].

Collingwood apparently changed this view, however. For in the manuscripts on cosmology he explicitly discusses the concept of "historical cause". Contrasting it with Mill's notion of an "unconditional antecedent" he contends:

> An historical cause is a fact or assembly of facts which, when an agent is aware that he stands in them as his circumstances, determine him through this awareness to act in a certain way. The causality is doubly conditional: 1) he cannot be acted upon by the facts unless he is aware of them as his circumstances 2) nor unless he freely and intelligently thinks out a line of appropriate response to them. He can truthfully say 'I *had* to retreat *because* of the enemy's strength', where *because* carries its full meaning; but 1) if he hadn't known the enemy's strength it wouldn't have thus affected him 2) nor if he had been a sufficiently incompetent officer to neglect it[91].

In his "Notes on the History of Historiography and Philosophy of History" Collingwood speaks of "the false assumption that the historical process consists of events in causal series: that the event E 1 *causes* the event E 2". He continues: "The cause of my buying ink (if it is insisted that I use that phrase at all) is not the fact that I have none left, but my resolve to buy new instead of borrowing, using a pencil, etc. The cause of my being involved in a traffic accident is not the fact of my being on the scene at the time, but my carelessness when there. A series of historical events is never a causal series in the sense of a series in which the earlier determine the later"[92].

These passages make clear that Collingwood is only willing to use the concept of causation in history if it is used in a special sense. He elaborated on this idea in his well-known analysis of three senses of the word "cause" in *An Essay on Metaphysics* (EM,285-327). He makes there a distinction between the concept of causation as used in history, in practical natural science and in theoretical natural science. Since the last of these is not relevant for history, we will only discuss the first two senses of causality[93].

1) Causation in history is defined by Collingwood as follows:

> In sense I of the word 'cause' that which is caused is the free and deliberate act of a conscious and responsible agent, and 'causing' him to do it means affording him a motive for doing it. For 'causing' we may substitute 'making', 'inducing', 'persuading', 'urging', 'forcing', 'compelling', according to differences in the kind of motive in question (EM,290).

It is called the *historical* sense of the word "cause", "...because it refers to a type of case in which both C and E are human activities such as form the subject-matter of history. When historians talk about causes, this is the sense in which they are using the word, unless they are aping the methods and vocabulary of natural science" (EM,286). As an example the following statement is given: "Mr. Baldwin's speech causes adjournment of House" (EM,290). Within the historical concept of causation Collingwood makes the following important distinction:

> A cause in sense I is made up of two elements, a *causa quod* or efficient cause and a *causa ut* or final cause. The *causa quod* is a situation or state or things existing; the *causa ut* is a purpose or state of things to be brought about. Neither of these could be a cause if the other were absent ... The *causa quod* is not a mere situation or state of things, it is a situation or state of things known or believed by the agent in question to exist... The *causa ut* is not a mere desire or wish, it is an intention. The *causa ut* of a man's acting in a certain way is not his wanting to act in that way, but his meaning to act in that way (EM,292-293).

2) Of causation in practical natural science Collingwood gives the following definition:

> In sense II that which is caused is an event in nature; but the word 'cause' still expresses an idea relative to human conduct, because that which causes is something under human control, and this control serves as means whereby human beings can control that which is caused. In this sense, the cause of an event in nature is the handle, so to speak, by which human beings can manipulate it... This sense of the word may be defined as follows. *A cause is an event or state of things which it is in our power to produce or prevent, and by producing or preventing which we can produce or prevent that whose cause it is said to be* (EM,296-297).

On this usage of the word 'cause' Collingwood observes that it "... can be recognized by two criteria: the thing described as a cause is always conceived as something in the world of nature of physical world, and it is always something conceived as capable of being produced or prevented by human agency". Examples he then gives are: "The cause of malaria is the bite of a mosquito; the cause of a boat's sinking is her being overloaded; the cause of books going mouldy is their being in a damp room ..." (EM, 299).

Of the three senses of the concept of causation, according to Collingwood only the first one is relevant for history. This view, however, is too limited. For Collingwood only speaks of two possible causal relations (leaving aside sense III): both cause and effect are human activities (sense I) and both cause and effect are natural events or things (sense II). What is omitted is the possibility that a cause is a natural event and the effect human actions. This would be a combination of sense I and sense II in the sense that in this case the *causa quod* would be a natural event to be handled by a human agent and the *causa ut* his intention. This handling, though, should be interpreted in a broad sense, since of most natural events playing a role in history one cannot say that they could be produced or prevented at will. One can say, however, that certain natural events or conditions can either be profited from or protected against. Strangely enough, Collingwood himself gives an example of this in an illustration of sense I: "... bad weather causes [a man] to return from an expedition" (EM,290). This is inconsistent, since in his view in this sense a cause can only be a human action.

It is obvious that a causal relation in the sense proposed here, which could be called sense I/II, is the one at issue in the question of the role of natural conditions or events in history. Though it is not discussed by Collingwood in his treatment of the three senses of the word "cause" in *An Essay on Metaphysics*, this sense is discussed by him elsewhere in various places. In the draft of his Inaugural lecture, for instance, Collingwood maintains:

> However important it may be for the historian to obtain a correct idea of the natural surroundings – geographical, climatic, biological and so forth – in which historical events took place, it is even more important for him to understand that these things are only the stage upon which historical events are enacted and which provides them with a raw material offering alternative directions for development. These natural facts condition the course of history but do not determine it; nothing determines it except itself; and hence the historian in seeking a reason for its proceeding in this direction and not in that can never find such a reason in its natural stage or background[94].

The same subject is discussed by Collingwood in his notes made during his journey to the East Indies. Under the heading "Historical

Naturalism" he maintains:

> I use this term as a name for that kind of failure to think historically which ends in either a) *substituting* natural facts for the historical facts about which one is trying to think (losing the distinction between them altogether), or else b) *superordinating* natural facts to historical facts, as the causes of which these historical facts are the effects... Only historical facts can be causes of historical facts... *Examples* of historical naturalism: 1) geographical (and climatic) history, where geographical facts (I include meteorology in geography) are taken as the causes of historical facts (Montesquieu and other 18th century writers). Here a long view would soon dissolve the illusion, by showing that the same geographical facts which are said to cause one type of historical development are compatible with developments of a just different kind; so that what causes a development of that type is not the geographical facts in question but the way in which people think of these facts and of themselves in relation to them[95].

That Collingwood, so far as the relation between natural conditions and human responses to them is concerned, focusses his attention on the latter may be demonstrated by his historical practice as well. Discussing the Roman conquest of Britain he says in *The Home-Reading Magazine*:

> They conquered because they saw; because they grasped the character of the country to which they had come, and realized what action it demanded of them. They seem to have understood the geography of Britain in its broad outlines as accurately as if they had possessed our physical maps and geographical text-books; for they planned their campaigns and developed their scheme of conquest with an evident attention to the distinctions between the main physical divisions of the country[96].

In *Roman Britain*, too, emphasis is put on the human element:

> In the Roman period there was a great difference in civilization between the south-east of England and the rest. In part this was due to differences of soil and climate; the south-east is more fertile and less wet than most other parts; but it is easy to over-emphasize these differences... The important difference lay less in the country than in the people[97].

With regard to the causal relation between natural events or conditions and human actions one could describe Collingwood's position as one which focusses attention not on the *causing* factor of nature, but on the *caused* one of human activity. It should be remembered that the *causa quod*, the reference to the situation in a causal explanation, in

Collingwood's view does not concern the (objective) causal influence of an "objective" nature, but how it is conceived by men. This is made clear by Collingwood's rejection of the accidental and contingent in history. For we have seen that these concepts are defined by him as an event "which happens to a person without his intending that it shall happen to him", and this can only refer to a natural event[98]. Collingwood does not mean by this, however, that since natural events are not intended by agents they do not have to be taken into account by the historian. What he does mean is that the historian is not interested in these events as such, but only in the human responses to them. This position is clarified in the following passage:

> Actually, accidental or contingent events do not occur *in* history at all: they form the background or scenery of history. A mariner is caught in a storm: this is an accident; but that storm appears in the history of navigation only if the historian is interested in the mariner's handling of the situation to which this accident gave rise... Nature as such is contingent... Nature may be called a background or scenery for history, but it does not figure in history as a constant. It affects the course of history in different ways according as man copes with it differently. Nature as the background of history is what man makes it[99].

An aspect not discussed by Collingwood is the fact that the *causa quod* may exhibit various degrees of causal "force". This is implied, though, by Collingwood's using, as equivalents for "causing", terms like "making" and "inducing", as well as "forcing" and "compelling. One can assume, for instance, that when a mariner is caught in a storm, it will be – beside other conditions – the nature of the storm which will determine the "compelling" character of the actions to be taken.

I have given some examples from Collingwood's historical work, in which the human responses to natural conditions are emphasized. Elsewhere attention is focussed on the "compelling" nature of these conditions. "The main lines of communication, in any given country, alter very little from age to age", Collingwood says in "The Fosse", "They are *dictated* by geography... The main Roman roads are laid out with such accurate attention to geographical facts, that the railway engineers of the nineteenth century were unconsciously *forced* to imitate their choice of track" (italics mine)[100]. In "Town and Country in Roman Britain" Collingwood contends that "... the 'ancient Britons' were *driven* to cultivate the miserable soils of the mountain-side because they could not face the capital expenditure of clearing the better soils of the valley" (italics mine)[101]. The importance Collingwood attaches to geographical factors is also demonstrated in the following passages: "... England is a country designed by nature to be invaded from the mainland of Europe"[102], and "The impact of the Roman Empire on Britain, and its consequences, can only be understood in the light of certain geographical considerations"[103].

In some places, however, Collingwood goes so far in emphasizing the compelling character of natural conditions that he is no longer consistent with his own principles. This is the case when he assumes a law-like relation between natural conditions and human responses. For, as we have seen, he is of the opinion that nature does not figure in history as a constant and that the same natural conditions are compatible with historical developments of different kinds. In the passage quoted above from "The Fosse", however, Collingwood points to certain constant geographical factors, which forced both Roman road builders and nineteenth century railway engineers. Similarly he asserts in *The Home-Reading Magazine*: " [I] f the population of England were annihilated to-morrow, and it was recolonized with savages, who gradually civilized themselves, in a few thousand years London would again stand out as the centre of trade, and there would again be some kind of junction corresponding with Reading". A little further on he says that "... geography compels men to build towns in certain places..."[104]. Likewise, Collingwood contends elsewhere that "... the focal position of London in the road-system was imposed on the Roman engineers by the *force majeure* of geographical facts..."[105].

As noted already in the beginning of this section, *Roman Britain and the English Settlements* starts with an analysis of Britain's geographical characteristics and their influence on British history. He makes a distinction between the highland zone, rough in landscape and climate, and the lowland zone with more gentle features. Because of this, the latter is adaptive to foreign influences, while the former is more conservative. Collingwood goes so far as to speak explicitly of causal laws to indicate the influence of Britain's geographical conditions: "The highland zone is unattractive to invaders, hard to invade, and hard to conquer in detail when invaded; its landscape and climate impose peremptory laws on any one, no matter whence he comes, who settles there; all these causes, therefore, combine to make it a region tenacious of its old customs, conservative in temperament, stubborn to resist any kind of change"[106].

The idea of a distinction in Britain between a highland and lowland zone and its influence on British history was taken over by Collingwood from the archaeologist Cyril Fox. Though he does not mention him in his first chapter of *Roman Britain and the English Settlements*, Collingwood quotes in his article on prehistory[107] the following passage from Fox: "In the lowlands of Britain new cultures of continental origin tend to be *imposed*. In the highland, on the other hand, these tend to be *absorbed*. In the lowlands you get *replacement*, in the highlands *fusion*". After this Collingwood observes: "I shall venture to refer to this principle as 'Fox's law'"[108].

I do not see how this idea of a law-like causal relation between natural conditions and human activities can be renconciled with Collingwood's views on causation in history as expounded elsewhere. We are here a long way from the discussion of causality in *An Essay on Metaphysics*, which allows only human activities to be causes in history[109]. As I have noticed already, the latter doctrine is too limited, since it does not take into account the possibility of nature playing a causal role in history. We have seen, however, that in other places Collingwood explicitly discusses the latter view. Yet in Collingwood's view nature's role in history is of a special kind. It is never determinative, but always dependent on man's interpretation. Any law-like vision with regard to nature's influence goes against this view. One can only note that there is here a discrepancy between Collingwood's theory and practice.

There are also other passages where one feels at a loss concerning Collingwood's view on causation. What type of causal relation would be meant, for instance, when he asserts that "... the newly established *pax Romana* was causing population to rise..."[110]? or, one could ask, which theory of causation has Collingwood in mind, when he contends, comparing peasant risings in Gaul with those in Britain: "Causes being identical, it is hardly to be doubted that effects were identical too"[111]?

Though Collingwood may not be consistent in his views on causality and the role of natural conditions in history, he certainly developed valuable and fertile ideas on these important subjects. The present discussion has amply shown, moreover, that the criticism that Collingwood placed human actions in a vacuum, is far from justified. That this is unjustified not only with regard to natural conditions, but also with regard to the "objective conditions" of economic life, may be demonstrated by a passage in the manuscripts on folklore, which even has a Marxist flavour. Discussing totemism, Collingwood maintains:

> ...[I]t would be well to reflect that totemism, like other systems of custom and belief, has a certain connexion with the economic life of the people among whom it exists. The classical examples of it have been found among food-gathering peoples; and although it may, and often does, survive the adoption of agriculture, its natural home is in a food-gathering civilization. For though all religions are at bottom attempts to solve one and the same problem, the problem of establishing a sound relation between man and the power that works in and behind the world he inhabits, this problem is never a purely 'metaphysical' one, in the sense of being divorced from his immediate practical interests. The god that satisfies his hunger and thirst after righteousness and eternal life is also the god that gives him his daily bread[112].

8.6. GENERAL KNOWLEDGE

Another aspect of his theory of history for which Collingwood has been criticized is his alleged failing to take into account the use of general knowledge by historians. In chapter 3.3.7 some of these criticisms have been discussed, as well as the few observations Collingwood made on the subject in *The Idea of History.* These observations, however, are too random to give a reliable picture of Collingwood's views.

Interpretations, therefore, which are only based on *The Idea of History*, necessarily suffer from this circumstance. This is true, for example, of the interpretations of Walsh and Martin, the authors who deal most explicitly with the question of the use of general knowledge in Collingwood's thought. When we take into account what Collingwood has written about the subject apart from *The Idea of History*, both in his published and unpublished work, we cannot avoid the conclusion that the traditional interpretations have to be reconsidered. For these writings show that general truths do play a role in various respects, not only in Collingwood's thought on history, but also in his historical practice.

It is obvious from the passages quoted from *The Idea of History* in chapter 3.3.7, that Collingwood is highly critical of a generalizing approach in history. That this viewpoint, like many of Collingwood's other mature views, is of an early date, may be demonstrated by a passage in *Religion and Philosophy*, where Collingwood says:

> To see uniformities is the mark of a superficial observer; to demand uniformities is characteristic of all the less vital and more mechanical activities. What we call uniformity in people, in society and history, is generally a name for our own lack of insight; everything looks alike to the person who cannot see differences (RPh,213).

In "The Nature and Aims of a Philosophy of History" Collingwood explicitly discusses generalizations in history. He rejects the idea of a philosophy of history, which "... aims at the erection of a superstructure of generalisations based upon historical facts". "It assumes", he continues, "that the facts have been finally settled by historians; and using these facts as material for inductions, it proceeds to determine the abstract and universal laws which govern their occurrence" (NAPhH, 35). The assumption, however, that the alleged historical facts would be secure enough to be used for generalizations is called false by Collingwood: "... there is no given fact upon which at any given moment historical research has said the last word..." (NAPhH, 35). In *The Idea of History* Collingwood expresses the same viewpoint. The only difference is that this time the fact is emphasized that history aims at the inner- or thought-side of human actions, while in a generalizing science "... the 'historical data'... upon which it is based are merely perceived and not historically known" (IH,223). Likewise, Collingwood speaks in this connection of a "false assumption". There is a similar parallel between Collingwood's views as expressed in "The Nature and Aims of a Philosophy of History" and *The Idea of History*: in the first he says that "generalisations which pretend to be true of all history are, as a matter of fact, true only of certain phases in history" (NAPhH,35), while in the latter he asserts that "... social orders are historical facts, and subject to inevitable changes, fast or slow" (IH,223)[113].

In "The Nature and Aims of a Philosophy of History", however, Collingwood adds after the passage quoted above an important observation, which is not to be found in *The Idea of History*:

> ...[W]ithin the body of historical thought itself, not erected upon it as a superstructure but contained within itself as a subordinate but necessary element, generalisation and inductive thinking have an important place. Historical research cannot proceed without using its own previous results as materials upon which to generalise in order thereby to help itself in the determination of fresh facts. Actual historical thinking is a constant alternation of the general and the individual, the individual as end and the general as means. No historical fact can be determined without the help of generalisations; thus it is only through inductive study of ancient pottery that a man can recognise the presence of a Roman villa in his garden... The determination of facts and the using of them as material for generalisations are not two separate and independent activities, one history and the other philosophy of history; they are two interlocking and interacting elements in history itself (NAPhH, 35-36).

Further on in the same article Collingwood contends: "...[T]he special activity of the scientist is to generalise; but the historian ... generalises too, only he generalises not for the sake of generalising, like the scientist, but for the sake of helping himself to determine historical fact" (NAPhH, 48).

In "The Philosophy of History" Collingwood expounds the same view:

> In science, the individual fact is of importance only so far as it illustrates a general law. The law is the end, the fact is the means to it ... In history, the opposite is true. The individual fact is the end, and the general law is of importance only so far as it enables us to determine the fact. Whether bad money invariably drives out good, is a question not for the historian, but for the economist, who is a kind of scientist; but the historian may appeal to this principle as an aid in discovering what happened on a certain occasion (PhH, 132)[114].

In the draft of his Inaugural Collingwood expands the idea of using generalizations into the principle of making use of any knowledge that may be possessed by an historian. After saying that the historian wishes to draw a conclusion from his evidence "which shall help him to build up his imaginary picture of the past", he maintains:

> Into the drawing of this conclusion everything that the historian knows may enter either as additional premisses or as controlling principles: knowledge about nature and man, mathematical

> knowledge, philosophical knowledge and so forth. The whole sum of his mental habits and possessions is active in determining the way in which he shall draw his conclusion[115].

The passages quoted above clearly show that the charge that Collingwood neglects the role of generalizations in history is unjustified. As was mentioned, the reason for this misinterpretation must be found in its being based only on the *Idea of History* (especially "Human Nature and Human History"). And Collingwood's short discussion of the use of generalizations in history as expressed there should in any case be seen in its proper context. For what is at issue in "Human Nature and Human History" is the question of the right approach to the study of human mind. In this context emphasis is naturally put on the differences between the scientific and historical approaches. Since generalizing is a typical feature of the former, it is no surprise to see Collingwood opposing it. It is no omission that the use of generalizations in history was not discussed in "Human Nature and Human History". For this paper is not about historical study as such. At one place, though, Collingwood does allude to the use of generalizations in history, when he says: "It is only when the particular fact cannot be understood by itself that such [general] statements are of value" (IH,223). This implies that generalizations can be useful. In his comparative studies on folklore Collingwood formulates and uses such generalizations. These studies, however, are pursued in order to interpret specific phenomena and do not have generalization as their aim. Or, as Collingwood puts it himself:

> I shall... confine my study to fairy-tales found in the British Isles, and among these to certain classes of theme which I think I know how to interpret. Illustrative material I shall use wherever I think needful, and from whatever sources are most helpful; but my task is interpretation, not comparison, and the piling-up of parallels... is a method of impressing readers which does not advance the inquiry I have undertaken[116].

Collingwood's alleged neglect of generalizations can be most obviously disproved, of course, by reference to his contributions to archaeology. We have seen how he considered archaeology (or "empirical methodology", as he called it in the 1926 lectures) explicitly abstract, classificatory and aiming at generalizations in order to interpret specific types of evidence[117]. In chapter 5.2.3 we have seen that *The Archaeology of Roman Britain* is a standard work in this field and that Collingwood did pioneer work on the classification of various types of archaeological remains.

General or pure methodology (equated in the 1928 lectures with philosophy of history) is conceived by Collingwood not as aiming at material generalizations concerning various types of sources, but at the formal principles of their interpretation[118].

The best example of the explicit elaboration of such principles is to be found in the manuscripts on folklore. We have seen how he opposes in them the naturalistic approach of anthropology and works out an historical method instead. The essence of the latter is that primitive people should not be seen as foreign "objects" – which makes their understanding impossible – but as human beings, who share a common rationality and emotional life with us. So Collingwood develops the view that a certain vision of human nature is a necessary precondition for understanding primitives. This principle, then, is materially worked out by Collingwood in his study on magic, in which he demonstrates that there is in fact a common human nature. It is interesting to note in this connection that this vision of human nature is not in line with the way Collingwood is usually interpreted. For it is more or less generally accepted that, in Collingwood's view, human nature is "liquidated" by becoming history[119].

In his study of magic, however, Collingwood assumes certain common characteristics in all men. The emphasis is on common emotional aspects, but as we have seen, he is far from denying that the primitive mind has a rationality of its own[120]. To give a few examples of Collingwood's referring to aspects of human life common to all men[121]: speaking about emotional attitudes to personal belongings he says: "Emotions of this kind have been felt *semper, ubique, ab omnibus*". Again, "In every society known to us, marriage bulks larger in the emotional life of a woman than in that of a man". In connection with the habit of wearing hats, he declares:

> These feelings are to some extent observable in all human beings. But they are developed in different ways by different peoples.
> The wearing of clothes gives one a feeling of security or self-confidence. To take off one's clothes in public is to 'give oneself away', to 'make an exhibition of oneself' that is, to forfeit one's dignity. It makes one ridiculous or contemptible. This is a universal human feeling, to be traced, I think, in every civilization.
> In every case we have found that the magical practice has its basis in emotions which are universally human and can be verified as existing, and even sometimes as giving rise to definite customs, in and among ourselves.
> Granted the general form of a taboo-system, that is, a system of prohibitions with purely emotional sanctions, as something always and everywhere found in every human society...

And his conclusion about magic is as follows:

> ...[T]he result of our inquiries in this chapter has been that the term magic, from being a term of reproach, has become a term of description: it no longer implies that the thing so denominated is foolish or in any way discreditable, it expresses scientifically the nature of that thing and assigns to it a permanent and necessary place in every department of human life.

These passages are not only of great interest because they throw another light on Collingwood's view of human nature, but also because they clearly show that he was far from avoiding generalizations. On the contrary, he even developed generalizations himself. He did this, however, in the context of his argument aimed at establishing the necessary preconditions for the proper study of primitive man. The corollary would be that the same idea holds for the study of the past. What vision, then, on human nature would be necessary for a proper understanding of the past? Or, alternatively, which vision makes such an understanding impossible? Collingwood did not explicitly deal with these questions. At one place in *The Idea of History* he does make the following observation: "There is a kind of pre-established harmony between the historian's mind and the object which he sets out to study; but this pre-established harmony, unlike that of Leibniz, is not based on a miracle – it is based on the common human nature uniting the historian with the men whose work he is studying" (IH,65). On the character of this common human nature, however, Collingwood does not elaborate[122].

What about generalizations and the re-enactment of past thought? Martin has expressed the view that these have nothing to do with each other. "...[T]he essence of Collingwood's position", he contends, "was to conceive a logical discontinuity between explanation by re-enactment and explanation by generalization"[123]; and he holds that "[r]ational actions ... are understood and explained by re-enactment ... without reference to generalizations of any sort"[124]. Elsewhere he asserts that "[g]eneralization is wholly extrinsic to the logic of re-enactment"[125], and that "... any argument designed to show that explanations by re-enactment do require *general* hypothetical premises has a certain unwelcome drift to it – unwelcome, that is, to Collingwood or to those who share his philosophical predispositions"[126].

I do not think this interpretation is correct. It would make the re-enactment doctrine quite incredible if it were. For how could one sensibly interpret human actions without referring them to their proper context? And how could such a context be conceived without reference to general propositions? The view that re-enactment has nothing to do with general knowledge of any kind seems to be influenced by the intuitive version of this doctrine[127]. However, just as Collingwood nowhere indicates that he endorses intuition, he similarly nowhere explicitly asserts that all general truths should be excluded from re-enactment. That he adheres to the opposite view is made clear by his example of the chess-players. For he emphasizes the fact that the actions of the players remain unintelligible without knowledge of the rules of the game[128]. For an historian, then, in order to understand human actions in the past similarly the use of general knowledge is needed. How this is done can be demonstrated by Collingwood's own historical practice. Here some examples will be given where Collingwood refers to general truths of various kinds.

In his discussion of the Roman invasion of Britain Collingwood mentions the suggestion, that Plautius might have sent one flotilla to each of three ports. One of the reasons he then gives for thinking this unlikely is that "... it is a maxim of strategy that forces should not be divided in the face of the enemy"[129]. Elsewhere Collingwood similarly refers to a law-like principle in his argument. The building of villas in Britain, he contends, has the remarkable feature of being spontaneous. He is of the opinion that this is not due to an influx of settlers from abroad. He gives three arguments for this view, of which the last one reads: "partly [it is shown] by general historical considerations, such as the improbability of foreign settlers at so early a date going anywhere except to the towns"[130]. Likewise, Collingwood contends on the purpose of Hadrian's Wall: "...[S]ound tactics demanded that the Romans should use the Wall precisely as a modern army should use a fortress: to delay the enemy, and to cover flank or rear in battle, but not to constitute the main defence. And no one who bears in mind the equipment and tactical traditions of the Roman army can doubt that the men responsible for designing and administering the Wall knew this well enough"[131].

There may also be references to general truths, which are limited in scope; for instance, when Collingwood asserts: "The true motives for the conquest of Britain were those which had been permanent factors in the British question ever since Julius had first raised it"[132]. Or when he says: "It seems to have been a constant principle of Roman engineering in Britain as in Gaul that, wherever possible, roads should be built along ridges"[133]. To give another example: discussing the purpose of the Vallum, Collingwood develops the theory that it served as a customs barrier under the supervision of the procurator, while the military function of the Wall was controlled by the governor. He bases his theory on the following argument: "...[A] Roman frontier had two functions, one military or defensive, the other financial, as a line where traffic passing between the province and the unconquered country outside it passed through supervised openings and paid duty. And it is a peculiar feature of Roman administration that the financial service under the procurators was entirely separated from the military service under the provincial governors"[134].

We have seen that Collingwood in his discussion of the interpretation of coins also explicitly refers to certain general principles concerning hoards, their relation to social and economic circumstances etc.[135]. A general truth of another kind is often implied by Collingwood's arguing from analogy. "Frontier raiding, which ... explicitly accounts for Commodus's works on the Danube", Collingwood asserts in his article on signal-stations on the Cumberland Coast, "is the true explanation of Hadrian's Wall; and the situation on the Solway is like enough to that on the Danube to suggest that in Cumberland, as in Pannonia, petty raiders might be provided with boats"[136].

Of all the general truths exemplified in the passages quoted above it can be said that they are based on historical insight and research, or on principles of a more general kind. Since they enter into the ordinary work of the historian, their validity is his responsibility. Collingwood would certainly agree with this contention and it is not denied by his re-enactment doctrine. It is another question, however, when generalizations are taken over "ready-made" and uncritically from the generalizing social sciences. On Collingwood's view such generalizations are useless for history if they are not properly based on historical knowledge. Paradoxically, it is Collingwood himself who provides an example of this faulty approach in his article on "Town and Country in Roman Britain". For, as we have seen, he based his argument on the estimation of the population of Roman Britain on the demographic studies carried out by Carr-Saunders[137]. The defects of his argument were in particular traceable to the generalizations he took over from this source. In his criticism of Collingwood's article Randall pointed out "... the danger of arguing from generalized propositions, without close attention to the limiting conditions"[138]. This contention indeed accords with Collingwood's own viewpoint. That does not mean, however, that generalizations do not play a role in history.

8.7. EXPLANATION AND UNDERSTANDING

Collingwood's ideas on explanation and understanding have already been implicitly dealt with at various places: in the discussion of the re-enactment doctrine, the role of objective conditions and causality. The subject comes up for most extensive and explicit consideration in "Reality as History"[139]. We have seen that he develops there the view that the flux of historical events can itself be made intelligible and does not have to be reduced to fixed types or the regularities of science. The essence of this view is that in history the continuity between individual things is shown, this having the character of a necessary relation. The necessity is rational, as illustrated by Collingwood's examples of a game of chess[140] and the changing of money after Great Britain went off the gold standard[141], as well as his account of the reason for the building of Hardknot Castle[142]. As we have seen in the last section, however, general truths are not excluded from his explanations.

The guiding principle of Collingwood's idea of historical understanding is, that history should be studied from within. Though this view was worked out in the re-enactment doctrine it was already held by Collingwood at an early date. In *Religion and Philosophy* he says with regard to the heresies of early Christianity that one should enter "... with some degree of sympathy into the problems which men wished to solve, and ... comprehend the motives which led them to offer their various answers" (RPh, 42). "...[T]he true task of historical theology", he observes, "is to find out not only what was said, but what was meant... Then we should be in a position to understand from within the new doctrines of Jesus, and really to place ourselves at the fountain-head of the faith. To speak of studying the mind of Jesus from within may seem presumptuous; but no other method is of the slightest value" (RPh,43). Further on he asserts on mind in general, that "[t]o know some one's mind is nothing more nor less than to see eye to eye with him, to look at reality as he looks at it, to know what he knows. His mind is not an object in itself; it is an attitude towards the real world, and to know his mind is to know and share that attitude" (RPh,156). It is Collingwood's conviction that it is always possible to know mind in this intimate way. For on the knowledge of a non-inherited tradition he contends in the draft of "Human Nature and Human History": "However remote from each other two historical traditions may be, each of them represents an achievement of mind; and mind is never wholly alien to mind, never a mere spectacle to be watched from outside, but always something that can be penetrated and seen from within by re-living its thoughts"[143]. That Collingwood really endorses the view that *nihil humanum a me alienum* is demonstrated by contentions he made on the understanding of Chinese civilization. Though there is no question in this case of an unconsciously inherited tradition, he maintains:

> Nevertheless, [the European] can find there certain ways of thinking which, though new to him, are intelligible. He is conscious, as he masters them, of a new growth of powers within himself... [F]or what Chinese art and literature have taught him is that he possessed powers of which hitherto he had been ignorant; he had fancied himself something altogether alien to Chinese civilization, and now he knows that he is not; he knows himself as heir to a tradition from which he had imagined himself cut off[144].

Here Collingwood expresses the same view as the one he stated so forcefully with regard to the understanding of primitive mind: that any mind is intelligible and has a rationality of its own. If understanding fails it is the student who is at fault. Or, as Collingwood puts it in *The Idea of History*: "It has been said that *die Weltgeschichte ist das Weltgericht*; and it is true, but in a sense not always recognized. It is the historian himself who stands at the bar of judgement, and there reveals his own mind in its strength and weakness, its virtues and its vices" (IH,219).

To study mind from within and to recognize its rationality could be considered the precondition of understanding it properly. Coming to the concrete interpretation of human actions, however, it is the conception of the logic of question and answer that plays a pivotal role. In the discussion of this concept I argued that it plays a role both in historical understanding and in historical inquiry[145]. Since the latter aspect has been dealt with in chapter 7.3, I will here focus attention on the former.

We have seen that Collingwood emphasizes the problem-solving character of historical inquiry and that his approach resembles Popper's in a remarkable way. The same can be said of his view of historical interpretation. For here again, both Collingwood and Popper emphasize the fact that history *a parte objecti* should be seen as problem-solving in nature. In *Religion and Philosophy* Collingwood already speaks of history "... as a process of the solution of problems and the overcoming of difficulties" (RPh, 166). Likewise, he contends in "The Theory of Historical Cycles" that the "... succession of problems... is the course of history" (THC, 86) and that "[t]he business of the historian is to discover what problems confronted men in the past, and how they solved them" (THC,85). In the *Autobiography* the same view is elaborated with regard to the history of philosophy. "...[A]ny one can understand any philosopher's doctrines if he can grasp the questions which they are intended to answer", Collingwood says (Aut, 55). Or, on the occasion of his meditation on the Albert Memorial he gives the rule: "'reconstruct the problem'; or, 'never think you understand any statement made by a philosopher until you have decided, with the utmost possible accuracy, what the question is to which he means it for an answer'" (Aut, 74). To give an example of this rule: Marx will be misunderstood, Collingwood argues, if his philosophy is not seen in the context of the "practical" problem he meant to solve (Aut, 152-153). "To think in that way about philosophies not your own", Collingwood observes, "is to think about them historically" (Aut, 58).

Though the problems of historical agents are mostly dealt with implicitly in Collingwood's historical writings, he sometimes refers to them explicitly. He says for instance that "[t]he chief tactical problem presented to [Caesar] by the British charioteers was the difficulty of pursuit"[146], and asserts elsewhere: "Even if Gaul were tranquil, the growth of a rich and powerful monarchy across the Channel would keep alive the unsolved problem bequeathed to posterity by Caesar"[147] (This problem was to be eventually solved by the conquest of Britain). Elsewhere Collingwood speaks of art in terms of problems to be solved[148].

Popper emphasizes as well that an historian should reconstruct the problem-situations of the past. This view is developed by his theory of the "logic of the situation". Though the idea is already discussed in *The Open Society and Its Enemies* and *The Poverty of Historicism*[149] it is more explicitly worked out by Popper in a paper read at a congress in Tübingen in 1961[150]. He speaks there of "... eine *rein objektive Methode* in den Sozialwissenschaften..., die man wohl als die *objectiv*-verstehende Methode oder als Situationslogik bezeichnen kann". After this he continues:

> Eine *objektiv*-verstehende Sozialwissenschaft kann unabhängig von allen subjektiven oder psychologischen Ideen entwickelt werden. Sie besteht darin, dass sie die *Situation* des handelnden Menschen hinreichend analysiert, um die Handlung aus der Situation heraus ohne weitere psychologische Hilfe zu erklären. Das objektive 'Verstehen' besteht darin, dass wir sehen, dass die Handlung

> objektiv *situationsgerecht* war. Mit anderen Worten, die Situation ist so weitgehend analysiert, dass die zunächst anscheinend psychologischen Momente, zum Beispiel Wünsche, Motive, Erinnerungen und Assoziationen, in Situationsmomente verwandelt wurden[151].

There is, surprisingly, a striking correspondence between this view and Dray's concept of a rational explanation. For according to the latter, which is based on an interpretation of Collingwood's re-enactment doctrine, what is asked for in an historical explanation is to show the appropriateness of certain actions in the light of an agent's beliefs, motives, principles etc. and his circumstances[152].

In "A Pluralist Approach to the Philosophy of History", however, Popper contrasts his logic of the situation with Collingwood's re-enactment doctrine. After quoting the passage in *The Idea of History*, in which Collingwood uses as an example the proper understanding of the Theodosian Code (IH, 283), he contends:

> Collingwood makes it clear that the essential thing in understanding history is not so much the analysis of the situation, but the historian's mental process of re-enactment. The analysis of the situation serves only as an indispensible help for this re-enactment. I, on the other hand, suggest that the psychological process of re-enactment is inessential, though I admit that it can be an excellent personal help for the historian, a kind of intuitive check of the success of his situational analysis. What is essential, I suggest, is not the re-enactment but the situational analysis: the historian's attempt to analyse and describe the situation is nothing else than his historical conjecture, his historical theory. And the question – 'what were the important or operative elements in the situation?' – is the central problem which the historian tries to solve. To the extent that he solves it, he has *understood* the historical situation, and that piece of history which he tries to recapture[153].

"The main significance of the difference between Collingwood's re-enactment method and my method of situational analysis", Popper concludes a little further on, "is that Collingwood's is a subjectivist method, while the method I advocate is objectivist. But this means that, for Collingwood, *a systematic rational criticism of competing solutions to historical problems is impossible*"[154]

We have seen that Popper's psychological interpretation of the re-enactment doctrine is incorrect[155]. The same can be said of his view that this doctrine would not allow for rational criticism. These claims can be demonstrated by an example from Collingwood's historical practice. Discussing the period between Caesar's expeditions to Britain and its actual conquest Collingwood reports that according to the ancient historian Dio the emperor Augustus developed serious plans to invade Britain. "If Dio's stories are true", Collingwood comments, "we must credit Augustus with the design of conquering and permanently occupying a part, at any rate, of Britain; Dio hints that he wished to outdo Caesar, and he must have learnt from Caesar's example that no permanent results could come from a mere raid. It is probable that the stories, as an account of his actions, are not true". After this, Collingwood continues:

> Britain was dangerous, but so was Parthia; and Augustus, who always had plenty to do nearer home, was inclined to shirk remote frontier problems. It was more characteristic of him to advertise an intention which he did not really entertain, than to abandon an enterprise he had once undertaken. For our present purpose it is not important to decide between these alternatives. The decision affects our view of Augustus' character, but not our view of the British question as it existed in his time. Whether he actually planned the conquest of Britain, only to be diverted from it by other tasks, or whether, recognizing from the first that these other tasks had a prior claim, he only allowed others to think he was planning it, in either case he was bearing witness to an unsolved problem on the north-western frontier, and the necessity of solving it, sooner or later, by conquest[156].

This passage clearly demonstrates that instead of giving a psychological interpretation in terms of Augustus' motives Collingwood focusses his attention on the objective "logic of the situation". This does not mean, of course, that motives never play a role in Collingwood's interpretations, for the examples given of the re-enactment doctrine show that they do[157]. In this case too, however, they are not interpreted by Collingwood in a subjective or psychological way, but are "objectively" analysed within the context of a certain problem-situation. The suggestion that these interpretations would not admit rational criticism would be fiercely rejected by Collingwood: it is indeed contrary to his view of historical science. In the passage quoted above Collingwood criticizes Dio's view. There is no reason at all to think that Collingwood would consider his own interpretation to be final, let alone beyond rational criticism.

Though Collingwood speaks at various places of "situations", his view of this concept is not easy to grasp. We have seen how in his discussion of re-enactment in the "Notes on the History of Historiography and Philosophy of History" he speaks of both the mind of William the Conqueror and the thoughts of his opponent Harold as being factors in a certain situation[158]. By a situation can also be meant, for instance, economic or natural conditions. Collingwood contends that we cannot speak of "objective" situations in either sense as determining human actions, but only of situations as interpreted by human agents. The word "situation", however, has a certain objective connotation and it is precisely this aspect that is referred to by Popper's concept of the "logic of the situation". At one place in *The Idea of History* Collingwood seems to endorse this view:

> For a man about to act, the situation is his master, his oracle, his god. Whether his action is to prove successful or not depends on whether he grasps the situation rightly or not. If he is a wise man, it is not until he has consulted his oracle, done everything in his power to find out what the situation is, that he will make even the most trivial plan. And if he neglects the situation, the situation will not neglect him. It is not one of those gods that leave an insult unpunished (IH, 316).

The way Collingwood continues, however, seems to contradict this position:

> The freedom that there is in history consists in the fact that this compulsion is imposed upon the activity of human reason not by anything else, but by itself. The situation, its master, oracle, and god, is a situation it has itself created... All history is the history of thought; and when an historian says that a man is in a certain situation this is the same as saying that he thinks he is in this situation. The hard facts of the situation, which it is so important for him to face, are the hard facts of the way in which he conceives the situation (IH, 317).

In the first instance, one has the inclination to dismiss this contention as a gross exaggeration, which is, moreover, belied by Collingwood's own historical practice. For, as we have seen, he ascribes an important role in history to geographical factors; the argument that these would be imposed upon the activity of human reason only by itself is hardly convincing. I will not try to conjure away this difficulty, but only attempt to clarify Collingwood's underlying idea by means of certain views he develops elsewhere. When these are taken into account his position is much more sensible.

We have seen how Collingwood in "Reality as History" identifies a man's character with his past[159]. This past is the determinate part of present actions. Beside this determinate part, Collingwood argues, an action has an indeterminate part assignable to free will. After its performance an action is past and done and will therefore accordingly determine future actions and so *ad infinitum*. Thus an action is never fully determined, but when performed it will be one of the determining factors of future actions.

In accordance with this conception Collingwood elsewhere asserts (in a few notes under the heading "History as the Understanding of the Present"[160]) that the given historical fact "... is always *more than what it is*": "It is this (i.e. the past) as to content; as to form, it is free activity. To understand it = understand what it is, its content; the form is not understood, it escapes the understanding"[161]. (This position reminds one of Kant's distinction between the realms of the phenomena and noumena, only the first of which can be known). A little further on Collingwood expands on this idea as follows:

> [T]he past does not possess the present, the present possesses its own past. The possessive act by which the present affirms itself affirms its own past. This *positedness* of the past is part of the selfpositingness of the present. This selfpositing of the present is what always and necessarily escapes the historian.
> For the historian, this limitation appears as *contingency*: i.e. an incomplete determination of the present by the past. The past event which he cites as explaining why the present is what it is does not determine the present: it only determines *possibilities* between which the present may choose. No historian can claim to have shown that a certain sequence of events *must*, have fallen out thus and no otherwise. The fall of a man's income may lead him to retrenchment or to bankruptcy: which it does, depends certainly on what kind of man he is, but what kind of man he is can never be finally determined: he determines it himself in his own action as he goes on. He goes on to bankruptcy and we say he *was* an extravagant and thriftless man, but this does not explain *why* he chose that alternative; it is only a way of saying that he *did* choose it. The 'free will' which chose this course only becomes a reality to the historian when it has ceased to be act and has passed into fact. When it has done so, the historian is no better off: because the free will is now in another situation and is confronted by another choice[162].

So the indeterminate or "free" part inherent in all action makes it impossible, in Collingwood's view, ever to give a complete explanation of the past. As he puts it in "History as the Understanding of the Present": "The past doesn't *actually* ever explain the present in its entirety. It only gives *part* of the present – not a complete determination of it. E.g. the peace treaty and German defeat partially explain Nazi mentality: but even if you add conditions *ad infinitum* (German socialism, fear of Russian influence, etc.) you never get a complete explanation"[163].

In "What 'Civilization' Means" Collingwood discusses the relation between man's actions and his situation in a clarifying way:

> The facts of the situation in which an act is done in one sense determine that act and in another sense do not. They determine it disjunctively, not categorically. They determine the agent's opportunities; they cannot determine his choice among those opportunities. Finding himself in that situation, he finds himself bound to act in one or another of certain ways. His awareness of himself as a free agent is awareness of these alternative possibilities for action which arise from the situation, and

> awareness of his power to choose between them. This choice is the work of his own will.
> The facts are what he inherits from his ancestry, who have bequeathed to him the situation in which he finds himself. They include economic facts, social facts, legal facts, and so forth. He does not inherit his will. Every man has to make his will for himself[164].

This passage clarifies the ambiguity involved in Collingwood's speaking in *The Idea of History* of a man's situation as "his master, his oracle, his god", and at the same time asserting that this situation is created by human reason itself (IH, 316-317). For, as I see it, what is referred to in the first case is the fact that one *has* to respond to a certain situation, while in the second it is to *how* this is done. It is confusing, though, that Collingwood suggests that in the latter case "situation" is used in the same sense as in the former. This cannot be correct, since a situation cannot be both a determining factor and a creation at the same time. One can say, however, that by responding to a situation a man creates another situation. It is the latter which is emphasized by Collingwood, in his opinion special attention being focussed upon this aspect of human action in history.

Summing up Collingwood's position, we may say that in his view the past can never be fully explained by its determining conditions, since the non-determined aspects of human actions always have to be taken into account. It is paradoxical, however, that an historian in Collingwood's view not only has to pay special attention to the latter, but has to show a necessary relation and continuity between them. The solution of this paradox is found in the fact that Collingwood conceived this relation as a rational one. Though human actions are not fully determined this does not mean that no rational connections can be shown. On the contrary, the possibility of showing them is in Collingwood's view a precondition of historical understanding. Such connections, however, are of another kind than determining relations in nature. They necessarily have a certain looseness, as has been pointed out by Dray[165].

The suggestion that Collingwood's view implies the neglect of determining conditions is nevertheless false. How could one speak of human actions as responses to certain situations without knowing what the latter are? Collingwood would only contend that they too have to be established by historical research. It would be rather odd to suggest the neglect of determining conditions by Collingwood, since the past itself is considered by him a determining condition.

8.8. HISTORICAL OBJECTIVITY

In "Reality as History" Collingwood discusses the objection that in an historical explanation something can be explained only in terms of something else which is itself left unexplained. Seeing this as an objection would amount to saying, Collingwood observes, that we cannot know any history unless we know all of it. If such a position were seriously maintained, he argues, it would bring all knowledge into question, not only history[166]. Collingwood strongly rejects the idea of final or complete knowledge: "Wherever thought is active and is achieving solid results, the solution of one problem leads to the statement of others"[167]. A little further on he asserts:

> There is no finality in any knowledge whatever. There is nothing about which we have any knowledge at all, about which there is not more to know... If it is truer in history than elsewhere (which I doubt) that here the incompleteness of all we know is peculiarly manifest, it follows, not that history is a peculiarly futile form of thought, but that it is a peculiarly priviliged one, where the thinker is more than commonly aware of what he is doing and more than commonly exempt from delusions about the nature and extent of his achievement[168].

This clearly refutes the "objectivist" interpretation of Collingwood, which ascribes to him a view of history aimed at indubitable and certain knowledge – especially by means of re-enactment[169].

We find the same idea expressed in the 1926 lectures, when Collingwood contends: "The work of collecting sources is as endless as is the work of interpreting them, and therefore every narrative that we can at any given moment put forward is only an interim report on the progress of our historical inquiries. Finality in such a matter is absolutely impossible. We can never say 'this is how it happened', but only and always 'this is how, as at present advised, I suppose it to have happened'"[170]. However, this does not mean, Collingwood argues, that there can be no advance in historical knowledge: he speaks in this connection of its "relative truth". "The only certainty that we can ever have in historical thinking", he maintains, "is the certainty of having made a definite advance on previous theories"[171].

The "subjectivist" interpretation of Collingwood is not true either. For we have seen how important he considers the sound interpretation of evidence – related, it should be added, to specific questions. The ambition to arive at "objective" historical knowledge is usually based on an implicit realist view of the past. This position, however, is epistemologically untenable. For this reason Collingwood's approach – based on an idealist view of the past – focussing attention on evidence and questions to be asked, is attractive and fruitful. His viewpoint is clearly expressed in the following passage from the 1928 lectures:

> The historian cannot have certain knowledge of what the past was in its actuality and completeness; but neither has he uncertain knowledge of this, or even conjecture or imagination of it. The past in its actuality and completeness is nothing to him; and, as it

> has finished happening, it is nothing in itself; so his ignorance of it is no loss. The only knowledge that the historian claims is knowledge of the answer which the evidence in his possession gives to the question he is asking... The certainty of history, then, is the certainty that *the evidence in our possession points to one particular answer to the question we ask of it*[172].

How do we know that a certain picture of the past is true? Collingwood asks in the draft of his Inaugural. His answer is: "There is only one way: by doing [the historian's] work over again for ourselves, that is, by reconsidering the evidence upon which his picture is based and, exercising upon this evidence our own historical imagination, finding that we are led to the same result"[173].

In his notes made during his journey to the East Indies Collingwood comes back to the question of finality in historical research under the heading "No Completeness in History":

> No piece of historical work ever exhausts either its subject, however small, or the evidence for its subject, however exiguous. 1) Its subject, because every profitable reconsideration of an historical subject re-defines the limits of the subject. Every competent historian who comes to the consideration of an old historical theme finds himself asking afresh what the theme is that he is dealing with, and giving a new and individual answer to this question. 2) Its evidence, because every advance in historical method is an advance in the power of historians to cite new kinds of evidence[174].

"But must not the historian *try* to be complete even though he knows he can't be? Is not historical completeness an 'idea of Reason'?", Collingwood asks. His answer is: "No. The historian should not aim at completeness, he should aim at relevance. He should aim at providing an answer that really is an answer to the question he is asking"[175].

If this viewpoint is to be termed relativistic it is a relativism that should be subscribed to by every historian. In fact, Collingwood's position with regard to the question of historical objectivity can be described neither as "subjectivist", nor as "objectivist". It turns on the notions of the historian's questioning activity and historical evidence, and their interrelation. Contrary to the view of some interpreters, the re-enactment doctrine is not involved in it.

CHAPTER NINE

HISTORY AND OTHER SCIENCES

9.1. HISTORY AND NATURAL SCIENCE

In his scheme for *The Principles of History* Collingwood mentions as topic for the second book the relation between history and natural science. The themes would be, first, that neither history nor natural science may be reduced to the other, and second, that there is a certain relation between them[1]. For a good understanding of these topics it is useful, however, first to consider Collingwood's views on the characteristics of a natural process and an historical process. For his opinion on the relation between natural science and history is to a great extent based on these views.

We have seen how Collingwood develops in his manuscript on cosmology a developmental view of reality, including not only mind, but also matter and life. He rejects, therefore, Hegel's position, that nature has no history. For Darwin had demonstrated, Collingwood argues, that in nature as well forms change, so that it, too, has a history[2]. He is not willing, however, to equate natural history with "history in the fullest sense", or, as he calls it, "historian's history". Their difference, however, is not yet fully grasped and Collingwood is content simply to distinguish between "history" in the wider sense, which would include natural development, and "history" in the narrower sense, which he calls "the stricter or more proper" one[3].

It is only in "Reality as History" that the distinction between a natural and historical process is worked out. Its essence is that nature in its development does not retain its own past, while the contrary is true of mind's development[4]. To put it differently: nature's past is dead, while that of mind is alive. This idea is eloquently described by Collingwood in his draft of "Human Nature and Human History": "Nature ceases to be what it was in becoming what it is; the phases of its process fall outside one another. Mind, in becoming something new, also continues to be what it was; the stages of its development interpenetrate one another"[5]. Though Collingwood agrees with regarding both nature and history as a process, he emphasizes that the latter is of a peculiar kind: "...[A]n historical process is a process of thought, and a mind which has developed its thought from one stage to another has not left the first stage behind when it reaches the second"[6].

In contrast with this, the past of a natural process is "over and gone"[7].

In his 1936 lectures Collingwood gives the example of the equilibrium of a pound of sugar and a pound weight in a pair of scales. This equilibrium is arrived at by adding to or taking away from the original amount of sugar put on the scales. For the equilibrium once established, however, it does not make any difference which procedure has been taken[8]. Collingwood then gives the following example of an historical process, where the past lives on in the present and is therefore relevant for it:

> ...[S]uppose that two people agree in thinking that the earth is round, but one of them has arrived at that belief through accepting it as something told to him by persons in a position of authority, the other through considering the evidence for himself. Because they have reached the same belief through different processes, there will be differences in the way in which they hold it, and a scrutiny of their present frame of mind will reveal these differences and show how the belief has been arrived at[9].

The reason that Collingwood discusses the differences between natural and historical processes is that in his time the evolutionary view of nature had become very influential. In this connection he refers at various places especially to the philosophies of Whitehead and Alexander. Though he thought highly of their ideas, he opposed their conception of all reality having the character of a process in a similar sense. Collingwood is of the opinion that this suggestion is misleading, since it does not take into account the essentially different nature of an historical process. Or, as it is tersely put by him: "... all process is not history and history [is] not a mere process"[10].

So the refusal to reduce natural science to history is derived from the distinction between a natural and an historical process. But the complementary refusal not to reduce history to natural science is similarly based on it. For it is the characteristic of an historical process – a process of thought – to be living in the present. This process, then, can only be known by studying it from within through the re-enactment of past thought. In contrast to this, natural events are studied from the outside, as mere phenomena. Based on this notion there are many other things which, in Collingwood's view, make history different from natural science. For instance, history does not reason inductively, nor does it search for invariable causes or general laws[11]. It is aimed at the individual situated at a definite time, while natural science looks for timeless universals[12]. History, moreover, tries to make the flux of events in itself intelligible, while in natural science this flux is reduced to fixed types, subsuming events under general laws[13]. In "Can Historians be Impartial" Collingwood also asserts that in history, in contrast to natural science, value judgments are necessary[14].

In his scheme for *The Principles of History* Collingwood refers to the similarities between natural science and history as well. For the former is not only based on certain principles, he contends, but also on "historical facts" in the form of certain observations. We find the same view expressed in "Are History and Science Different Kinds of Knowledge", *Speculum Mentis*, the 1929 lectures, "Reality as History" and *The Idea of Nature*[15]. Another similarity between natural science and history, pointed out by Collingwood in his lectures of 1931 and 1932, is the fact that both deal with abstract and ideal objects[16]. In his manuscript on cosmology he enlarges on this idea. The historian also uses abstract concepts, he asserts there, which are separate from actual experiences. These concepts, however, are not pure essences, as in natural science, but are existentialised[17]. Finally, the questioning activity could be mentioned as a point of similarity between natural science and history[18].

Though, on Collingwood's view, there are in theory certain similarities between natural science and history, he usually emphasizes their dissimilarities in practice. He thinks recognition of the latter the more pressing, since he believes that scientific methods have more than once been wrongly copied – in various ways – by history. Examples are the way in which Hegel distinguishes the periods of the past in terms of logical formulae[19], or Spengler's historical morphology[20]. The "comparative method" is accordingly rejected by Collingwood, and is called by him "the apotheosis of anti-historicism in a positivistic interest". "You cease to care about what a thing *is*", he observes, "and amuse yourself by saying what it is *like*"[21].

The practice of copying the methods of natural science is called by Collingwood positivism or naturalism. We have seen how he rejects this approach in anthropology[22]. Collingwood also uses in this connection the term "pseudo-history": "...[N]aturalistic methods", he asserts in his manuscripts on folklore, "instead of yielding historical knowledge, yield a pseudo-history which is merely a magnified projection of the would-be historian's desires upon the blank screen of the unknown past"[23]. That Collingwood's rejection of this approach is based on his distinction between a natural process and an historical one, is made clear by the definition of pseudo-history given by him elsewhere: "Pseudo-historicity then shall be my name for what Alexander, Whitehead, etc. call historicity. A pseudo-history is an account of changes, whether geological, astronomical, social, or any other kind, where the person giving the account does not re-enact in his own mind the thoughts of the person or persons by whose action these changes came about"[24].

9.2. HISTORY AND THE SOCIAL SCIENCES

Discussing the general causes of Collingwood's intellectual isolation Johnston mentions as "one of the major reasons why Collingwood has fallen into neglect since the 1930's" the fact that history relied in the twentieth century more on social science than on philosophy. "In ignoring the rise of social science", he comments, "he [Collingwood] failed to contribute to what seems to be the most influential movement in social thought in the first half of the twentieth century"[25]. Though it is true, as Johnston observes, that Collingwood did not take notice of leading social scientists, such as Durkheim or Weber[26], I disagree with the contention that he ignored the social sciences. I would even contend that he contributed to them and that the views he developed in this field, especially with regard to what he considered the historical approach, are still of great interest. The reason for their neglect is, that they are mostly worked out in the unpublished manuscripts. In what follows I will not only briefly discuss Collingwood's view of the social sciences and their relation to history, but also the attention he paid to social factors in his historical practice and the relativism that would be implied by his position.

In discussing the 1929 lectures we have seen how Collingwood fiercely opposes the positivistic theory of history developed by people like Taine and Buckle. In their view the historian's task was to provide the "facts", which would then be explained by the "scientific" historian or sociologist. This would lead, Collingwood argues, to a tampering with the historical facts, since they would be adapted to certain theories, while he also argues that the positivistic conception is based on a wrongly conceived separation of facts from causes[27].

The positivistic approach is discussed by Collingwood in "Reality as History" in a slightly different context. He speaks there of the influence of scientific thought on "... a number of hybrid sciences such as anthropology, Völkerpsychologie, comparative philology, etc. whose general principle lies in extracting historical facts from the context in which alone they are truly, that is historically, intelligible, reassembling them in a classificatory system according to their likenesses and unlikenesses, and attempting to lay down general laws governing their relations"[28]. "These sciences have always been regarded with distaste by historians", Collingwood observes, "because the historian, trained as he is to think of facts in their concrete actuality, cannot tolerate the substitution for any one fact whatever of another more or less like it"[29]. "Hence the attempt to reduce history to a science (as it was called)", he contends a little further on, "which means to renounce history as such and substitute sociology in the wide Comtian sense of that word, was an attempt whose futility and viciousness every historian recognised..."[30]. Collingwood then makes an important, though surprising, observation:

> This is not to deny the utility of the studies which in general may be called sociological. They are vicious only when they invert their own true relation to historical studies. History can find room within itself for studies of a scientific kind; indeed, the relation between historical and scientific thought is in this way not unlike that between scientific and mathematical[31].

After pointing out that, although mathematical thinking is the indispensable groundwork of scientific thought, nature cannot be described "without residue" in mathematical terms, Collingwood continues:

> Human activity as it takes place in history ... cannot be described without residue in scientific terms, yet the ability to think scientifically is indispensable to the historian. All the sociological sciences are useful to him; what he cannot allow, if he understands anything about his own work, is that this relation should be inverted and that the sociological sciences should represent themselves as the end to which historical thought is the means[32].

In his manuscripts on folklore Collingwood gives examples of the use of social science. For he contends that "[a]rchaeology, helped by comparative ethnology, is coming to realize that the old classification of Stone, Bronze, and Iron Ages ... gives a superficial view of the earliest developments of human life"[33], and a little further on: "It is easy to imagine, and here again comparative ethnology gives confirmation, that the traditional tales of food-gathering peoples will differ from those of agricultural peoples"[34]. Discussing the fact that a theme contained in a fairy-tale must be seen as a fragment of ancient custom or belief, such as a stone implement is a fragment of ancient technical skill, Collingwood similarly maintains: "In each case the historian uses the fragment by reconstructing in his mind the life and thought of the people who have left him this sample of their work. The archaeologist is helped in this attempt by the ethnologist, who may be able to point out similar implements now used in a particular way by this or that primitive society. The student of fairy-tales must avail himself of the same kind of help"[35].

Though these passages plainly show that Collingwood does not neglect the social sciences, it is far from clear how the latter are conceived by him. For, as we have seen, he is not only against sociology "in the wide Comtian sense of that word", but also against a science such as anthropology, which extracts "historical facts from the context in which alone they are truly, that is historically, intelligible". But how else, one could ask, can comparative ethnology collect its data than by extracting them from historical contexts? It is thus a puzzling question how these data can be made useful again for concrete historical or archaeological studies. In the discussion of the use of generalizations in history we have met with the same difficulty with regard to Collingwood's use of certain demographical theories[36]. When Collingwood speaks, therefore, of the use of social sciences for history, his conception of their status is not altogether clear.

The question what Collingwood means by social science is the more puzzling in view of his argument in his manuscripts on folklore that anthropology should practise the historical method. He even goes so far as to call anthropology "an historical science"[37]. In these manuscripts this claim is based on two considerations. On the one hand Collingwood means by it that anthropology should study its object from within and not as something external to the inquirer. On the other hand, he insists that it should not abstract, but study the concrete details. Collingwood observes on the interpretation of magical practices: "The method is to reconstruct, from all the evidence at our disposal, the social structure in which they grew up. It is thus an historical method: one proceeding not by abstraction and generalization, but by the reconstruction of fact in all its detail"[38].

In his notes made during his journey to the East Indies Collingwood even goes so far as to deny that the social sciences form a separate discipline. For in the scheme for *The Principles of History* it is stated that the human sciences "...are crypto-history or just history"[39]. Crypto-history is defined as "... a name for sciences which *are* historical sciences but profess not to be"[40], and the first example given is that of teaching military tactics, while its historical dimension is "concealed or denied". Collingwood enlarges on this idea as follows:

> The so-called classical economists have written of an 'iron law of wages', meaning that a certain theorem about wages must always be true under any kind of social system. Actually, this theorem was true of the social system under which they wrote: but under a different social system it would not necessarily remain true. The so-called classical economics is thus a crypto-historical science, describing a certain set of transient historical conditions under the belief that it was stating eternal truths. – Anthropology is crypto-history. A number of different historical complexes are lumped together under the name of 'primitive life' or something like that: and their characteristics, all of which have their own historical contexts, are thus by a fiction abstracted from these contexts and treated as a kind of matrix ... within which historical formations arise[41].

We may conclude from this that, in Collingwood's view, the social sciences cannot claim a separate status. For they are either pseudo-history or crypto-history. In the first case an historical process is positivistically studied on the analogy of the study of a natural process[42], while in the second the historical dimension of human conduct is denied. As has been noticed already, it therefore remains obscure what type of social science Collingwood has in mind, when he refers to the use an historian can make of it – for instance, when he speaks of a certain economic "principle" the historian may appeal to, "as an aid in discovering what happened on a certain occasion"[43].

Notwithstanding this difficulty, we have seen from the treatment of Collingwood's historical writings that he pays ample attention to social and economic aspects of Roman Britain[44]; and we have also seen how the anthropologist Evans-Pritchard considers the topics discussed in *Roman Britain* highly relevant for ethnology[45]. In other reviews of Collingwood's historical work we similarly find the opinion expressed that its special value lies in its treatment of social and economic matters. Since Collingwood's theory of history has more than once been accused of being relevant only to accounts of the actions of individuals in political and military history, it is rather paradoxical to find one reviewer of *Roman Britain and the English Settlements* asserting that "Dr. Collingwood has ... been more concerned with the state of the country and of its people than with military and political history"[46], and that he "is the first to draw attention to the agrarian problems of Roman Britain..."[47]. Another reviewer contends on the same book that "[t]he discussion of urban and country life and of agriculture in particular forms a real contribution to the social and economic history of the province..."[48]. And the political scientist Hammond gives the following judgment on *Roman Britain and the English Settlements*: "The book ... should prove valuable not merely to historians of Rome or England but to the political scientist and the sociologist because of the attention which is paid to the character and development of the two societies, in each case the result of the imposition of a new culture on a different and well-established one but under wholly different conditions"[49]. There is also other evidence which seriously undermines Collingwood's alleged one-sided interest for the political aspects of history – for instance, when he points out the important influence of new inventions, and maintains that "[t]o the average medieval household, the advent of the spinning-wheel was a far more significant revolution than the signing of Magna Carta"[50].

When we do not take into account Collingwood's reduction of the social sciences to history, but think of these sciences in their common-sense meaning, we can only conclude that, rather than neglecting them, Collingwood's views are highly relevant to them. They revolve around the following notions: that society should be studied from within and not from without as a mere spectacle; that attention should be paid to its appearances in detail and that it should not be reduced to abstractions and generalizations; that its historical dimension should be taken into account and weighed against any analysis in terms of static "laws".

It is obvious that these views are opposed to any positivistic, naturalistic or behaviouristic conception of social science. Collingwood's own view of social science is implied in the various views he developed with regard to history. Since these views are the subject of this study it suffices to declare that, in my opinion, they are highly relevant for the social sciences and still of great current interest. There is one aspect of them, however, that I should like to discuss separately. This is Collingwood's "historicist" position and the complete relativism or scepticism that is implied by it, according to some interpreters[51].

Of Hegel's philosophy of history Collingwood asserts that it has "one supreme merit": "It is based on a firm grasp of the fact ... that different periods of history are really different – not only chronologically different, but different in their fundamental characteristics"[52]. This contention certainly expresses an historicist and thus a relativistic view. That Collingwood was conscious, however, of the dangers of a complete relativism is made clear by what he writes about the subject in "What 'Civilization' Means". In this essay the concept of civilization is treated as a philosophical one, in the sense developed in *An Essay on Philosophical Method.* Civilization, Collingwood argues, should be seen as a process, but not as one that is always and everywhere of the same nature. This "historical monism" he considers "a chief defect of nineteenth-century historical thought"[53]. Against this he develops the following view:

> Twentieth-century historians take for granted the idea of an 'historical pluralism', the idea that at different periods and among different peoples different ideals have been envisaged and correspondingly different processes enacted. Hence we speak of Chinese civilization and European civilization, for example, as different things, realizing or attempting to realize different ideals, and not capable of being described as merely differences in the degree to which one single ideal has been realized. They are different, we think, in kind ...
> Thus the present-day conception of the relation between civilization and barbarism is that any given society at any given time has its own standard of civilized life, and thinks of itself as civilized in so far as it recognizes that standard, and of other societies as barbarous in so far as they do not[54].

Collingwood observes that this may seem "a dangerous opinion". For it may seem to imply, he argues, that the definition of civilized conduct is reduced to the actual conduct in a given society at a certain time. He comments on this as follows:

> This is called 'historical relativism', and is rightly regarded with suspicion, because it really amounts to denying what it professes to explain. It amounts to denying that there is any such thing as an ideal of civilized conduct: not merely that there is one single ideal valid for all societies and all times, but that there are many ideals each valid for one society at one time. For if 'civilized conduct' as a phrase in the mouth of certain persons at a certain time merely means 'the way in which we behave', the ideal element in the meaning of the word 'civilized' has vanished, and only a factual element is left. In that case the verb 'to civilize' has lost all meaning[55].

Collingwood denies, therefore, that the notion of civilization proposed by him implies "historical relativism": "It does not imply the negation of all ideals and the substitution for social ideals of social facts. All it asserts is that the social facts which are called civilizations are orientated towards different ideals"[56]. In order for the concept of civilization to be meaningful, however, there must be some similarity between its various realizations, Collingwood argues. He concludes, therefore, that "... the historical pluralism of the present day does not exclude a certain kind of historical monism. The plurality of civilizations does not exclude a sense in which civilization is one"[57]. Collingwood develops this idea in the rest of his essay, in which he distinguishes three orders of ideal of civilization. The last, then, would be "an ideal of universal civility"[58].

The questions of historicism and relativism come most explicitly to the fore with regard to the various civilizations in past and present, especially as to the question if and how far they should be judged. From Collingwood's discussion of this problem it is clear that he does not provide a simple and one-sided solution, which could be described as "complete relativism". On the contrary, he is not only conscious of the pitfalls involved in this delicate question, but also develops a highly sophisticated argument. As with Collingwood's other views it is not only relevant for history, but also for the social sciences at large.

In this section the science of psychology has not been discussed. The reason is, that Collingwood developed specific ideas on this science, which are especially relevant for the practical dimension of his theory of mind. This question will therefore be dealt with in the next chapter.

CHAPTER TEN

HISTORY AND PRACTICE

10.1. INTRODUCTION

In his *Autobiography* Collingwood declares that "[i]n the kind of history ... I have been practising all my life, historical problems arise out of practical problems" (Aut, 114). Further on he says that besides the *rapprochement* between philosophy and history, he "... was also working at a *rapprochement* between theory and practice" (Aut, 147), while he ends the book with the statement: "I know that all my life I have been engaged unawares in a political struggle, fighting against these things [i.e. Fascism and irrationalism] in the dark. Henceforth I shall fight in the daylight" (Aut, 167). As with regard to other aspects of his views and development as described in the *Autobiography*, Collingwood's view on the practical dimensions of history and philosophy has not been taken seriously by his interpreters. Johnston, for instance, asserts:

> If one defines an intellectual as a man of trained intellect who tries to guide society toward a specific goal or goals, then Collingwood was not an intellectual. He never supported a specific program until very late in life, and then only in an ineffectual manner. Collingwood never lobbied for legislation of any kind. He never became a journalist in the manner of George Orwell or R.H.S. Crossman. He preferred to regard himself as a thinker who stood above the storm, who from a vantage point above controversy might urge all combatants to take a broader view of the issues at stake. He tried, of course, to relate philosophy to life, but by life he meant the search of the individual for self-fulfillment. Society could take care of itself, so long as the individual knew what he was seeking. And Collingwood was there to help the individual persevere in this quest for self-fulfillment[1].

The same Crossman who is referred to by Johnston gives a similar judgment:

> Among the dry bones in the valley of Oxford, he [Collingwood]

> built an ivory tower of Hegelian aesthetics, and listened in delicate disdain to the wind among the skeletons.
> Though he did not realize it at the time, the new philosophy of history which he constructed was as remote from the world as the realism he despised. Viewing all history as the history of thought and the clash of purposes, Collingwood denied that there were such things as historical 'events' ... Philosophy for him was the selfconsciousness of the contemplative mind[2].

In my opinion these judgments seriously misinterpret Collingwood. For, I would contend, there is ample evidence to endorse Collingwood's statement that the practical dimension played an important role in his theory of history and philosophy in general. I would go even further, and declare that his views concerning this subject are still of great current (practical) interest. Johnston is right, however, if he asserts that Collingwood never supported a specific program, lobbied for any legislation, or became a journalist. At least there is no evidence that he ever engaged in these endeavours. He was, however, less conspicuously engaged in social affairs of his time in his own, a philosophical way. In that respect one could even say the he tried "to guide society toward a specific goal or goals", as is denied by Johnston. As examples one could mention his essay "Man Goes Mad"[3] and his views on "primitive" and "civilized" cultures developed in the manuscripts on folklore. The "practical" dimension of his thought comes more explicitly to the fore in the chapter "Theory and Practice" of the *Autobiography* and certain parts of *An Essay on Metaphysics*[4], while we have seen that Collingwood wrote *The New Leviathan* on the occasion of the outbreak of the war[5]. In contrast to the views of Johnston and Crossman, I would contend, therefore, that Collingwood was more interested in and worried about the social and political developments of his time – at least in the thirties – than were most (Oxford) philosophers[6].

We have seen how, in Collingwood's view, thought is always involved in action[7]. The first sentence of *Speculum Mentis*, on the other hand, reads: "All thought exists for the sake of action" (SM, 15). At various places in *The New Leviathan* Collingwood even asserts that thought is primarily practical (NL, 5, 100, 102, 125). We will discuss here only the practical aspects of history as conceived by Collingwood. For a good understanding of these it is necessary first to deal with his view of psychology.

10.2. PSYCHOLOGY

History, in Collingwood's view, has as its object mind's activities. Since the nineteenth century, however, psychology has come to the fore as a formidable rival to this claim. For it claims to be the only scientific study of mind. All his life Collingwood strongly objected to this pretension. In his opinion, only special aspects of mind are studied by psychology, while others remain the domain of history and philosophy. We find this view already expressed in 1913, in the notes for the first lectures Collingwood gave, when he was 24 years old. He lectured on Aristotle's *De Anima* and remarks in an Introduction that this work "... is commonly described as Aristotle's contribution to psychology". He makes it clear, however, that Aristotle's study should be distinguished from psychology in the present-day sense of the word:

> The psychology of modern psychologists is an a posteriori, inductive, generalising and as a rule experimental science. It is an empirical as opposed to a philosophical science. It studies the will or the understanding not a priori or universally as do ethics and logic, but by empirical examination of a number of instances. Thus ethics would ask what is the distinction, a universal and a priori distinction, between a good and bad act; psychology would never ask such a question but would ask for instance whether men commit crimes more in the morning than the afternoon, or more in winter than summer. Logic tells us how a cogent argument is formed and how it differs from a fallacious one; psychology collects statistics and averages which tell us that on the whole this or that class of men is most likely to form valid rather than invalid arguments. In short psychology is a science which deals in averages and inductive generalisations, and thus differs sharply from philosophical sciences for which the conceptions of an average or of induction are meaningless[8].

It would be a misconception, however, to think that Collingwood was against the science of psychology. In the manuscript on cosmology he assigns to it the task of studying a special aspect of mind: not mind "as spirit", but "mind as psyche, the psychologist's notion of mind, consisting not of an activity of clear thought and pure activity but of conations and 'ideas' which are subjective and, regarded as thought, mainly illusory". "At bottom perhaps what psyche is", Collingwood adds, "could be identified as desire or even appetite – it is in substance a complex of emotions"[9]. We have seen that the same field is assigned to psychology in the notes of 1936[10]; and the same thing is done in *The Idea of History*, when feelings and the irrational elements of mind are said to be the subject-matter of psychology (IH, 231). That Collingwood does not object to psychology if it confines itself to the study of these aspects of mind may be illustrated by the fact that he expresses his admiration for Freud's work and speaks of psychologists as "experts (all honour to them) in explaining dreams"[11].

What Collingwood objects to is the claim of psychology to study mind in its entirety. The spiritual aspects of mind, he argues, do not belong to the field of psychology. Yet this area is claimed by psychology as well. In *Religion and Philosophy*, Collingwood says in his *Autobiography*, he attacked the notion of a psychology of religion. As a crucial statement he quotes from the former book: "the mind, regarded in this way, ceases to be a mind at all" (Aut, 93). This quotation however, is not complete. For the passage in *Religion and Philosophy* reads: "The mind, regarded in this external way, really ceases to be a mind at all"; it continues: "To study a man's consciousness without studying the thing of which he is conscious is not knowledge of anything, but barren and trifling abstraction" (RPh, 42). Before this passage Collingwood similarly asserts with regard to the psychology of knowing that it "... differs from logic or the philosophical theory of knowledge in that it treats a judgment – the act of knowing something – as an event in the mind, a historical fact. It does not go on to determine the relation of this mental event to the 'something' known, the reality beyond the act which the mind, in that act, apprehends" (RPh, 40). Likewise, Collingwood asserts in *The Idea of History*: "Psychology is thought of the first degree; it treats mind in just the same way in which biology treats life. It does not deal with the relation between thought and its object, it deals directly with thought as something quite separate from its object, something that simply happens in the world, as a special kind of phenomenon, one that can be discussed by itself" (IH, 2).

In Collingwood's rejection of the claim of psychology to study thought or the rational aspects of mind, we come to a crucial element in his philosophy of mind. For we have seen that thought, in contrast to feeling, in Collingwood's view has the characteristic of being "criteriological" and critical[12]. That is, a thought always implies the object that is thought about and is judged according to certain criteria. It can only be studied, therefore, "from within" and not from the outside as a mere event. We have seen how crucial this notion is also for a good understanding of the re-enactment doctrine[13]. What psychology does, is to reduce thoughts to certain experiences, which can be studied from the outside as "objective" events. By reducing thoughts to psychic phenomena in this way, Collingwood argues, they are fundamentally distorted.

Collingwood's position is well illustrated by a review he wrote of Jung's *Psychological Types, or the Psychology of Individuation* (London,1923). Jung makes a distinction between conscious minds being introverted (turned upon itself) or extraverted (turned upon the object). Collingwood offers the following comment:

> Now the value of the theory, Jung says more than once, is that it explains the differences between different people's attitudes to life; for instance, it reveals the conflict between medieval nominalism (extravert thinking) and realism (introvert thinking) as a conflict of temperaments. If this is so, all scientific doctrines are the products of temperamental peculiarities, and therefore no such doctrine is true, and no conflict between two doctrines can be settled. Any given theory is of interest not as telling us the truth about the object, but as revealing to us the psychological

> peculiarities of its inventor; and this applies to Jung's own theories. Hence the theory taught in this book discredits all theory whatever[14].

Another, most interesting, example is Collingwood's review of Spearman's *The Nature of 'Intelligence' and the Principles of Cognition* (London, 1923). In this book Collingwood discusses the use of intelligence tests which would make it possible, according to Spearman, to establish "ultimate laws" of psychology. Collingwood's criticism is important enough to be quoted in full:

> The original error, we think, lies in the hope of using intelligence-tests as a basis for the psychology of cogniton. It might be supposed that anything done in a laboratory is a scientific experiment and a firm basis for any amount of theoretical superstructure; but this is a mere idol of the theatre. Intelligence-tests are meant to test intelligence, and intelligence, as Professor Spearman's opening chapter shows, is not scientifically definable. The word denotes not a scientific concept but a vaguely-defined and fluctuating mass of attributes which we wish to find in persons who are to be entrusted with certain vaguely-defined responsibilities. To pretend, in such inquiries, to scientific accuracy is like trying to plot the edge of a fog with a theodolite[15]. We can see, normally, when we are in a fog and when we are not; so we can, after ordinary experience of a person, tell whether he is or is not a person of 'intelligence', and suitable for positions of responsibility. We might invent an instrument which should inform us whether or not we were in a fog; this might be useful at night, much as intelligence-tests are useful when we cannot have prolonged practical experience of a person's character. But such an instrument would not revolutionise meteorology, because, though in some cases indispensable, it would be less reliable than the unaided senses of the normal man. The 'success' of intelligence-tests, of which Professor Spearman speaks so highly, consists not in telling us something we could not otherwise know, but simply in telling us that those people are intelligent who our unaided common-sense tells us are intelligent. Thus intelligence-tests can never widen the sphere of our accurate knowledge; for when they tell us something which our unaided common-sense does not confirm, we call them unsuccessful; and so does Professor Spearman[16].

This passage is of great interest for several reasons. It makes clear, in the first place, that, in Collingwood's view, the concept of intelligence should not be seen *in vacuo* as a pure psychic phenomenon, but rather as related to certain behaviour. The criteriological aspect is apparent here. It is further a striking anticipation of Ryle's description of intelligence as "knowing how", as developed in *The Concept of Mind*[17].

Finally, the general observation could be made that, in view of the fact that the notion of intelligence tests is still highly debatable, it is remarkable that Collingwood was already making in 1923 this fundamental criticism.

In *An Essay on Metaphysics* Collingwood pays extensive attention to psychology. His criticism of its pretensions is severe: it is called by him the 'pseudo-science' of thought. "[W]hen we are told that psychology is the science which tells us how we think", Collingwood asserts, "we must never forget that the word 'think' is being used in a rather special sense. It has lost all suggestion of self-criticism. It has lost all suggestion of an attempt to think truly and avoid thinking falsely. In fact, since this is at bottom what distinguishes thinking from feeling, the word 'think' here simply means feel" (EM, 117). In another passage Collingwood makes clear that psychology is valued by him if it keeps to its own field. What he opposes is its trespassing on others:

> As the science of feeling, psychology is not only a science of respectable antiquity; it is a science with great triumphs to its credit...
> My suspicions are not about this; they are about the status of psychology as the pseudo-science of thought which claims to usurp the field of logic and ethics in all their various branches, including political science, aesthetics, economics, and whatever other criteriological sciences there may be, and finally of metaphysics (EM, 141-142).

Collingwood's objections against the pretensions of psychology as developed in *An Essay on Metaphysics* are primarily of a practical nature. Since critical thinking is set aside, Collingwood argues, psychology plays a vital role in "the propaganda of irrationalism"[18]. This Collingwood considers extremely dangerous for European civilization. In his view the rational dimensions of mind cannot be studied by psychology, but only by philosophy and history. Thus we are warned against the negative practical corollaries of a wrongly-conceived approach of the study of mind in psychology; but Collingwood still regards this study, given the right approach, as important for the opportunity it offers of solving certain problems, based on a real understanding.

10.3. THE USE OF HISTORY

With regard to Collingwood's views concerning the practical dimension of history three aspects may be distinguished: the role of history in achieving human self-knowledge, the use of history in understanding the present and the contribution of history to a proper understanding of human nature. All three will be briefly dealt with[19].

The great value Collingwood attaches to history is demonstrated by the fact that in his *Autobiography* he even goes so far as to assert that the "revolution in historical method... had brought into existence a genuine, actual, visibly and rapidly progressing form of knowledge which now for the first time was putting man in a position to obey the oracular precept 'know thyself'..." (Aut, 116). Though I think the claim that it is only through history that man can know himself to be an exaggeration, Collingwood elsewhere shows what he has in mind. For in the draft of "Human Nature and Human History" he offers the example of becoming conscious of the political tradition that has formed one's mind by studying the political history of one's country. "...[I]n thus coming to know himself he is altering the self that he knows", Collingwood argues. "He is disentangling the various threads of thought which were at first confused into a single pattern; in so doing, he will inevitably discover that some represent ideas no longer valid for the situation in which he stands: self-knowledge becomes self-criticism..."[20]. Likewise, Collingwood contends that by getting acquainted with Chinese civilization "... there is a new self-knowledge which is a development or reconstruction of the self"[21].

The claim that history is the unique way of arriving at human self-knowledge is only understandable within the context of Collingwood's theory of history in general. In this connection the close relation between history *a parte subjecti* and *a parte objecti* should first be mentioned. Both, for Collingwood, are part of the same historical process: the past for him is therefore not a dead past to be contemplated from the outside, but one living in the present and to be understood from within. This living past, however, may only be living unconsciously. It is by historical thought that it can be made conscious. The consciousness realized by historical knowledge makes man aware of the historical process he is part of and enlarges his understanding of the situation he finds himself in. Because of the inseparable unity of thought and action this has the practical corollary that his actions can be made more rational and effective. It is thus only when the various aspects of Collingwood's theory of history are seen in their interrelation that his claim that history provides human self-knowledge can be properly assessed. In the draft of "Human Nature and Human History" this claim is stated in the following clarifying way:

> Historical knowledge ... is the way, and the only way, in which mind knows itself; it is the theory, and the only theory, of human nature, in so far as human nature means mind or rationality. But it is more than mere theory; for theory... means the contemplation by mind of an object presented to it ready-made; and mind, in knowing itself, raises itself to a higher power of rationality or mindhood by becoming self-conscious, and thus possessing in a new manner the self of which previously it was unaware. The self-awareness of mind, which is history, is at the same time the self-enlargement of mind by developing in itself this new power[22].

The practical dimension of history is most clearly discussed by Collingwood in the chapter entitled "The Foundations of the Future" in his *Autobiography*. His starting-point is the contrast that has developed since the sixteenth century between man's power to control nature and his inability to control human situations. For Collingwood this contrast was most poignantly illustrated by the First World War. "The War was an unprecedented triumph for natural science", he observes cynically. "Bacon had promised that knowledge would be power, and power it was: power to destroy the bodies and souls of men more rapidly than had ever been done by human agency before" (Aut, 90). On the other hand, nobody wanted it: "It happened because a situation got out of hand. As it went on, the situation got more and more out of hand. When the peace treaty was signed, it was more out of hand than ever" (Aut, 90). Collingwood is not without reason afraid that the growing gap between the capability to control nature and the incapability to control human situations will eventually lead to disaster. The only way this can be averted, he argues, is not by a change of heart – "more goodwill and human affection" – but by a change in the head – "more understanding of human affairs and more knowledge of how to handle them" (Aut, 92). This understanding and knowledge, Collingwood claims, can only be provided by history. This may seem an extravagant pretension. Yet I am of the opinion that Collingwood's argument is not only in line with his theory of history, but worthy of serious consideration.

Many of our actions are done, Collingwood argues, according to certain rules. The attempt to apply rules is unsuccessful, however, in two cases: when one finds oneself in a situation that is unlike previously known types and when one is not content to refer a situation to a known type, though it could be done (Aut, 103-104). If one believes that actions should always be related to certain rules, one is in these cases at a loss. Collingwood's solution to the problem with which one is confronted in cases like these, could be described as the maxim not to lose one's nerve in such circumstances. For he asserts: "there are situations which, for one reason or another, can be handled without appeal to any ready-made rules at all, so long as you have insight into them. All you need in such cases is to see what the situation is, and you can then extemporize a way of dealing with it which will prove satisfactory. This ... type of case ... [is] of great importance in moral and political life..." (Aut, 101-102)[23].

This insight in a situation, then, is provided by historical knowledge. Though Collingwood does not enlarge on this in his *Autobiography*, he refers to it at various places in his unpublished manuscripts. We have seen, for instance, how he declares in his 1926 lectures that "[h]istory is nothing but the attempt to understand the present by analysing it into its logical components of necessity, or the past, and possibility, or the future"[24]. In "History as the Understanding of the Present" it is likewise asserted that any present possesses the past as its determined part, while free activity is its undetermined part; or, as Collingwood calls it, the past determines the possibilities of the present[25]. Seen from the standpoint of the free agent, he speaks in "What 'Civilization' Means" of the opportunities provided for by the past[26]. The choices made from these opportunities are not determined by the past. "What a man can bequeath is only opportunity", Collingwood contends, "and opportunity is indifferent how it shall be used"[27].

Collingwood's view concerning the importance of history in coping with practical problems can be summed up, therefore, as follows. In cases where one is confronted with situations that cannot be related to certain types, it is only by historical knowledge that a solution can be found. For it may provide insight into the present unique situation by analysing it into its determined and undetermined components, that is, its aspects of possibility and opportunity. The historical approach is therefore a necessary condition to warrant a decision for present action. In this case one is not alarmed, moreover, by the uniqueness of the situation, since the historical dimension of the present is by definition as unique as the present.

I think Collingwood's view is valuable and has not lost its importance. On the contrary, since the present world is changing with an ever growing speed, caused not least by developments in science and technology, it is more and more difficult to relate present situations to known and more general types. The suggestion that historical knowledge can provide insight into present unique situations by distinguishing their determined and undetermined aspects is important. For by pointing out the former, decisions that will lead to blind alleys may be prevented, while by indicating the latter opportunities that are left may be shown, and the corresponding choices that are still open.

Collingwood's theory of history shows its practical relevance most fundamentally, perhaps, in the view of human nature that is implied by it. We have seen how in "Reality as History" the idea is rejected that the historical process is underpinned by a permanent substance of human nature[28]. We find this view already expressed in *Religion and Philosophy*, when Collingwood asserts: "In fact there is no such thing as human nature in the sense of a definite body of characteristics common to every one..." (RPh, 164). In "Reality as History" this view is demonstrated by means of an analysis of the concept of human character. For analogous to human nature, this is often seen as a fixed and permanent entity. We have seen how Collingwood considers this a false conception and how he argues that human character should be equated with man's past[29]. Since this past is never the same and changes in time, man's character changes as well. Man's actions, we have seen, consist of two elements, according to Collingwood: a determined one in the form of his past or character, and an undetermined one in the form of his free will. So present actions are partly free and never fully determined. Since these actions are the past actions of the future, this implies that man's character is partly the product of his own choice.

The same theory holds for the concept of human nature, which could be described as the corporate character of man. In the draft of "Human Nature and Human History" it is asserted, therefore, that "... human nature determines itself, creates itself as this or that kind of human nature, in the perpetuation and development of determinate historical traditions"[30]. In this connection it may seem confusing that Collingwood in his manuscripts on folklore explicitly speaks of certain characteristics common to all men[31]. If these are interpreted, however, as indicating certain common aspects of the pasts of all societies known to us, their assertion is consistent with his theory. We have seen that Collingwood in "Reality as History" asserts that the phrase human nature "... may be *loosely* used as a collective name for those sets or patterns of human activity which we regard, at any given moment, as permanent, and accept as things beyond our power to change"[32] (italics mine). This is not in opposition to the assertion made elsewhere, that "... *strictly* speaking, there is no such thing as human nature; that what has gone by that name is, properly described, not human nature but human history"[33] (italics mine). For the latter assertion implies that non-strictly, that is, loosely speaking, the term human nature might be used[34].

What Collingwood attacks in his criticism of the concept of human nature is its interpretation as a group of permanent and fixed characteristics, analogous to the laws of nature. He thinks it the more urgent to resist this notion, since the science of psychology is based on it. We have seen how Collingwood considers the latter fundamentally misleading as a science of mind. In "Reality as History" psychology is criticized for its attempt to understand man by the same methods as natural science. For the latter, Collingwood argues, is based on the assumption that man as the knower and controller of nature is intelligent, while nature as the known and controlled is unintelligent and mere mechanism. When psychology studies man by means of the methods of natural science, Collingwood points out, the same distinction is presupposed. The consequences of this approach he considers disastrous:

> They [the methods] therefore assume that human nature, as the object upon which they are exercised, is unintelligent. The result is that intelligence itself is converted into unintelligence. For mind, we are given mechanisms which merely enjoy the honorary title of mental; for activity we are given passive reaction to stimulus; for thought, ideas associated automatically in accordance with fixed laws. Since the wealth and dignity of modern man are constituted by his superiority as mind to the mere Nature over which he rules, this reduction of his own mental nature to the level of unintelligent mechanism marks his bankruptcy and degradation[35].

While this criticism is primarily of a moral kind, further on in the same essay Collingwood refers to the practical consequences of the belief in fixed human characteristics as presupposed by psychology. As we have seen, he does this by giving the example of the distinction between considering the habit of smoking as "natural" and as historically conditioned: the first shuts the gates of the future, Collingwood argues, while the latter leaves them open[36]. This difference is then illustrated by another example:

> ...[I]f war as we know it is an institution that has grown up in the course of human history, and if its recurrent appearance as a fact in our world is due to the way in which our political systems have been organized and our political habits shaped, it follows that since history (in this case the history of corporate human action) has created war, history can abolish it; and since in principle it is a thing that can be abolished, the task of men who think it an evil is to discover some way of reorganizing our political life so as to do without it. If war is due to some natural instinct of pugnacity, it cannot be abolished, and the attempt to abolish it will be attended by mischievous consequences.
> The converse does not hold. The failure of one, or two, or a hundred attempts to abolish war goes no way towards proving that war is due to a natural instinct, any more than a man's failure to give up smoking proves that the craving for tobacco is natural[37].

This passage is not only a clear illustration of the practical dimension of Collingwood's historical view of human nature. It is also of great current interest in connection with the views on human culture and society developed by present-day fashionable biology and ethology, culminating in so-called "sociobiology". One can only come to the sad conclusion, however, that these views express an even worse form of naturalism than Collingwood could imagine. For in this case human conduct is reduced, not to permanent and fixed psychological laws, but to neurological and genetical ones. If Collingwood resists the idea of a psychology of religion, he would certainly be horrified by the idea of relating religion to "... an inherited need, represented in gene pools, for some kind of reverence for the sacred..." – to quote from Stuart Hampshire's review[38] of *On Human Nature* by the leading sociobiologist E.O. Wilson[39].

Collingwood's position is eloquently expressed in a letter to the editor of *Philosophy*[40]. This letter is an answer to another by Sir Herbert Samuel, who complains that philosophers no longer discuss urgent practical questions "of personal and social morality, of economic organization, of international relationship"[41]. Collingwood answers that the solution of these problems presupposes two things: "a conviction that the problems can be solved, and a determination that they shall be solved". "Of these two", he adds, "the first is, I think, capable of being provided, in a reasoned form, by philosophy". Without such a reasoned conviction, he argues, "... the will to solve them is so handicapped by doubts within and opposition without, that its chance of success dwindles to vanishing-point". He is especially afraid of "... the defeatist spirit which fears that what we are aiming at is no more than a Utopian dream"[42]. This defeatism Collingwood considers especially fatal if it is based on the notion of a fixed human nature:

> And this fear becomes paralysing when, not content with the status of a natural timidity or temporary loss of nerve, it calls in the help of philosophical ideas, and argues that the evils admittedly belonging to our moral, social, and political life are essential elements in all human life, or in all civilization, so that the special problems of the modern world are inherently insoluble[43].

Against this idea Collingwood develops his own view as follows:

> As the seventeenth century needed a reasoned conviction that nature is intelligible and the problems of science in principle soluble, so the twentieth needs a reasoned conviction that human progress is possible and that the problems of moral and political life are in principle soluble. In both cases the need is one which only philosophy can supply ... What would correspond to the Renaissance conception of nature as a single intelligible system would be a philosophy showing that ... social and political institutions are creations of the human will, conserved by the same power which created them, and essentially plastic to its hand; and that therefore whatever evils they contain are in principle remediable. In short, the help which philosophy might give to our 'dissatisfied, anxious, apprehensive generation' would lie in a reasoned statement of the principle that there can be no evils in any human institution which human will cannot cure[44].

This passage not only shows Collingwood's interest in the practical problems of his time, but also the role he assigned to history in connection with them. This role does not only consist, however, in the equation of human nature with history, but also in the fact that, in Collingwood's view, it is history alone that can provide the proper insight necessary for the solution of present human problems[45]. He is convinced, moreover, that any study of human conduct based not on historical method, but on natural science, is on the wrong track[46]. These views have not lost their relevance. They may all be considered elaborations of Collingwood's notion of a *rapprochement* between philosophy and history.

NOTES

NOTES TO CHAPTER ONE

1. D. Emmet, review of A. Donagan, *The Later Philosophy of R.G. Collingwood* (Oxford, 1962), in *The Philosophical Quarterly* 13 (1963), 371.
2. C.V. Wedgwood, review of R.G. Collingwood, *The Idea of History* (Oxford, 1946), in *The Observer*, 8-9-1946.
3. M. Beloff, review of *The Idea of History*, in *Time and Tide*, 28-9-1946.
4. See chapter 3.3.
5. R.W. Harris, "Collingwood's *Idea of History*", *History* 37 (1952), 1.
6. (Oxford, 1957).
7. Ibid., 122.
8. (Oxford, 1962).
9. (Bloomington and London, 1969).
10. (Toronto, 1970). Besides the books mentioned, there is A. Shalom, *R.G. Collingwood: Philosophe et Historien* (Paris, 1967). This book mainly gives a detailed description of Collingwood's writings and does not try to give a general interpretation of Collingwood as the other works do. In spite of the suggestion made by the title, Collingwood's historical writings are not discussed.
11. Mink, *Mind*, 1.
12. L.J. Goldstein, *Historical Knowing* (Austin, 1976).
13. R. Martin, *Historical Explanation: Re-enactment and Practical Inference* (Ithaca and London, 1977).
14. P. Munz, *The Shapes of Time: A New Look at the Philosophy of History* (Middletown, 1977).
15. See, for instance, F.A. Hanson, *Meaning in Culture* (London, 1975). The anthropologist Thompson asserts that Collingwood "...was isolated and neglected as the tide of interest turned towards Wittgenstein and the elaboration of 'Oxford philosophy'. In this way an enormously attractive and promising line of enquiry was lost and is only now, thirty or more years later, being painfully rediscovered" (M. Thompson, *Rubbish Theory: The Creation and Destruction of Value* (Oxford, 1979). I am thankful to Professor Schuyt for calling these works to my attention.
16. Donagan, *Later Philosophy*, 42-43, 45; H.V. White, review of Donagan, *Later Philosophy*, in *History and Theory* 4 (1965), 246-247.
17. Mink, *Mind*, 265; S. Toulmin, "Introduction", in R.G. Collingwood, *An Autobiography* (Oxford, 1978), XV. One could also refer in this connection to Foucault's notion of "episteme", as developed in *Les Mots et les Choses* (Paris, 1966). The subject of absolute presuppositions and its possible relevance to the science of history will not systematically be discussed in this study. A satisfactory treatment of this notion would require a thorough study of its own. In this connection, moreover, the manuscript on

"Metaphysics" (LM, 1937-38) should also be consulted, since it seems to contain important information on the subject. The scope of this study, however, does not allow a full discussion of this matter.

18. Mink, *Mind*, 7-12. With regard to existentialism Mink refers especially to the conception of freedom. One could also mention in this connection Heidegger's notion of "*geschichtlichkeit*".
19. See chapter 7.3.
20. See chapter 8.7. This has also been pointed out by P. Skagestad, *Making Sense of History: The Philosophies of Popper and Collingwood* (Oslo, 1975), 18-19, 93.
21. Toulmin, "Introduction", XI.
22. Ibid., XIV.
23. In his Introduction to the German translation of Collingwood's *Autobiography* Gadamer maintains: "Auf eine überraschende, fast rätselhafte Weise ist dieser fremde Denker und Schriftsteller für uns kein Fremder, und wenn er jetzt auf deutsch zu uns zu reden beginnt, ist er fast wie ein Heimgekehrter, der seine geistige Heimat draussen, wo er lebte und kämpfte, nie vergass. Diese Heimat is die grosse weiträumige Landschaft der deutschen Romantik und der 'Historischen Schule', Hegels und Schellings, Humboldts, Rankes und Droysens, Schleiermachers und Diltheys, von der das deutsche Philosophieren unserer Jahrzehnte noch immer in unverkennbarer Weise zeugt" (H.G. Gadamer, "Einleitung" in R.G. Collingwood, *Denken: Eine Autobiographie* (Stuttgart, 1955), VII).
24. On 7 April 1931 Collingwood's father wrote to the archaeologist F.G. Simpson: "The last thing is that Robin is down with chicken-pox!" That Collingwood did not recover quickly from this illness is shown by the letter he wrote to Simpson on 3 June 1931: "Come for Monday night by all means. We shall be delighted to see you. But I must warn you, that you will not find me a very bright host. I am not recovering my strength with any great rapidity and at present I can do no kind of work – no reading or writing, no talk of any kind that requires concentration or intelligence. If I try these things, I have fits of complete exhaustion that last a day or more, and my doctor takes a very serious view of my condition, threatening me with a permanent loss of strength following a complete breakdown, unless I take steps to repair my reserves of strength by several months' rest. I tell you this so that you may know that it will be useless to try to talk to me when you come here, and that you can't expect me to do any work for months to come" (Letters in the possession of Dr. Grace Simpson).
25. LM, 1933-1, 17.
26. See chapter 4.10.
27. LM, 1939-40-1.
28. See the end of chapter 9.2.
29. LM, 1934-1.
30. See the beginning of chapter 4.12.
31. LM, 1933-34.
32. Donagan, *Later Philosophy*, 17.
33. See p. 63.
34. Mink, *Mind*, 3-4.
35. Donagan begins his book with the contention: "The first obstacle to understanding Collingwood's later philosophy is his own narrative of its development, which beyond doubt is untrue" (*Later Philosophy*, 1). For Knox, see IH, X-XI, and Mink, *Mind*, 14.

36. I disagree, though, with Mink's view that *Religion and Philosophy* is "the least illuminating of his works" (*Mind*, 16). For one may find in this book many anticipations of Collingwood's mature views.

NOTES TO CHAPTER TWO

1. He had previously translated B. Croce, *The Philosophy of G. Vico* (London, 1913).
2. T.M. Knox, "Collingwood, Robin George", *The Dictionary of National Biography, 1941 - 1950* (Oxford, 1959), 169.
3. See "List of work done", 15 (LM, 1933-1).
4. Reprinted in A. Donagan ed., *Essays in the Philosophy of Art by R.G. Collingwood* (Bloomington, 1964) 5 - 41. The original edition (Kendal, 1922) is also re-edited, with a short "prefatory note" by T.M. Knox (Chichester, 1971). Here Donagan's edition is used.
5. The resemblance of this view to the theory of "absolute presuppositions" as developed by Collingwood in his *Essay on Metaphysics* (Oxford, 1940) is striking. Both Donagan, in *Later Philosophy*, 264, and Rubinoff, in *Collingwood*, 224, have noticed this.
6. The term "historicism" has been the subject of much confusion. It is the translation of the German "*Historismus*", but may also refer to the well-known concept of Popper, to which he gave "the somewhat unfamiliar label 'historicism'" (*The Poverty of Historicism* (London, 1957), 3). When Popper wrote this in 1944, however, the translation of the German "*Historismus*" was just changing from "historism" into "historicism". Lee and Beck mention the fact that in the 1918 edition of Baldwin's *Dictionary of Philosophy and Psychology* "historism" appeared, but not "historicism", while in D.D. Rune's *Dictionary of Philosophy* (New York, 1942) the reverse is the case (L.E. Lee and R.N. Beck, "The Meaning of 'Historicism'", *American Historical Review* 59 (1954), 568) Mandelbaum says that "...[t]he term 'historicism' was adopted into the English language in the late 1930's and the 1940's both in the United States and in England" (M. Mandelbaum, "Historicism", in P. Edwards ed., *The Encyclopaedia of Philosophy* (New York, 1967), Vol. 4, 24). Donagan comes to the same conclusion, saying that de Ruggiero used the term "historicism" for the first time in 1932 (A. Donagan, "Popper's Examination of Historicism" in P.A. Schilpp ed., *The Philosophy of Karl Popper*, Vol. 2 (La Salle, 1974), 906). The commentators mentioned here apparently have not noticed that Beard had already in 1935 explicitly used the term "historicism", saying that: "... it may be well to Anglicize *Historismus*". (Ch. A. Beard, "That Noble Dream", *American Historical Review* 41, (1935), 77, reprinted in R.N. Nash ed., *Ideas of History*, Vol. 2 (New York, 1969), under the title "The Case for Historical Relativism", 162 - 176, the quotation being on p. 165). According to Donagan, however, the same Beard again translated "*Historismus*" as "historism" in 1937 (Donagan, *Popper*, 906). Anyhow, when Collingwood used the term "historicism" in *Ruskin's Philosophy* in 1919 he seems to be the first one to do so, which was not noticed before.
   The present confusion about the term "historicism" is due to the fact that it may refer both to the German "*Historismus*" and to Popper's concept. Donagan defends Popper against his critics by saying that Popper was justified in thinking when he wrote *The Poverty of Historicism*, that

"*Historismus*" was translated by "historism", it being therefore incorrect to blame him for using the term "historicism". "I contend", Donagan says, "that it is anachronistic to take his account of historicism as a false image of *Historismus.* It is true that, since Popper wrote and published *The Poverty of Historicism*, the word 'historicism' has replaced 'historism' as the usual rendering of '*Historismus*': Friedrich Engel-Janosi's *The Growth of German Historicism* (1944) was perhaps the turning point" (Donagan, *Popper*, 908). It so happened, however, that Popper reviewed this same book by Engel-Janosi. In it he makes a remark, which not only undermines Donagan's contention, but adds to the confusion: "Historicism, as the author defines it (and I have no quarrel with his definition *since I have used the term in a similar sense*) has an importance far beyond the problems of historiography" (italics mine) (K.R. Popper, review of F. Engel-Janosi, *The Growth of German Historicism* (Baltimore, 1944), in *Economica*, N.S. 12 (1945), 259).

To make the confusion complete, the German "*Historismus*" recently seems to be translated again by "historism". For the standard work on the subject, F. Meinecke's *Die Entstehung des Historismus* (München und Berlin, 1936), has been translated with the title *Historism: The Rise of a New Historical Outlook* (London, 1972).

7. Rubinoff, *Collingwood*, 27.
8. Mink, *Mind*, 16.
9. N. Rotenstreich, in *Philosophy, History and Politics: Studies in Contemporary English Philosophy of History* (The Hague, 1976), 6, though noticing the difference between *Religion and Philosophy* and *Speculum Mentis*, does not mention dialectic in this connection. According to him "*Religion and Philosophy* stresses the identity of history and philosophy in terms of the programme [i.e. the claim of history to be philosophy], while *Speculum Mentis* stresses the difference between the two realms in terms of the *realization* of the common programme".
10. Mink, *Mind*, 20.
11. See for instance Collingwood's article "Croce's Philosophy of History", which appeared in 1921 and is reprinted in W. Debbins ed., *R.G. Collingwood: Essays in the Philosophy of History* (Austin, 1965), 3 - 22.
12. H.S. Harris, "Introduction", in G. Gentile, *Genesis and Structure of Society* (Urbana, 1960), 14 - 20.
13. See LM, Correspondence.
14. Harris, "Introduction", 8 - 11; J. Passmore, *A Hundred Years of Philosophy* (Harmondsworth, 1968), 299, says about Smith that he " ...turned south for his Idealism, to the writings of B. Croce and G. Gentile.
15. R.G. Collingwood, *An Autobiography* (Oxford, 1939), 18.
16. "List of work done", 16 (LM, 1933-1).
17. LM, Correspondence.
18. Review of R.F. Alfred Hoernlé, *Matter, Life, Mind and God: Five Lectures on Contemporary Thought* (London, 1923), in *The Oxford Magazine* 41 (1923), 423. This review by Collingwood is not signed, but can be found in a list of reviews in his "List of work done", 67 -69 (LM, 1933-1).
19. A. Greppi Olivetti, *Due saggi su R.G. Collingwood: Con un' appendice di lettere inedite di Collingwood a G. de Ruggiero* (Padova, 1977), 97 (see also LM, Correspondence).
20. Rotenstreich, *Philosophy*, 19.
21. Ibid.
22. Rubinoff, *Collingwood*, 113.

23. Ibid., 115.
24. L.J. Goldstein, review of L.O. Mink, *Mind* and L. Rubinoff, *Collingwood*, in *Man and World* 6 (1973), 88.
25. L.J. Goldstein, "Collingwood's Theory of Historical Knowing", *History and Theory* 9 (1970), 4.
26. Rubinoff, *Collingwood*, 113.
27. R.G. Collingwood, "The Devil" in B.F. Streeter et al. (eds.), *Concerning Prayer* (London, 1916), reprinted in L. Rubinoff, *Faith and Reason: Essays in the Philosophy of Religion by R.G. Collingwood* (Chicago, 1968), 212 - 233. The quotation is from p. 231.
28. Ibid., 232.
29. Review of J.N. Figgis, *The Will to Freedom: or the Gospel of Nietzsche and the Gospel of Christ* (London, 1917), in *The Oxford Magazine* 36 (1918), 299. See "List of work done", 67 (LM, 1933-1).
30. According to Collingwood he had destroyed the manuscript after writing his *Autobiography* (Aut, 99). Chapter 2 of it, however, has survived (see LM, 1917).
31. LM, Correspondence.
32. Ibid. Collingwood refers here to *Libellus de Generatione*, an essay he wrote in 1920 (Aut, 99). See also p. 53 of this study.
33. Greppi Olivetti, *Collingwood*, 91 (see also LM, Correspondence).
34. LM, Correspondence.
35. According to Mink, *Mind*, 111 - 113, Collingwood's epistemology is "beyond realism and idealism". Though he also emphasizes Collingwood's dialectical approach he does not associate it with his explicit rejection of realism. For an extensive discussion by Collingwood of the subject, see his manuscript on "Realism and Idealism" (LM, 1936-7).
36. Greppi Olivetti, *Collingwood*, 93 (see also LM, Correspondence).
37. They are reprinted in W. Debbins ed., *Essays*. This edition will be used.
38. Debbins ed., *Essays*, 3 - 22.
39. Ibid., 16.
40. Ibid., 8.
41. One has to keep in mind that when writing *Speculum Mentis* Collingwood had already been active in both archaeological and historical research for at least ten years. Goldstein seems to forget this fact, when he says about Collingwood: "...it is his anti-realism which opens his mind to an appreciation of what historical knowing is, once he has determined to devote himself to the actual study of it" (L.J. Goldstein, "Collingwood's Theory of Historical Knowing", *History and Theory* 9 (1970), 10). Collingwood was already engaged in the actual study of history before being an anti-realist. This corresponds with the description Collingwood himself gave of his development in the *Autobiography*, assigning to his historical and archaeological experience an instrumental role in his attack on realism. There is therefore no reason to endorse Goldstein's view that "[i] t may well be that the order of development is the reverse of that recounted in the *Autobiography*, and that his already discrediting realism made it easier for him to discern the characteristic features of historical knowing" (Ibid.).

    When writing *Speculum Mentis* Collingwood's anti-realism was already developed. On history, however, he still held a realistic point of view. So we come to the paradoxical conclusion that history apparently functioned as a leaven in Collingwood's rejection of realism, while he still held a realistic view on history itself. Though working already for many years in the field of archaeology and history, it is only after 1925, when he concentrated on the epistemological problems concerning history, that his anti-realism extended to the science of history.

42. Debbins ed., *Essays*, 34 - 56.
43. "[N] o single historical problem is ever finally solved", Collingwood writes. "All history at its actual best is the provisional and tentative answer to a question which remains at bottom unanswered. The actual and the ideal do converge; the historian does get nearer to a real knowledge of the infinite world of fact; but they converge asymptotically. The nearer the actual comes to the ideal, the greater becomes the force, generated by this very approach, to prevent a still closer convergence. The more the historian knows, the more acutely he becomes aware that he will never really know anything, and that all his so-called knowledge is to an unverifiable extent erroneous" (NAPhH, 55).
44. Greppi Olivetti, *Collingwood*, 99 (see also LM, Correspondence).
45. Donagan, *Later Philosophy*, 315. Collingwood refers in this letter to an application of his for a chair in Oxford. On 25 March 1928 he writes about it to de Ruggiero: "I hoped to get more time for philosophical work if I could get elected to the vacant chair of Moral Philosophy at Oxford; so I applied for it, Croce kindly giving me a very handsome testimonial; but they did not elect me. They elected a man 20 years older, who richly deserves it: so I have no cause for complaint" (LM, Correspondence).
46. Debbins ed., *Essays*, 90 - 103.
47. See "List of work done", 24 (LM, 1933-1).
48. See note 41.
49. The questions discussed by Collingwood in "The Limits of Historical Knowledge", have recently been made of current interest again by L.J. Goldstein, in his *Historical Knowing* (Austin, 1976). In this book Goldstein argues with vigour against the realist position on historical knowledge, linking it to scepticism, as did Collingwood (see esp. ch. 2). Goldstein's view has started a discussion on the "Constructionist Thesis" versus the "Realist Thesis" in *The Constitution of the Historical Past, History and Theory* 16 (1977) (Beiheft 16), with contributions by P.H. Nowell-Smith, L.J. Goldstein and W.H. Walsh.

    In "Collingwood on the Constitution of the Historical Past", in M. Krausz ed., *Critical Essays on the Philosophy of R.G. Collingwood* (Oxford, 1972), 241 - 267, Goldstein had already put forward the view – crediting it to Collingwood – that historians are only concerned with evidence. When he contends however that "...historical constitution has the function of explaining evidence ..." (Ibid., 266), he seems to go to far, putting in fact the cart before the horse.

    P. Munz, in *The Shapes of Time: A New Look at the Philosophy of History* (Middletown, 1977), also rejects the idea of trying to grasp the "real" past as an illusion: "History will appear to every historian", he quotes from G. Barraclough, "as the story that emerges from the sources he is most familiar with. History is not what happened but what people think happened" (Ibid., 208). "*Res gestae* are unknowable, unsurveyable, and beyond grasp" (Ibid., 209).
50. The classical example is Charles Beard; see among others his "That Noble Dream", *American Historical Review*, 41 (1935), 74 - 87, reprinted in R.H. Nash ed., *Ideas of History*, Vol. 2 (New York, 1969), 162 - 176.
51. Debbins, *Essays*, XXXI - XXXII.
52. E.E. Harris, review of Debbins, *Essays*, in *History and Theory* 5 (1966), 202.

53. E.E. Harris, "Collingwood's Theory of History", *The Philosophical Quarterly* 7 (1957), 35.
54. N. Rotenstreich, "From Facts to Thoughts: Collingwood's Views on the Nature of History", *Philosophy* 35 (1960), 125.
55. See also Rotenstreich, *Philosophy*, 49 - 60.
56. L. Rubinoff, review of Debbins, *Essays*, in *Dialogue* 5 (1966), 472.
57. Ibid., 474.
58. Rubinoff, *Collingwood*, 108 - 112.
59. L.J. Goldstein, "Collingwood's Theory of Historical Knowing", *History and Theory* 9 (1970), 16.
60. *Proceedings of the Aristotelian Society* 26 (1925 - 1926), 135 - 150.
61. Ibid., 149.
62. Ibid.
63. Ibid., 150.
64. Rotenstreich, *Philosophy*, 65 -66, does not come to this conclusion and gives another interpretation. According to him Collingwood wanted to "...transform the time differences into modal ones without realizing, it seems, that it is precisely time that makes this transformation possible" (66). Rotenstreich sees this as an example of Collingwood's wish to "get rid of time".
65. Debbins ed., *Essays*, 121 - 139.
66. L.O. Mink, "Collingwood's Historicism: A Dialectic of Process", in Krausz ed., *Essays*, 157 - 161. In Mink, *Mind*, 261, it is already mentioned. For the distinction between empirical and philosophical concepts, see also chapter 3.1.
67. *The International Journal of Ethics* 36 (1925 - 1926), 162 - 185.
68. Ibid., 162.
69. Ibid., 163.
70. Ibid., 167.
71. Ibid., 166.
72. *Proceedings of the Aristotelian Society* 29 (1928 - 1929), 155 - 176.
73. Ibid., 157.
74. Ibid., 165.
75. This corresponds with Collingwood's own statement in his *Autobiography*, that his ideas on history "...were being worked out for nearly twenty years after I became a teacher of philosophy" (Aut,116) (he became a fellow of Pembroke College in 1912) and that "[t] his train of thought was not complete until about 1930" (Aut,115). Mink has said about this statement of Collingwood that "...it has seemed completely incredible to his interpreters that this 'train of thought' was 'complete' as early as 1930" (Mink, *Mind*, 14). In my opinion, however, there is no reason not to give credence to Collingwood's own interpretation.
76. Debbins ed., *Essays*, 23 - 33.
77. The similarity of this view with the thesis developed by Goldstein in *Historical Knowing* is remarkable. Goldstein distinguishes the "superstructure" of history from its "infrastructure". By the first he means "...that part of the historical enterprise which is visible to nonhistorian consumers of what historians produce. That means the literary product of the historian's work, the final form in which his conclusions are cast ..." By the infrastructure of history he refers "...to the aspect ... of historical work which, for the most part, remains beneath the visibility line of most nonhistorian consumers of what historians produce", it being "...that range of intellectual activities whereby the historical past is constituted in

historical research" (141). He is also of the opinion that "...if one turns to the infrastructure of history, one finds a situation rather like that which obtains in science" (142).

78. Greppi Olivetti, *Collingwood*, 96 (see also LM, Correspondence).
79. Ibid., 96 - 97.
80. *The Vasculum* 9 (1923), 52 - 59. This article is not included in the collection of essays by Debbins on the philosophy of history by Collingwood and has not been noticed before.
81. Ibid., 52.
82. Ibid., 53.
83. Ibid., 54.
84. Ibid., 55.
85. Ibid., 55 - 56.
86. Ibid., 57.
87. Ibid., 57 - 58.
88. Ibid., 58.
89. Ibid.
90. "Science and History" was published in January 1923 and Collingwood began work on *Speculum Mentis* in December 1922 (see "List of work done", 16, (LM, 1933-1).
91. "Science and History", 59.
92. Rotenstreich, *Philosophy*, 68 - 72, says that Collingwood "[i] n the first phase of his system" (referring to *Speculum Mentis*) emphasized the difference between science and history (68). According to Rotenstreich Collingwood came to a "...new understanding of the relation between science and history ... which abolished the former clearcut distinction" (69). He even calls it "the Copernican turn in Collingwood's system" (72). This claim is a rather strange one, because Collingwood's "new understanding", then, would be reached in 1922, in comparison to a position he actually reached at a later date (in 1924). Rotenstreich apparently realizes the inconsistency of his interpretation: "Actually the article which considers the similarity of the two kinds of knowledge is earlier than the book which sées them as different" (69).
93. Debbins ed., *Essays*, 57 - 75.
94. See p. 36.
95. The first sentence of Spengler's book reads: "In diesem Buche wird zum erstenmal der Versuch gewagt, Geschichte vorauszubestimmen".
96. Guido de Ruggiero "Science, History and Philosophy", *Philosophy* 6 (1931), 166 -179.
97. Greppi Olivetti, *Collingwood*, 100 (see also LM, Correspondence).
98. de Ruggiero, "Science", 171.
99. Ibid.
100. Ibid., 171 - 175.
101. Ibid., 176.
102. Ibid., 177.
103. Ibid., 178.
104. "It is all very well to ask for an impartial judge; there are no impartial judges ... We must have an impartial decision, but no one is qualified to give it" (SM,43).
105. Greppi Olivetti, *Collingwood*, 101.
106. R.G. Collingwood, review of Harold Temperley ed., *Selected Essays of J.B. Bury* (Cambridge, 1930), in *The English Historical Review* 46 (1931), 461 - 465. In the "Editor's Preface" of *The Idea of History*,VII, T.M. Knox mentions

that he made use of this review in Part IV, Par. 1 (iv) of *The Idea of History* (147 - 151). The text, however, is not the same as the original review. Here the original review is used.

107. Ibid., 462.
108. Ibid., 464.
109. Ibid., 465.
110. Krausz ed., *Essays*, 154 - 178.
111. Ibid., 161 - 164.
112. Rotenstreich, *Philosophy*, 67.
113. Ibid., 80.
114. Debbins ed., *Essays*, 76 - 89.
115. Ibid., 104 - 120.

NOTES TO CHAPTER THREE

1. "List of work done", 17 (LM, 1933- 1).
2. Letter dated 7 February 1934 (LM,Correspondence).
3. The letters are in the possession of the Oxford University Press. In the letter of 18 October 1939 the Press suggested to Collingwood that two stars should be put on the spine of *An Essay on Metaphysics* as a way of indicating the second number of a series. *An Essay on Philosophical Method* would accordingly get one star.
   In his answer of 19 October Collingwood agrees with this proposal, and continues: "The same problem will arise over *Principles of Art*, when *Principles of History* gets finished. Perhaps it should be considered now in case any more *Principles of Art* want binding before its successor reaches the press. Here the series will run: Philosophical Principles, by R.G.C., I. *The Principles of Art*, II. *The Principles of History*. I don't know how far we go in the announcement of books in preparation: but actually one-third of II is written. Similarly II of Studies in the History of Ideas (*The Idea of History*) is in preparation: I (*The Idea of Nature*) being now in its lecture stage, i.e. being tried on the dog". I am grateful to the Oxford University Press for permission to quote from this letter.
   Since it was arranged that *An Essay on Philosophical Method* (EPhM) and *An Essay on Metaphysics* (EM) would be number one and two in the same series, with respectively one and two stars on the spine, it is surprising to find on the reprints of EPhM and EM respectively two and three stars (the original edition of EM has two stars). Inside the wrapper of EM is written: "This volume continues the series of Philosophical Essays begun in *An Essay on Philosophical Method* (1933), and is to be regarded as Volume III in that series". Two questions may be asked. First, which book could be meant to be Volume II of the series? It cannot be *The Principles of Art*, because we have seen that this book would be Volume I of the series "Philosophical Principles". Second, why has EPhM, as Volume I of the series "Philosophical Essays", two stars on the spine?
4. See LM,1936 - 2.
5. Greppi Olivetti, *Collingwood*, 103 (see also LM, Correspondence).
6. Ibid., 104.
7. See "Log of a Journey in the East Indies in 1938-9", LM, 1938 - 39-1.
8. He visited there, among others, the Dutch and German artists Bonnet and Spies, who lived in Bali.
9. "Historiography", LM,1938- 39- 2, 20.

10. The letters are in the possession of Mrs. V. Collingwood.
11. Letter in the possession of Dr. Grace Simpson (See Appendix IV).
12. See note 3.
13. Bodleian Library, Oxford, Ms. Crawford, 4, 118.
14. *Antiquity*, 1 (1927), 117.
15. The version Knox there gave of it has been an object of controversy among Collingwood's interpreters. For a short description of the main interpretations see chapter 1.2.
16. In his Preface Knox says that for the section on Bury (IH, 147-151) he made use of a book review by Collingwood in *The English Historical Review* (IH, VII). The text in *The Idea of History*, however, differs quite considerably from te original review (see also note 106 of chapter 2). In the contents of the 1936 lectures Bury is also mentioned (pp. 114-118) (see LM, 1936-2).
17. *Proceedings of the British Academy* 22 (1936), 97-127.
18. (Oxford, 1935). This was Collingwood's Inaugural as Waynflete Professor of Metaphysical Philosophy.
19. Inquiries made by the present writer suggest that it was possibly destroyed by the Oxford University Press, as is usual with manuscripts once they are published, not realizing that in this case only a part of it was used.
20. See also chapter 2.5 of this study.
21. In connection with Oakeshott he says: "The third alternative is that it should be a living past, a past which, because it was thought and not mere natural event, can be re-enacted in the present and in that re-enactment known as past" (IH,158).
22. Debbins ed., *Essays*, 15.
23. Ibid., 71.
24. Elsewhere Collingwood says about the activity of thinking in general: "The act of thinking, then, is not only subjective but objective as well. It is not only a thinking, it is something that can be thought about. But, because... it is never merely objective, it requires to be thought about in a peculiar way, a way only appropriate to itself. It cannot be set before the thinking mind as a ready-made object, discovered as something independent of that mind and studied as it is in itself, in that independence. It can never be studied 'objectively', in the sense in which 'objectively' excludes 'subjectively'. It has to be studied as it actually exists, that is to say, as an act" (IH,292).
25. *Time and Tide*, 28-9-1946.
26. C.V. Wedgwood, *The Observer*, 8-9-1946; A.L. Rowse, *The Sunday Times*, 29-9-1946. Rowse had also reviewed Collingwood's *Autobiography*, calling it "the most interesting book that has come out of Oxford for some time" (*The Spectator* 163 (1939), 262). Discussing *The Idea of History*, however, in his *The Use of History* (London, 1946) his assessment was much less kind, considering it remarkable "how obtuse clever men can be" (148).
27. *The English Historical Review* 62 (1947), 84, 85. Not all the reviews were consistent with each other. A striking example is a review in *The Contemporary Review* 171 (1947), 187 - 188, by R. Aris, which comments on the essays in *The Idea of History* that the one "on historical evidence should be read by every historian" (187), while another reviewer contends: "The editor confesses doubt as to the inclusion of some passages; and the reader may regret that Part V, section 3, at least was not omitted" (*The Oxford Magazine* 65 (1945-1946), 113). This section is the one on historical evidence.

28. See for instance the review by Charles Beard in *The American Historical Review* 52 (1947), 704-708.
29. "Explanation in History and Philosophy", *Aristotelian Society, Supplementary Volume* 21 (1947), 33-77.
30. G.J. Renier, *History: Its Purpose and Method* (London, 1950), 45.
31. P. Smith, *The Historian and History* (New York, 1960), 85.
32. T.R. Tholfsen, *Historical Thinking: An Introduction* (New York, 1967), 229.
33. D.H. Fischer, *Historian's Fallacies: Toward a Logic of Historical Thought* (New York, 1970), 195.
34. G.R. Elton, *Political History: Principles and Practice* (New York and London, 1970), 133.
35. K.G. Faber, *Theorie der Geschichtswissenschaft* (München, 1971), 35.
36. A. Marwick, *The Nature of History* (London, 1970), 83.
37. A.J. Toynbee, "R.G. Collingwood's View of the Historian's Relation to the Objects that he Studies", in *A Study of History*, Vol. 9 (London, 1954), 718-737.
38. Ibid., 720-721.
39. Ibid., 732.
40. Ibid., 737.
41. A.E. Murphy, review of *The Idea of History*, in *The Philosophical Review* 56 (1947), 590.
42. C.K. Grant, "Collingwood's Theory of Historical Knowledge", *Renaissance and Modern Studies* 1 (1957), 69.
43. P. Winch, *The Idea of a Social Science and its Relation to Philosophy* (London, 1958), 131.
44. K.O. Apel, *Transformation der Philosophie*, Band 1 (Frankfurt am Main, 1973), 51.
45. W.H. Walsh, "The Character of a Historical Explanation", *Aristotelian Society, Supplementary Volume* 21 (1947), 55.
46. See for instance A.M. MacIver, "The Character of a Historical Explanation", *Aristotelian Society, Supplementary Volume* 21 (1947), 33, 47; Ph. Bagby, *Culture and History: Prolegomena to the Comparative Study of Civilizations* (London, 1958), 68; Renier, *History*, 41 ff.; K. Acham, *Analytische Geschichtsphilosophie: Eine Kritische Einführung* (Freiburg und München, 1974), 31; M. White, *Foundations of Historical Knowledge* (New York, 1965) 3; P.K. Conkin and R.N. Stromberg, *The Heritage and Challenge of History* (New York, 1971), 88ff.

    W.H. Walsh calls the view of Croce and "his follower, R.G. Collingwood" "the standard idealist account of historical knowledge" (*An Introduction to Philosophy of History* (London, 1951, 42). While Walsh is generally sympathetic towards the idealist position (see pp. 48 ff.), this cannot be said of Renier, who speaks of "the pan-idealism of Croce-Collingwood" and Collingwood's "mysticism" (*History*, 75, 81).
47. A. Child, "History as Imitation", *The Philosophical Quarterly* 2 (1952), 200.
48. T.A. Roberts, *History and Christian Apologetic* (London, 1960), 13.
49. J.N. Hartt, "Metaphysics, History and Civilization: Collingwood's Account of their Interrelationships", *The Journal of Religion* 33 (1953), 209.
50. P. Gardiner, *The Nature of Historical Explanation* (Oxford, 1952), 49.
51. Fischer, *Historical Fallacies*, 197.
52. Walsh, *Introduction*, 49.
53. Ibid., 50.
54. Ibid., 58-59. Walsh seems here less positive with regard to Collingwood's theory than in his earlier assessment (see p. 82).

55. B. Baumann, *Imaginative Participation: The Career of an Organizing Concept in a Multidisciplinary Context* (The Hague, 1975), 131.
56. M. Ginsberg, "The Character of a Historical Explanation", *Aristotelian Society, Supplementary Volume* 21 (1947), 70-71.
57. Gardiner (*Historical Explanation*, 49) says of Collingwood's theory that "the behaviour of human beings *en masse* rather than *qua* individuals is not easily covered by it", while White (*Foundations*, 148) contends: "Insofar as he deals with social behavior, with social events, with group action, the historian can hardly be described as someone who seeks the thoughts in the mind of *society*. The Collingwoodian scheme is too rationalistic as a model for the explanation of all individual behavior, and it leads to difficulties if we wish to analyze the distinctive task of the social historian".
58. E.H. Carr, *What is History* (London, 1961), 46.
59. G. Leff, *History and Social Theory* (Garden City, 1971), 24, 25.
60. Fischer, *Historical Fallacies*, 197.
61. Donagan, *Later Philosophy*, 206.
62. Mink, *Mind*, 174-178.
63. *Philosophy* 27 (1952), 211-220.
64. Ibid., 213.
65. Ibid., 214.
66. Ibid., 213.
67. Ibid., 216. It is obvious that Gardiner's criticism implicitly relies on Ryle's theory of mind, as developed in *The Concept of Mind* (London, 1949).
68. Gardiner, *Historical Explanation*, 47-48.
69. J. Cohen,"A Survey of Work in the Philosophy of History, 1946-1950", *The Philosophical Quarterly* 2, (1952), 173. How differently Collingwood's philosophy of mind may be interpreted is demonstrated by the fact that Skagestad in his treatment of the subject comes to a conclusion which is exactly opposite that of Gardiner and Cohen. He sums up Collingwood's position as saying that "[t]he life of the mind belongs to the public sphere; there is no privileged access and no right of privacy to protect against intrusion". He speaks in this connection of "Collingwood's monism" and adds: "like Ryle's, though perhaps even more radically" [sic]. (P. Skagestad, *Making Sense of History: The Philosophies of Popper and Collingwood* (Oslo, 1975)), 65-66. It should be added, though, that Skagestad's view is not based on *The Idea of History*, but on passages from *Religion and Philosophy*. As will be shown hereafter, the views of Dray, Walsh and Grant already go in the direction of Skagestad's.
70. *Revue Internationale de Philosophie* 11 (1957), 420-432.
71. Ibid., 424-425.
72. Ibid., 425.
73. Ibid., 426.
74. Ibid., 427. Walsh is also of the opinion that Collingwood's theory need not contradict Ryle's (*Introduction*, 55-56) and according to Grant "Collingwood's theory can ... be shown to be consistent with Ryle's main thesis" ("Collingwood's Theory of Historical Knowledge", 79).
75. Walsh, *Introduction*, 53-54.
76. Ibid., 54.
77. W.H. Dray, "Historical Understanding as Re-thinking", *University of Toronto Quarterly* 27 (1958), 204-205.
78. Debbins ed., *Essays*, 86.
79. Rotenstreich, *Philosophy*, 13-14.
80. Renier, *History*, 47.

81. Roberts, *History*, 13.
82. Grant, "Collingwood's Theory of Historical Knowledge", 69.
83. (London, 1978).
84. Ibid., 26.
85. Ibid., 26-27. Against this objection one could retort, however, that nothing would prevent people from trying to climb them. Though it may objectively not be possible to conquer the world, it is not precluded that the actions of a megalomaniac should still be guided by this idea.
86. W.H. Walsh, "R.G. Collingwood's Philosophy of History", *Philosophy* 22 (1947), 155. Immediately following this passage Walsh expresses his doubt, however, "as to what exactly it is that Collingwood is trying to express". Since Collingwood is obviously against the idea of cause-and-effect relations in history he wonders if it is more in the direction of Toynbee's notion of challenge-and-response, but then immediately rejects this hunch because of the quasi-biological background to Toynbee's theory. It being also implausible that Collingwood would maintain that all human actions are the autonomous product of human reason, Walsh remains in doubt as to the right interpretation of his ideas on this subject. We will see that, notwithstanding all the subsequent discussions, there is still reason for this doubt.
87. E.E. Harris, "Collingwood's Theory of History", *The Philosophical Quarterly* 7 (1957), 40.
88. Donagan, *Later Philosophy*, 204.
89. Mink, *Mind*, 171-172.
90. R. Bultmann, *Geschichte und Eschatologie* (Tübingen, 1958), 166. "These conditions and events within nature, so far as they are relevant for human life and history, may be called *encounters, sufferings.* Not only human actions but also human sufferings belong to history. It may be wondered whether sufferings do not always set in motion actions. Yet sufferings not only set in motion actions, but being human and distinct from mere natural or mechanical encounters, in a certain sense they are also actions, an action as reaction." This passage is incomplete in the English edition (R. Bultmann, *History and Eschatology* (Edinburgh, 1957), 140).
91. For a criticism of Bultmann's treatment of Collingwood see J. Hopkins, "Bultmann on Collingwood's Philosophy of History", *The Harvard Theological Review* 58 (1965), 227-233.
92. W.H. Dray, "R.G. Collingwood et la Connaissance Historique", *Dialogue: Revue Canadienne de Philosophie* 17 (1978), 673.
93. Ibid.
94. W.H. Walsh, "The Character of a Historical Explanation", 55.
95. J. Cohen, "A Survey of Work in the Philosophy of History, 1946-1950", 177.
96. Hayden V. White, "Collingwood and Toynbee: Transitions in English Historical Thought", *English Miscellany* 8 (1957), 166.
97. Marwick, *History*, 83. Similar interpretations may be found for instance in E.C. Rust, *Evolutionary Philosophies and Contemporary Theology* (Philadelphia, 1969), 30; L.B. Rasmussen, *Two Essays on the Scientific Study of History* (Bern and Frankfurt/M, 1975), 30; Conkin and Stromberg, *Heritage and Challenge*, 88.
98. W.B. Gallie, *Philosophy and the Historical Understanding* (New York, 1964), 18.
99. P. Munz, *The Shapes of Time: A New Look at the Philosophy of History* (Middletown, 1977), 68.

100. Van Austin Harvey, *The Historian and the Believer: The Morality of Historical Knowledge and Christian Belief* (London, 1966), 92. On the same page he says of the doctrine that its adherents "sometimes talk as if the historian had some special intuitive powers by virtue of which he could 'get inside' other minds in a self-authenticating fashion".
101. M.C. D'Arcy, *The Sense of History: Secular and Sacred* (London, 1958), 31.
102. Bagby, *Culture and History*, 6.
103. R.G. Shoemaker, "Inference and Intuition in Collingwood's Philosophy of History", *The Monist* 53 (1969), 101, 103.
104. Walsh, *Introduction*, 57.
105. Ibid., 58. Responding to the objections by Donagan and Dray against the intuitive version Walsh changed his view on the subject in a later edition (London, 1967). In an "additional note" (70-71) he declares: "I should wish myself to stress that Collingwood's aim was to lay bare the peculiar character of historical knowledge, and would admit that his doctrine could be reconstructed without any reference being made to intuition ..." He is still of the opinion, though, that the inference that Collingwood thought that historical understanding must be immediate is "entirely natural". Nielsen reports that in conversations with her in 1975 both Walsh and Gardiner have abandoned their intuitionistic positions (M.H. Nielsen, review of Skagestad, *Making Sense of History*, in *Inquiry* 22 (1979), 488).
106. A. Donagan, "The Verification of Historical Theses", *The Philosophical Quarterly* 6 (1956), 193-208.
107. Ibid., 197.
108. Ibid., 196.
109. Donagan, *Later Philosophy*, 216.
110. W.H. Dray, "R.G. Collingwood and the Acquaintance Theory of Knowledge", *Revue Internationale de Philosophie* 11 (1957), 420.
111. W.H. Dray, review of Donagan, *Later Philosophy*, in *The Canadian Historical Review* 45 (1964), 130.
112. L.J. Goldstein, "Collingwood on the Constitution of the Historical Past", in M. Krausz ed., *Essays*, 247, 253-254. See also Goldstein's remark in *Man and World* 6 (1973), 85, reviewing Mink, *Mind*, and Rubinoff, *Collingwood*.
113. Atkinson, *Knowledge and Explanation*, 27. For similar rejections of the intuitive version see Debbins' "Introduction", in Debbins ed., *Essays*, XXVIII, W.G. Runciman, *A Critique of Max Weber's Philosophy of Social Science* (Cambridge, 1972), 19, and R. Martin, *Historical Explanation: Re-enactment and Practical Inference* (Ithaca, 1977), 49-51, 57-58, 61-62.
114. Munz, *The Shapes of Time*, 66.
115. Ibid., 312.
116. Ibid., 67.
117. Ibid., 312.
118. Ibid., 313.
119. L.B. Cebik, "Collingwood: Action, Re-enactment, and Evidence", *Philosophical Forum* 2 (1970), 68.
120. N. Sykes, "Some Current Conceptions of Historiography and Their Significance for Christian Apologetic", *The Journal of Theological Studies* 50 (1949), 32. Sykes continues that "its relevance may be succinctly indicated by reference to the movement in biblical study usually known as 'the quest of the historical Jesus'". "If ... the validity of Collingwood's argument be accepted", he says further, "the historian is not concerned to ask for fact without interpretation ... the historian of Christianity must be interested in the trial and crucifixion of Jesus in relation to the claim to

Messiahship which the evangelists advance as the cause of that episode ... From this standpoint it may be possible to see more clearly the fundamental issues involved in the claim of Christianity to rest upon a reliable historical tradition preserved in the gospels. First, there is no *necessary* contradiction between what former critics sought as 'a purely historical picture of Jesus' and the gospel portrait 'always of Jesus regarded as the Christ', since it may well prove to be the case that the purely historical Jesus did in fact believe himself to be the Messiah, and that therefore the gospels would be historically justified in presenting the facts of his life in the framework of this Messianic interpretation" (33-34). It is remarkable how far Sykes' contention, based on his interpretation of *The Idea of History*, corresponds to Collingwood's view, as expressed in a passage in *Religion and Philosophy*: "The 'historical Jesus' can never solve the problem of Christianity, because there never was a 'historical' Jesus pure and simple; the real Jesus held definite beliefs about God and himself and the world; his interest was not historical but theological. By considering him as a mere fact in history, instead of also an idea in theology, we may be simplifying our task, but we are cutting ourselves off from any true understanding and sharing of his consciousness ... the true task of historical theology is to find out not only what was said, but what was meant; what current Judaism, to begin with, meant by its formulae, and how far its meaning was a satisfactory theology. Then we should be in a position to understand from within the new doctrines of Jesus, and really to place ourselves at the fountain-head of the faith. To speak of studying the mind of Jesus from within may seem presumptuous; but no other method is of the slightest value" (RPh, 43). We find here already *in nuce* Collingwood's mature view of history. At that time, however, he still adhered to a realist conception of history and the quoted passage is part of an argument in which "a history of the Church" is opposed to "philosophical theology" and the limits of "pure history" – it being equated to "historical positivism" – are pointed out.

121. A.E. Murphy, review of *The Idea of History*, in *The Philosophical Review* 56 (1947), 590.
122. L. Strauss, "On Collingwood's Philosophy of History", *The Review of Metaphysics* 5 (1952), 566.
123. Ibid., 575.
124. Ibid.,
125. Toynbee, *A Study of History*, Vol. 9, 717-737.
126. Ibid., 733.
127. Ibid., 736.
128. Renier, *History*, 48. Fischer calls Collingwood's argument in the Epilegomena of *The Idea of History* "a sore trial to a serious reader" (*Historical Fallacies*, 196).
129. Gardiner, "The 'Objects' of Historical Knowledge", 212.
130. Gardiner, *Historical Explanation*, 39.
131. Ibid., 48, 49. C.E. Kanichai, *R.G. Collingwood's Philosophy of History: A Critical Exposition and Evaluation* (Bangalore, 1972), 47, also interprets the re-enactment doctrine as an example of the acquaintance theory: "It is interesting to notice that Collingwood, who constantly fought against the compresence or acquaintance theory in historical knowledge has, as if unawares, come to employ it when he describes historical knowledge as the re-enactment of past thoughts in the mind of the historian".
132. M. Mandelbaum, review of *The Idea of History*, in *The Journal of Philosophy* 44 (1947), 186.

133. R. Flenley, review of *The Idea of History*, in *The Canadian Historical Review* 28 (1947), 71.
134. G. Buchdahl, "Logic and History: An Assessment of R.G. Collingwood's *Idea of History*", *Australasian Journal of Philosophy* 26 (1948), 109.
135. K.M. Martin, "Caesar and Collingwood as Historians", *Latomus: Revue d'Etudes Latines* 28 (1969), 164.
136. Ibid., 167.
137. Roberts, *History*, 9.
138. Grant, "Collingwood's Theory of Historical Knowledge", 74.
139. Ibid., 76.
140. L.B. Cebik, "History's Want of Authority - Some Logical and Historical Speculations", *Southern Journal of Philosophy* 8 (1970), 150.
141. Ibid., 151.
142. L.J. Goldstein, "Collingwood on the Constitution of the Historical Past", 241-267.
143. Ibid., 245.
144. Ibid., 250.
145. Ibid., 261.
146. Ibid., 266. In his review of Mink, *Mind* and Rubinoff, *Collingwood* in *Man and World* 6 (1973), 89, Goldstein also speaks of the re-thinking of past thoughts as a method.
147. See p. 95.
148. Donagan, "The Verification of Historical Theses", 199-200.
149. Ibid., 203.
150. Ibid.
151. Ibid., 205.
152. Ibid., 206.
153. Ibid.
154. (Oxford, 1957), 118-155.
155. W.H. Dray, "R.G. Collingwood and the Acquaintance Theory of Knowledge", 420-432.
156. Ibid., 430.
157. Ibid., 431.
158. Ibid., 432.
159. *University of Toronto Quarterly* 27 (1958), 200-215.
160. Ibid., 211-212.
161. Ibid., 212.
162. "Such an agent would have to be regarded by the historian as 'irrational' in a stronger sense than the agent who, although he misconceived his situation, nevertheless acted in the way required by the situation as he conceived it", Dray says.
163. Ibid., 213-214.
164. Dray, *Laws and Explanation*, 121-122.
165. Ibid., 128. Dray emphasizes that a rational explanation is based on empirical research: " ... it has an inductive, empirical side, for we build up to explanatory equilibrium *from the evidence*" (Ibid., 129). For the concept of rational explanation see also Dray's *Philosophy of History* (Englewood Cliffs, 1964), 10-12.
166. W.H. Dray, "R.G. Collingwood et la Connaissance Historique", 659-682.
167. Ibid., 676-677.
168. Ibid., 679-680.
169. Shoemaker, "Inference and Intuition in Collingwood's Philosophy of History", 112-113.

170. S. Toulmin, *Human Understanding*, Vol. 1, (Oxford, 1972), 491.
171. Rubinoff, *Collingwood*, 282-283.
172. P. Gardiner, "Historical Understanding and the Empiricist Tradition", in B. Williams and A. Montefiore eds., *British Analytical Philosophy* (London, 1966), 277.
173. Ibid., 277-278. See also note 105 of this chapter.
174. R. Martin, *Historical Explanation: Re-enactment and Practical Inference* (Ithaca, 1977), 50.
175. Ibid., 51. Grant also distinguishes "two complementary stages of inference and understanding", but for him only the latter is re-enactment (see p. 98). Martin would certainly object, however, to re-enactment being described as "aesthetic", as Grant does. Donagan, on the other hand, makes a distinction between the discernment and discovery of past thoughts, re-enactment being connected with the first one (see p. 101).
176. Ibid., 51-52. See also pp. 53-54, where he explicitly refers to Dray.
177. Walsh, *Introduction*, 92-93.
178. In *Objective Knowledge: An Evolutionary Approach* (Oxford, 1972), 153-190.
179. Ibid., 154.
180. Ibid., 170.
181. Ibid., 176-177.
182. Ibid., 177.
183. Ibid., 178.
184. Ibid.
185. Ibid., 187.
186. Ibid., 188.
187. Ibid. In "A Pluralist Approach to the Philosophy of History", in E. Streissler et al.(eds.), *Roads to Freedom: Essays in Honour of F.A. von Hayek* (London, 1969), 181-200, Popper develops the same argument against Collingwood.
188. Skagestad, *Making Sense of History*, 55.
189. See note 69.
190. Skagestad, *Making Sense of History*, 87.
191. Ibid., 89.
192. Shoemaker, "Inference and Intuition in Collingwood's Philosophy of History", 110.
193. Ibid., 110-111.
194. W. Debbins, "Introduction", in Debbins ed., *Essays*, XXVI.
195. Munz, *The Shapes of Time*, 44-45.
196. Ibid., 177.
197. Ibid., 177-178.
198. Collingwood's philosophy of mind is most fully worked out by Mink in chapter 4 of his *Mind, History and Dialectic*. A shorter description of it may be found in "Collingwood's Dialectic of History", *History and Theory* 7 (1968), 7-17. (See also *Mind, History and Dialectic*, 162-170, and "Collingwood's Historicism: A Dialectic of Process", in Krausz ed., *Essays*, 164-167).
199. Mink, "Collingwood's Historicism", 167.
200. Mink, "Collingwood's Dialectic of History", 14.
201. R. Stover, *The Nature of Historical Thinking* (Chapel Hill, 1967), 106.
202. Ibid.
203. W.G. Runciman, *A Critique of Max Weber's Philosophy of Social Science* (Cambridge, 1972), 20.
204. Martin, "Caesar and Collingwood as Historians", 167.

205. Dray, "Historical Understanding as Re-thinking", 206.
206. A.L. Burns, "Ascertainment, Probability and Evidence in History", *Historical Studies: Australia and New Zealand* 101 (1951), 330.
207. Mink, *Mind*, 189.
208. Dray, "Historical Understanding as Re-thinking", 206.
209. Ibid., 209.
210. Ibid.
211. Donagan, *Later Philosophy*, 182.
212. Ibid., 182-185.
213. Ibid., 185.
214. Ibid.
215. Ibid., 191.
216. W.H. Dray, "Singular Hypotheticals and Historical Explanation", in Llewellyn Gross ed., *Sociological Theory: Inquiries and Paradigms* (New York, 1967), 192.
217. Ibid., 193.
218. Ibid., 194-197.
219. Martin, *Explanation*, 14-15.
220. Ibid., 55.
221. Ibid., 71.
222. Ibid., 97.
223. Ibid., 148.
224. Ibid., 149.
225. Ibid.
226. Ibid., 150.
227. Ibid., 154.
228. Ibid., 150.
229. W.H. Dray, review of Martin, *Explanation*, in *The American Historical Review* 83 (1978), 1220.
230. Donagan, *Later Philosophy*, 200.
231. Walsh, "The Character of a Historical Explanation", 54.
232. Donagan, "The Verification of Historical Theses", 194.
233. "I should ... no longer wish to rely on the passage from *The Idea of History* ... to show that Collingwood believed thought to be self-explanatory, as Donagan has convinced me that the word 'it' at the end of the second sentence was intended to refer back to 'event' in the first" (Walsh, *Introduction* (London, 1967)), 71).
234. Dray, "Singular Hypotheticals and Historical Explanation", 188. Collingwood says that the historian investigates "not mere events (where by a mere event I mean one which has only an outside and no inside) but actions, and an action is the unity of the outside and inside of an event" (IH,213).
235. Dray, "Historical Understanding as Re-thinking", 207.
236. Dray, "Singular Hypotheticals and Historical Explanation", 188.
237. Ibid., 190.
238. Ibid., 190-191. For a recent discussion by Dray of the "what is why" question see his "R.G. Collingwood et la Connaissance Historique", 667-671.
239. Donagan, *Later Philosophy*, 203.
240. Ibid.
241. Ibid., 268.
242. Ibid.
243. See pp. 111 - 112.
244. J.W.N. Watkins, "On Explaining Disaster", *The Listener*, 10 -1- 1963, 69-70.

245. J.W.N. Watkins, "Imperfect Rationality", in R. Borger and F. Cioffi eds., *Explanation in the Behavioural Sciences* (Cambridge, 1970), 167-217.
246. Ibid., 209-210. Watkin's discussion of Collingwood is dealt with by I.C. Jarvie, *Concepts and Society* (London, 1972), 24-29.
247. Harvey, *The Historian and the Believer*, 70-89, discusses, among others, Bradley's and Collingwood's positions.
248. Walsh, "The Character of a Historical Explanation", 60.
249. Ibid.
250. Ibid., 62.
251. Ibid., 65.
252. Walsh, *Introduction*, 64-71. See also Walsh, "R.G. Collingwood's Philosophy of History", 158-160.
253. Ibid., 57.
254. Ibid., 57-58. A few pages further on Walsh refers again to the more comprehensive generalizations: [I] n addition to the specific generalizations which historians assume, each for his particular purposes, there is also for each a fundamental set of judgments on which all his thinking rests. These judgments concern human nature: they are judgments about the characteristic responses human beings make to the various challenges set them in the course of their lives, whether by the natural conditions in which they live or by their fellow human beings" (65).
255. Buchdahl, "Logic and History", 113.
256. Acham, *Analytische Geschichtsphilosophie*, 114-115. That Acham speaks of "psychic phenomena" with regard to motives clearly demonstrates that he has not understood at all what the theories of Collingwood and Dray amount to.
257. Grant, "Collingwood's Theory of Historical Knowledge", 74.
258. Martin, *Explanation*, 46.
259. Ibid. In contrast with this statement, however, Martin asserts on Collingwood's position two pages before: "The historian's move is not in the direction of generalization at all, even where generalization might be restricted to a nonuniversal scope" (44).
260. Ibid., 47, 63.
261. Walsh, *Introduction*, 94.
262. Renier, *History*, 215.
263. D.M. Mackinnon, review of *The Idea of History*, in *The Journal of Theological Studies* 48 (1947), 252.
264. Donagan, "The Verification of Historical Theses", 200.
265. Cohen, "A Survey of Work in the Philosophy of History, 1946-1950", 172.
266. Walsh, *Introduction* (1967), 71.
267. Ibid., 58.
268. H. Fain, *Between Philosophy and History: The Resurrection of Speculative Philosophy of History within the Analytic Tradition* (Princeton, 1970), 151, 155.
269. M. Mandelbaum, review of *The Idea of History*, in *The Journal of Philosophy* 44 (1947), 187.
270. Hartt, "Metaphysics, History and Civilization", 209.
271. B. Verhaegen, *Introduction à l'Histoire Immédiate: Essai de Méthodologie Qualitative* (Gembloux, 1974), 39.
272. Walsh, "The Character of a Historical Explanation", 65. See also Walsh, "R.G. Collingwood's Philosophy of History", 159. In his *Introduction* Walsh asserts: "Whether any reputable philosopher advocates a thoroughgoing scepticism about historical knowledge I do not know. But Collingwood,

however inconsistent it might be with the rest of his theory, came near to doing it ..." (109).

273. It is interesting to note that this passage is not to be found in the manuscripts. In fact it is a quotation from a private letter of Collingwood to Knox. This information the present writer got from Knox. The passage is used by Knox in his argument that Collingwood became a relativist in his later years. In this connection he also quotes Collingwood as saying that "philosophy as a separate discipline is liquidated by being converted into history" (IH,X) and that "history is the only kind of knowledge" (IH,XII). Knox's interpretation has become very influential. It is contested, among others, by Rubinoff, "Collingwood's Theory of the Relation Between Philosophy and History: A New Interpretation", *Journal of the History of Philosophy* 6 (1968), 363-380. According to Rubinoff Collingwood "at no time did ... subscribe to the doctrines of radical historicism" (365-366).

274. Carr, *What is History*, 20.

275. A.M. MacIver, "The Character of a Historical Explanation", *Aristotelian Society, Supplementary Volume* 21 (1947), 33.

276. Ibid., 48.

277. Strauss, "Collingwood's Philosophy of History", 575.

278. Munz, *The Shapes of Time*, 195.

279. Strauss, "Collingwood's Philosophy of History", 579.

280. Munz, *The Shapes of Time*, 195.

281. Martin, "Caesar and Collingwood", 170.

282. Elton, *Political History*, 133.

283. Martin, *Explanation*, 63. In *The Idea of History* Collingwood says that "Bradley rightly saw that the historian's criterion is something which he brings with him to the study of the evidence, and that this something is simply himself" (IH,139). In another place Collingwood declares that "... so far from relying on an authority other than himself, to whose statements his thought must conform, the historian is his own authority and his thought autonomous, self-authorizing, possessed of a criterion to which his so-called authorities must conform and by reference to which they are criticized" (IH,236). In this connection Collingwood speaks of "the autonomy of historical thought" (IH,236) and "the historian's autonomy" (IH,237).

284. Ibid., 64.

285. Goldstein, "Collingwood's Theory of Historical Knowing", 22.

286. Ibid., 23.

287. Dray, "Historical Understanding as Re-thinking", 211.

288. Ibid.

289. Dray, "Collingwood and the Acquaintance Theory of Knowledge", 431; Shoemaker "Inference and Intuition in Collingwood's Philosophy of History", 113. Shoemaker contends: "Once the historian has re-thought another man's thought, he has no positive assurance ... that he has done it correctly" (Ibid, 108). He is therefore of the opinion that "Collingwood did *not* see historians as infallible" (ibid., 113). R. Martin is of the same opinion: "There is no way in which re-enactment can ever verify, in the sense of make true infallibly or beyond doubt, the facts that the investigator has adduced in his explanation ... Collingwood's analysis was not designed to show that a particular explanatory re-enactment could ever be beyond doubt or conclusive in any final sense" (*Explanation*, 56-57).

1. Quotations may be made with the permission of Mrs. Teresa Smith, daughter of R.G. Collingwood, who retains ownership of them, but publication or photocopying is not allowed. The reason for this is to be found in Collingwood's will, in which he declares himself against publication of his unpublished manuscripts.
2. Besides these, the Bodleian Library has in its possession another group of manuscripts that has been available for consultation only from May 1980. These contain, among others, Collingwood's lectures on the ontological proof, and his notes and lectures on ethics of the years 1921, 1923, 1928, 1932, 1933 and 1940. At the time of writing these manuscripts could not yet be consulted. They are included, however, in the List of Manuscripts.
3. LM, 1933-34-A, 1b.
4. LM, 1919-3, 2-3.
5. Ibid., 4.
6. Ibid., 5-7. It is surprising in this connection that, according to Knox, Collingwood "used to say that his favourite philosopher was Plato" (IH, VIII).
7. Ibid., 9.
8. Ibid., 10.
9. See pp. 25-26.
10. LM, 1920-21-2, 1-2.
11. Ibid., 2.
12. Ibid., 3.
13. Ibid., 4-5.
14. Ibid., 6.
15. Ibid., 7.
16. Ibid., 8.
17. In *Speculum Mentis* Collingwood adheres to the second alternative in the objectivist view, which he had refuted in his earlier essay: "But the real prejudice against which we have to guard ... is the more dangerous for the scientific rigour upon which we plume ourselves when reconstructing history ... It is precisely our own prejudice ... We know that just as we remember not what happened but what we want to remember, as we perceive not what is 'there' but what we attend to, so we reconstruct history not as it was but as we choose to think it was. We are all ready to see this fault in other historians, but we all commit it ourselves, none of us more so than those who pride themselves on the 'objectivity' of their methods. And yet to surrender this 'objectivity', to confess frankly that our histories are nothing but an expression of personal points of view, this is the very cynicism of history, the conscious acquiescence in what has now become a deliberate fraud. If every history is an historical romance, we may leave out the adjective and frankly identify history with art. As long as we pretend to write history, we must claim access to the fact as it really was. This fact, we have seen, is inaccessible. History as a form of knowledge cannot exist" (SM, 237-238). With the inaccessibility of historical fact in the one but last sentence Collingwood refers to an argument he had made a few pages before. He argues there that historical facts must be seen in their context, which is a universal one. Since this context, however, is always incomplete "we can never know a single part as it actually is" (SM,231). This argument is in fact a third possibility in the objectivist view. So in the above quoted passage Collingwood combines two of them.
18. Debbins ed., *Essays*, 44.

19. Ibid., 46.
20. *Proceedings of the Aristotelian Society*, New Series 26(1925-1926), 135-150.
21. LM,1925-2, 14-15.
22. Ibid., 15.
23. See pp 35-36.
24. *Proceedings*, 150.
25. Ibid., 146.
26. For the lectures given by Collingwood, see Appendix II.
27. In April 1927 Collingwood stayed a few weeks in Rome, in the house of de Ruggiero. See LM, Correspondence, letters to de Ruggiero of 2-1-1927, 16-1-1927, March 1927, 16-4-1927.
28. LM,1927-1, XVIII.
29. Ibid., XIX.
30. Ibid., XXV.
31. Ibid., XXII-XXIII.
32. Debbins ed., *Essays*, 128.
33. LM,1927-1, XXIII-XXV. Mink has also pointed out the arbitrariness and therefore futility of all attempts to define the empirical concept of history. The question "What is history", he contends, "cannot be answered if it is taken to mean 'What characteristics are common and peculiar to everything that is called history (or that has been done by people called historians, or that is done by people who are professional academic historians, etc.)?' A question of this form is impossible to answer not just because such a wide variety of things have been called history – and even by people called historians – but rather because no matter how circumspect and carefully qualified your generalization, it can be confuted in an instant by a university president with original ideas and a post to fill – or even, for that matter, by an historian with original ideas" (*Mind*, 161).
34. Ibid., XII-XIII.
35. Ibid., XIII-XIV.
36. Ibid., XIV.
37. Ibid., XIV-XV.
38. Ibid., XV-XVI.
39. Ibid., XVI.
40. LM,1926-2, 2.
41. Ibid., 5.
42. Ibid., 6.
43. Ibid.
44. Ibid., 8. There is a difference between the distinction Collingwood makes here between the material and formal elements with that made in his "Preliminary Discussion". In the latter he discusses the concept of philosophy of history, and methodological questions are considered there the matter, while in his lectures of 1926 the science of history is discussed, in which context methodological questions are the form and the sources the matter.
45. Ibid., 18.
46. Ibid., 18-19.
47. Ibid., 21.
48. Ibid., 22-23.
49. Ibid., 23-24. Collingwood's conception proves how much beside the mark is Gallie's allegation that Collingwood neglects the achievements of predecessors in history: "... Collingwood's account appears to neglect one of the most obvious facts about the general context in which the need for fresh

historical thinking, the need for problem-solving history, commonly arises. This context is, of course, the narratives which the historian inherits from his predecessors or sources and which it is both natural and inevitable for him to regard – in a first provisional attitude – as witnesses of truth from a certain limited point of view. He must first adopt this attitude, whatever his final judgment may be, if he is to know, if he is to be able to follow, what his predecessors' narratives amount to ... But Collingwood, with his detestation of all 'authorities', not only neglects but appears to deny the existence of this obvious fact" (W.B. Gallie, *Philosophy and the Historical Understanding* (New York, 1964), 18). "Second-order history" is also discussed by Collingwood in his *Autobiography* (Aut,132), so this conception could have been known before the manuscripts were available.

50. Ibid., 26.
51. Ibid., 27-28.
52. Ibid., 29.
53. Ibid., 30.
54. Ibid., 32.
55. Ibid., 33.
56. Ibid.
57. Ibid., 33-34.
58. Ibid., 35.
59. Ibid., 52.
60. Ibid.
61. Ibid., 53-54.
62. Ibid., 51.
63. Ibid., 63.
64. Ibid., 69-71.
65. Ibid., 71.
66. LM,1928-1, Preface.
67. Ibid., 1.
68. Ibid., 2.
69. Ibid., 3.
70. Ibid., 4.
71. Ibid., 4-6.
72. Ibid., 6.
73. Ibid., 6-8.
74. Ibid., 9-11.
75. Ibid., 12-13.
76. Ibid., 6-7.
77. Ibid., 7.
78. Ibid., 8.
79. Ibid., 13-14.
80. Ibid., 5.
81. Ibid., 9.
82. Ibid., 14.
83. Ibid., 8-9.
84. Ibid., 9.
85. Ibid., 15.
86. Ibid., 19.
87. Ibid., 21-22.
88. Ibid., 24.
89. Ibid., 24-25.

90. Ibid., 28.
91. Ibid.
92. Ibid., 29.
93. Ibid., 30.
94. Ibid., 39-40.
95. Ibid., 40-41. This conception is in line with Collingwood's view of the concept of truth in the context of his "logic of question and answer", as developed in his *Autobiography*: "What is ordinarily meant when a proposition is called 'true', I thought, was this: (a) the proposition belongs to a question-and-answer complex which as a whole is 'true' in the proper sense of the word; (b) within this complex it is an answer to a certain question; (c) the question is what we ordinarily call a sensible or intelligent question, not a silly one, or in my terminology it 'arises'; (d) the proposition is the 'right' answer to that question ... Whether a given proposition is true or false, significant or meaningless, depends on what question it was meant to answer; and any one who wishes to know whether a given proposition is true or false, significant or meaningless, must find out what question it was meant to answer" (Aut,38-39).
96. Ibid., 45.
97. Ibid.
98. Ibid., 44.
99. Ibid., 41.
100. Ibid.
101. 290-295. See also p. 331 of this study.
102. LM,1928-1, 41-42
103. Ibid., 49.
104. Ibid.
105. Ibid., 50.
106. Ibid.
107. Ibid., 50-51.
108. Ibid., 52.
109. Ibid., 53.
110. Ibid., 54.
111. Ibid.
112. Ibid., 56-57.
113. Ibid., 58.
114. Ibid., 60.
115. Ibid., 62.
116. Ibid.
117. See pp. 3-5.
118. See Appendix III.
119. See Appendix III and LM.
120. See Appendix II
121. See p. 4.
122. LM,1929-1, 12.
123. Ibid., 14.
124. Ibid., 16-17.
125. Ibid., 18.
126. Ibid., 19.
127. Ibid., 20. On the same page Collingwood says of Marx: "The practical consequences of his teaching have been, perhaps, the most important feature in general politics in the last hundred years. But it does not follow that his philosophy of history is likely to satisfy a critical historian. That,

indeed, it has never done". The point Collingwood criticizes Marx for is that capitalism and socialism are abstractly put against each other by him, while they should be seen as "co-existent tendencies". In *The New Leviathan* Collingwood develops the same view with regard to the concepts of democracy and aristocracy: "...[D]emocracy and aristocracy, properly understood, are not hostile to each other. They are mutually complementary...Each of them gives a partial answer to the question: 'How shall we make the ruling class as strong as possible?'" (192). Democracy and aristocracy are, according to Collingwood, "positive and negative elements" in a dialectical process of " [t]he inevitable recruitment of a ruling class from its correlative ruled class" (ibid.).
"[W]hen these elements are considered in *false abstraction* from the process to which they belong", we get " [t]he rise of doctrinaire democracy or doctrinaire aristocracy" (ibid).

128. Ibid., 22.
129. Ibid.
130. Ibid., 23.
131. Ibid., 26.
132. Ibid., 36.
133. "(1) What kind of reality has the object of historical knowledge? in what sense are historical facts real? (2) What is the scope of historical knowledge? how much material ought the historian to master, and what limits are there to the collection of material? (3) How are historical facts interrelated with each other and bound up into logical or historical wholes? What is historical causation or necessity? (4) What kind of certainty does historical knowledge possess, and in what sense is it susceptible of demonstration?".
134. LM,1931-2,2.
135. Cf.IH,1-3, where the same is said.
136. LM,1931-2, 2.
137. Ibid., 2a.
138. Ibid.
139. LM,1932-2, 3.
140. Ibid., 4.
141. Ibid., 6.
142. Ibid., 7.
143. Ibid., 7-8.
144. Ibid., 8-9.
145. Ibid., 9.
146. LM,1935-8,4.
147. Ibid., 4-5.
148. Ibid., 8.
149. Ibid., 10.
150. Ibid., 12.
151. Ibid., 13.
152. Ibid.
153. LM,1936-3, 7. It should be added, though, that the distinction between science and history is conceived here within the context of the difference between a natural process as "outward" and an historical process as "inward".
154. LM,1935-8, 14.
155. *The Journal of Philosophy* 39 (1942), 35-48; reprinted in P. Gardiner ed., *Theories of History* (Glencoe, Illinois, 1959), 344-356, and C.G. Hempel,

Aspects of Scientific Explanation and Other Essays in the Philosophy of Science (New York, 1965), 231-243.

156. Dray, *Laws and Explanation*, 1.
157. LM,1935-8, 14. It is interesting to note that Dray's wording is almost the same: "To put it in a summary way", he says, "what the theory maintains is that explanation is achieved, and only achieved, *by subsuming what is to be explained under a general law*" (*Laws and Explanation*, 1).
158. Ibid., 14-15.
159. Dray, *Laws and Explanation*, ch. 5.
160. LM,1935-8, 15.
161. Ibid.
162. Ibid.
163. Ibid., 15-16.
164. Ibid., 17.
165. Ibid.
166. Ibid.
167. Ibid., 18.
168. Ibid.
169. Ibid.
170. Ibid., 19-20. For the same point see p. 110.
171. Ibid., 21.
172. Collingwood does not say here that a character may similarly cut off certain possibilities. Further on, however, he calls a character "a power at once liberating and constraining" (Ibid., 35).
173. Ibid., 22.
174. Ibid.
175. Ibid.
176. Ibid., 23.
177. Ibid.
178. Ibid., 24.
179. Ibid., 25.
180. Ibid.
181. Ibid.
182. Ibid., 26.
183. Ibid., 27.
184. Ibid., 26.
185. Ibid., 27.
186. Ibid., 28.
187. Ibid., 26.
188. "I spell Nature with a capital", Collingwood says, "when I mean by it the totality or system of what are called natural things" (Ibid., 29).
189. Ibid., 32-33. Collingwood adds to this argument, however, that these points must not be over-emphasized, because we know so little about the behaviour of individual electrons. We can only give generalized descriptions of them in the nature of "statistical laws".
190. Ibid., 34.
191. Ibid., 34-35.
192. Ibid., 35.
193. Ibid.
194. Ibid., 35-36.
195. Ibid., 36.
196. Ibid.
197. This manuscript will not be treated separately.

198. See Appendix II.
199. LM,1936-5, 10-11.
200. Ibid., 12.
201. LM,1936-3, 21-22.
202. LM,1936-5, 27-28.
203. Ibid., 28.
204. Ibid., 16-17.
205. Ibid., 17.
206. Cf. what Collingwood says in the published version of "Human Nature and Human History": "If the discovery of Pythagoras concerning the square on the hypotenuse is a thought which we to-day can think for ourselves, a thought that constitutes a permanent addition to mathematical knowledge, the discovery of Augustus, that a monarchy could be grafted upon the Republican constitution of Rome by developing the implications of *proconsulare imperium* and *tribunicia potestas*, is equally a thought which the student of Roman history can think for himself, a permanent addition to political ideas" (IH,217-218).
207. LM,1936-5, 18.
208. Ibid., 18-19.
209. Ibid., 19.
210. Ibid.
211. Ibid., 19-20.
212. Ibid., 21-22.
213. See Appendix I.
214. LM,1938-39-2, 24.
215. LM,1938-39-1, 29.
216. See p. 64.
217. This information the present writer got from Knox. In his preface, however, Knox speaks of quotations "from a series of notes written early in 1939 for *The Principles of History*" (IH,X).
218. LM,1938-39-2, 11.
219. Ibid., 16.
220. Ibid., 17.
221. On the cover of the notebooks on folklore (LM,1936-6) is written "Magdalen College", Collingwood's college after he was elected Waynflete Professor of Metaphysical Philosophy in 1935.
222. (Oxford, 1938), 57-77.
223. LM,1936-37-2, 1.
224. Ibid., 19.
225. Ibid., 17.
226. Ibid., 5.
227. Ibid., 19.
228. Ibid., 1.
229. LM,1936-37-4, 4.
230. Ibid., 6.
231. LM,1936-37-3, 12.
232. Ibid., 23.
233. Ibid., 25.
234. LM,1936-6,I-78.
235. LM,1936-37-3, 38.
236. Ibid., 48.
237. LM,1936-37-4, 7.
238. Ibid.

239. Ibid.
240. Ibid., 3-4.
241. LM, 1936-37-3, 26.
242. LM,1936-37-4, 8.
243. Ibid., 12.
244. Ibid., 20-21.
245. Ibid., 5.
246. LM,1936-37-5, 9.
247. Ibid., 2. In another passage Collingwood says: "All historical knowledge involves the re-creation in the historian's mind of the past experience which he is trying to study. If magic were a form of belief or custom peculiar to primitive peoples and absolutely foreign to the mind of civilized man, the civilized historian could never understand it" (LM,1936-37-2, 18).
248. Ibid., 3.
249. Ibid., 14.
250. Ibid., 15. There is an interesting parallel between this view and the theme of the suppression of emotions and irrational behaviour in general in European civilization, as elaborated by N. Elias, *Über den Prozess der Zivilisation* (Basel, 1936, 1939) and M. Foucault, *Histoire de la Folie à l'Age Classique* (Paris, 1972).
251. Ibid., 15-16.
252. Ibid., 16.
253. Ibid., 18.
254. Ibid.
255. Ibid.
256. Ibid., 28.
257. Ibid., 29.
258. Ibid., 30.
259. Ibid., 30-31.
260. Ibid., 33.
261. LM, 1936-37-6, 2. The concept of rationality in connection with primitive societies is of current interest, especially with regard to the aspect of understanding (see for instance B.R. Wilson ed., *Rationality* (Oxford, 1974)). The aspect of the irrational aspects in European civilization, however, does not get similar attention.
262. LM, 1936-37-5, 33.
263. LM, 1936-37-6, 3. Collingwood sums up his view by contending: "It is 'civilization' not 'savagery', that takes explaining" (LM, 1936-37-5, 34). The Dutch historian Jan Romein has developed a similar view by maintaining that European history from 1600 should be considered rather a deviation from a "common human pattern" than a "normal" one. ("The Common Human Pattern", *Cahiers d'Histoire Mondiale* 4 (1958), 449-463).
264. (Oxford, 1945).
265. *Mind* 80 (1971), 151-152.
266. See Appendix II.
267. Prefatory note to *The Idea of Nature*.
268. LM, 1933-34-A, 1b.
269. Ibid., 1d.
270. "Metaphysics is the attempt to find out what absolute presuppositions have been made by this or that person or group of persons, on this or that occasion or group of occasions, in the course of this or that piece of thinking" (EM, 47). "...[M] etaphysics ... is primarily at any given time an attempt to discover what the people of that time believe about the world's

general nature...Secondarily, it is the attempt to discover the corresponding presuppositions of other peoples and other times, and to follow the historical process by which one set of presuppositions has turned into another" (Aut, 65-66).

271. LM, 1934-1,2.
272. Ibid., 9.
273. Ibid., 14.
274. Ibid., 14-15.
275. Ibid., 15.
276. Ibid., 16.
277. Ibid., 17, 18.
278. Ibid., 18.
279. Ibid., 18-19.
280. Ibid., 25.
281. Ibid., 27.
282. Ibid., 28.
283. Ibid., 30.
284. Ibid., 30-31.
285. "Such a phrase as 'inquiry into the truth of an absolute presupposition' is nonsense" (*EM*, 54).
286. Ibid., 77.
287. LM, 1934-1, 16.
288. LM, 1933-34-A, 1d.
289 Ibid.
290. Ibid., 3.
291. Ibid., 3-4.
292. Ibid., 8.
293. LM, 1933-34-E,82.
294. LM, 1933-34-A,4-5.
295. Ibid., 5-6.
296. Ibid., 6.
297. Ibid., 15.
298. Ibid., 58.
299. LM, 1933-34-E, 89.
300. LM, 1933-34-C, 80.
301. Ibid., 84.
302. Ibid.
303. LM, 1933-34-D, 4.
304. Ibid., 5.
305. Ibid., 5-6.
306. Ibid., 77.
307. Ibid., 87.
308. LM, 1933-34-E, 27.
309. Ibid, 104.
310. Ibid.
311. Ibid., 105.
312. Ibid.
313. LM, 1933-34-D, 68.

1. F.D. Schneider, "Collingwood and the Idea of History", *The University of Toronto Quarterly* 22 (1953), 172 - 183; A. Donagan, "The Verification of Historical Theses", *The Philosophical Quarterly* 6 (1956), 193 - 208.
2. L.J. Goldstein, "Collingwood's Theory of Historical Knowing", *History and Theory* 9 (1970), 3 - 36 and "Collingwood on the Constitution of the Historical Past", in Krausz ed., *Essays*, 241 -267. In the latter article, though, Goldstein makes the strange remark that "...even [sic] the treatment of such stuff as archaeology deals with can be treated in accordance with Collingwood's conception of historical method ..." (247).
3. L.O. Mink, "Collingwood's Historicism: a Dialectic of Process", in Krausz ed., *Essays*, 154.
4. *Proceedings of the British Academy* 29 (1943), 476 - 480.
5. *Later Philosophy*, 196 - 200.
6. "Collingwood's Theory", 28.
7. *The Dictionary of National Biography, 1941 - 1950* (Oxford, 1959), 168.
8. "Collingwood's Theory", 28.
9. See "R.G. Collingwood: Bibliography of Writings on Ancient History and Archaeology", *Proceedings of the British Academy* 29 (1943), 481 - 485.
10. T.A. Roberts, *History and Christian Apologetic* (London,1960), 6.
11. See the "In Memoriam" R.G. Collingwood wrote of his father in *CW* 33 (1933), 308 - 312. For Collingwood's family background see W.M. Johnston, *The Formative Years of R.G. Collingwood* (The Hague, 1967), 3 - 30.
12. *CW* 66 (1966), V - VII.
13. See I.A. Richmond, "Frank Gerald Simpson", *Archaeologia Aeliana*-4th ser., 34 (1956), 219 - 221; E.B. Birley, "In Memoriam", *CW* 55 (1955), 359 - 365; G. Simpson, "Editor's Introduction: A Short Survey of Gerald Simpson's Life and Work", in: G. Simpson ed., *Watermills and Military Works on Hadrian's Wall. Excavations in Northumberland 1907 - 1913 by F. Gerald Simpson* (Kendal, 1976), 1 - 23.
14. "I remember with gratitude that it was he [Collingwood] who sent me to the Wall when I was still an undergraduate", Birley says, "and I believe that Professor Richmond, too, owed to Collingwood his first introduction to Simpson's programme of research, and so to the many years of close collaboration which served to transform the whole position of mural studies" (*Research on Hadrian's Wall* (Kendal, 1961), 68). At another place Birley declares that one of Richmond's "main inspirations" had been "the teaching of R.G. Collingwood at Oxford" ("Sir Ian Archibald Richmond", *Proceedings of the British Academy* 52 (1966), 293).
15. R.G. Collingwood, *The Archaeology of Romain Britain* (London, 1930).
16. LM, Correspondence.
17. Before this passage Toulmin says: "It is worth recalling that, when Collingwood first encountered ancient history, the subject was in some disarray. The work of Schliemann at Troy, and of Arthur Evans at Knossos, had made the recovery of Antiquity a matter of popular interest, a craze, even 'big business'. In the excitement of discovery, serious intellectual questions were thrown into shade by the beauty of Minoan frescos and the glitter of Agamemnon's Mask. Digs were conducted in a manner that appears to us, in retrospect, undiscriminating and needlessly destructive. Crates of material unearthed from the ancient sites of the Aegean and Levant were shipped off to the museums of western Europe, where some of them remained unopened for many years. At times, it was hard to tell

whether the most fashionable archaeologists were moved by the enthusiasm of children or by the cupidity of grave-robbers" (Aut,XVII).

18. *PSAN* 9 (1919), 118. Many years later Collingwood wrote: "It was Haverfield who placed the study of Roman Britian on the *sicheren Gang einer Wissenschaft*" (*JRS* 23 (1933), 100).
19. "The Milecastle on the Wall of Hadrian at Poltross Burn", *CW* 11 (1911), 390-461.
20. Birley, *Research*, 65.
21. *The Archaeological News Letter* 2 (1949), 42.
22. *CW* 55 (1956), 360 - 361.
23. Birley, *Research*, 67 - 68.
24. Richmond, "Appreciation of R.G. Collingwood", 478.
25. Ibid., 478 - 479.
26. See p. 192.
27. Donagan also refers to this point (*Later Philosophy*, 198 - 199).
28. Simpson "Introduction", 21.
29. Debbins ed., *Essays*, 43.
30. Richmond, "Appreciation of R.G. Collingwood", 479.
31. A.H.A. Hogg, "Sample Excavation", *Current Archaeology* 63 (1978), 126.
32. Donagan, Later Philosophy, 200.
33. R.G. Collingwood, "Explorations in the Roman Fort at Ambleside (fourth year, 1920) and at other sites on the Tenth Iter", *CW* 21 (1921), 30 - 31.
34. R.G. Collingwood, "Report of the Excavations at Papcastle, 1912", *CW* 13 (1913), 131- 141.
35. Ibid., 134.
36. Ibid., 141.
37. Ibid.
38. Professor Haverfield and R.G. Çollingwood, "Report on the Exploration of the Roman Fort at Ambleside, 1913", *CW* 14 (1914), 435.
39. Richmond, "Appreciation of R.G. Collingwood", 476.
40. Collingwood, "Report, 1913", 449.
41. Ibid., 451 - 452.
42. R.G. Collingwood. "The Exploration of the Roman Fort at Ambleside: Report on the second year's Work (1914)", *CW* 15 (1915), 8.
43. Ibid., 31.
44. Ibid.
45. R.G. Collingwood, "The Exploration of the Roman Fort at Ambleside: Report on the third year's Work (1915)", *CW* 16 (1916), 79- 81.
46. T. Garlick, *Ambleside Roman Fort* (Whitehaven, 1975), 9, 26.
47. Richmond, "Appreciation of R.G. Collingwood", 479.
48. R.G. Collingwood, "King Arthur's Round Table: Interim Report on the Excavations of 1937", *CW* 38 (1938), 1 - 31.
49. Richmond, "Appreciation of R.G. Collingwood", 479.
50. G. Bersu, "King Arthur's Round Table: Final Report, including the excavations of 1939, with an appendix on the Little Round Table", *CW* 40 (1940), 169 - 206.
51. Ibid., 201.
52. Ibid., 202.
53. See pp. 140,155.
54. LM, 1926 - 2, 30.
55. LM, 1928-1, 56.
56. Ibid.
57. Ibid., 57.

58. LM, 1929 - 1, last page.
59. LM, 1926 - 2, 30 - 32.
60. Collingwood, *Archaeology*, V - VI.
61. Ibid., 2 - 3. This passage is an example of what Collingwood calls in his 1926 lectures the "practical side" of the science of archaeology, "consisting of general recommendations as to the search for the special kind of evidence in question" (LM, 1926 - 2, 30).
62. Collingwood calls this the "theoretical side" of the science of archaeology, "consisting of general propositions concerning such things as the period of history at which this or that moulding or piece of ornament was used" (LM, 1926 - 2, 30).
63. Collingwood, *Archaeology*, 7.
64. Ibid., 25.
65. Ibid., 35.
66. Ibid., 39 - 40.
67. Ibid., 42 - 44.
68. Ibid., 44.
69. Ibid., 92.
70. Ibid., 244. The metaphor of a family reminds one of Wittgenstein's conception of "family resemblances" (*Philosophical Investigations* (Oxford, 1958), I, 66-67). There are, however, some differences. A group of brooches identified by Collingwood with a family are genealogically related to each other, which is not the case with Wittgenstein's examples. Collingwood further speaks of a close relation between groups (families) of brooches producing certain likenesses, while Wittgenstein only speaks of individuals (for instance games) forming one family.
There is another place where Collingwood also uses the concept of family likeness: "Of a Celtic pantheon or group of gods worshipped *semper, ubique, ab omnibus*, Britain yields no trace. What uniformity these cults possessed was due, not to the existence of dominating figures like the Juppiter Optimus Maximus of Rome, but to family likeness among the cults and festivals themselves" (*CAH*, Vol.12 (Cambridge, 1939), 295)
71. Ibid., 26 - 27.
72. Ibid., 153.
73. Ibid., 155.
74. As far as they could be traced by the present writer.
75. I.A. Richmond, *Antiquity* 5 (1931), 504.
76. E.B. Birley, *CW* 31 (1931), 210.
77. R.E.M. Wheeler, *The Antiquaries Journal* 11 (1931), 173.
78. R.C. Bosanquet, *JRS* 23 (1933), 100.
79. Ibid., 101.
80. J.P. Droop, *Annals of Archaeology and Anthropology*, 18 (1931), 56.
81. S. Casson, *The London Mercury* 24 (1931), 282.
82. Wheeler, *Antiquaries*, 173.
83. Richmond, *Antiquity*, 506.
84. Bosanquet, *JRS*, 101; G.M. Churchill, *Art and Archaeology* 33 (1932), 112.
85. Birley, *CW*, 210.
86. Casson, *Mercury*, 282.
87. According to R.I.W. Westgate "Collingwood has done pioneer work of fundamental importance in classifying Forts, Town-houses, Villas, Coarse Pottery, and Brooches" (*American Journal of Archaeology* 35 (1931), 361). For similar appraisals see M. Cary, *History* 15 (1930 - 1931), 348; C.A. R. Radford, *Journal of the British Archaeological Association* 36 (1930), 427; *Nature* 127 (1931), 9.

88. (London).
89. Ibid., XXIII - XXIV.
90. F.H. Thompson, *The Antiquaries Journal* 50 (1970), 133.
91. R.E.M. Wheeler, *Antiquity* 43, (1969), 239.
92. Ibid.
93. Vol.1, *Inscriptions on Stone* (Oxford).
94. See Wright's Preface in *Inscriptions*, V - VII.
95. G. Macdonald says in a "Biographical Notice" to Haverfield's *The Roman Occupation of Britain* (Oxford, 1924), speaking about Haverfield's plans to publish a complete collection of Romano-British inscriptions, that " [o] ne of the scholars whose collaboration he [Haverfield] intended to secure was Rostovtzeff, whom the turmoil in Russia had driven to England. In the early autumn of 1919 he revisited in his company several familiar Roman sites ...." (32). See also W.M. Johnston, *Formative Years*, 38.
96. See Aut,145.
97. Wright, *Inscriptions*, VI. In his 1932 Report as University Lecturer Collingwood mentions as his first project: "Corpus of Roman Inscriptions in Britain, begun in 1920 and planned to take 20 years; every inscription to be illustrated with a facsimile drawing by myself. ... The drawings made have risen from about 600 to about 1100. The whole material has been arranged and indexed" (See Appendix III).
98. Wright gives the following description: "After a sheet of flexible paper has been strapped to, or held against, the inscribed face, the epigraphist follows the exact form of the letters and decoration. The resultant drawings are then fair-copied and photographically reduced in making the line-blocks to a fixed scale, usually 1/4, 1/8 or 1/12. This method gives a faithful record of the spacing and letter-forms of the inscription and is superior, not only to the facsimile in capitals adopted by CIL [Corpus Inscriptionum Latinarum], but, in most cases, even to the photograph, in which the problem of illuminating all parts of the lettering equally and without distortion is often insuperable". (*Inscriptions*,XIV).
99. By the time he wrote his *Autobiography* in 1938 Collingwood seems to have lost his interest in inscriptions. He contends that " ... the inscriptions themselves were not of very great service to my Romano-British studies". "...[T] he epigraphic historian as such can never be wholly Baconian in spirit. Regarded as documents, inscriptions tell you less, under critical scrutiny, than literary texts; regarded as relics, they tell you less than archaeological material proper. And on the questions which I particularly wanted to ask, it happened that inscriptions threw hardly any light" (Aut,145-146). This assertion is in sharp contrast to that made in *The Archaeology of Roman Britain*: "The value of inscriptions as historical material is so great that it can hardly be exaggerated. Apart from modern forgeries ... they are contemporary and authoritative documents, whose text if legible cannot be corrupt, and whose cumulative value, in the hands of scholars accustomed to handling them in the mass, is astonishing. They are the most important single source for the history and organisation of the Roman Empire" (162). Similarly, in *Roman Britain and the English Settlements* it is stated: "For every part of the Roman Empire, Britain not excepted, inscriptions constitute a source of high importance" (464).

It is difficult to explain these striking differences in judgment on a subject on which Collingwood had worked for so many years of his life. When writing his *Autobiography* he was perhaps too much obsessed by the cleavage he had constructed between "scissors-and-paste" and "Baconian" history. He does not make clear, however, why exactly inscriptions could not be studied in a Baconian spirit.

100. See p. 163.
101. R.G. Collingwood, "An Introduction to the Prehistory of Cumberland, Westmorland and Lancashire north of the Sands", *CW* 33 (1933), 163 - 200.
102. See Appendix III.
103. Collingwood, "Introduction", 163.
104. Ibid., 165.
105. Ibid., 165 - 166.
106. Ibid., 189 - 190.
107. Ibid., 193.
108. Ibid.
109. Ibid., 194.
110. Ibid.
111. Ibid., 195.
112. Ibid.
113. Ibid., 196.
114. Ibid., 197. Compared with the questions to be asked by fieldwork those of excavation are more specific.
115. Ibid., 197 - 198.
116. The archaeologist Crawford wrote an appreciative review of the article (*Antiquity* 8 (1934), 361 - 363).
117. Collingwood's article "Science and History" was published in the same journal (see pp. 43-47).
118. *The Vasculum* 8 (1921), 4 - 9.
119. Ibid., 4.
120. Ibid., 5.
121. Ibid., 6.
122. Ibid., 7.
123. Ibid., 8.
124. Ibid., 8 - 9.
125. Birley, *Research*, 269 - 270.
126. D.J. Breeze and B. Dobson, *Hadrian's Wall* (London, 1976), 39.
127. *JRS* 11 (1921), 37 - 66.
128. Ibid., 37.
129. LM, 1928-1, 27.
130. Ibid., 28.
131. Collingwood, "Hadrian's Wall", 37.
132. Ibid., 43.
133. Ibid., 44 - 45.
134. Ibid., 45.
135. Ibid., 47.
136. Ibid., 52.
137. Ibid., 53.
138. Ibid.
139. Ibid., 55.
140. Ibid., 61 - 62.
141. Ibid., 62 - 63.
142. LM, 1926 - 2,58
143. Collingwood, "Hadrian's Wall", 66.
144. According to Birley "[t]here is much to be said for Collingwood's logical order of subjects, and in some respects his paper is of abiding value", though he is also of the opinion that the facts given by Collingwood are not always accurate (*Research*, 48).

145. R.G. Collingwood, "Hadrian's Wall: 1921 - 1930", *JRS* 21 (1931), 36 -64.
146. Ibid., 37.
147. Ibid. This view corresponds in a remarkable way to Kuhn's observation on people who usually bring about a paradigm change: "Almost always the men who achieve these fundamental inventions of a new paradigm have been either very young or very new to the field whose paradigm they change. And perhaps that point need not have been made explicit, for obviously these are the men who, being little committed by prior practice to the traditional rules of normal science, are particularly likely to see that those rules no longer define a playable game and to conceive another set that can replace them" (Th. S. Kuhn, *The Structure of Scientific Revolutions*, 2nd ed. (Chicago, 1970), 90).
148. Ibid.
149. Ibid., 37 - 38.
150. It would have been more in accordance with his own interpretation if Collingwood had mentioned Simpson in this connection. In a note at the beginning of his article, moreover, commenting on the work done on Hadrian's Wall, Collingwood says: "Of the men responsible for directing the work, I must mention first and foremost Mr. F.G. Simpson, who has done more than any other one person to lay down the lines on which archaeological study of the Wall must proceed, and to bring about all the advances which have been made in that study during the present century" (36).
151. Ibid., 38.
152. Ibid., 39.
153. Ibid., 36.
154. *Archaeologia Aeliana*-4th ser., 15 (1938), 1 - 42.
155. See pp. 230-231.
156. Collingwood, "John Horsley", 2 - 3.
157. Ibid., 25 - 26.
158. Ibid., 39.
159. Ibid., 39 - 40. This passage is important for several reasons. The type of scientific revolutions Collingwood discusses here is based on revolutions in technique and is therefore different in character from paradigm changes as discussed by Kuhn. It must also be distinguished from the revolutionary change from "complete excavation" to "selective excavation", which indeed can be compared with a paradigm change (see note 147).
Half a year after his lecture on Horsley Collingwood developed his theory of absolute presuppositions (see LM, 1937 - 38). As several interpreters have observed, this conception closely corresponds with Kuhn's conception of a paradigm. In the above quoted passage, however, Collingwood is not discussing a change in absolute presupposition. On the contrary, he emphasizes rather the continuity in what Toulmin has called the "disciplinary aim" between Horsley and archaeology of the present day (see S. Toulmin, "Conceptual Change and the Problem of Relativity ", in Krausz ed., *Essays*, 214-216).
There is also an interesting parallel to be drawn between Collingwood's emphasis on the need for sound theoretical thinking on the occurrence of revolutions in technique and his admonition not to rely too much on the scientific tools man has invented, but rather to handle them with wisdom (see p. 191).
160. Birley speaks of "the pedestal on which Collingwood's imagination set him" ("John Horsley and John Hodgson", *Archaeologia Aeliana*-4th ser., 36 (1958), 5).

161. Collingwood, "John Horsley", 41.
162. Ibid., 42.
163. R.G. Collingwood, "Roman Signal-Stations on the Cumberland Coast", *CW* 29 (1929), 138 - 165.
164. Ibid., 139.
165. Ibid., 140.
166. Ibid., 143.
167. Ibid.
168. Collingwood does not mention the eighteenth century antiquary Gordon, who, according to Birley, was the first to refer to them (*Research*, 126).
169. Collingwood, "Signal-Stations", 148.
170. Ibid., 164.
171. See Birley, *Research*, 126 - 131, and Ch. Daniels ed., J. Collingwood Bruce, *Handbook to the Roman Wall*, 13th ed. (Newcastle upon Tyne, 1978), 260 - 286.
172. R.G. Collingwood, "A System of numerical references to the parts of Hadrian's Wall and the structures along its line", *CW* 30 (1930), 108 - 115; R.G. Collingwood "A System of Numerical References", *PSAN* 4 (1929 - 1930), 179 - 187. Birley reports that the numbering-system was "...worked out by Collingwood in consultation with Simpson in 1929 ..." (*Research*, 70).
173. R.G. Collingwood, *A Guide to the Roman Wall* (Newcastle upon Tyne, 1926, 2nd ed., 1932).
174. (Newcastle upon Tyne, 1933).
175. In his 1932 Report as University Lecturer Collingwood mentions as one of his projects "A large-scale work on Hadrian's Wall, with special reference to the problems of archaeological method there arising". In his 1935 Report he refers again to this work, saying that it, like the work on Roman inscriptions, "...at the present rate of progress, will not be finished for several years" (See Appendix III). This projected study on Hadrian's Wall never appeared.
176. (London). In the Preface Collingwood says that the book "...represents a set of lectures given at the Oxford Summer Meeting in 1921". In his *Autobiography*, however, Collingwood reports that the book was written in 1921 "at the invitation of the Delegates of the Clarendon Press" (Aut,120). He further says about it: "It was a short book; I wrote it in two days; it was designed to be elementary, and it was full of faults. However, it served to lay down once for all my general attitude towards the problems, and, even more important, my general conception (partly due to Haverfield, but partly different from his) of what the problems were; it gave me a first opportunity of finding out, more clearly than was possible within the limits of a short article, how my conception of historical research was developing" (Aut,120 - 121).
177. (Oxford).
178. *RB* (1932), 3 - 4.
179. *RB* (1923), 12.
180. *RB* (1932), 3.
181. Ibid., 2.
182. *RB* (1923), 15 - 16.
183. *RB* (1932), 4.
184. Ibid., 5.
185. Ibid.
186. Ibid., 6.
187. *RB* (1923), 18 - 19.

188. *RB* (1932), 12 - 13.
189. See pp. 57-59.
190. *RB* (1932), 91.
191. Ibid., 92.
192. Ibid., 93.
193. *RB* (1923), 68.
194. A. Grenier, *Revue des Etudes Anciennes* 34 (1932), 333.
195. E. Albertini, *Revue Historique* 173 (1934), 626.
196. J.N.L. Myres, *JRS* 22 (1932), 252.
197. Chr. Hawkes, *The Antiquaries Journal* 13 (1933), 67.
198. *Man* 32 (1932), 220 - 221.
199. That this view is indeed Collingwood's is clearly demonstrated by what he contends on the subject in the manuscripts on folklore: "...[C]ultural borrowings such as those insisted upon by the diffusionist theory imply not only the power to lend but also the power to borrow: that is, they imply that whatever was borrowed was something capable of incorporation in the existing culture of the borrowers. There must be some kind of organic relation between a given culture and the elements which it borrows from elsewhere... The incorporation of what is borrowed into the culture of the borrowers always implies a certain modification adapting it to its new context; as thus adapted, it becomes a new and original thing, the product of that culture's spontaneous energy and the expression of its peculiar needs" (LM,1936 -37-2, 12). In *Roman Britain and the English Settlements* Collingwood similarly maintains "...that the people upon whom this [Roman] system was imposed were people with an historical tradition of their own, who learnt from Rome not the lessons she was able to teach but the lessons their previous training enabled them to learn" (VI).
200. *Man*, 220. Evans-Pritchard was teaching in Oxford in the thirties. It is probable that he and Collingwood not only knew each other, but were acquainted with each other's work. Collingwood mentions Evans-Pritchard's *Witchcraft, Oracles and Magic among the Azande* (Oxford, 1937) in his *Principles of Art* (8). According to Kuper, Evans-Pritchard turned from 1950 to an historical orientation, having been from 1937 a functionalist: "...in line with his belief that social systems should be studied by the methods of the humanities rather than those of the natural sciences... Evans-Pritchard insisted upon the use of history. He argued that there was little essential difference between the techniques and aims of social anthropology and those of history, particularly social history" (A. Kuper, *Anthropologists and Anthropology: The British School 1922 -1972* (Harmondsworth, 1975), 163). Apparently Evans-Pritchard was interested in history in the beginning of his career as well. Ions reports on him that "...at the end, he returned to his first passion – history" (E.Ions, *Against Behaviouralism: A Critique of Behavioural Science* (Oxford, 1977), XII). Thompson even suggests that Evans-Pritchard and Collingwood had corresponding views, an idea which is made plausible by the former's review of *Roman Britain.* After discussing Whitehead, Thompson asserts: "Whitehead's ideas were taken up by two men who were contemporaries at Oxford: the anthropologist Evans-Pritchard and the philosopher and historian Collingwood" (M. Tompson, *Rubbish Theory: The Creation and Destruction of Value* (Oxford, 1979), 75).
201. See Appendix III.
202. *RBES*, V.
203. See note 1 and 2.

204. I.A. Richmond, *Journal of the British Archaeological Association* 3rd ser., 2 (1937), 197.
205. R.E.M. Wheeler, *JRS* 29 (1939), 87 - 88.
206. G. Clark, *The Times*, 15 -11- 1961.
207. This is not to say, of course, that the view that all history is the history of thought necessarily leads to the neglect of evidence.
208. *RBES*, 82.
209. P.K. Bailly Reynolds, *The Antiquaries Journal* 17 (1937), 451.
210. Ibid., 452.
211. Wheeler, *JRS*,89.
212. 36 (1924), 6 - 8, 37 - 39, 71 - 73; 36 (1925), 101 - 104, 133 - 135, 165 - 168.
213. Ibid., 39.
214. Ibid., 104.
215. R.G. Collingwood, *Roman Eskdale* (Whitehaven, 1929)
216. R.E.M. Wheeler, *Archaeological Journal* 87 (1930), 371.
217. J.B. Bury, "The Notitia Dignitatum", *JRS* 10 (1920), 131 - 154; R.G. Collingwood, "The Roman Evacuation of Britain", *JRS* 12 (1922), 74 - 98.
218. Vol. 10 (Cambridge, 1934), 790 - 802; Vol. 11 (Cambridge, 1936), 511 - 525; Vol. 12 (Cambridge, 1939), 282 - 296.
219. Vol. 3 (Paterson, 1937), 1 - 118.
220. *Antiquity* 3 (1929), 261 - 276.
221. A.M. Carr-Saunders, *The Population Problem* (Oxford, 1922) and *Population* (Oxford, 1925).
222. Collingwood "Town and Country", 263.
223. Ibid., 261.
224. Ibid., 262.
225. Ibid., 263.
226. Ibid.
227. Ibid., 262.
228. "The Roman legions were far away in the north and west, and Boudicca's blow was struck before they could return; when they did, her armies met them and were wiped out to the number, it is said, of 80,000, a number even exceeding the 70,000 Romans and Romanized Britons whom they had massacred" (*Roman Britain* (1923), 23). But for a few words in the first sentence the same description is given in the second edition (*Roman Britain* (1932), 20 -21).
229. Collingwood, "Town and Country", 262.
230. While Collingwood gives an estimate of 50 to 70 million for the total population of the Roman Empire he asserts that "...one-seventh of the whole population of the Empire is accounted for by Egypt alone ..." (Ibid., 261).
231. Ibid., 270.
232. H.J. Randall, "Population and Agriculture in Roman Britain", *Antiquity* 4 (1930), 80 - 90.
233. Ibid., 89.
234. R.E.M. Wheeler, "Mr. Collingwood and Mr. Randall: A Note", *Antiquity* 4 (1930), 91 - 95.
235. Ibid., 95.
236. Collingwood, "Roman Britain", in Frank ed., *Economic Survey*, 10.
237. Collingwood, "Town and Country", 276.

1. See Appendix I.
2. See p. 68.
3. Under the heading "the object of history" Collingwood describes history in *The Idea of History* as "the science of *res gestae*, the attempt to answer questions about human actions done in the past" (IH,9).
4. Logic and ethics are given as examples (EM,108). In *The Principles of Art* Collingwood also mentions logic and ethics as "criteriological" sciences, that is, they are "...concerned not only with the 'facts' of thought but also with the 'criteria' or standards which thought imposes on itself" (PA,171).
5. See especially "Human Nature and Human History" (IH,205-231).
6. LM, 1927-1, XVIII.
7. Ibid., XIX.
8. *The International Journal of Ethics* 36(1925-1926), 162-185.
9. Ibid., 162. See also p. 37.
10. See pp. 57-59.
11. Mink, *Mind*, 79-118, and *History and Theory* 7 (1968), 3-37.
12. Skagestad, *Making Sense of History*, 63-70.
13. LM, 1933-34.
14. See pp. 245-246.
15. LM, 1933-34-D, 1 b.
16. LM, 1933-34-C, 84.
17. This passage makes clear that Collingwood rejects precisely those ideas ascribed to him by Gardiner, who asserts that "Collingwood's compromise suggestion is to treat thoughts as entities...which two different persons may 'have' at different times" ("The 'Objects' of Historical Knowledge", *Philosophy* 27(1952), 214). Elsewhere Gardiner says on the "'inside-outside' theory of historical events", that it is misleading, "...because the introduction of a spatial metaphor gives the impression that what are called the 'insides' of events are queer objects, invisible engines that make the wheels go round" (*Historical Explanation*, 47).
18. See pp. 82-83.
19. LM, 1933-34-A, 12.
20. See pp. 108-109.
21. LM, 1933-34-B, 12.
22. See p. 198.
23. See p. 199.
24. Walsh, *Introduction*, 49-50.
25. "Not only is the history of thought possible", Collingwood says in the 1928 lectures, "but, if thought is understood in its widest sense, it is the only thing of which there can be history" (LM,1928-1,7).
26. See p. 83.
27. Mink, *Mind*, 96.
28. *RB* (1923), 82; *RB* (1932), 115.
29. See pp. 187-191.
30. LM, 1936-4.
31. Ibid., 29.
32. Ibid., 30.
33. Ibid., 31.
34. Ibid.
35. Ibid., 32.
36. See p. 53.

37. See pp. 48-49, 66-67.
38. LM, 1933-34-A, 60.
39. Ibid., 61.
40. Ibid., 62.
41. Ibid. Compare what Collingwood states in *The Idea of History*: "...[I]n human affairs... there is no...fixed repertory of specific forms. Here, the process of becoming ... [involves] not only the instances or quasi-instances of the forms, but the forms themselves" (IH,210).
42. Ibid., 63-64.
43. *RB* (1923), 17.
44. *RBES*, 83.
45. LM, 1936-37-3, 33.
46. LM, 1939-40-1, 10.
47. Ibid., 11.
48. *RB* (1932), 47.
49. *JRS* 21 (1931), 36-64. See also pp. 233-235.
50. Ibid., 61.
51. Ibid., 62-63. The phrase "the real is the rational and the rational is the real" has been wrongly borrowed from Hegel, since the latter says in his Preface to the *Grundlinien der Philosophie des Rechts*: "Was vernünftig ist, das ist wirklich; und was wirklich ist, das ist vernünftig" ("The rational is real; and the real is rational").
52. Debbins ed. *Essays*, 17.
53. LM, 1926-2, 46-49.
54. LM, 1929-1, 12.
55. See p. 165.
56. LM, 1935-8, 8.
57. LM, 1933-34-A, 68-69.
58. Ibid., 69-70.
59. Collingwood would have made the distinction between the second and third form clearer if he had mentioned also the feature of rationality in the case of the third form. For, as we have seen, he considers this an essential element of the human past. In this connection Collingwood's appraisal of Bergson in the *Idea of History* is illustrative. After praising him for conceiving mental life as a process Collingwood contends: "But the process which Bergson is describing, although it is a mental process, is not a rational process" (IH, 187-188).
60. LM, 1926-2, 64-65.
61. LM, 1928-1,41.
62. See pp. 153-54.
63. LM, 1928-1,45.
64. Reviewing W. Benett, *Freedom and Liberty* (Oxford, 1920) Collingwood says: "Mr. Benett's book is conspicuous for its philosophical plot: that is, its concentration on the working-out of a single concept, that of Freedom, as a synthesis of opposing moments, which he calls Liberty and Law. The development of this plot carries the interest through the book in a remarkable way". Collingwood also speaks of "a real dialectical drama" and says that "Mr. Benett's Hegelism is a sign of the times: the more so as it appears to be quite unconscious". G.L. Raymond, *Ethics and Natural Law* (London, 1920) is called by Collingwood "of the opposite kind": "it has no plot at all, and consists of a mass of disconnected detail strung together on an all too slender thread" (*The Oxford Magazine* 39(1921), 264).

65. What Collingwood means by historical plots closely resembles Walsh's conception of "colligation" in history, being "...the procedure of explaining an event by tracing its intrinsic relations to other events and locating it in its historical context" (*Introduction*, 59).
66. Aut,38-39.
67. This statement echoes Ranke's well-known dictum "jede Epoche ist unmittelbar zu Gott" (*Weltgeschichte*, Vol. 9 (Leipzig, 1888),5).
68. LM, 1936-37-2, 15, 16.
69. LM, 1938-39-2, 22. Collingwood also observes that "[t]he practice of confining the use of the term to cases where the event is untoward is a vulgarism which has established itself only in certain contexts, e.g. in the vocabulary of insurance".
70. Ibid., 23.
71. This standpoint is related to the question of the role of objective conditions in history, which will be discussed later in chapter 8.5.
72. *The English Historical Review* 46(1931), 465. For Bury see also p. 52.
73. That the word "revolution" is not used any more by historians is of course far from true. The same may be said of the word "chance" with physicists. Collingwood's contention is clearly the product of wishful thinking. What he asserts, however, on the French Revolution is interesting and a good illustration of his point in repudiating the concept: "The 'people' on which the Revolution aimed at bestowing power was not the population as a whole...It was not a rabble; it was the *bourgeoisie*; and the *bourgeoisie* was already an organized body corporately possessed of economic power. The problem of the revolutionaries was to bestow political power where economic power already lay". (NL,198-199). Of political thought in the nineteenth century on the French Revolution he says that it "failed in particular to apprehend its continuity with the long historical process out of which it had grown; failed to see it as the legitimate offspring of that process, a development of tendencies long visibly at work, predictable long beforehand by any intelligent observer and predicted by many" (NL,199).
74. LM, 1939-40-1,10.
75. Mink, "Collingwood's Historicism", in Krausz ed., *Essays*, 154-157, 161-164.
76. See pp. 143-144.

NOTES TO CHAPTER SEVEN

1. Donagan, "The Verification of Historical Theses", 207.
2. See L.B. Cebik, "Collingwood: Action, Re-enactment, and Evidence", *Philosophical Forum* 2(1970), 68-90; G.S. Couse, "Neglected Implications of R.G. Collingwood's Attack on 'Scissors-and-Paste History'", *The Canadian Historical Association. Historical Papers*, 1972, 23-38; C.A.J. Coady, "Collingwood and Historical Testimony", *Philosophy* 50(1975), 409-424.
3. See Appendix III.
4. It is interesting to see how closely this statement corresponds with Collingwood's view of Vico as expressed nine years before: "Vico's chosen field, as an historian, was the history of remote antiquity. He studied distant and obscure periods precisely because they were distant and obscure; for his real interest was in historical method, and, according as the sources are scanty and dubious and the subject-matter strange and hard to understand, the importance of sound method becomes plain" (PhH, 127).

5. Collingwood, "Hadrian's Wall: 1921-1930", 62.
6. LM, 1936-3, 1.
7. Ibid., 1-2. See also IH, 206-207.
8. Ibid., 1.
9. Ibid., 2. See also IH, 209.
10. LM. 1936-2, 11.
11. Ibid.
12. Ibid., 11-12.
13. Ibid., 12.
14. Ibid.
15. LM, 1935-6, 1-19.
16. See p. 139.
17. LM, 1935-6, 9.
18. LM, 1931-2, 10.
19. See p. 33.
20. Oakeshott's view corresponds remarkably closely to that of Collingwood: "'What really happened' ...as the end in history must, if history is to be rescued from nonentity, be replaced by 'what the evidence obliges us to believe'. All that history has is 'the evidence'; outside this lies nothing at all" (*Experience and Its Modes* (Cambridge, 1933), 107-108). The same wording, however, had already been used before by Collingwood. For in the table of contents of the 1928 lectures one section is entitled: "historical truth = *what evidence obliges us to believe*".
21. In *The Idea of History* he observes that "[i] f history means scientific history, for 'source' we must read 'evidence'" (IH, 279). This statement is made in the context of a discussion where "source" seems to be identified with the testimonies of scissors-and-paste history. Neither in theory, nor in practice, however, is this distinction made in the rest of his work.
22. On evidence he asserts that it is "[c]*reated*, not *discovered*, because evidence is not evidence until it makes something evident" (NAPhH, 52).
23. LM, 1928-1, 22.
24. See pp. 155-156.
25. "A history of the Romano-British people has to be written with attention to the somewhat meagre literary sources, eked out by analogy from those of other provinces; but its material comes mainly from archaeology", Collingwood contends (*RBES*, VI). In the bibliography of the same book he asserts: "Our knowledge of Roman Britain is based only to a very small extent on texts: most of it depends on archaeological material, and a bibliography can only refer a reader to descriptions and discussions of this material, not to the material itself" (*RBES*, 462).
26. See p. 284.
27. LM, 1926-2, 12-13.
28. R.G. Collingwood, review of R.E.M. Wheeler, *Prehistoric and Roman Wales* (Oxford, 1925), in *The English Historical Review* 42(1927), 109.
29. J.B. Bury, "The Notitia Dignitatum", *JRS* 10(1920), 131-154.
30. R.G. Collingwood, "The Roman Evacuation of Britain", *JRS* 12(1922), 75.
31. See pp. 216-218.
32. Collingwood, *Archaeology*, 185-186.
33. Ibid., 186-187.
34. Ibid., 187-189.
35. Ibid., 189.
36. Ibid., 191.
37. Ibid., 192.

38. Ibid., 192-193.
39. Ibid., 193.
40. R.G. Collingwood, "Britain and the Roman Empire", in F.S. Marvin ed., *England and the World* (Oxford, 1925), 55.
41. Collingwood, "Roman Signal-Stations", 155.
42. R.G. Collingwood, "A newly-discovered Roman Site in Cumberland", *Antiquity* 4(1930), 472-477.
43. LM, 1935-6, 42-43. See also IH, 247.
44. Ibid.
45. See p. 184.
46. Couse, "Neglected Implications", 28, 30; Coady, "Testimony", 417.
47. Coady, "Testimony", 420.
48. Collingwood, "Roman Britain", in Frank ed., *Economic Survey*, 7-118.
49. Ibid., 54.
50. See pp. 229-230.
51. Couse, "Neglected Implications", 31.
52. Coady, "Testimony", 420.
53. Ibid., 421.
54. Elton, *Political History*, 133.
55. Fain, *Between Philosophy and History*, 144.
56. Ibid., 155.
57. See p. 184.
58. (New York, 1969)
59. Ibid., 165.
60. Ibid., 14.
61. See pp. 203-204.
62. See pp. 21, 201.
63. See pp. 21-22.
64. See Aut, 29-43.
65. See for instance Donagan, *Later Philosophy*, 56-63; Mink, *Mind*, 123-131. Collingwood's logic of question and answer is discussed by the German philosopher Gadamer within the context of his own theory of hermeneutics (H. G. Gadamer, *Wahrheit und Methode* (Tübingen, 1972), 351-360).
66. Mink, *Mind*, 124, 138.
67. Ibid., 132.
68. Ibid. Compare this with Collingwood's description of the past and future as the two elements of necessity and possibility within the present (see p. 36).
69. Compare the distinction made with regard to archaeology between "backward thinking" and "forward thinking" (see pp. 221-222).
70. Mink, *Mind*, 192.
71. "The Logic of Question and Answer is not a theory of logic at all, in any ordinary sense of that term, nor is it even a theory of semantics; it is a hermeneutics" (Ibid., 131).
72. Ibid., 139.
73. LM, 1928-1, opposite p.54.
74. LM, 1935-6, 15.
75. See pp. 43-47, 51-52, 69-71, 163-164.
76. LM, 1934-1, 17.
77. See note 75. In chapter 2.4 Collingwood's views on the relation between science and history have also been discussed. We have seen there how he emphasizes in "Are History and Science Different Kinds of Knowledge?" (1922) the similarity between science and history from another viewpoint than the questioning activity. Both science and history, he contends in this

article, make use of generalizations but are focussed on the individual. In "Science and History" (1923) the differences between science and history are worked out. Likewise, coming back to the subject in the thirties, Collingwood also develops certain ideas both on the similarity and dissimilarity between science and history. This time, however, other aspects are discussed: the similarity between science and history is seen in their questioning activity and their dissimilarity in their being an inferential and reasoned form of knowledge in different ways.

78. See for instance K. R. Popper, *Conjectures and Refutations* (New York, 1963), 198-200; idem, "A Pluralist Approach to the Philosophy of History", in E. Streissler et al. (eds.), *Roads to Freedom: Essays in Honour of F. A. von Hayek* (London, 1970), 188-192; idem, *Objective Knowledge: An Evolutionary Approach* (Oxford, 1972), 118-119, 180-182, 260-261, 290.
79. Compare Popper's contention that "[w]e are not students of some subject matter but students of problems" (*Conjectures*, 67).
80. Popper, "Pluralist Approach", 191. Elsewhere Popper declares that "...we are led back to the view of science and of the growth of knowledge as always starting from, and always ending with, problems - problems of an ever increasing depth, and an ever increasing fertility in suggesting new problems" (*Conjectures*, 222).
81. Popper, *Objective Knowledge*, 246.
82. K.R. Popper, *The Logic of Scientific Discovery* (London, 1959), 22.
83. Popper, "Pluralist Approach", 191.
84. See p. 151. It is interesting to compare the principle with Popper's contention that "...you should study the *problem situation* of the day. This means that you pick up, and try to continue, a line of inquiry which has the whole background of the earlier development of science behind it; you fall in with the tradition of science" (*Conjectures*, 129).
85. See pp. 151-152.
86. LM, 1928-1, 51.
87. Ibid., 52.
88. Ibid., 53. See also p. 155.
89. LM, 1935-6, 27.
90. LM, 1936-5, 43. In history "fact" may both refer to a past event and the evidence for it. Though Collingwood does not make this distinction in the quoted passage, one could say that his contention is relevant for both meanings.
91. Compare the similar criticism Popper levels against the "essentialist" definitions of the Greeks (*The Open Society and its Enemies*, Vol. 2 (London, 1945), 9-21).
92. Collingwood, "Hadrian's Wall: 1921-1930", 37-38.
93. See p. 54. Donagan seems to interpret the "questions of principle", Collingwood refers to, as his (Baconian) "principles of historical thinking" ("The Verification of Historical Theses", *The Philosophical Quarterly* 6(1956), 199). This interpretation does not appear to be correct. For the question of principle Collingwood refers to is the principle of incapsulation of a cultural inheritance. This principle serves as a presupposition for the right question to be asked about the problem of the "Celtic Revival".
94. *RBES*, 252-253.
95. Ibid., 252.
96. *RB* (1932), 143.
97. Ibid., 145.
98. Ibid., 145-149.

99. (Oxford, 1926).
100. *Antiquity* 1(1927), 369.
101. Ibid.
102. *RBES*, 170.
103. Ibid. A similar example, where the presuppositions of a question are discussed by Collingwood, is to be found in his review of R. E. Zachrisson, *Romans, Kelts and Saxons in Ancient Britain* (Uppsala, 1927). Commenting on Zachrisson's discussion of the survival of Romano-British towns, Collingwood argues: "If we asked what we meant by saying 'a town survived' or 'did not survive', it would become clear that the word survival, in such a context, has various meanings. For the social, economic or political historian a town is an institution. Its life is a complex of social habits, economic facts and political ideas. A town survives, in this sense, as long as these habits and facts and ideas have an effective existence within it. Now, says Z., 'when a Saxon town is built on the very site of its Roman predecessor it is difficult to believe that there was not some kind of continuity'. Perhaps; but *what* kind? A handful of de-Romanized Britons, squatting among the ruins of a Roman town, represent a continuity of race and to some extent a continuity of language. But from the point of view of the social, economic and political historian they are discontinuity incarnate" (*JRS* 18(1928), 118).
104. R.G. Collingwood, "The Romans and Britain", in S.A. Cook et al. (eds.), *CAH*, Vol. 10 (Cambridge, 1934), 797.
105. A methodological interpretation of the re-enactment doctrine does not entail, though, an intuitionistic one. Goldstein is an example of an interpreter, who takes the re-enactment doctrine methodologically while rejecting the intuitionistic version (see pp. 95,99).
106. See pp. 94-95.
107. LM, 1926-2, 26.
108. Ibid., 34.
109. Ibid., 35.
110. See p. 285.
111. LM, 1929-1, 24.
112. Ibid., 26.
113. Ibid., 28.
114. Ibid., 33. For Collingwood's judgment on the German view of history see also p. 161.
115. LM, 1936-2, 106-107.
116. LM, 1929-2, 3.
117. Ibid., 3-4.
118. LM, 1936-5, 23.
119. See pp. 95-96.
120. L. Rubinoff ed., *Faith and Reason*, 278-279. The article "Can the New Idealism Dispense with Mysticism?" originally appeared in *Proceedings of the Aristotelian Society*, Supplementary Volume 3 (1923), 161-175.

NOTES TO CHAPTER EIGHT

1. See pp. 68-69, 71.
2. See p. 54.
3. See pp. 68-69.
4. Though he had not yet developed his theory of re-thinking Collingwood already links, however, in passing in his articles on Croce (1921) and

Spengler (1927) the conceptions of re-thinking and a living past (see pp. 72-73).

5. See pp. 28-34, 138, 141-142, 144-148.
6. See pp. 35-36.
7. See pp. 141-142.
8. See pp. 144-147.
9. For the close interrelation between the historical process and the thought about it see also pp. 71-72.
10. LM, 1938-39-2, 19.
11. See pp. 170-171.
12. LM, 1934-3, 1.
13. See pp. 37-38, 133-135, 143.
14. LM, 1933-34-D, 38.
15. Ibid., 39.
16. See pp. 147-149.
17. See p. 143.
18. For a discussion of their views see pp. 97-99, 106.
19. See pp. 99-101. Beside the passages Donagan refers to, one could mention in support of this viewpoint the following statement by Collingwood: "History cannot be scientifically written unless the historian can re-enact in his own mind the experience of the people whose actions he is narrating" (IH, 39).
20. See pp. 101-105.
21. See pp. 69-70.
22. LM, 1936-3, 4.
23. See pp. 134-135.
24. LM, 1936-3, 5.
25. Ibid., 7.
26. Ibid.
27. (Oxford, 1936).
28. *The English Historical Review* 52(1937), 143-144.
29. See p. 66.
30. See pp. 203-204.
31. See pp. 155-156.
32. LM, 1928-1, 58.
33. Ibid.
34. See p. 261.
35. See pp. 74-76.
36. For Collingwood's statement that thought is *de jure* and *de facto* "common property", see p. 149.
37. See p. 149.
38. LM, 1933-34-E, 33-34.
39. It is remarkable how much Dray's interpretation of the re-enactment doctrine is in line with Collingwood's view as expressed in the passage quoted above. "...[A]lthough the thoughts which two people have may be identical", Dray maintains, "the experience which shows that each has the relevant thought may, on Collingwood's view, be entirely different. In terms of experience, there are many ways of thinking the 'same thought'" ("R.G. Collingwood and the Acquaintance Theory of Knowledge", 430).
40. See pp. 178-179.
41. In his manuscripts on folklore Collingwood asserts: "The historian can only understand a thought by thinking it over again for himself" (LM, 1936-37-4, 20). See also p. 187 of this study.
42. Walsh, *Introduction*, 93.

43. See pp. 105-106.
44. Skagestad, *Making Sense of History*, 54. For Skagestad's position see pp. 106-107 of this study.
45. Ibid., 54-55, 67-68.
46. Collingwood's views on psychology will be discussed in chapter 10.2.
47. The same view is expressed by Shoemaker and Debbins, and it is implied by Dray's interpretation (see p. 107).
48. For the notion of "incapsulation" see pp. 77-78.
49. See p. 250.
50. *The Home Reading Magazine* 36(1925), 7.
51. Ibid., 8.
52. Collingwood, *Roman Eskdale*, 38.
53. Collingwood, *A Guide to the Roman Wall*, 9-10.
54. See p. 258.
55. See p. 78.
56. R.G. Collingwood, "Hadrian's Wall", *History* 10 (1925), 201.
57. *RBES*, 51-52.
58. Ibid., 157.
59. Ibid., 291-292.
60. Ibid., 99.
61. R. G. Collingwood, "The Bewcastle Cross", *CW* 35(1935), 2.
62. Collingwood, "Britain and the Roman Empire", 59; *RB* (1932), 123.
63. Letter to Hughes, 8 -12- 1939, in Bodleian Library, Oxford, Ms. Eng. misc., c 516, 38.
64. *RB* (1932), 109.
65. See p. 85.
66. *RB* (1923), 87.
67. Reprinted in Rubinoff ed., *Faith and Reason*, 226.
68. Collingwood, "Some Perplexities about Time", 136.
69. Collingwood, "Political Action", 168-169.
70. Elsewhere Collingwood asserts: "The body of human thought or mental activity is a corporate possession, enjoyed both severally and collectively by individual men" (LM, 1936-3, 17). Dray's article "Collingwood's Historical Individualism" (*Canadian Journal of Philosophy* 10(1980), 1-20) was published after the completion of this study. Dray comes to the conclusion that Collingwood "...embraced a form of methodological holism, while clinging still to an *ontological* individualism, the precise warrant for which, in the context of his own thought, he left unclear"(20).
71. LM, 1933-34-B, 15.
72. LM, 1936-37-1, 3.
73. That Collingwood is also far from over-emphasizing the role of individuals - as his view of history is so often thought to imply - is made clear by the following passage: "The conception of an original work of science or art, created in its entirety by a single man, is a mere fiction. We owe it to Descartes, who undertook in his own philosophical system to give us an example of such a thing: but actually that system contained large elements of medieval scholasticism, Renaissance Platonism, and so forth, which he found ready-made in the philosophical tradition to which he was heir, and incorporated in his own work as he saw fit". After also giving Shakespeare, Voltaire and Gibbon as examples, he concludes: "The absolutely unprejudiced thinker, the original creator who makes an absolutely new work of art, is no actual human being; he is an imaginary creature, like the purely economic man of the early economists. Every actual thinker, every actual

artist, works by adding his quota to a tradition, a common stock of ideas already current" (LM, 1936-37-1, 7-8).

74. See p. 260.
75. In *The New Leviathan* Collingwood rejects the notion of the "unconscious", especially as elaborated by Freud (NL, 36-38). He agrees, however, with Freud's notion of the "preconscious". Since Collingwood himself also uses in practice the word "unconscious" (apparently in its common-sense meaning), I will follow him in this.
76. *RBES*, 252-253.
77. LM, 1936-5, 18-19.
78. See p. 54. It is a moot point, however, how the notion of tradition as unconscious historical process can be reconciled with Collingwood's idea that "...the historical past...is a living past, kept alive by the act of historical thinking itself..." (IH,226) , while historical thinking is seen as an activity "...which is...a form of thought possible only to a mind which knows itself to be thinking in that way" (IH,289) and as being reflective (IH,307).
79. A clear example is the one given by Collingwood of the rationality implicit in Paris dressmakers (see p. 278).
80. See p. 88.
81. LM, 1936-5, 4.
82. LM, 1931-2, 8.
83. LM, 1933-34-E, 110.
84. LM, 1935-6, 20-21.
85. Ibid., 21.
86. At one place, though not in connection with economic history, Collingwood even explicitly deserts the re-enactment doctrine. "*So far as* the historian's thought is a re-thinking of past thought" (italics mine), Collingwood asserts in his "Notes on the History of Historiography and Philosophy of History", "what was important for the persons whose thought he re-thinks is important for his own too. The state of the weather at the battle of Trafalgar is important to the naval historian because it was important to Villeneuve and Nelson. For them, it determined the conditions under which they had to fight: for him, the knowledge of it is essential to any understanding of the battle. Still, what was unimportant to the historical agent may be important to the historian, and then he is right to say that it was important although no one saw its importance at the time. A statesman may think it unimportant that his policy is opposed by a small minority; to the historian, the existence of this minority may be an important fact because of the light it throws on later developments" (LM, 1936-5, 50-51).
87. LM, 1935-6, 39. Though a system of social or economic relations of which contemporaries could not have been conscious is studied from the outside, one could observe that in the interpolations made by an historian questions of principle are involved which do not conflict with the re-enactment doctrine.
88. M. Hammond, review of *RBES*, in *Political Science Quarterly* 53(1938), 456.
89. Rubinoff ed., *Faith and Reason*, 218-219.
90. See p. 153.
91. LM, 1933-34-A, 66-67.
92. LM, 1936-5, 49.
93. The third sense of cause is the tight relation of a necessary and sufficient condition in nature. Discussing its problems Collingwood comes to the conclusion that it should be seen in anthropomorphic terms as compulsion (EM, 321-322).

94. LM, 1935-6, 40-41.
95. LM, 1938-39-2, 5-6. Likewise, Collingwood asserts in *An Essay on Metaphysics*: "It became evident that when eighteenth-century historians spoke of nature as the cause of historical events what they ought to have said, and would have said if their grasp on historical method had been firmer, was that man's historical activities were conditioned not by nature itself but by what he was able to make of nature. And since what man makes of nature depends on man's own historical achievements, such as the arts of agriculture and navigation, the so-called conditioning of history by nature is in reality a conditioning of history by itself" (EM, 98).
96. Collingwood, "Rome in Britain", 71.
97. *RB* (1923), 41; *RB* (1932), 48.
98. See p. 279.
99. LM, 1938-39-2, 23.
100. Collingwood, "The Fosse", *JRS* 14(1924), 252. When Collingwood speaks of people unconsciously being forced to do something he contradicts his own principle of causation in history that "...which is caused is the free and deliberate act of a conscious and responsible agent..." (EM, 290).
101. Collingwood, "Town and Country in Roman Britain", 265.
102. Collingwood, "The Roman Signal Station", in A. Rowntree ed., *History of Scarborough* (London, 1931), 40.
103. Collingwood, "Britain", in *CAH*, Vol. 11 (1936), 511.
104. Collingwood, "Rome in Britain", 7.
105. Collingwood, "Britain", in *CAH*, Vol. 11 (1936), 520.
106. *RBES*, 3.
107. See chapter 5.2.4.
108. Collingwood, "An Introduction to the Prehistory of Cumberland etc.", 171. For other references to Fox, see LM, 1936-37-8, 32, and "Roman Britain", in Frank ed., *Economic Survey*, 7-9. In I. L. L. Foster and L. Alcock eds., *Culture and Environment: Essays in Honour of Sir Cyril Fox* (London, 1963), the editors say in their preface, that "[t]he study of environment and distribution and their influence on material culture has been a fundamental and inspiring theme in the many contributions of Sir Cyril Fox to archeology". The distinction between the highland and lowland zone in Britain and its influence on British history has been worked out by Fox in *The Personality of Britain* (Cardiff, 1932). Collingwood refers to this book several times. In *Culture and Environment* Wheeler says of Fox that already at an early date " ...he was enunciating what became widely known as 'Fox's law': the basic cultural differentiation of the Highland from the Lowland Zone of the British Isles" (3). Wheeler does not mention that it had been Collingwood who coined the term "Fox's law" for the theory.
109. Besides this Collingwood contends that "[i]n sense I every causal proposition is an individual proposition" (EM, 308).
110. *RBES*, 181.
111. Ibid., 304.
112. LM, 1936-37-8, 23-24.
113. For the complete passage see p. 117.
114. In the same article Collingwood discusses the element of universality contained in individual judgments (see p. 49 of this study).
115. LM, 1935-6, 44. See also IH, 248.
116. LM, 1936-37-2, 21.
117. See pp. 213-214.
118. See pp. 140-141, 155-156.

119. Martin, for instance, asserts: "The essence of his [Collingwood's] criticism of the science of human nature was that, due to the development and increasing differentiation of mind in historical process, there are no universally recurrent phenomena. What we do have is a transhistorical heterogeneity in the phenomena of human thought and action" (*Explanation*, 29).
120. See p. 190.
121. All the following quotations are from LM, 1936-37-5, respectively 5, 12, 22, 24, 36, 41, 48.
122. Of the maxims of Spinoza "neither to condemn nor to deride the feelings and actions of men, but to understand them", and of Butler "that every thing is what it is, and not another thing", mentioned by Collingwood in connection with the study of primitives (see p. 187), one could contend that they also hold for the study of the past. It should be added, though, that the first one would be highly debatable.
123. Martin, *Explanation*, 45.
124. Ibid., 47.
125. Ibid., 63.
126. Ibid., 104.
127. Walsh adheres to both views. Martin, however, rejects the intuitive version of the re-enactment doctrine (see note 113 to chapter 3).
128. See p. 154.
129. *RBES*, 79-80.
130. Ibid., 215.
131. Collingwood, "Hadrian's Wall", *History* 10(1925), 202.
132. Collingwood, "The Conquest of Britain", in *CAH*, Vol. 10 (1934), 797.
133. Collingwood, "Britain", in *CAH*, Vol. 11 (1936), 519.
134. *RBES*, 133.
135. See pp. 290-292.
136. Collingwood, "Roman-Signal Stations", 143.
137. See pp. 250-252.
138. Randall, "Population and Agriculture", 81.
139. See chapter 4.8.
140. See p. 153.
141. See pp. 167-168.
142. See p. 321.
143. LM, 1936-3, 19.
144. Ibid.
145. See pp. 296-297.
146. *RBES*, 41.
147. Collingwood, "The Conquest of Britain", in *CAH*, Vol. 10 (1934), 793.
148. *RBES*, 252, 254; *CAH*, Vol. 12 (1939), 293.
149. Popper, *Open Society*, Vol. 2, 97-99, 265; idem, *Poverty*, 147-152.
150. K. R. Popper, "Die Logik der Sozialwissenschaften", in Th. W. Adorno et. al., *Der Positivismusstreit in der deutschen Soziologie* (Neuwied und Berlin, 1969), 103-123. This paper originally appeared in *Kölner Zeitschrift für Soziologie und Sozialpsychologie* 14(1962), 233-248.
151. Ibid., 120. "...*a purely objective method* in the social sciences which may well be called the method of *objective* understanding, or situational logic. A social science orientated towards objective understanding or situational logic can be developed independently of all subjective or psychological ideas. Its method consists in analysing the social *situation* of acting men sufficiently to explain the action with the help of the situation, without any

further help from psychology. Objective understanding consists in realizing that the action was objectively *appropriate to the situation.* In other words, the situation is analysed far enough for the elements which initially appeared to be psychological (such as wishes, motives, memories, and associations) to be transformed into elements of the situation" (Th.W.Adorno et al., *The Positivist Dispute in German Sociology* (London, 1976), 102).

152. Dray, *Laws and Explanation*, 124-125. Dray says: "[T]hat what we want to know when we ask to have the action explained is in what way it was *appropriate*" (124). Dray's "rational explanation" model for history was set up as an alternative to Popper's and Hempel's "covering law model". Recently, however, Popper has expressed the view that he did not consider the latter important for historical explanation and has made a remarkable switch to the "logic of the situation". Referring to the "covering law" model he asserts: "I did not ...regard this particular analysis as especially important for historical explanation, and what I did regard as important needed some further years in which to mature. It was the problem of rationality (or the 'rationality principle' or the 'zero method' or the 'logic of the situation'). But for years the unimportant thesis - in a misinterpreted form - has, under the name 'the deductive model', helped to generate a voluminous literature. The much more important aspect of the problem, the method of situational analysis ... was developed from what I had previously called the 'zero method'. The main point here was an attempt *to generalize the method of economic theory (marginal utility theory) so as to become applicable to the other theoretical social sciences.* In my later formulations, this method consists of constructing a *model of the social situation,* including especially the institutional situation, in which an agent is acting, in such a manner as to explain the rationality (the zero-character) of his action. Such models, then, are the testable hypotheses of the social sciences; and those models that are 'singular', more especially, are the (in principle testable) singular hypotheses of history" (K.R. Popper, "Intellectual Autobiography", in P.A. Schilpp ed., *The Philosophy of Karl Popper*, Vol. 1, 93-94).

153. Popper, "Pluralist Approach", 197-198.

154. Ibid., 198.

155. See pp. 319-320.

156. Collingwood, "The Romans and Britain", in *CAH*, Vol. 10 (1934), 793-794.

157. See p. 323.

158. See p. 180.

159. See pp. 170-171.

160. LM, 1934-3.

161. Ibid., 1. With "the given historical fact" Collingwood refers to a present fact, namely Nazi mentality, mentioned by him before. Collingwood's point, though, also holds for past facts, insofar as they are seen as facts which were present once.

162. Ibid., 2.

163. Ibid., 1.

164. LM, 1939-40-1, 24.

165. Dray, *Laws and Explanation*, 133; idem, *Philosophy of History* (Englewood Cliffs, 1964), 44.

166. LM, 1935-8, 19. (See also pp. 169-170 of this study). Burns and Mink have pointed out the same corollary, while Dray has discussed it with regard to history (see p.110).

167. Ibid.

168. Ibid., 20.
169. See chapter 3.3.8.
170. LM, 1926-2, 36.
171. Ibid., 37.
172. LM, 1928-1, 53.
173. LM, 1935-6, 41.
174. LM, 1938-39-2, 14.
175. Ibid., 14-15.

NOTES TO CHAPTER NINE

1. See Appendix I.
2. LM, 1933-34-A, 85, 90-91. See also IH, 211.
3. Ibid., 91.
4. See pp. 172-174.
5. LM, 1936-3, 16.
6. Ibid., 15.
7. Ibid.
8. LM, 1936-2, 9-10.
9. Ibid., 10.
10. Ibid., 11.
11. LM, 1929-2, 3.
12. See p. 285.
13. See p. 165.
14. "Judgments of value are nothing but the ways in which we apprehend the thought which is the inner side of human action" (LM, 1936-1, 12). Similarly, Collingwood contends elsewhere: "When a man is dead, the world has judged him, and my judgment does not matter; but the mere fact that I am rethinking his history proves that he is not dead, that the world has not yet passed its judgment. In my person, indeed, it is now about to pass judgment" (Debbins ed., *Essays*, 15). Collingwood is not consistent, however, when he asserts at another place: "Moses is dead, and there is no need to get in sweat about him ... There is no sense in using terms like good and bad except of persons or things, that come into practical relations with one's own will" (THC, 76). In his 1926 lectures Collingwood even explicitly rejects moral judgments: "We are not called upon to pass moral judgments at all. Our business is simply to face the facts ... The real holocaust of history is the historian's holocaust of his emotional and practical reaction towards the facts that it presents to his gaze. True history must be absolutely passionless, absolutely devoid of all judgments of value, of whatever kind". A little further on he sums up his position as follows: "[T]he dead must be left to bury their dead and to praise their virtues and lament their loss" (LM, 1926-2, 50-51). Collingwood does not keep to this principle, though, when he asserts in *Roman Britain and the English Settlements*: "The story of Boudicca from first to last is a story of horror and shame, a story of things that ought never to have happened" (103-104). Collingwood's ambivalence towards the question of value judgments in history may be shown by his review of Rostovtseff's *The Social and Economic History of the Roman Empire*. For after saying that Grote's and Mommsen's views of the past had been inspired by their political ideas, he contends: "Dry-as-dust historians, who believe that all modern interests and party passions should be forgotten by the suppliant on entering the temple of Clio, may deplore it; but they

cannot alter it; unless the past is seen in the light of the present, it cannot be seen at all. It can only be fumbled with. Yet the dangers are obvious. To read one's own political passions into the politics of the past is certainly the only way of bringing the past to life; but one may endow it with a life too like that of the present. And this is a danger which Professor Rostovtseff has not, perhaps, wholly escaped" (*Antiquity* 1(1927), 368).

15. See respectively pp. 42, 17, 160-161, 166 of this study. For *The Idea of Nature* see pp.176-177.
16. See pp. 162, 164.
17. LM, 1933-34-E, 110-114.
18. See p. 298.
19. See p. 159.
20. See p. 48. "[D] amnosa hereditas of the positivistic idea of fact", Collingwood asserts in his Notes of 1936 on Spengler, "The isolation of fact persists: only the facts are now large chunks with a uniform and rigidly-repeated internal morphology ... The relations between fact and fact remain as they are for positivism purely external, i.e., non-historical, they are related naturalistically (morphologically) not historically" (LM, 1936-5, 23).
21. LM, 1938-39-2, 10. After this he remarks: "Imagine a 'comparative pathology'. This condition is *like* nasal catarrh (but it *is* scarlet fever)".
22. See pp. 184-185.
23. LM, 1936-37-4, 6.
24. LM, 1938-39-2, 19.
25. Johnston, *Formative Years*, 142.
26. Ibid., 141.
27. See pp. 160-161.
28. LM, 1935-8, 11.
29. Ibid.
30. Ibid.
31. Ibid.
32. Ibid., 11-12.
33. LM, 1936-37-2, 5-6.
34. Ibid., 6.
35. Ibid., 17.
36. See pp. 342-343.
37. See p. 186.
38. LM, 1936-37.-4, 19-20.
39. See Appendix I.
40. LM, 1938-39-2, 18.
41. Ibid. For similar contentions see pp. 339, 356.
42. See p. 355.
43. See p. 338.
44. See chapter 5.4.
45. See p. 247.
46. P. H. Blair, review of *RBES*, in *The English Historical Review* 52(1937), 684.
47. Ibid., 685.
48. G. C. Whittick, review of *RBES*, in *The Classical Review* 51(1937), 78.
49. M. Hammond, review of *RBES*, in *Political Science Quarterly* 53(1938), 456.
50. LM, 1936-37-2, 6.
51. See pp. 3-5 of this study; M. Mandelbaum, review of *The Idea of History*, in *The Journal of Philosophy* 44(1947), 188; R. Aris, review of *The Idea of History*, in *Contemporary Review* 171(1947), 188.
52. LM, 1929-1, 15.

53. LM, 1939-40-1, 13.
54. Ibid.
55. Ibid., 14.
56. Ibid., 14-15.
57. Ibid., 15.
58. Ibid., 20.

NOTES TO CHAPTER TEN

1. Johnston, *Formative Years*, 143.
2. R.H.S. Crossman, "When Lightning struck the Ivory Tower: R.G. Collingwood", *The New Statesman and Nation* 17(1939), 223; reprinted in *The Charm of Politics and Other Essays in Political Criticism* (London, 1958), 107.
3. LM, 1936-4.
4. See especially EM, 133-142.
5. See p. 63.
6. Discussing in his *First Mate's Log* (Oxford, 1940) the accusations of his American shipmates that English education is not interested in the broader human issues, Collingwood asserts: "The accusation against English life in general, and Oxford life in particular ... is to a certain extent valid, as indicating faults in English education, and in the English idea of the relation between learned men and the public ... for the gulf between the learned specialist and the general public is a great evil, a menace to the welfare both of the public and of the learned" (86-87). In a letter to the editor of *Philosophy* Collingwood maintains with regard to the contribution philosophy might make to the solution of modern social and political problems: "It cannot descend like a *deus ex machina* upon the stage of practical life and, out of its superior insight into the nature of things, dictate the correct solution for this or that problem in morals, economic organization, or international politics". "If the philosopher is no pilot", he asserts a little further on, "neither is he a mere spectator, watching the ship from his study window. He is one of the crew" (*Philosophy* 9 (1934), 262-263).
7. See pp. 257-258.
8. LM, 1913, 1.
9. LM, 1933-34-A, 24.
10. See p. 177.
11. LM, 1936-37-3, 43, 46.
12. See pp. 258, 263.
13. See pp. 319-320.
14. *The Oxford Magazine* 41(1923), 425-426. This review is unsigned, but is mentioned in Collingwood's "List of work done" (LM, 1933-1).
15. A portable surveying instrument.
16. *The Oxford Magazine* 42(1923), 118. This review, too, is unsigned but mentioned in the "List of work done". The same book by Spearman is discussed in *An Essay on Metaphysics*, 126-132.
17. G. Ryle, *The Concept of Mind* (London, 1949), 27-32.
18. See *An Essay on Metaphysics*, 133-142.
19. Self-knowledge will be interpreted here as self-knowledge of the individual. In the Introduction of *The Idea of History* the notion of self-knowledge is used in a broader sense. As an answer to the question "what is history for?", Collingwood divides it into three aspects: "Knowing yourself means

knowing, first, what it is to be a man; secondly, knowing what it is to be the kind of man you are; and thirdly, knowing what it is to be the man *you* are and nobody else is". (IH, 10). For Collingwood's assessment of the Greek precept of human self-knowledge, see *The First Mate's Log*, 61-63.

20. LM, 1936-3, 18.
21. Ibid., 19. A little before that Collingwood asserts: "In acquiring historical knowledge we are enriching our personality by actively incorporating into it historical traditions which hitherto we have possessed only in a confused form or not at all" (ibid., 18).
22. Ibid., 22.
23. Sometimes, however, Collingwood refers to analogous historical conditions or developments. For instance, when he asserts: "Of all past historical periods, the Roman Empire is that which most closely resembles our own; and we are most of us aware that our civilization is exposed to forces which seem bent upon its destruction ... History never repeats itself; but its processes may resemble one another so closely that, so long as we duly attend to the features peculiar to each, it is not impossible to argue from one to another, and use Antiquity as a lantern to explore Futurity" (review of S. Dill, *Roman Society in Gaul in the Merovingian Age* (London, 1926), in *Antiquity* 1(1927), 117). Likewise, Collingwood elsewhere maintains: "...[T]he history of the Roman Empire is the history of the most important experiment known to us in conscious political co-operation among peoples widely differing in race, language, traditions and civilisations. At the present crisis in the history of the world, it concerns us to know how such an experiment succeeded and how it failed" ("Hadrian's Wall", *JRS* 21(1931), 61). That Collingwood is conscious, however, of the dangers of using analogies too concretely, may be shown by his contention elsewhere: "Some people are never tired of holding up the decline and fall of the Roman Empire as a warning of ourselves ... No doubt, resemblances may be found. A sufficiently determined seeker can always find them. But there are also differences, and the differences are fatal to the parallel" ("Town and Country", *Antiquity* 3(1929), 274). In the following example, finally, Collingwood shows how the past may determine certain aspects of the present: "...[T]o this day the English character blends the law-abiding Roman's love of sound government with the self-reliance of the seafaring Saxon, and is unintelligible and unmanageable except by people who realize this fact. So close the lessons of ancient history stand to the problems of modern life" (*The Roman Signal Station on Castle Hill, Scarborough* (Scarborough, 1925), 2). In a letter of 17 June 1925 to F. G. Simpson, Collingwood refers to this pamphlet (see Appendix IV).
24. See p. 143.
25. See p. 349.
26. See p. 349.
27. LM, 1939-40-1, 25.
28. See p. 170.
29. See pp. 170-171.
30. LM, 1936-3, 17.
31. See p. 340.
32. See p. 171.
33. LM, 1936-3, 13.
34. Broadly interpreted, one could contend, of course, that Collingwood's philosophy of mind already implies a common human nature in the sense that it describes certain characteristics of mind common to all men. His theory of

history similarly holds for the history of all societies. Interpreted in this way Collingwood's speaking of a "common human nature uniting the historian with the men whose work he is studying"(IH, 65) is understandable (see p. 341).

35. LM, 1935-8, 5-6.
36. See p. 171.
37. LM, 1935-8, 25-26.
38. Stuart Hampshire, "The Illusion of Sociobiology", *New York Review of Books*, 12-10-1978, 64. For a critical discussion of sociobiology see also M. Sahlins, *The Use and Abuse of Biology* (Ann Arbor, 1976).
39. (Cambridge, Mass., 1978).
40. R.G. Collingwood, "The Present Need of A Philosophy", *Philosophy* 9(1934), 262-265.
41. Ibid., 121.
42. Ibid., 264.
43. Ibid.
44. Ibid., 264-265.
45. In a letter to the sinologist Hughes, dated 8 December 1939, Collingwood speaks of "the idea of a union of philosophy and history" as "the point to which I have devoted and shall devote my life". He continues on it: "I am quite clear that this idea could save Europe, and believe that nothing else can. The present developments of European philosophy are, broadly speaking, nothing but a series of attempts to push Europe more quickly and more conclusively over the precipice ... you seem to see these truths, which I had almost fancied no one in England saw except myself!" (Bodleian Library, Oxford, Ms. Eng. misc., c 516).
46. Collingwood not only has epistemological arguments for this position, but also moral ones (see the scheme for the third book of *The Principles of History*, in Appendix I).

APPENDIX I

SCHEME FOR *THE PRINCIPLES OF HISTORY**

Scheme for a book, 9-2-39, "*The Principles of History*". *Main* topics will include, 1) a simple account of the most obvious characteristics of history *as a special science*, 2) Relations between this and others, 3) Relation of history as thought to practical life. These could be Books I, II, III.

I.1. State and expound the conception of *Evidence*. Contrast this with the conception of *Testimony* and the *Scissors-and-paste history* which that implies. I.2. State and expound the conception of *Action* (res gestae). Contrast this with the conception of *Process* or *Change* and the pseudo-history which that implies. I.3. Conception of *Re-enactment*, and contrast the *Dead Past* and *Completeness*. I.4. History as the *self-knowledge of mind*. Exclusion of other sciences of mind.

(there might be an *introduction* on the word "history", showing how originally it = investigation of *any* kind, e.g., historia naturalis: and how it gradually focussed itself on the kind of meaning which it had for Herodotus, Thucydides, and Polybius. Survivals of the original generalised sense in modern usages. This book deals with the specialised sense).

II.1. History and natural science. A. Difference between them irreducible: A 1. History not to be reduced to natural science; A 2. Natural science not to be reduced to history. B. Relation between them: any given natural science an historical achievement: its dependence on "facts" and "principles", i.e. the *historical* fact that someone has observed something, and thinks in a certain way about it. II.2. History and the human sciences. These are crypto-history or just history. II.3. History and philosophy...

III. The main idea here is that history is the negation of the traditional distinction between theory and practice. That distinction depends on taking, as our typical case of knowledge, the contemplation of nature, where the object is presupposed. In history the object is enacted and is therefore not an *object* at all. If this is worked out carefully, there should follow without difficulty a characterisation of an historical morality and an historical civilization, contrasting with our "scientific" one. Where "science" = of or belonging to *natural* science. A scientific morality will start from the idea of *human nature* as a thing to be conquered or obeyed: a historical one will deny that there is such a thing, and will resolve what we are into and what we do. A scientific society will turn on the idea of *mastering* people (by money or war or the like) or alternatively *serving* them (philanthropy). A historical society will turn on the idea of *understanding* them.

* LM, 1938-39-2, 20-21.

APPENDIX II

LECTURES GIVEN BY R. G. COLLINGWOOD *

1914	TT:	Theory of Knowledge
	MT:	Aristotle: De Anima
1915	TT:	Theory of Knowledge (Elementary)
	MT:	Philosophy of Religion
1920	HT:	The Ontological Proof of the Existence of God
	TT:	Introduction to the Theory of Knowledge
	MT:	Guido de Ruggiero
1921	HT:	1) Ontological Proof of the Existence of God; 2) The Roman Wall: History and Archaeology
	TT:	Introduction to the Theory of Knowledge
	MT:	Moral Philosophy (intended primarily for students in their last year)
1922	HT:	1) Ontological Proof of the Existence of God; 2) The Roman Wall: History and Archaeology
	TT:	Introduction to the Theory of Knowledge
	MT:	Moral Philosophy
1923	HT:	1) Philosophy of Religion; 2) The Roman Wall: History and Archaeology
	TT:	Theory of Knowledge (Introductory)
	MT:	Moral Philosophy
1924	HT:	Philosophy of Religion
	TT:	1) Introduction to the Theory of Knowledge; 2) Philosophy of Art
	MT:	Moral Philosophy
1925	HT:	Philosophy of Religion
	MT:	Moral Philosophy
1926	HT:	Philosophy of History
	TT:	1) Moral Philosophy; 2) Philosophy of History; 3) Roman Britain
	MT:	Moral Philosophy
1927	HT:	Roman Britain

* From the *Oxford University Gazette*, vols. 44-71 (1913-1914/1940-1941).

MT = Michaelmas Term (October-December)
HT = Hilary Term (January-March)
TT = Trinity Term (April-June)

	TT: Philosophy of History
	MT: Moral Philosophy
1928	HT: Roman Britain
	TT: Philosophy of History
1929	HT: Roman Britain
	TT: 1) Introduction to Ancient Philosophy (for the Regius Professor of Greek); 2) Philosophy of History
	MT: Moral Philosophy
1930	HT: Roman Britain
	TT: Philosophy of History
	MT: Moral Philosophy
1931	HT: Roman Britain
	TT: Philosophy of History
1932	MT: Moral Philosophy
1933	HT: Roman Britain
	TT: Philosophy, its Nature and Method
	MT: Moral Philosophy
1934	HT: Roman Britain
	MT: Nature and Mind
1935	HT: Roman Britain
	MT: 1) Nature and Mind; 2) Informal Instruction
1936	HT: 1) Philosophy of History; 2) Informal Instruction
	TT: 1) Philosophy of History (continued); 2) Informal Instruction
	MT: 1) Central Problems in Metaphysics; 2) Informal Instruction
1937	HT: 1) Nature and Mind; 2) Informal Instruction; 3) Roman Britain
	TT: 1) Philosophy of History; 2) Informal Instruction
	MT: 1) Philosophy of History (continued); 2) Philosophy of Art; 3) Informal Instruction
1938	HT: 1) Central Problems of Metaphysics; 2) Philosophy of Art (continued); 3) Informal Instruction; 4) Roman Britain
1939	TT: 1) Metaphysics; 2) Informal Instruction
	MT: 1) Nature and Mind; 2) Informal Instruction
1940	HT: 1) The Idea of Nature in Modern Science; 2) Goodness, Rightness, Utility
	TT: 1) The Idea of History; 2) Informal Instruction
	MT: 1) Philosophical Theory of Society and Politics; 2) Informal Instruction
1941	HT: 1) Theory of Society and Politics (continued); 2) Informal Instruction

APPENDIX III

REPORTS BY R. G. COLLINGWOOD TO THE FACULTY OF LITERAE HUMANIORES

Report on quinquennium as University Lecturer, from summer 1927 to time of writing in January 1932 *

By appointing me Lecturer in Philosophy and Roman History, I understand the University to mean, not only that I am to study and teach these two subjects, but also that I am to study and teach them in their mutual connexions: i.e. in philosophy, to investigate the philosophy of history, and, in history, not to neglect the methods and logic of historical work, and to emphasize the relation between history and its sources. Apart, therefore, from my ordinary work as a College tutor in philosophy, of which I say nothing, I have devoted my leisure to the following five projects:

1. Corpus of Roman Inscriptions in Britain, begun in 1920 and planned to take 20 years; every inscription to be illustrated with a facsimile drawing by myself.
2. An archaeological textbook of the materials used in the study of Roman Britain.
3. A large-scale work on Hadrian's Wall, with special reference to the problems of archaeological method there arising.
4. A history of Roman Britain with special reference to the relation between Roman and Celtic elements in its civilisation to form vol. I of the Oxford History of England.
5. A study of the philosophical problems arising out of history: especially a) logical and epistemological problems connected with the question "how is historical knowledge possible?", b) metaphysical problems concerned with the nature and reality of the objects of historical thought.

During this period, the above projects have advanced as follows:

1. The drawings made have risen from about 600 to about 1100. The whole material has been arranged and indexed.
2. The book is now published (no. 13 in the annexed list).
3. The Clarendon Press has agreed to publish the book; all the excavators and local archaeologists are assisting in a co-ordinated scheme of research; several preliminary studies have been written, e.g. no. 14 in the annexed list. In this connexion I have travelled in

* Oxford University Archives, Reports of the Board of the Faculty of Literae Humaniores, 1912-41, LH/R/1/5, 83.

Germany to study the similar remains there and to compare our methods of research with those of the Germans.

4. This book will gather up all the results of my studies on Roman Britain. In order to equip myself for it I have embarked on two new lines of study:
a) prehistoric Britain, with special reference to the Celtic Iron Age; b) the Roman occupation of other Western provinces, for purposes of comparison. I have chosen Spain for special study, and have been travelling there and reading on the subject for the last two years.

5. This I regard as my chief work, involving the whole of my philosophical and historical studies in their mutual connexions. I am of opinion that there is important work to be done here, and that it cannot be done except by a trained and practising historian who is also in constant work as a philosopher. This opinion has been strengthened by much reading in the last four years, and by gradually reaching, in that time, a provisional solution of most of the chief problems. Some results of this study are published in nos. 1, 4, 5, 6, 10 of the annexed list. No. 1 is in effect the synopsis of a complete treatise, but I do not intend to begin writing such a treatise until I have done several years' work on various aspects of the subject.

In conclusion, I may state that project no. 1 was my legacy from the late Professor Haverfield; the others have been rendered possible only by my appointment as University Lecturer and my consequent command of leisure.

Appendix

I List of Lectures, 1927-1931

A. Philosophical

Moral Philosophy: general course, given yearly: 16 lectures. Revised yearly and rewritten on a new plan in 1929.

Philosophy of History: 16 lectures, given yearly. Completely rewritten 1928; revised yearly.

Early Greek Philosophy: A single lecture.

B. Historical

Roman Britain: A general course, given yearly: 16 lectures. Revised yearly.

The above are public lectures under the Faculty of Lit.Hum. The following are the most important lectures, unpublished, given outside Oxford:

Recent advances in the study of Roman Britain. The J. H. Gray lectures at Cambridge, 1930.

Aesthetic Theory and Artistic Practice. Address to the British Institute of Philosophical Studies, 1931.

II List of Publications, 1927-1931

A. Philosophical

Books and pamphlets:

1. The Philosophy of History (Historical Association)
2. Faith & Reason (Benn's "Affirmations")

Articles:

3. Aesthetics (in the Volume "The Mind", ed. R. J. C. McDowall)
4. The Theory of Historical Cycles (*Antiquity*)
   I Oswald Spengler
   II Cycles and Progress
5. The Limits of Historical Knowledge (*J.Phil.Stud.*)
6. Progress and History (*The Realist*)
7. Political Action (*Proc.Arist.Soc.*)

Translations:

8. G. de Ruggiero, History of European Liberalism

8a. G. de Ruggiero, Science, History & Philosophy (in *Philosophy*)

9. B. Croce, art. Aesthetics in Encycl. Brit.

Reviews of books:
about 20, the most important being

10. J. B. Bury, Posthumous Essays (*Eng.Hist.Rev.*: on the philosophy of history)
11. A. E. Taylor, Plato, the man & his work (*Criterion*)
12. G. Santayana, The Realm of Essence (*Adelphi*)

B. Roman History and Archaeology

Books and pamphlets:

13. The Archaeology of Roman Britain
14. The Book of the Pilgrimage of Hadrian's Wall
15. Roman Eskdale

Articles (omitting a number of short articles on matters of detail):

a) General, on Roman Britain

16. Town & Country in Roman Britain (*Antiquity*)
17. Romano-Celtic Art in Northumbria (*Archaeologia*)
18. Hadrian's Wall: 1921-1930 (J.Rom.St.)
19. Ten Years' Work on Hadrian's Wall (Cumb.&West.Trans.)

b) Topographical studies

20. Hardknot Castle (*C.&W.Trans.*)
21. Roman Ravenglass (*C.&W.Trans.*)
22. Old Carlisle (*C.&W.Trans.*)
23. Roman Fort at Watercrook, Kendal (*C.&W.Trans.*)

24. Roman signal-stations on the Cumberland coast (*C.&W.Trans.*)

c) Catalogues

25. Scaleby Castle Roman Antiquities (*C.&W.Trans.*)
26. Roman objects from Stanwix and Thatcham (*Antiq.Journ.*)
27. Roman objects in the Craven Museum (*C.&W.Trans.*)

d) Excavation reports

28. Excavations at Brough-by-Bainbridge (Leeds Phil.Soc.)
29. Roman Fortlet on Barrock Fell (*C.&W.Trans.*)

Reviews of books
about 30, the most important being:

30. Rostovtseff, Social & Economic History (*Antiquity*)
31. Dill, Roman Society in Gaul in the Merovingian Age (*Antiquity*)
32. Bury, Invasion of Europe by the Barbarians (*Antiquity*)
33. Zachrisson, Romans, Kelts & Saxons in Ancient Britain (*J.R.S.*)
34. Fabricius, Der Obergermanisch-rätische Limes, 47 (*J.R.S.*)
35. Stähelin, Die Schweiz in röm. Zeit (J.R.S.)

In collaboration (the most important items only):

36. Roman London (Royal Commission on Historical Monuments: by an *ad hoc* Committee, of which I was chairman)
37. Roman Britain in 1927 (with Miss M. V. Taylor: *J.R.S.*)
38. Roman Britain in 1928 (with Miss M. V. Taylor: *J.R.S.*)
39. Roman Britain in 1929 (with Miss M. V. Taylor: *J.R.S.*)
40. Roman Britain in 1930 (with Miss M. V. Taylor: *J.R.S.*)
41. Review of Schulten, *Numantia*, vol. II (with Miss M. I. Munro: *J.R.S.*)

R. G. Collingwood

12-1-32

*Report as university lecturer in Philosophy and Roman History for the three years Michaelmas 1932 - Michaelmas 1935 **

I. *General observations*

1. Time has been spent about equally on the two subjects: the academic year 1932-3 chiefly on philosophy with some Roman history, 1933-4 almost exclusively on philosophy, 1934-5 almost exclusively on Roman history.
2. In philosophy I have been working a) at the problem of method, b) at metaphysical and in particular cosmological problems, upon which my main interest is at present concentrated.
3. In Roman History I have been concentrating upon a) the economic

* Oxford University Archives, Reports of the Board of the Faculty of Literae Humaniores, 1912-41, LH/R/1/6, 76.

aspects of provincial history (always with special reference to Britain), b) the relation between Romanized provincial life and its pre-Roman substratum; since I find that current conceptions of Romanization are vitiated by lack of contact between Roman historians and "prehistoric" archaeologists, which has hitherto made it impossible to form a just estimate of the work done by the Roman Empire in (at any rate) the Celtic provinces.

4. Publications during this period have been relatively few, because I have been, almost the whole time, preparing material for certain large works not yet written: these should be completed in the next 3 years or so if I continue to have leisure for them.

II. *Philosophy*

a. *Lectures given in Oxford*
   1. *Moral Philosophy.* In the long vacation of 1932 I completely rewrote my lectures on this subject and delivered them MT. 1932. In the long vacation 1933 I again thoroughly revised them, rewriting about half, adding a good deal of new material and omitting some old, and delivered them in this shape MT. 1933.
   2. *Philosophical Method.* A new course delivered TT. 1933, covering roughly the same ground as my book (see below).
   3. *The Nature and Scope of Metaphysics.* Two lectures given HT. 1934 as part of a "circus" course of metaphysics.
   4. *Nature and Mind.* A new course delivered MT. 1934. An historical review of philosophical theories of Nature, with special reference to its relation with Mind (as knower, as creator, as evolutionary product, etc.), from the pre-Socratics to the present day; concluding with a short statement of certain results reached in my own study of the subject (see below).

b. *Publications*
   The only one of importance has been my *Essay on Philosophical Method,* the writing of which occupied me more or less continuously from autumn 1932 to early summer of 1933. It was intended to serve as preface to a series of philosophical works based on the conception of method there expounded. This intention I am now carrying out as time permits.

c. *Unpublished work*
   In the summer of 1933, on finishing the above Essay, I began applying the ideas there stated to certain metaphysical problems connected with the ideas of Nature, Matter, Life, Evolution etc. in the hope of doing something to advance a branch of philosophical thought which has lately been brought again into prominence by Alexander, Whitehead and others. I spent a year exclusively on this

(except for the routine of teaching) and put together about 700 pages of rough MSS. containing what I hope will prove to be the groundwork of a future treatise. One by-product of this year's work, viz. the historical review of theories on the subject, formed my lectures in M.T. 1934; another by-product has been co-operation with various psychologists, biologists, physicists etc. in and out of Oxford, with regard to the philosophical aspects of their own work.

III. *Roman History*

a. *Lectures given in Oxford*

1. *Roman Britain*, annual course each Hilary Term. For HT 1935 the lectures have been completely rewritten, incorporating the results of my recent studies in provincial economic history and the relation between Romanized provincial life and its pre-Roman substratum.
2. Two lectures on *Britain* contributed to a "circus" on Roman provincial civilization, HT. 1935, summarizing general characteristics in the light of the new material already mentioned.

b. *Publications* (a select list of the most important only, but including MSS. not yet actually published).

1. Bruce's *Handbook of the Roman Wall*, more or less completely rewritten and forming the most complete account of the Wall as yet published.
2. Revised edition of my small Roman Britain.
3. Roman Britain down to Nero, in Cambridge Anc.Hist. vol. X.
4. Roman Britain in the Flavian-Antonine period, ibid. XI.
5. *Economic Survey of Roman Britain*, a work of 60-70,000 words intended as a contribution to Tenney Frank's *Economic Survey of Rome* (will probably have to be abridged for publication). This is divided into sections on land and population, finance, communications, agriculture, mining, industry, education, etc., and in most of these I have found that an exhaustive review of the evidence yields new and sometimes interesting conclusions.
6. Annual reports on Britain in *J.R.S.*, with Miss M. V. Taylor.
7. *Introduction to the Prehistory of Cumberland and Westmorland*, in Cumb. & West. Trans. XXXIII 163-200. I mention this as one example of the "prehistoric" studies which I have lately been undertaking in order to arrive at a more adequate conception of the substratum upon which the fabric of Romanized provincial life was erected. It is illustrated with several distribution-maps, and occupied me for much of the summer and early autumn of 1932.
8. The Bewcastle Cross, ibid. XXXV (forthcoming). This is really a monograph on Northumbrian carved crosses of the

Anglian period, reconsidering the highly controversial questions of their origin and dating. I mention it because I have approached the subject from a Romano-British standpoint, studying the connexions between Anglian civilization and the Roman and Celtic elements surviving in northern Britain.

c. *Unpublished work* (again I mention only the most important and tangible things).

1. I continue to amass and organize material for two large works projected many years ago: the corpus of Roman inscriptions in Britain and a book on Hadrian's Wall. These, at the present rate of progress, will not be finished for several years.
2. I have been in touch, as usual, with most of the excavations going on in this country, either in an advisory capacity or at least making myself familiar with their aims and results, and often assisting in the preparation of their published reports.
3. I have not travelled abroad during this period, but I have continued to keep in touch with work done in other Celtic provinces, especially Gaul.
4. Above all, I have been collecting material and arranging my ideas for writing vol. I of the *Oxford History of England*, which will be the fullest account of Roman Britain yet written. I intend to begin writing it within the next few months, and hope to publish it during 1936.

R. G. Collingwood

18 January 1935

APPENDIX IV

LETTERS FROM R.G. COLLINGWOOD

Letters to F.G. Simpson:

(i)

6-X-22

There are some observations on history and the characteristics of historically-minded people in the essay on Ruskin, so I send that too. My visit to you left me with a great deal to think about, and especially with the problem *why* history should be so important for the happiness and salvation of man. I doubt if it is primarily because history enables one to control the present – that seems to me too narrow a view, and not wholly true: one doesn't control the will of God, one learns to bow to it and make it one's own: to understand the world is perhaps no less valuable than to control it. Spinoza said that to be happy, to love God intelligently, meant to see everything *sub specie aeternitatis* – stripping off the passions and distorting interests of the moment and acquiring the 'point of view of eternity'. Now the truth seems to me to be that only the historian sees things from the point of view of eternity: because the evolutionary biologist, the astronomer, the mathematician etc. only see from the point of view of the momentarily fashionable biological or other theory: the scientist never sees *himself*. But the historian sees from the point of view of eternity because his history is the history of himself, and he achieves eternity not by ignoring time but just by recognising time and recognising himself as the heir of the past. Therefore to understand history is to understand oneself, which is the Delphic oracle's formula for salvation. That doesn't mean, I think, that if everyone was a historian there would be no wars: I think there still would be: but it does mean that wars wouldn't destroy anyone's faith or the inmost spring of their happiness...

(ii)

17.vi.25

I enclose MS.2075 words long for your perusal, wishing you to check it and pass it for press...
It was easy to work from your admirably clear notes. It has been less

easy to obey your request for something about the *aim* of such studies; I have tried to do this by pointing out, not in abstract general terms but in a concrete form, the peculiar way in which an understanding of modern English character is improved by an understanding of this particular phase in history. You will realise how impossible a fuller treatment is, within the limits of the pamphlet, when you see how everything is compressed. But it won't, I fear, be much good to the I.L.P. [Independent Labour Party]; yet I do mean to suggest to them that political doctrines may work in Russia or Germany, and may fail to work in England owing to the Roman element which we still inherit, and that – here I would strongly oppose certain 19th-century notions, including some of Marx's – there can be no such thing as an international culture, proletarian or any other, that ignores differences of historical background...

[Collingwood refers in this letter to the pamphlet *The Roman Signal Station on Castle Hill, Scarborough* (Scarborough, 1925) (see note 23 to chapter 10).]

(iii)

10 July 34

... It was splendid news (which I received after posting a letter to Richmond yesterday) that a Turf Wall turret had been found in site 79b. I feel like some personage at a desk in Scotland Yard, hearing that a murderer, wanted for close on 30 years, has been caught at last – or shall I say some skeleton-haunted feaster of medieval legend, whose spectre has been converted by the light of dawn into a long-lost friend in disguise? I would sing paeans to the glory of The Cumberland Excavation Committee, if it weren't such an awkward phrase to fit into the metre of a paean.

(iv)

Easter Monday 1939

... Coming back to a stack of letters two feet high, I am writing briefly. But I must tell you that in the last 6 months I have written two books and begun a third. One of them is already in galley proof: this is my autobiography, which was written to give an outline account of my life's work in case it should not be God's will that I should report to the public on that work in detail. The second, which is complete in MS., is a treatise on Metaphysics. The third, of which I wrote some 40.000 words in Java, is called *The Principles of History* and is the book which my whole life has been spent in preparing to write. If I can finish that, I shall have nothing to grumble at. – I say all this in order to show you that I have by no means lost my strength. In fact, as soon as the pressure of routine work was lifted, the strength came back very rapidly...

Letter to E.B. Birley:

(v)

Nov.10.1930

... I am very glad indeed that your appointment has gone through at last and seems, so far as I can judge, to promise a satisfactory arrangement. When I think of myself teaching for 30-40 hours a week from the age of 23 to that of 40, barring the War, I rejoice to think how much you will be able to do that none of us could ever hope to do. And my only apprehension about you, in fact, is that you will do too much: I mean, that you will devote yourself so exclusively to historical and archaeological studies that you will lose (as many archaeologists do) the quality of a man totus teres atque rotundus. People who lose that quality suddenly find their historical and archaeological work dry and loathsome and become discontented, unhappy and half-hearted in their job. I have always escaped that fate by doing other things primarily, and keeping my archaeology as an amateur's hobby, but then of course mine *is* amateurish, and yours aims at a higher standard. But there is no reason why you should not reach that standard while still keeping much of your mind free for other interests and occupations...

BIBLIOGRAPHY I

LIST OF MANUSCRIPTS OF R.G. COLLINGWOOD*

A. PHILOSOPHICAL

1908 "The Devil in Literature", (27 pp., 5500 words). R.G.C.: "an essay upon the Mythology of the Evil One; Read before 'Etanos', 1908".

1913 "Aristotelis 'De Anima', libri tres. Translation and commentary", (92 pp., 35.000 words). R.G.C.: "The ms. is intended for a lecture to cover 2 hours a week for 8 weeks given in 1913 and enlarged in 1914".

1916 "Lectures on the Philosophy of Religious Evolution", (59 pp., 16.000 words). R.G.C.: "Delivered at the Foyer d'Etudiants, Kingsway, March-April, 1916".

1917 *Truth and Contradiction, Chapter II*, (21 pp., 9500 words). Collingwood writes about this manuscript in his *Autobiography* (Oxford, 1939); on p. 99 he says about it in a note: "The original manuscript [speaking about *Libellus de Generatione*], like the only manuscript of *Truth and Contradiction*, was destroyed after I wrote this book". Chapter II however has apparently survived.

1918 1) "Words and Time", (14 pp., 5500 words). Dated Sept. 13, 1918.

2) "Lecture on the Philosophy of St. Paul", (11 pp., 3000 words). R.G.C.: "Somerville College, Oxford 1918".

1919 1) "The Spiritual Basis of Reconstruction", (16 pp., 4000 words). Dated May 7, 1919. R.G.C.: "Address to the Belgian Students' Conference at Fladbury, May 10, 1919".

2) "Money and Morals", (12 pp., 3000 words). Dated May 24, 1919. Lecture to the Student Movement London branch on May 27, 1919 and the Indian Students' Hostel in June 1919.

3) "A Footnote to Future History", (11 pp., 1600 words). Dated "Vigil of All Saints 1919".

4) "Lectures on the Ontological Proof of the Existence of God",

* The number of words given is only approximate. If Collingwood has written something on the cover-sheet, it is mentioned after his initials (R.G.C.).

(96 pp., 42.000 words). R.G.C.: "Written December MCMXIX for delivery Hilary Term MCMXX".

1920 1) "The Church", (21 pp., 5000 words). Dated April 13, 1920. R.G.C.: "paper read before 'The Group', Oxford, April 29, 1920".

2) "Croce's Philosophy", (6 pp., 1300 words). Dated July 3, 1920.

3) "Notes on Hegel's Logic", (9 pp., 4000 words). Dated Sept. 19, 1920.

4) "The Philosophy of the Christian Religion", (14 pp., 4000 words). R.G.C.: "First draft - not very much altered in revision". At the back: "Sept. 29-1920; sent to *Theology*, Oct. 1, 1920".

5) "Notes on Formal Logic", (19 pp., 3000 words). Dated 1920. R.G.C.: "These notes on Formal Logic are to be recast into a chapter of the complete work. There ought to be notes on the Logic of Becoming as well".

1920-21 1) "Draft of opening chapter of a 'Prolegomena to Logic' (or the like)", (60 pp., 15.000 words). On p.3 is written "1920-1" and on p.29 "III-1921".

2) *An illustration from historical thought*, (8 pp., 1500 words). Is attached to the above-mentioned manuscript.

1921 1) "Fred. Rauh", (16 pp., 5000 words). Dated April 28, 1921. Notes on F. Rauh, *Essai sur le Fondement Métaphysique de la Morale* (1890) and *L'Experience Morale* (1903) and books about him.

2) "Lectures on Moral Philosophy", (121 p., 37.000 words). R.G.C.: "for M.T. 1921, written at various times, May-Oct. 1921".

3) "Jane Austen", (19 pp., 4500 words). Dated Nov. 27, 1921. Lecture for the "Johnson Society".

1922-23 "Fragment on Neo-Realism", (2 pp., 400 words). Dated 1922-23.

1923 "Action", (91 pp., 36.000 words). R.G.C.: "A course of lectures (16 lectures) on Moral Philosophy. Written September 1923, for delivery in Michaelmas term 1923". Added at a later date: "Much rewritten and expanded from about 70 to about 100 pages (i.e. 35.000 to 40.000 words) in Mich. term 1926, with some very considerable alterations on points of theory. This expansion brought it up to about the limit for 16 lectures".

1924 "Lectures on the Philosophy of Art", (44 pp., 13.000 words). R.G.C.: "Delivered at noon on Saturdays, TT 1924" and on inside of the cover: "These Lecture-notes, written during Trinity Term 1924, represent a much abbreviated and systematized version of the notions worked out at Avignon in the spring of 1924 and noted down, in a rough and chaotic form, in the notebook entitled 'Rough Notes for a book on the

Philosophy of Art"'.

1925 1) "Economics as a Philosophical Science", (34 p., 8000 words). Draft of article in *The International Journal of Ethics* 36 (1925-26).

2) "Some Perplexities about Time", (17 pp., 5500 words). Dated July 4, 1925. R.G.C.: "First draft of Ar. Soc. paper 1925" (printed in *Proceedings of the Aristotelian Society* 26 (1925-26)).

1926 1) "The Philosophy of History", (5 pp., 1400 words). R.G.C.: "fragment intended as introduction to lectures: 1926".

2) "Lectures on the Philosophy of History", (75 pp., 30.000 words). R.G.C.: "written January 9-13, 1926, for delivery in Hilary Term, 1926".

1927 1) "Preliminary Discussion. The Idea of a Philosophy of Something, and, in Particular, a Philosophy of History", (26 pp., 10.500 words). Dated April 1927. Is added to the above-mentioned manuscript.

2) Analysis of *Art and Instinct* by S. Alexander, (7 pp., 2700 words).

1928 1) "Outlines of a Philosophy of History" (69 pp., 28.000 words). The Preface ends with: "April 1928, Le Martouret, Die, *Drôme*". (Collingwood mentions this manuscript in his *Autobiography*, p.107).

2) "Stray Notes on Ethical Questions", (46 pp., 10.500 words). R.G.C.: "1928 - I. Moral Standards, II. A Political Antinomy, III. Punishment".

1929 1) "Lectures on Philosophy of History", (38 pp., 11.000 words). R.G.C.: "II (T.T. 1929)". These lectures deal with Kant, Hegel, nineteenth-century positivism and Croce.

2) "Oswald Spengler", (15 pp., 4000 words). R.G.C.: "Lecture delivered at the London School of Economics, 13-V-29, Written 12-V-29".

1930 1) "The Good, the Right, and the Useful", (18 pp., 4000 words). R.G.C.: "Exeter College Dialectical Society, March 3, 1930". Separate: "The Good and the Useful", (1 p., 250 words) and "Promise", (2 pp., 600 words).

2) "Historical Background of N.T. Thought", (15 pp., 4000 words). R.G.C.: "The Group, Oct. 1930".

3) "Science, Religion and Civilization", (14 pp., 2500 words). R.G.C.: "The third of a series of lectures under that title delivered in Coventry Cathedral, Oct.-Dec. 1930, by Joseph Needham, B.H. Streeter and R.G.C.".

1930-31 "Notes on aesthetic", (13 pp., 1800 words); for the paper "Aesthetic Theory and Artistic Practice".

1931 1) "Aesthetic Theory and Artistic Practice", (44 pp., 12.000 words). R.G.C.: "This is the complete form of the paper which was to be delivered, in abbreviated form, as a lecture before the British Institute of Philosophical

Studies on March 17, 1931. Feb. 23-28, 1931".
2) "The Origin and Growth of the Idea of a Philosophy of History", (24 pp., 6000 words). Notes for lectures.

1932 1) "Report on Work Done During Quinquennium as University Lecturer, from summer 1927 to the time of writing, January 1932", (7 pp., 1200 words). Is a draft; for the official report see Appendix III.
2) "The Philosophy of History", (12 pp., 4500 words). Lectures.
3) "Moral Philosophy Lectures", (172 pp., 65.000 words). R.G.C.: "I. Beginning of 1932 course, including a) methodological introduction b) discussion of matter, life and mind – all cut out in 1933 as overloading the course. II. Complete course as written 1929 and superseded (after not being delivered 1931) in 1932. III. Complete course as written 1923, with alterations of 1926, 1927".
4) "War in its relation to Christian Ethics with special reference to the Lambeth Report, 1930", (10 pp., 4000 words). R.G.C.: "paper read to the Group, on 17 November 1932".

1933 1) "List of Work Done", (24 pp., 1500 words). R.G.C.: "This record got hopelessly into arrears about 1929".
2) "Correspondence between Collingwood and H.A. Prichard about 'claims'", (6000 words). In photocopy.
3) "Lectures on Moral Philosophy", (130 pp., 55.000 words). Dated 1933. R.G.C.: "This is substantially the complete new MS. written in the long vacation of 1932; but there are alterations...".
4) "Notes on Morals", (7 pp., 1000 words). R.G.C.: "This notebook (begun Nov. 27, 1933) is to contain notes on important works concerned with moral philosophy, re-read (or read) for the special purpose of a projected book on the subject and not for the general purpose of following their doctrine". Notes on Green's *Prolegomena* and Rousseau's *Contrat Social*. At the back notes on Spinoza (26 pp., 4000 words).
5) "Outline of a Theory of Primitive Mind", (8 pp., 2800 words). Dated December 1933.

1933-34 "Notes towards a Metaphysic"; 5 notebooks: A: 91 pp., 24.000 words; begun Sept. 1933, till Sept. 23, 1933. B: 90 pp., 26.000 words; begun Sept.23, 1933; on p.78 "March 1934". C: 93 pp., 24.500 words; from March 23, 1934 till April 5, 1934. D: 91 pp., 23.500 words; April 6-22, 1934. E: 157 pp., 32.500 words; from April 22, 1934 till May 8, 1934; added: December 27, 1937 "Postscript to causation" (2 pp., 300 words).

1934 1) "The Nature of Metaphysical Study", (31 pp., 7500

words). R.G.C.: "Two lectures, opening a course of 16 lectures on Metaphysics by various speakers, to be delivered on 15 and 17 January 1934".

2) "Rough draft on Existence, Space and Time, Matter", (25 pp., 7500 words).

3) "History as the Understanding of the Present", (2 pp., 700 words). This manuscript is not dated. Since it refers to Oakeshott's *Experience and its Modes*, which was published in 1933, it is estimated to be from 1934.

1935 1) "Central Problems in Metaphysics", (1 p., 40 words). R.G.C.: "Central Problems in Metaphysics. Lectures written April 1935, for delivery T.T. 1935. Begun 8 April (after nearly a week's work on preliminary notes); finished 18 April: Tetbury"; added on a later date: "if given in M.T. (as in 1936) should be cut down – turned out of shortly after the lectures". Only the cover sheet of this manuscript has survived.

2) "Correspondence between Collingwood and G. Ryle", (16.500 words). This correspondence has developed in connection with Ryle's article "Mr. Collingwood and the Ontological Argument", *Mind* 44 (1935).

3) "Rule-making and Rule-breaking", (12 pp., 3500 words). R.G.C.: "Sermon preached in St. Mary the Virgin's Church, Oxford, 5 May 1935".

4) Analysis of P. Leon, *The Ethics of Power* (1935), (13 pp., 2500 words). Dated June 10, 1935.

5) "Method and Metaphysics", (27 pp., 7000 words). R.G.C.: "Paper read before the Jowett Society, 19 June 1935".

6) "Inaugural. Rough Notes", (53 pp., 10.500 words). Is completely different text from the one reprinted in *The Idea of History*, 231-249.

7) "Manuscript on 'Sense-data", (4 pp., 2500 words). Dated October-November 1935. Is incomplete.

8) "Reality as History", (38 pp., 17.000 words). Dated December 1935. R.G.C.: "An experimental essay designed to test how far the thesis can be maintained that all reality is history and all knowledge historical knowledge".

9) "Experiment in New Realism", (6 pp., 3000 words).

1936 1) "Can Historians be Impartial?", (14 pp., 4500 words). R.G.C.: "Paper read to the Stubbs Historical Society, 27 Jan. 1936".

2) "Lectures on the Philosophy of History", (24 pp., 12.000 words). R.G.C.: "Second term's course, containing, I: History of the idea of history in England, Germany, France and Italy, since about 1870, II: Metaphysical Epilegomena to the course, written April 1936"; added

on a later date on p.1: "Written Jan.-March 1936 for delivery at the same time". A "contents of 2nd term's lectures" is given, beginning with p.105 "English Thought on History since 1870" and ending with p.194. "Part II Metaphysical Epilegomena" begins on p.154. Of this manuscript only the following pages are left: 1,2,8-12, 19d, 37, 42, 68, 84, 94, 104-107, 114, 116, 125, 133, 144a-c. Pages 144a-c are from 1940; on p.144a is written: "16th lecture-14 June 1940". In the text are sometimes corrections of a later date, probably 1940. A note is attached from T.M. Knox: "Passages from the Ms. of the Idea of History, either not used or used in a different form in the published work".

3) "Human Nature and Human History", (25 pp., 8500 words). Dated March 1936. R.G.C.: "First draft of paper written May 1936 and sent up for publication by the British Academy". The latter publication is reprinted in *The Idea of History*, 205-231.

4) "Man Goes Mad", (38 pp., 12.500 words). R.G.C.: "Rough Ms. begun 30 Aug. 36".

5) "Notes on History of Historiography and Philosophy of History", (66 pp., 13.000 words).

6) "English Folklore"; 3 notebooks; I: 91 pp., 18.000 words; II: 89 pp., 17.000 words; III: 48 pp., 8000 words.

7) "Realism and Idealism", (125 pp., 55.000 words). This manuscript is not dated; added to it is a printed table of contents. The manuscript is probably from 1936, for Collingwood lectured in Michaelmas Term of that year on "Central Problems in Metaphysics" (see Appendix II), while the manuscript begins as follows: "In these lectures I propose to deal with some of the central problems in metaphysics; but the method by which I shall approach them is through the consideration of certain central types of metaphysical theory. I begin by dividing such theories, according to a distinction which has become traditional, into realistic and idealistic; and roughly speaking the first half of the lectures will be about realism and the second half about idealism".

1936-37 1) "Folklore and Folk-tale", (21 pp., 6000 words).

2) "Folklore I", (21 pp., 7000 words).

3) "Folklore II. Three Methods of Approach: Philological, Functional, Psychological", (52 pp., 17.000 words).

4) "Folklore III. The Historical Method", (21 pp., 7000 words).

5) "Folklore IV. Magic", (52 pp., 16.000 words).

6) "Folklore" (concluding chapter), (11 pp., 3500 words). Is incomplete.

7) "Fairy-tales", (4 pp., 1500 words).
8) "Cinderella", (32 pp., 10.000 words).

1937 "Notes on Causation", (22 pp., 3600 words).

1937-38 "Metaphysics", (24 pp., 10.500 words). Is incomplete: only pp. 29-52 are left. On p.41: "9-1-38".

1938 "Racine", (20 pp., 9000 words). Analysis of different works by Racine.

1938-39 1) "Log of a Journey in the East Indies, 1938-39", (39 pp., 7500 words). In photocopy.
2) "Historiography", (29 pp., 6500 words). R.G.C.: "Notes on Historiography written on a voyage to the East Indies, 1938-9".

1939-40 1) "What 'Civilization' Means", (44 pp., 14.000 words). Probably an early draft of the *New Leviathan.*
2) "Moral Philosophy Lectures, 1940", (77 pp., 37.000 words). R.G.C.: "Goodness, Rightness, Utility. Lectures delivered in H.T. 1940 and written as delivered, forming a continuation of those on Feeling, Appetite, Desire and Will, delivered in the previous term. Written Dec. 1939-Feb. 1940".

1940 1) "Historiography", (6 pp., 700 words). On p.2 is written : "8-3-40. The *Idea of History* (Notes for lectures, on discovering that the Ms. which contains the results of my last 15 years' work on the subject has disappeared)".
2) "Fascism and Nazism", (10 pp.). Manuscript of article published in *Philosophy* 15 (1940).

1939-41 *The New Leviathan.* Manuscript of the printed work (Oxford, 1942). At 40.97 is written: "1-2 Jan. 1941".

Correspondence

1) 36 Letters (in photocopy) of Collingwood to de Ruggiero, from 1920 to 1938 (34 in English, 2 in Italian). Of these letters 9 are printed in A. Greppi Olivetti, *Due saggi su R.G. Collingwood: Con un' appendice de lettere inedite di Collingwood a G. de Ruggiero* (Padova, 1977), 89-104.
2) 6 Letters (in photocopy) of Collingwood to Alexander, from 1925 to 1938.

Manuscripts of which no date is given or can be given

1) "Economics as a Philosophical Science", (25 pp., 4000 words). R.G.C.: "For a section of a comprehensive ethical treatise; or alternatively as a small book under the above title".
2) "Rough notes on Politics", (3 pp., 600 words).
3) "Notes towards a Theory of Politics as a Philosophical Science", (7 pp., 1700 words).
4) "The Breakdown of Liberalism", (2 pp., 300 words).

5) "Outlines of a Concept of the State", (1 p., 200 words).
6) "Ms. about 'mind and thought'", (2 pp., 600 words).
7) "Ms. about 'action'", (3 pp., 600 words).
8) "The Confusion of Sense", (19 pp., 5500 words).
9) "Observations on Language", (4 pp., 1000 words).
10) "Hieronimo and Hamlet", (17 pp., 1400 words).
11) "The Electra-story", (2 pp., 350 words).
12) "Aeschylus", (4 pp., 2000 words).
13) "Notes on Euripides", (45 pp., 8800 words).
14) "Jane Austen", (18 pp., 5000 words).
15) "Notes on Mill, Taylor,Schlegel, Windelband, Voltaire", (4 pp., 1000 words).
16) "Notes on Descartes' 'Principia'", (13 pp., 3000 words).
17) "Translation of the Preface to the 'Critique of Pure Reason' of Kant", (32 pp., 8500 words). Translation of both editions (1781, 1787).
18) "Comment on the Preface of Kant's 'Critique of Pure Reason'", (34 pp., 4500 words).
19) "Notes on Voltaire's 'Siècle de Louis XIV'", (2 pp., 600 words).
20) "Causation in Spinoza", (7 pp., 1500 words). Is incomplete.
21) "Ms. on freedom and the relation between body and mind", (10 pp., 2500 words). Notes on MacTaggart, James, Hodgson, Gibson, Galloway, Birdson[?]. R.G.C.: "MacTaggart, James, Will to Believe. Hodgson, Phil. of Experience, ch. 14. Boyce Gibson, Personal Idealism, problem of freedom. Galloway, Studies in Phil. of Religion. Birdson, Free will". The name Birdson does not occur in the National Union Catalog.
22) "Notes on Ducasse 'The Philosophy of Art' (1931) and Reid 'Study in Aesthetics'" (1931), (17 pp., 4000 words).
23) "Art and Instinct", (7 pp., 2000 words). Is incomplete.
24) "Art and the Machine", (15 pp., 5000 words).
25) "Religious Intolerance", (5 pp., 1000 words). R.G.C.: "Unfinished address for Lady Margaret Hall. They rejected the subject and I substituted an extempore address on the Christian view of pain".
26) "Notes on MacIver, *The Modern State*", (17 pp., 3500 words). This book was published in 1926.
27) "Utility, Right and Duty", (50 pp., 14.000 words). This ms. is probably from around 1920, since it is written on paper of the "Admiralty War Staff", where Collingwood worked during the First World War.
28) "Lectures on Moral Philosophy", (96 pp., 25.000 words). Ms. is incomplete.

B. ROMAN BRITAIN

1921 "Notes for the Extension Lectures", (28 pp., 4500 words).

R.G.C.: "Notes for the Extension Lectures (1921) on which *Roman Britain* was based".

1922 "Roman Frontiers", (25 pp., 5000 words). R.G.C.: "a lecture delivered before the Newcastle Literary and Philosophical Society; Oct. 1, 1922".

1923 1) "The Romans on the Wall", (4 pp., 800 words). R.G.C.: "for 'Yorkshire Post', slightly altered in copying, 23-VIII-23".

2) "Roman Inscriptions", (8 pp., 1000 words). R.G.C.: "Soc. for Reform of Latin Teaching". This lecture was given in August 1923 (see *The Classical Review* 37 (1923), 145. I am thankful to Dr. Grace Simpson for having given me this information).

1925 "The Roman Wall", (11 pp., 3000 words). R.G.C.: "address to the English Historical Association at Newcastle; January 9, 1925".

1931 "The Vallum Crossings", (17 pp., 4000 words). R.G.C.: "an excerpt from a paper on 'Ten Years' Work on Hadrian's Wall, 1920-1930', to be published in Cumb. and West. A. and A.S. Trans. N.S. XXXI: to be read at Kendal, April 22, 1931".

1933 "The State of Britain at the time of the Anglo-Saxon settlements", (16 pp., 3000 words). R.G.C.: "contribution to a discussion at a meeting of the Archaeological Institute, 1 February 1933".

1935 1) "The Truth about the Vallum", (6 pp., 2000 words). R.G.C.: "by Infelix".

2) "Notebook", (16 pp., 1000 words). R.G.C.: "C & W Prehistory, Cartmel 1935". Many drawings.

3) "Review of G. Macdonald *The Roman Wall in Scotland* (1934^2)", (15 pp., 4500 words). Printed in the *Journal of Roman Studies* 26 (1936).

1936 1) "Who was King Arthur?", (15 pp., 4000 words). R.G.C.: "paper to the Martlets, Univ. Coll. Oxon., June 1936".

2) "Mayborough and King Arthur's Round Table", (9 pp., 2500 words). R.G.C.: "read 8 July 1936".

1937 1) "King Arthur's Round Table, (8 pp., 1400 words).

2) "Notebook", (56 pp., 6000 words). R.G.C.: "Excavations at King Arthur's Round Table, Eamont Bridge, Westmorland, July 1937".

3) "Notebook", (30 pp., 4000 words). R.G.C.: "Parallel sites and illustrative material".

Manuscripts of which no date is given or can be given

1) "Dr. Hugh Todd's Ms. at St. Edmund Hall" "Bronsted's Early English Ornament", (56 pp., 2500 words).
2) "C. & W. Britt. Setts", (11 pp., 150 words). Drawings.
3) "The Roman Tombstone at Fordington", (3 pp., 700 words).
4) "Lancaster", (3 pp., 750 words). Is incomplete (only pp. 19-

21).

5) "Evidences of W. Saxons along historical route: Salisbury, Mildenhall, odd burials Dorset", (2 pp., 500 words).

6) "Ms. on Notitia Dignitatum", (15 pp., 4500 words). Is incomplete (only pp.3-17).

7) "Ms. on 5th cent. Roman Britain", (4 pp., 2000 words). Is incomplete (only pp. 285-288). Probably ms. of *Roman Britain and the English Settlements*.

8) "Notes on Hambledon", (2 pp., 400 words).

9) "Britain and the Roman Empire", (26 pp., 6500 words).

10) "The Bewcastle Cross", (7 pp., 1700 words).

BIBLIOGRAPHY II

WORKS BY R.G. COLLINGWOOD *

1913 "Report of the Excavations at Papcastle, 1912", *CW* 13 (1913), 131-141.

1914 With Professor Haverfield: "Report on the Exploration of the Roman Fort at Ambleside, 1913", *CW* 14 (1914), 433-464.

1915 "The Exploration of the Roman Fort at Ambleside: Report on the second year's Work (1914)", *CW* 15 (1915), 1-62.

"Roman Ambleside", *The Antiquary* 51 (1915), 91-96.

1916 *Religion and Philosophy* (Oxford, 1916).

"The Devil", in B.F. Streeter et al. (eds.), *Concerning Prayer: Its Nature, its Difficulties and its Value* (London, 1916); repr. in Rubinoff ed., *Faith and Reason*, 212-233.

"The Exploration of the Roman Fort at Ambleside: Report on the third year's work (1915)", *CW* 16 (1916), 57-90.

1918 Review of J.N. Figgis, *The Will to Freedom: Or the Gospel of Nietzsche and the Gospel of Christ* (London, 1917), in *The Oxford Magazine* 36 (1917-1918), 229.

1919 Obituary Notice of Professor Haverfield, *PSAN* 9 (1919), 117-118.

1921 "Croce's Philosophy of History", *The Hibbert Journal* 19 (1921), 263-278; repr. in Debbins ed., *Essays*, 3-22.

Review of W. Benett, *Freedom and Liberty* (Oxford, 1920) and G.L. Raymond, *Ethics and Natural Law* (London, 1920), in *The Oxford Magazine* 39 (1920-1921), 264.

"The Purpose of the Roman Wall", *The Vasculum* 8 (1921), 4-9.

"Hadrian's Wall: A History of the Problem", *JRS* 11 (1921), 37-66.

"Explorations in the Roman Fort at Ambleside (fourth year, 1920) and at other sites on the Tenth Iter", *CW* 21 (1921), 1-42.

1922 "Are History and Science Different Kinds of Knowledge?", *Mind* 31 (1922), 433-451; repr. in Debbins ed., *Essays*, 23-33.

Ruskin's Philosophy (Kendal, 1922); repr. in Donagan ed., *Essays*, 5-41.

"The Roman Evacuation of Britain", *JRS* 12 (1922), 74-98.

1923 *Roman Britain* (Oxford, 1923).

*This bibliography is not complete; only those works are included which have been used in this study. For a survey of Collingwood's writings on archaeology and history, see the bibliography compiled by I.A. Richmond, in *Proceedings of the British Academy* 29(1943), 481-485.

"Science and History", *The Vasculum* 9 (1923), 52-59.
"Can the New Idealism Dispense with Mysticism?", *Proceedings of the Aristotelian Society*, Supplementary Volume 3 (1923), 161-175; repr. in Rubinoff ed., *Faith and Reason*, 270-282.
Review of R.F.A. Hoernlé, *Matter, Life, Mind and God: Five Lectures on Contemporary Thought* (London, 1923), in *The Oxford Magazine* 41 (1922-1923), 423.
Review of C.G. Jung, *Psychological Types, or the Psychology of Individuation* (London, 1923), in *The Oxford Magazine* 41 (1922-1923), 425-426.
Review of C. Spearman, *The Nature of 'Intelligence' and the Principles of Cognition* (London, 1923), in *The Oxford Magazine* 42 (1923-1924), 117-118.

1924 *Speculum Mentis* (Oxford, 1924).
"Rome in Britain", *The Home-Reading Magazine* 36 (1924-1925), 6-8, 37-39, 71-73.
"The Fosse", *JRS* 14 (1924), 252-256.

1925 *Outlines of a Philosophy of Art* (London, 1925); repr. in Donagan ed., *Essays*, 45-154.
"The Nature and Aims of a Philosophy of History", *Proceedings of the Aristotelian Society* 25 (1924-1925), 151-174; repr. in Debbins ed., *Essays*, 34-56.
"Rome in Britain", *The Home-Reading Magazine* 36 (1924-25), 101-104, 133-135, 165-168.
"Britain and the Roman Empire", in F.S. Marvin ed., *England and the World* (London, 1925), 35-64.
"Hadrian's Wall", *History* 10 (1925), 193-202.
The Roman Signal Station on Castle Hill, Scarborough (Scarborough, 1925).

1926 "Some Perplexities About Time: With an Attempted Solution", *Proceedings of the Aristotelian Society*, New Series 26 (1925-1926), 135-150.
"Economics as a Philosophical Science", *The International Journal of Ethics* 36 (1925-1926), 162-185.
Review of F.J. Teggart, *Theory of History* (London, 1925), in *Journal of Philosophical Studies* 1 (1926), 255-256.
A Guide to the Roman Wall (Newcastle-upon-Tyne, 1926) (2nd ed. 1932).

1927 "Oswald Spengler and the Theory of Historical Cycles", *Antiquity* 1 (1927), 311-325; repr. in Debbins ed., *Essays*, 57-75.
"The Theory of Historical Cycles", *Antiquity* 1 (1927), 435-446; repr. in Debbins ed., *Essays*, 76-89.
Review of S. Dill, *Roman Society in Gaul in the Merovingian Age* (London, 1926), in *Antiquity* 1 (1927), 117-119.
Review of R.E.M. Wheeler, *Prehistoric and Roman Wales* (Oxford, 1925), in *The English Historical Review* 42 (1927), 109-110.

Review of M. Rostovtseff, *The Social and Economic History of the Roman Empire* (Oxford, 1926), in *Antiquity* 1 (1927), 367-369.

1928 "The Limits of Historical Knowledge", *Journal of Philosophical Studies* 3 (1928), 213-222; repr. in Debbins ed.,*Essays*, 90-103.

Review of R.E. Zachrisson, *Romans, Kelts and Saxons in Ancient Britain* (Uppsala, 1927), in *JRS* 18 (1928), 117-119.

1929 "A Philosophy of Progress", *The Realist* 1 (1929), 64-77; repr. in Debbins, ed. *Essays*, 104-120.

"Political Action", *Proceedings of the Aristotelian Society*, New Series 29 (1928-1929), 155-176.

"Town and Country in Roman Britain", *Antiquity* 3 (1929), 261-276.

"Roman Signal-Stations on the Cumberland Coast", *CW* 29 (1929), 138-165.

Roman Eskdale (Whitehaven, 1929).

1930 *The Archaeology of Roman Britain* (London, 1930); revised edition by I.A. Richmond (London, 1969).

The Philosophy of History (London, 1930); repr. in Debbins ed., *Essays*, 121-139.

"A System of numerical references to the parts of Hadrian's Wall and the structures along its line", *CW* 30 (1930), 108-115.

"Hadrian's Wall: A System of Numerical References", *PSAN* 4 (1929-1930), 179-187.

"A newly-discovered Roman Site in Cumberland", *Antiquity* 4 (1930), 472-477.

1931 Review of H. Temperley ed., *Selected Essays of J.B. Bury* (Cambridge, 1930), in *The English Historical Review* 46 (1931), 461-465.

"Hadrian's Wall: 1921-1930", *JRS* 21 (1931), 36-64.

"The Roman Signal Station" in A. Rowntree ed., *The History of Scarborough* (London and Toronto, 1931), 40-50.

1932 *Roman Britain* (Oxford, 1932) (revision of 1923 edition).

1933 *An Essay on Philosophical Method* (Oxford, 1933).

Edition of J. Collingwood Bruce, *The Handbook to the Roman Wall*, 9th ed. (Newcastle-upon-Tyne and London, 1933).

"An Introduction to the Prehistory of Cumberland, Westmorland and Lancashire north of the Sands", *CW* 33 (1933), 163-200.

In Memoriam of W.G. Collingwood, *CW* 33 (1933), 308-312.

Review of G. MacDonald, *Roman Britain: 1914-1918* (London, 1931), in *JRS* 23 (1933), 99-100.

1934 "The Present Need of a Philosophy", *Philosophy* 9 (1934), 262-265.

"The Romans and Britain" and "The Conquest of Britain", in S.A. Cook et al. (eds.), *CAH*, Vol. 10 (Cambridge, 1934), 790-802.

1935 *The Historical Imagination* (Oxford, 1935); repr. in IH, 231-

249.

"The Bewcastle Cross", *CW* 35 (1935), 1-29.

1936 With J.N.L. Myres: *Roman Britain and the English Settlements* (Oxford, 1936).

"Human Nature and Human History", *Proceedings of the British Academy* 22 (1936), 97-127; repr. in IH, 205-231.

"The Latin West: Britain", in S.A. Cook et al. (eds.), *CAH*, Vol. 11 (Cambridge, 1936), 511-525.

1937 Review of R. Klibansky and H.J. Paton eds., *Philosophy and History: Essays presented to Ernst Cassirer* (Oxford, 1936), in *The English Historical Review* 52 (1937), 141-146.

"Roman Britain", in T. Frank ed., *An Economic Survey of Ancient Rome*, Vol. 3 (Paterson, 1937), 1-118.

1938 *The Principles of Art* (Oxford, 1938).

"John Horsley and Hadrian's Wall", *Archaeologia Aeliana* 4th ser., 15 (1938), 1-42.

"King Arthur's Round Table: Interim Report on the Excavations of 1937", *CW* 38 (1938), 1-31.

1939 *An Autobiography* (Oxford, 1939).

"Britain" in S.A. Cook et al. (eds.), *CAH*, Vol. 12 (Cambridge, 1939), 282-296.

1940 *An Essay on Metaphysics* (Oxford, 1940).

The First Mate's Log of a Voyage to Greece in the Schooner Yacht 'Fleur de Lys' in 1939 (London, 1940).

1942 *The New Leviathan* (Oxford, 1942).

1945 *The Idea of Nature* (Oxford, 1945).

1946 *The Idea of History* (Oxford, 1946).

1965 With R.P. Wright: *The Roman Inscriptions of Britain*, Vol. 1, *Inscriptions on Stone* (Oxford, 1965).

BIBLIOGRAPHY III

REVIEWS OF WORKS BY R.G. COLLINGWOOD

Religion and Philosophy (1916)

Eliot, T.S., *International Journal of Ethics* 27 (1917), 543
Galloway, G., *Mind* 28 (1919), 365-367
Moffat, J., in "Survey of Recent Theological Literature", *The Hibbert Journal* 15 (1917), 677-682
Webb, C.C.J., *The Oxford Magazine* 35 (1916-1917), 281-282

"The Devil" (in B.F. Streeter et al. (eds.), *Concerning Prayer: Its Nature, its Difficulties, and its Value* (London, 1916))

Drummond, J., *The Hibbert Journal* 15 (1917), 327-331
Stewart, H.F., *The Journal of Theological Studies* 18 (1917), 79-82

Roman Britain (1923)

Lamer, H., *Philologische Wochenschrift* 44 (1924), 664-665

Speculum Mentis (1924)

Anonymous, *The Church Times* (11-9-1925)
Anonymous, *The Guardian* (14-11-1924)
Anonymous, *Jewish Guardian* (12-9-1924)
Anonymous, *Revue de Métaphysique et de Morale* 33 (1926), supplem., 8
Anonymous, *The Scotsman* (10-11-1924)
Anonymous, *The Tablet* (28-3-1925)
Anonymous, *The Times Literary Supplement* 23(30-10-1924), 676
Burns, C.D., *International Journal of Ethics* 35(1925), 323
Croce, B., *La Critica* 23 (1925), 55-59
F., W.R., *The Irish Statesman* (19-11-1924)
Joad, C.E.M., "What is Left of Modern Philosophy", *The Spectator*, no. 5025 (18-10-1924), 550-553.
Laird, J., *Mind* 34 (1925), 235-241
Marvin, F.S., *Nature* 115 (1925), 78-79
Stebbing, L.S., *The Hibbert Journal* 28 (1925), 566-569
Windle, B.C.A., *The Catholic World* 122 (1925), 128-129

Outlines of a Philosophy of Art (1925)

Anonymous, *Nature* 116 (1925), 94
Anonymous, *The Times Literary Supplement* 24 (16-4-1925), 269
Jahn, J., *Zeitschrift für Bildende Kunst 62* (1928), *Beil., 32*

Wébert, J., *Revue des Sciences Philosophiques et Théologiques* 16 (1927), 70

Roman Eskdale (Whitehaven, 1929)

Wheeler, R.E.M., *Archaeological Journal* 87 (1930), 371-372

The Archaeology of Roman Britain (1930)

Anonymous, *The Archaeological Journal* 87 (1930), 381-382
Anonymous, *Nature* 127 (1931), 9
Anonymous, *Journal of the Royal Institute of British Architects* 38(1930), 51
Anonymous, *The Times Literary Supplement* 29 (18-12-1930), 1076
Birley, E.B., *CW* 31 (1931), 210-211
Bosanquet, R.C., *JRS* 23 (1923), 99-101
Cary, M., *History* 15 (1930-1931), 348
Casson, S., *The London Mercury* 24 (1931), 282
Churchill, G.M., *Art and Archaeology* 33 (1932), 112
Droop, J.P., *Annals of Archaeology and Anthropology* 18 (1931), 55-56
MacNeill, E., *Studies: An Irish Quarterly Review* 20 (1931), 144-145
Mattingly, H., *The Classical Review* 45 (1931), 84-85
Radford, C.A.R., *Journal of the British Archaeological Association* 36 (1930), 426-428
Richmond, I.A., *Antiquity* 5 (1931), 504-506
Vulliamy, C.E., *Man* 31 (1931), 13-14
Westgate, R.I.W., *American Journal of Archaeology* 35 (1931), 361
Wheeler, R.E.M., *The Antiquaries Journal* 11 (1931), 173-175
Winbolt, S.E., *The Geographical Journal* 77 (1931), 261-263

Second Edition, revised by I.A. Richmond (1969):
Thompson, F.H., *The Antiquaries Journal* 50 (1970), 133-134
Webster, G., *JRS* 60 (1970), 245-247
Wheeler, R.E.M., *Antiquity* 43 (1969), 239-240

Roman Britain (1932)

Anonymous, *Scottish Geographical Magazine* 49 (1932), 297
Anonymous, *The Times Literary Supplement* 31 (7-4-1932), 240
Albertini, E., *Revue Historique* 173 (1934), 626-627
Anderson, J.C.C. *The Oxford Magazine* 50 (1931-1932), 771-772
Evans-Pritchard, E.E., *Man* 32 (1932), 220-221
Grenier, A., *Revue des Etudes Anciennes* 34 (1932), 332-333
Hawkes, Chr., *The Antiquaries Journal* 13 (1933), 67-69
Myres, J.N.L., *JRS* 22 (1932), 252-253
Weiss, J., *Deutsche Geographische Gesellschaft in Wien* 76 (1932), 83

"An Introduction to the Prehistory of Cumberland, Westmorland

and Lancashire north of the Sands" (*CW* 33 (1933), 163-200)

Crawford, O.G.S., *Antiquity* 8 (1934), 361-363

An Essay on Philosophical Method (1933)

Anonymous, *The Guardian* (5-1-1934)
Anonymous, *The Times Literary Supplement* 33 (1-3-1934), 136
B., G., *The Tablet* (23-12-1933)
Cross, F.L., *The Church Quarterly Review* 118 (1934), 303-305
D'Arcy, M.C., *The Criterion* 13 (1933-1934), 500-503
Ducasse, C.J., "Mr. Collingwood on Philosophical Method", *The Journal of Philosphy* 33 (1936), 95-106
G., T., *Nature* 134 (1934), 648
Hartshorne, Ch., *International Journal of Ethics* 44 (1934), 357-358
Ronayne, C.F., "Conclusions as Premises", *The American Review* 4 (1934-1935), 627-633
Russell, L.J., *Philosophy* 9 (1934), 350-352
Ryan, J.K., *The New Scholasticism* 3 (1934), 172-174
Schiller, F.C.S., *Mind* 43 (1934), 117-120
Williams, H.R., *The Dublin Review* 98 (1934), 150-153

Roman Britain and the English Settlements (1936)

Anonymous, *The Geographical Review* 29 (1939), 524-525
Anonymous, *The Times Literary Supplement* 36 (23-1-1937), 52
Bailly Reynolds, P.K., *The Antiquaries Journal* 17 (1937), 451-453
Blair, P.H., *The English Historical Review* 52 (1937), 683-687
Brandl, A., *Deutsche Literaturzeitung* 58 (1937) 968-972
Cronne, H.A., *History* 22 (1937), 350-351
Darby, H.C., *The Cambridge Review* 58 (1937), 239
Fox, C., *Nature* 139 (1937), 213-214
Gray, L.H., *The American Historical Review* 43 (1937-1938), 89-90
Hammond, M., *Political Science Quarterly* 53 (1938), 455-457
Johnson, A.Ch., *Classical Weekly* 30 (1936-1937), 236-237
McGuire, M.R.P., *The Catholic Historical Review* 23 (1937), 482-483
Richmond, I.A., *Journal of the British Archaeological Association* 3rd ser., 2 (1937), 197-200
Tischler, F., *Offa: Berichte und Mitteilungen aus dem Schleswig-Holsteinischen Landes des Museums vorgeschichtlicher Altertümer in Kiel* 3 (1938), 156-160
Wheeler, R.E.M., *JRS* 29 (1939), 87-93
Whittick, G.C., *The Classical Review* 51 (1937), 77-78
Winbolt, S.E., *The Geographical Journal* 89 (1937), 66-67
Zeiss, H., *Historische Zeitschrift* 159 (1938-1939), 111-112

"Human Nature and Human History" (*Proceedings of the British Academy* 22 (1936), 97-127) de Burgh, WG., *Philosophy* 12 (1937), 233-236

"Roman Britain" (in T. Frank ed., *An Economic Survey of Ancient Rome*, Vol. 3 (Paterson, 1937), 1-118

Cary, M., *Classical Philology* 33 (1938), 430-432

The Principles of Art (1938)

Anonymous, *The Times Literary Supplement* 37(13-8-1938), 533
B., J., *Gazette des Beaux-Arts* 20 (1938), 329-330
Carritt, E.F., *Philosophy* 13 (1938), 492-496
Carritt, E.F., *The Manchester Guardian* (27-5-1938)
Cournos, J., *The Sunday New York Times* (17-7-1937)
McMahon, A.Ph., *Parnassus* 11 (1939), 29-30
Oakeshott, M., *The Cambridge Review* 59 (1938), 487
Price-Jones, G., *Burlington Magazine For Connoisseurs* 73 (1938), 185
Tomlin, E.W.F., *The Criterion* 18 (1938-1939), 118-124
Vivas, E., *Nation* 148 (1939), 98

An Autobiography (1939)

Anonymous, *The Manchester Guardian* (15-8-1939)
Anonymous, *The Times Literary Supplement* 38 (5-8-1939), 464
Hannay, H., *International Journal of Ethics* 51 (1941), 369-370
Hicks, G.D., in "Survey of Philosophical Literature", *The Hibbert Journal* 38 (1939), 128-131
Lamprecht, S.P., *The Journal of Philosophy* 36 (1939), 717-718
Marcham, F.G., *The Philosophical Review* 50 (1941), 546
Muirhead, J.H., *Philosophy* 15 (1940), 89-91
Rowse, A.L., *The Spectator* 163 (18-8-1939), 262
Schlenke, M., *Historische Zeitschrift* 184 (1957), 594-597 (review of German translation: *Denken: Eine Autobiographie* (Stuttgart, 1955))

An Essay on Metaphysics (1940)

Anonymous, *The Manchester Guardian* (6-5-1940)
Anonymous, *The Times Literary Supplement* 39 (18-5-1940), 240
Baldwin, B.M., *Nature* 147 (1941), 7-8
Buchler, J., *The Sunday New York Times* (1-12-1940)
Ducasse, C.J., *The Philosophical Review* 50 (1941), 639-641
Goheen, J., *The Journal of Philosophy* 38 (1941), 48-50
Lion, A., *Philosophy* 16 (1941), 74-78
Stebbing, L.S., *Mind* 50 (1941), 184-190

The First Mate's Log of A Voyage to Greece in the Schooner Yacht 'Fleur de Lys' in 1939 (1940)

Anonymous, *The Times Literary Supplement* 39 (6-4-1940), 172

The New Leviathan (1942)

Anonymous, *The Glasgow Herald* (10-9-1942)
Anonymous, *The Guardian* (5-6-1942)
Anonymous, *The Scotsman* (20-8-1942)
Anonymous, *The Spectator* 169 (17-7-1942), 70
Anonymous, *The Times Literary Supplement* 41 (11-7-1942), 340
Barker, E., *The Oxford Magazine* 61 (1942-1943), 162-163
Carew-Hunt, R.N., *The Nineteenth Century and After* 132 (1942), 213-216
Catlin, G., *Political Science Quarterly* 58 (1943), 435-436
Dewar, W., *Time and Tide* (15-8-1942)
Laird, J., *Philosophy* 18 (1943), 75-80
Laski, H.J., *The New Statesman and Nation* 24 (8-8-1942), 97-98
MacCormac, J., *The Sunday New York Times* (1-8-1943)
MacIver, R.M., *The Annals of the American Academy of Political and Social Science* 228 (1943), 181-182
Ritchie, A.D., *The Manchester Guardian* (9-8-1942)
Sibley, M.Q., *The American Political Science Review* 37 (1943), 724-725
Thomas, I., *The Observer* (12-7-1942)

The Idea of Nature (1945)

Anonymous, *The Times Literary Supplement* 44 (10-3-1945), 111
Ambrosino, G., *Critique* 19 (1945), 546-552
Dewar, W., *Time and Tide* (19-5-1945)
Laird, J., *Mind* 54 (1945), 274-279
Murphy, A.E., *The Philosophical Review* 55 (1946), 199-202
Myres, J.L., *Man* 5 (1945), 133
Paton, H.J., *The Hibbert Journal* 43 (1945), 276-277
Price, H.H., *The Sunday Times* (11-3-1945)
Ritchie, A.D., *Nature* 155 (1945), 741-742
Ritchie, A.D., *The Manchester Guardian* (2-3-1945)
Truron, J.W., *The Guardian* (27-4-1945)
Webb, C.C.J., *Journal of Theological Studies* 46 (1945), 248-251
Whittaker, E., *Philosophy* 20 (1945), 260-261

The Idea of History (1946)

Anonymous, *The Times Literary Supplement* 45 (17-8-1946), 385-386
Aris, R., *The Contemporary Review* 171 (1947), 187-188
Beard, Ch.A., *The American Historical Review* 52 (1947), 704-708
Beloff, M., Time and Tide (28-9-1946)
Flenley, R., *Canadian Historical Review* 28 (1947), 68-72
Lefebvre, G., *Revue Historique* 202 (1949), 132-133

MacKinnon, D.M., *Journal of Theological Studies* 48 (1947), 249-253
Mandelbaum, M., *The Journal of Philosophy* 44 (1947), 184-188
Murphy, A.E. *The Philosophical Review* 56 (1947), 586-592
Oakeshott, M., *The English Historical Review* 62 (1947), 84-86
Powell, A., *The Spectator* 177 (16-8-1946), 172
Rowse, A.L., *The Sunday Times* (29-9-1946)
Schlenke, M., *Historische Zeitschrift* 184 (1957), 594-597 (review of German translation: *Philosophie der Geschichte* (Stuttgart, 1955))
T., J.M., *The Oxford Magazine* 65 (1946-1947), 113-114
Webb, C.C.J., *The Hibbert Journal* 45 (1946), 83-86
Wedgwood, C.V., *The Observer* (8-9-1946)
Wunberg, G., *Philosophischer Literaturanzeiger* 9 (1956), 156-161 (review of German Translation: *Philosophie der Geschichte* (Stuttgart, 1955))

Additional reviews of *Roman Britain and the English Settlements* (1936):

Etienne, R., *Revue des Etudes Anciennes* 53 (1951), 391
Faider- Feytmans, G., *L'Antiquité Classique* 21 (1952), 230-231
Piganiol, A., *Revue Historique* 66 (1941), 307-308

BIBLIOGRAPHY IV

WORKS BY OTHERS

Acham, K., *Analytische Geschichtsphilosophie: Eine kritische Einführung* (Freiburg und München, 1974).

Apel, K.O., *Transformation der Philosophie*, 2 vols. (Frankfurt am Main, 1973).

Atkinson, R.F., *Knowledge and Explanation in History: An Introduction to the Philosophy of History* (London, 1978).

Bagby, Ph., *Culture and History: Prolegomena to the Comparative Study of Civilizations* (London, 1958).

Bauman, Z., *Hermeneutics and Social Science: Approaches to Understanding* (London, 1978).

Baumann, B., *Imaginative Participation: The Career of an Organizing Concept in a Multidisciplinary Context* (The Hague, 1975).

Bersu, G., "King Arthur's Round Table: Final Report, including the excavations of 1939, with an appendix on the Little Round Table", *CW* 40 (1940), 169-206.

Birley, E.B., "In Memoriam of F.G. Simpson", *CW* 55 (1955), 359-365.

-, "John Horsley and John Hodgson", *Archaeologia Aeliana*,4th ser., 36 (1958), 1-46.

-, *Research on Hadrian's Wall* (Kendal, 1961).

-, "Sir Ian Archibald Richmond, 1902-1965", *Proceedings of the British Academy* 52 (1966), 293-302.

Bouch, C.M.L., "In Memoriam of R.G. Collingwood", *CW* 43 (1943), 211-214.

Breeze, D.J. and Dobson, B., *Hadrian's Wall* (London, 1976).

Buchdahl, G., "Logic and History: An Assessment of R.G. Collingwood's *Idea of History*", *Australasian Journal of Philosophy* 26 (1948), 94-113.

-, "Has Collingwood been Unfortunate in his Critics?", *Australasian Journal of Philosophy* 36 (1958), 95-108.

Bultmann, R., *Geschichte und Eschatologie* (Tübingen, 1958).

Burns, A.E., "Ascertainment, Probability and Evidence in History", *Historical Studies: Australia and New Zealand* 101 (1951), 327-339.

Carr, E.H., *What is History?* (London, 1961).

Casserley, J.V.L., *The Christian in Philosophy* (London, 1949).

Cebik, L.B., "History's Want of Authority – Some Logical and Historical Speculations", *Southern Journal of Philosophy* 8 (1970), 143-155.

-, "Collingwood: Action, Re-enactment, and Evidence", *Philosophical Forum* 2 (1970), 68-90.

Child, A., "History as Imitation", *The Philosophical Quarterly* 2 (1952), 193-207.

Coady, C.A.J., "Collingwood and Historical Testimony", *Philosophy* 50 (1975), 409-424.

Cohen, L.J., "A Survey of Work in the Philosophy of History, 1946-1950", *The Philosophical Quarterly* 2 (1952), 172-186.

-, "Has Collingwood been Misinterpreted?", *The Philosophical Quarterly* 7 (1957), 149-150.

Conkin, P.K. and Stromberg, R.N., *The Heritage and Challenge of History* (New York, 1971).

Corder, Ph., "Another Pioneer", *The Archaeological News Letter* 2 (1949), 42.

Couse, G.S., "Neglected Implications of R.G. Collingwood's Attack on 'Scissors-and-Paste History'", *The Canadian Historical Association. Historical Papers*, 1972, 23-38.

Crossman, R.H.S., "When Lightning struck the Ivory Tower: R.G. Collingwood", *The New Statesman and Nation* 17 (1939), 222-223; repr. in idem, *The Charm of Politics and Other Essays in Political Criticism* (London, 1958), 105-109.

Daniels, Ch. ed., J. Collingwood Bruce, *Handbook to the Roman Wall*, 13th ed. (Newcastle-upon-Tyne, 1978).

D'Arcy, M.C., *The Sense of History: Secular and Sacred* (London, 1959).

Debbins, W. ed., *R.G. Collingwood: Essays in the Philosophy of History* (Austin, 1965).

Donagan, A., "The Verification of Historical Theses", *The Philosophical Quarterly* 6 (1956), 193-208.

-, *The Later Philosophy of R.G. Collingwood* (Oxford, 1962).

- ed., *Essays in the Philosophy of Art by R.G. Collingwood* (Bloomington, 1964).

Dray, W.H., *Laws and Explanation in History* (Oxford, 1957).

-, "R.G. Collingwood and the Acquaintance Theory of Knowledge", *Revue Internationale de Philosophie* 11 (1957), 420-432.

-, Historical Understanding as Re-thinking", *University of Toronto Quarterly* 27 (1958), 200-215.

-, R.G. Collingwood on Reflective Thought", *The Journal of Philosophy* 57 (1960), 157-163.

-, Historical Causation and Human Free Will", *University of Toronto Quarterly* 29 (1960), 357-369.

-, *Philosophy of History* (Englewood Cliffs, 1964).

-, Review of Donagan, *The Later Philosophy of R.G. Collingwood*, in *The Canadian Historical Review*45 (1964), 130-132.

-, "Singular Hypotheticals and Historical Explanation", in L. Gross ed., *Sociological Theory: Inquiries and Paradigms* (New York, 1967), 181-203.

-, "R.G. Collingwood et la Connaissance Historique", *Dialogue: Canadian Philosophical Review* 17 (1978), 659-682.

-, Review of Martin, *Historical Explanation*, in *American Historical Review* 83 (1978), 1219-1220.

-, "Collingwood's Historical Individualism", *Canadian Journal of Philosophy* 10 (1980), 1-20.

-, *Perspectives on History* (London, 1980).
Elton, G.R., *Political History: Principles and Practice* (New York and London, 1970).
Emmet, D., Review of Donagan, *The Later Philosophy of R.G. Collingwood*, in *The Philosophical Quarterly* 13 (1963), 371-372.
Faber, K.G., *Theorie der Geschichtswissenschaft* (München, 1971).
Fain, H., *Between Philosophy and History: The Resurrection of Speculative Philosophy of History within the Analytic Tradition* (Princeton, 1970).
Fischer, D.H., *Historical Fallacies: Toward a Logic of Historical Thought* (New York, 1970).
Fruchon, P., "Signification de l'Histoire de la Philosophie selon l'Autobiographie de Collingwood", *Les Etudes Philosophiques* 13 (1958), 143-160.
Gadamer, H.G., "Einleitung", in R.G. Collingwood, *Denken: Eine Autobiographie* (Stuttgart, 1955), V-XIV.
Gallie, W.B., *Philosophy and the Historical Understanding* (New York, 1964).
Gardiner, P., *The Nature of Historical Explanation* (Oxford, 1952).
-, "The 'Objects' of Historical Knowledge", *Philosophy* 27 (1952), 211-220.
-, "Historical Understanding and the Empiricist Tradition", in B. Williams and A. Montefiore eds., *British Analytical Philosophy* (London, 1966), 267-284.
Garlick, T., *Ambleside Roman Fort* (Whitehaven, 1975).
Ginsberg, M., "The Character of a Historical Explanation", *Aristotelian Society*, Supplementary Volume 21 (1947), 69-77.
Goldstein, L.J., "Collingwood's Theory of Historical Knowing", *History and Theory* 9 (1970), 3-36.
-, "Collingwood on the Constitution of the Historical Past", in Krausz, ed., *Essays*, 241-267.
-, Review of Mink, *Mind, History and Dialectic* and Rubinoff, *Collingwood and the Reform of Metaphysics*, in *Man and World* 6 (1973), 83-99.
-, *Historical Knowing* (Austin, 1976).
Grant, C.K., "Collingwood's Theory of Historical Knowledge", *Renaissance and Modern Studies* 1 (1957), 65-90.
Hampshire, St., "The Illusion of Sociobiology": Review of E.O. Wilson, *On Human Nature* (Cambridge, Mass., 1978), in *New York Review of Books*, (12-10-1978), 64-69.
Harris, E.E., "Collingwood's Theory of History", *The Philosophical Quarterly* 7 (1957), 35-49.
-, Review of Debbins ed., *R.G. Collingwood: Essays in the Philosophy of History*, in *History and Theory* 5 (1966), 202-207.
Harris, H.S., "Introduction", in G. Gentile, *Genesis and Structure of Society* (Urbana, 1960), 14-20.
Harris, R.W., "Collingwood's *Idea of History*", *History* 37 (1952), 1-7.
Hartt, J.N., "Metaphysics, History, and Civilization: Collingwood's

Account of their Interrelationship", *The Journal of Religion* 33 (1953), 198-211.
Harvey, V.A., *The Historian and the Believer: The Morality of Historical Knowledge and Christian Belief* (London, 1966).
Hempel, C.G., "The Function of General Laws in History", *The Journal of Philosophy* 39 (1942), 35-48.
Hogg, A.H.A., "Sample Excavation", *Current Archaeology* 63 (1978), 125-127.
Holmes, S., "*Nisi Credidero non Intelligam*: The Neo-Anselmian Theory of Categories in S. Körner and the Later Collingwood", *Rivista di Studi Crociani* 10 (1973), 28-42.
Hopkins, J., "Bultmann on Collingwood's Philosophy of History", *The Harvard Theological Review* 58 (1965), 227-233.
Jarvie, I.C., *Concepts and Society* (London, 1972).
Johnston, W.M., *The Formative Years of R.G. Collingwood* (The Hague, 1967).
Kanichai, C.E., *R.G. Collingwood's Philosophy of History: A Critical Exposition and Evaluation* (Bangalore, 1972).
Kennedy, D.E., "The Wood and the Trees: The Philosophical Development of R.G. Collingwood", *The Australian Journal of Politics and History* 10 (1964), 245-248.
Knox, T.M., "Prof. R.G. Collingwood, F.B.A.", *Nature* 151 (1943), 163-164.
-, "Notes on Collingwood's Philosophical Work: With a Bibliography", *Proceedings of the British Academy* 29 (1943), 469-475.
-, "Editor's Preface", in *The Idea of History*, V-XXIV.
-, "R.G. Collingwood", *Hamburger Akademische Rundschau* 2 (1947-1948), 491-496.
-, Collingwood, Robin George", in *The Dictionary of National Biography, 1941-1950* (Oxford, 1959), 168-170.
Krausz, M. ed., *Critical Essays on the Philosophy of R.G. Collingwood* (Oxford, 1972).
Kuhn, Th.S., *The Structure of Scientific Revolutions*, 2nd ed. (Chicago, 1970).
Leff, G., *History and Social Theory* (Garden City, 1971).
Legutko, R., "Mind and History. On R.G. Collingwood's Philosophy", *Reports on Philosophy* 2 (1978), 41-51 (Jagiellonian University; Polish Scientific Publishers (Cracow, 1978)).
Levine, J.M., "The Autonomy of History: R.G. Collingwood and Agatha Christie", *Clio* 7 (1978), 253-264.
-, "Collingwood, Vico and the *Autobiography*", *Clio* 9 (1980), 379-392.
MacIver, A.M., "The Character of a Historical Explanation", *Aristotelian Society*, Supplementary Volume 21 (1947), 33-50.
Martin, K.M., "Caesar and Collingwood as Historians", *Latomus: Revue d'Etudes Latines* 28 (1969), 162-174.
Martin, R., *Historical Explanation: Re-enactment and Practical Inference* (Ithaca, 1977).
Marwick, A., *The Nature of History* (London, 1970).

McCall, D.F., *Africa in Time-Perspective: A Discussion of Historical Reconstruction from Unwritten Sources* (New York, 1969).
McCallum, R.B., "Obituary: R.G. Collingwood (1889-1942) [sic], *The Oxford Magazine* 61 (1942-1943), 160-161.
-, "Robin George Collingwood", *Proceedings of the British Academy* 29 (1943), 463-468.
Mink, L.O., "Collingwood's Dialectic of History", *History and Theory* 7 (1968), 3-37.
-, "Collingwood's Historicism: A Dialectic of Process", in Krausz ed., *Essays*, 154-178.
Munz, P., *The Shapes of Time: A New Look at the Philosophy of History* (Middletown, 1977).
Neilson, G., "Obituary Notice of J.P. Gibson", *Archaeologia Aeliana*, 3rd ser., 8 (1912), 37-45.
Nielsen, M.H. and Shearmur, J.F.G., Review of Skagestad, *Making Sense of History: The Philosophies of Popper and Collingwood*, in *Inquiry* 22 (1979), 459-489.
Oakeshott, M., *Experience and its Modes* (Cambridge, 1933).
Olivetti, A.G., *Due saggi su R.G. Collingwood: Con un' appendice di lettere inedite di Collingwood a G. de Ruggiero* (Padova, 1977).
O'Sullivan, N.K., "Irrationalism in Politics: A Critique of R.G. Collingwood's *New Leviathan*", *Political Studies* 20 (1972), 141-151.
Passmore, J., *A Hundred Years of Philosophy* (Harmondsworth, 1968).
Popper, K.R., *The Open Society and its Enemies*, 2 vols. (London, 1945).
-, *The Poverty of Historicism* (London, 1957).
-, *The Logic of Scientific Discovery* (London, 1959).
-, *Conjectures and Refutations* (New York, 1963).
-, "A Pluralist Approach to the Philosophy of History", in E. Streissler et al. (eds.), *Roads to Freedom: Essays in Honour of F.A. von Hayek* (London, 1969), 181-200.
-, "Die Logik der Sozialwissenschaften", *Kölner Zeitschrift für Soziologie und Sozialpsychologie* 14 (1962), 233-248; repr. in Th. W. Adorno et al., *Der Positivismusstreit in der deutschen Soziologie* (Neuwied und Berlin, 1969), 103-123. English translation: *The Positivist Dispute in German Sociology* (London, 1976), 87-104.
-, *Objective Knowledge: An Evolutionary Approach* (Oxford, 1972).
-, "Intellectual Autobiography" in P.A. Schilpp ed., *The Philosophy of Karl Popper* (La Salle, 1974), 3-181.
Randall, H.J., "Population and Agriculture in Roman Britain", *Antiquity* 4 (1930), 80-90.
Rasmussen, L.G., *Two Essays on the Scientific Study of History* (Bern and Frankfurt am Main, 1975).
Renier, G.J., *History: Its Purpose and Methods* (London, 1950).
Richmond, I.A., "Appreciation of R.G. Collingwood as an Archaeologist" and "R.G. Collingwood: Bibliography of Writings on Ancient History and Archaeology", *Proceedings of the British Academy* 29

(1943), 476-485.
-, Robin George Collingwood", *Archaeologia Aeliana*,4th ser., 21 (1943), 254-255.
-, "Obituary Notice: Robin George Collingwood. Born 1889: Died 9 January 1943", *The Antiquaries Journal* 23 (1943), 84-85.
-, "Frank Gerald Simpson", *Archaeologia Aeliana*,4th ser., 34 (1956), 219-221.
Roberts, T.A., *History and Christian Apologetic* (London, 1960).
Rotenstreich, N., "Historicism and Philosophy: Reflections on R.G. Collingwood", *Revue Internationale de Philosophie* 11 (1957), 401-419.
-, "From Facts to Thoughts: Collingwood's Views on the Nature of History", *Philosophy* 35 (1960), 122-137.
-, *Philosophy, History and Politics: Studies in Contemporary English Philosophy of History* (The Hague, 1976).
Rowse, A.L., *The Use of History* (London, 1946).
Rubinoff, L., Review of Debbins ed., *R.G. Collingwood: Essays in the Philosophy of History*, in *Dialogue: Canadian Philosophical Review* 5 (1966), 471-475.
-, ed., *Faith and Reason: Essays in the Philosophy of Religion by R.G. Collingwood* (Chicago, 1968).
-, "Collingwood's Theory of the Relation between Philosophy and History: A New Interpretation", *Journal of the History of Philosophy* 6 (1968), 363-380.
-, "History and Perception: Reflections on Collingwood's Theory of History", *Philosophical Forum* 2 (1970), 91-102.
-, *Collingwood and the Reform of Metaphysics: A Study in the Philosophy of Mind* (Toronto, 1970).
Ruggiero, G. de, "Science, History and Philosophy", *Philosophy* 6 (1931), 166-179.
Runciman, W.G., *A Critique of Max Weber's Philosophy of Social Science* (Cambridge, 1972).
Rust, E.C., *Evolutionary Philosophies and Contemporary Theology* (Philadelphia, 1969).
Ryle, G., *The Concept of Mind* (London, 1949).
Schneider, F.D., "Collingwood and the Idea of History", *The University of Toronto Quarterly* 22 (1953), 172-183.
Shalom, A., *R.G. Collingwood: Philosophe et Historien* (Paris, 1967).
Shoemaker, R.G., "Inference and Intuition in Collingwood's Philosophy of History", *The Monist* 53 (1969), 100-115.
Simpson, G., "Editor's Introduction: A Short Survey of Gerald Simpson's Life and Work", in G. Simpson ed., *Watermills and Military Works on Hadrian's Wall: Excavations in Northumberland 1907-1913 by F. Gerald Simpson* (Kendal, 1976), 1-23.
Skagestad, P., *Making Sense of History: The Philosophies of Popper and Collingwood* (Oslo, 1975).
Smith, P., *The Historian and History* (New York, 1960).
Stover, R., *The Nature of Historical Thinking* (Chapel Hill, 1967).

Strauss, L., "On Collingwood's Philosophy of History", *The Review of Metaphysics* 5 (1952), 559-586.

Sykes, N., "Some Current Conceptions of Historiography and their Significance for Christian Apologetic", *The Journal of Theological Studies* 50 (1949), 24-37.

Tholfsen, T.R., *Historical Thinking: An Introduction* (New York, 1967).

Tomlin, E.W.F., *R.G. Collingwood* (London, 1953).

-, "The Philosophy of R.G. Collingwood", *Ratio* (1958), 116-135.

Toulmin, S., *Human Understanding*, Vol. 1 (Oxford, 1972).

-, "Conceptual Change and the Problem of Relativity", in Krausz ed., *Essays*, 201-221.

-, "Introduction", in Collingwood, *An Autobiography* (Oxford, 1978), IX-XIX.

Toynbee, A., "R.G. Collingwood's View of the Historian's Relation to the Objects that he Studies", in idem, *A Study of History*, Vol. 9 (Oxford , 1954), 718-737.

Van der Dussen, W.J., "Collingwood's Unpublished Manuscripts", *History and Theory* 18 (1979), 287-315.

Verhaegen, B., *Introduction à l'Histoire Immédiate: Essai de Méthodologie Qualitative* (Gembloux, 1974).

Walsh, W.H., "R.G. Collingwood's Philosophy of History", *Philosophy* 22 (1947), 153-160.

-, "The Character of a Historical Explanation", in *Aristotelian Society*, Supplementary Volume 21 (1947), 51-68.

-, *An Introduction to Philosophy of History* (London, 1951).

Watkins, J.W.N., "On Explaining Disaster", *The Listener* (10-1-1963), 69-70.

-, "Imperfect Rationality", in R. Borger and F. Cioffi eds., *Explanation in the Behavioural Sciences* (Cambridge, 1970), 167-217.

Wells, G.A., *Herder and After: A Study in the Development of Sociology* (The Hague, 1959)..

Wheeler, R.E.M., "Mr. Collingwood and Mr. Randall: A Note", *Antiquity* 4 (1930), 91-95.

White, H.V., "Collingwood and Toynbee: Transitions in English Historical Thought", *English Miscellany* 8 (1957), 147-178.

-, Review of Donagan, *The Later Philosophy of R.G. Collingwood*, in *History and Theory* 4 (1965), 246-247.

White, M., *Foundations of Historical Knowledge* (New York, 1965).

Winch, P., *The Idea of a Social Science and its Relation to Philosophy* (London, 1958).

INDICES

I. SUBJECTS

II. NAMES